AF598016

Methods in Molecular Biology

Series Editor
John M. Walker
School of Life Sciences
University of Hertfordshire
Hatfield, Hertfordshire, AL10 9AB, UK

For further volumes:
http://www.springer.com/series/7651

Spliceosomal Pre-mRNA Splicing

Methods and Protocols

Edited by

Klemens J. Hertel

Department of Microbiology & Molecular Genetics, University of California, Irvine, CA, USA

Editor
Klemens J. Hertel
Department of Microbiology & Molecular Genetics
University of California
Irvine, CA, USA

ISSN 1064-3745 ISSN 1940-6029 (electronic)
ISBN 978-1-62703-979-6 ISBN 978-1-62703-980-2 (eBook)
DOI 10.1007/978-1-62703-980-2
Springer New York Heidelberg Dordrecht London

Library of Congress Control Number: 2014931088

Printed on acid-free paper

Humana Press is a brand of Springer
Springer is part of Springer Science+Business Media (www.springer.com)

Preface

The splicing of nuclear pre-mRNAs is carried out by the spliceosome, which recognizes splicing signals and catalyzes the removal of noncoding intronic sequences to assemble protein coding sequences into mature mRNA prior to export and translation. Of the approximately 25,000 genes encoded by the human genome, more than 90 % are believed to produce transcripts that are alternatively spliced. Thus, alternative splicing of pre-mRNAs can lead to the production of multiple protein isoforms from a single pre-mRNA, significantly enriching the proteomic diversity of higher eukaryotic organisms. Because regulation of this process can determine the timing and location that a particular protein isoform is produced, changes in alternative splicing patterns modulate many cellular activities. Consequently, the process of splicing must occur with a high degree of specificity and fidelity to ensure the appropriate expression of functional mRNAs.

Mutations in RNA splicing regulatory elements or in genes encoding splicing regulators that bind splicing regulatory elements can cause or modify the severity of disease. Early estimates, based on the identification of mutations within splice sites, suggested that ~15 % of all single base mutations change splicing patterns. However, it is now clear that many more mutations affect splicing by disrupting other important RNA elements, such as splicing enhancers or silencers binding sites. New estimates suggest that up to 60 % of known mutations could cause disease through changes in pre-mRNA splicing. A significant step towards identifying some of these disease-causing mutations has been made recently by combining novel high-throughput experimental and bioinformatic approaches to define splicing patterns and splicing regulatory elements. The advent of novel methods to analyze the activities of the spliceosome has led to the merging of different analytical disciplines. The goal of this book is to provide the reader with a guide to classical experimental approaches to decipher splicing mechanisms and to provide experimental strategies that rely on novel multidisciplinary approaches.

This book was written with graduate and medical students, clinicians, and postdoctoral researchers in mind. It describes the theory of alternative pre-mRNA splicing in seven introductory chapters and then introduces protocols and their theoretical background relevant for a variety of experimental research. These protocol chapters cover basic methods to detect splicing events, analyses of alternative pre-mRNA splicing in vitro and in vivo, manipulation of splicing events, and high-throughput and bioinformatic analyses of alternative splicing. Each chapter provides a theoretical introduction and a practical guide for molecular biologists, geneticists, clinicians, and every researcher interested in alternative splicing. In general, the protocols require a basic knowledge of molecular biology and/or RNA methods.

The protocols in this book are a collection of commonly used methods in the field of alternative splicing. These protocols should be viewed as guides for experiments that allow investigators to understand basic procedures. It is hoped that the chapters will allow readers to quickly find the experimental tools necessary for their projects and that it will stimulate their interest in trying out other techniques. As such, I hope that this compendium of methods and protocols will help newcomers and seasoned molecular biologists to understand the fascinating world of alternative splicing with the ultimate goal of paving the way for many new discoveries to come.

Irvine, CA, USA *Klemens J. Hertel*

Contents

Preface . . . . *v*
Contributors . . . . *ix*

PART I INTRODUCTORY CHAPTERS

1 The Pre-mRNA Splicing Reaction 3
Somsakul Pop Wongpalee and Shalini Sharma

2 Diversity and Evolution of Spliceosomal Systems 13
Scott William Roy and Manuel Irimia

3 Mechanisms of Spliceosomal Assembly 35
Ni-ting Chiou and Kristen W. Lynch

4 Alternative Pre-mRNA Splicing 45
Stacey D. Wagner and J. Andrew Berglund

5 Regulation of Alternative Pre-mRNA Splicing 55
Miguel B. Coelho and Christopher W.J. Smith

6 Introduction to Cotranscriptional RNA Splicing 83
Evan C. Merkhofer, Peter Hu, and Tracy L. Johnson

7 Chromatin and Splicing 97
Nazmul Haque and Shalini Oberdoerffer

PART II METHODS CHAPTERS

8 Preparation of Splicing Competent Nuclear Extracts 117
Chiu-Ho T. Webb and Klemens J. Hertel

9 Preparation of Yeast Whole Cell Splicing Extract 123
Elizabeth A. Dunn and Stephen D. Rader

10 Efficient Splinted Ligation of Synthetic RNA Using RNA Ligase 137
Martha R. Stark and Stephen D. Rader

11 In Vitro Assay of Pre-mRNA Splicing in Mammalian Nuclear Extract . . . 151
Maliheh Movassat, William F. Mueller, and Klemens J. Hertel

12 Kinetic Analysis of In Vitro Pre-mRNA Splicing in HeLa Nuclear Extract 161
William F. Mueller and Klemens J. Hertel

13 In Vitro Systems for Coupling RNAP II Transcription to Splicing and Polyadenylation 169
Eric G. Folco and Robin Reed

14 Isolation and Accumulation of Spliceosomal Assembly Intermediates........ 179
Janine O. Ilagan and Melissa S. Jurica

15 Complementation of U4 snRNA in *S. cerevisiae* Splicing Extracts for Biochemical Studies of snRNP Assembly and Function 193
Martha R. Stark and Stephen D. Rader

16 Expression and Purification of Splicing Proteins from Mammalian Cells 205
Eric Allemand and Michelle L. Hastings

17 Single Molecule Approaches for Studying Spliceosome Assembly and Catalysis 217
Eric G. Anderson and Aaron A. Hoskins

18 Cell-Based Splicing of Minigenes................................. 243
Sarah A. Smith and Kristen W. Lynch

19 Quantifying the Ratio of Spliceosome Components Assembled on Pre-mRNA 257
Noa Neufeld, Yehuda Brody, and Yaron Shav-Tal

20 Antisense Methods to Modulate Pre-mRNA Splicing 271
Joonbae Seo, Eric W. Ottesen, and Ravindra N. Singh

21 Using Yeast Genetics to Study Splicing Mechanisms 285
Munshi Azad Hossain and Tracy L. Johnson

22 Medium Throughput Analysis of Alternative Splicing by Fluorescently Labeled RT-PCR................................ 299
Ryan Percifield, Daniel Murphy, and Peter Stoilov

23 Chromatin Immunoprecipitation Approaches to Determine Co-transcriptional Nature of Splicing....................... 315
Nicole I. Bieberstein, Korinna Straube, and Karla M. Neugebauer

24 Computational Approaches to Mine Publicly Available Databases 325
Rodger B. Voelker, William A. Cresko, and J. Andrew Berglund

25 Approaches to Link RNA Secondary Structures with Splicing Regulation 341
Mireya Plass and Eduardo Eyras

26 Methods to Study Splicing from High-Throughput RNA Sequencing Data 357
Gael P. Alamancos, Eneritz Agirre, and Eduardo Eyras

27 Global Protein–RNA Interaction Mapping at Single Nucleotide Resolution by iCLIP-Seq 399
Chengguo Yao, Lingjie Weng, and Yongsheng Shi

28 Predicting Alternative Splicing 411
Yoseph Barash and Jorge Vaquero Garcia

Index *425*

Contributors

ENERITZ AGIRRE • *Computational Genomics, Universitat Pompeu Fabra, Barcelona, Spain*
GAEL P. ALAMANCOS • *Computational Genomics, Universitat Pompeu Fabra, Barcelona, Spain*
ERIC ALLEMAND • *Unité de Régulation Epigénétique, Département de Biologie du Développement, Institut Pasteur, Paris, France*
ERIC G. ANDERSON • *Biochemistry and Molecular Pharmacology, University of Massachusetts Medical School, Worcester, MA, USA*
YOSEPH BARASH • *Department of Genetics, University of Pennsylvania, Philadelphia, PA, USA*
J. ANDREW BERGLUND • *Department of Chemistry and Institute of Molecular Biology, University of Oregon, Eugene, OR, USA*
NICOLE I. BIEBERSTEIN • *Max Planck Institute of Molecular Cell Biology and Genetics, Dresden, Germany*
YEHUDA BRODY • *The Mina & Everard Goodman Faculty of Life Sciences & Institute of Nanotechnology, Bar-Ilan University, Ramat Gan, Israel*
NI-TING CHIOU • *Department of Biochemistry and Biophysics, Perelman School of Medicine, University of Pennsylvania, Philadelphia, PA, USA*
MIGUEL B. COELHO • *Department of Biochemistry, University of Cambridge, Cambridge, UK*
WILLIAM A. CRESKO • *Department of Biology and Institute of Ecology and Evolution, University of Oregon, Eugene, OR, USA*
ELIZABETH A. DUNN • *Department of Biochemistry and Molecular Biology, University of British Columbia, Vancouver, BC, Canada*
EDUARDO EYRAS • *Computational Genomics, Universitat Pompeu Fabra, Barcelona, Spain; Catalan Institution for Research and Advanced Studies (ICREA), Barcelona, Spain*
ERIC G. FOLCO • *Department of Cell Biology, Harvard Medical School, Boston, MA, USA*
JORGE VAQUERO GARCIA • *Department of Genetics, University of Pennsylvania, Philadelphia, PA, USA*
NAZMUL HAQUE • *Laboratory of Ribonucleoprotein Biochemistry, National Heart, Lung, and Blood Institute, National Institutes of Health, Bethesda, MD, USA*
MICHELLE L. HASTINGS • *Department of Cell Biology and Anatomy, The Chicago Medical School, Rosalind Franklin University of Medicine and Science, North Chicago, IL, USA*
KLEMENS J. HERTEL • *Department of Microbiology & Molecular Genetics, University of California, Irvine, CA, USA*
AARON A. HOSKINS • *Department of Biochemistry, University of Wisconsin-Madison, Madison, WI, USA*
MUNSHI AZAD HOSSAIN • *Molecular Biology Section, Division of Biological Sciences, University of California, San Diego, La Jolla, CA, USA*
PETER HU • *Molecular Biology Section, Division of Biological Sciences, University of California, San Diego, La Jolla, CA, USA*

Janine O. Ilagan • *Department of Molecular, Cell and Developmental Biology and Center for Molecular Biology of RNA, University of California, Santa Cruz, CA, USA*
Manuel Irimia • *The Donnelly Centre, University of Toronto, Toronto, ON, Canada*
Tracy L. Johnson • *Department of Molecular, Cell and Developmental Biology, University of California, Los Angeles, Los Angeles, CA, USA*
Melissa S. Jurica • *Department of Molecular, Cell and Developmental Biology and Center for Molecular Biology of RNA, University of California, Santa Cruz, CA, USA*
Kristen W. Lynch • *Department of Biochemistry and Biophysics, Perelman School of Medicine, University of Pennsylvania, Philadelphia, PA, USA*
Evan C. Merkhofer • *Molecular Biology Section, Division of Biological Sciences, University of California, San Diego, La Jolla, CA, USA*
Maliheh Movassat • *Department of Microbiology & Molecular Genetics, University of California, Irvine, CA, USA*
William F. Mueller • *Department of Microbiology & Molecular Genetics, University of California, Irvine, CA, USA*
Daniel Murphy • *Department of Biochemistry, West Virginia University, Morgantown, WV, USA*
Noa Neufeld • *The Mina & Everard Goodman Faculty of Life Sciences & Institute of Nanotechnology, Bar-Ilan University, Ramat Gan, Israel*
Karla M. Neugebauer • *Max Planck Institute of Molecular Cell Biology and Genetics, Dresden, Germany*
Shalini Oberdoerffer • *Laboratory of Receptor Biology and Gene Expression, National Cancer Institute, National Institutes of Health, Bethesda, MD, USA*
Eric W. Ottesen • *Department of Biomedical Sciences, College of Veterinary Medicine, Iowa State University, Ames, IA, USA*
Ryan Percifield • *Department of Biochemistry, West Virginia University, Morgantown, WV, USA*
Mireya Plass • *Department of Biology, The Bioinformatics Centre, University of Copenhagen, Copenhagen, Denmark*
Stephen D. Rader • *Department of Chemistry, University of Northern British Columbia, Prince George, BC, Canada; Department of Biochemistry and Molecular Biology, University of British Columbia, Vancouver, BC, Canada*
Robin Reed • *Department of Cell Biology, Harvard Medical School, Boston, MA, USA*
Scott William Roy • *Department of Biology, San Francisco State University, San Francisco, CA, USA*
Joonbae Seo • *Department of Biomedical Sciences, College of Veterinary Medicine, Iowa State University, Ames, IA, USA*
Shalini Sharma • *Department of Basic Medical Sciences, University of Arizona, College of Medicine-Phoenix, Phoenix, AZ, USA*
Yaron Shav-Tal • *The Mina & Everard Goodman Faculty of Life Sciences & Institute of Nanotechnology, Bar-Ilan University, Ramat Gan, Israel*
Yongsheng Shi • *Department of Microbiology and Molecular Genetics, School of Medicine, University of California, Irvine, CA, USA*
Ravindra N. Singh • *Department of Biomedical Sciences, College of Veterinary Medicine, Iowa State University, Ames, IA, USA*
Christopher W. J. Smith • *Department of Biochemistry, University of Cambridge, Cambridge, UK*

SARAH A. SMITH • *Department of Biochemistry and Biophysics, Perelman School of Medicine, University of Pennsylvania, Philadelphia, PA, USA*
MARTHA R. STARK • *Department of Chemistry, University of Northern British Columbia, Prince George, BC, Canada*
PETER STOILOV • *Department of Biochemistry, West Virginia University, Morgantown, WV, USA*
KORINNA STRAUBE • *Max Planck Institute of Molecular Cell Biology and Genetics, Dresden, Germany*
RODGER B. VOELKER • *Institutes of Molecular Biology and Ecology and Evolution, University of Oregon, Eugene, OR, USA*
STACEY D. WAGNER • *Department of Chemistry and Institute of Molecular Biology, University of Oregon, Eugene, OR, USA*
CHIU-HO T. WEBB • *Department of Microbiology & Molecular Genetics, University of California, Irvine, CA, USA*
LINGJIE WENG • *Department of Computer Science, Institute for Genomics and Bioinformatics, University of California, Irvine, CA, USA*
SOMSAKUL POP WONGPALEE • *Department of Immunology and Molecular Genetics, Molecular Biology Institute, University of California, Los Angeles, CA, USA*
CHENGGUO YAO • *Department of Microbiology and Molecular Genetics, School of Medicine, University of California, Irvine, CA, USA*

Part I

Introductory Chapters

Chapter 1

The Pre-mRNA Splicing Reaction

Somsakul Pop Wongpalee and Shalini Sharma

Abstract

In eukaryotic organisms, nascent transcripts of protein-coding genes contain intronic sequences that are not present in mature mRNAs. Pre-mRNA splicing removes introns and joins exons to form mature mRNAs. It is catalyzed by a large RNP complex called the spliceosome. Sequences within the pre-mRNA determine intron recognition and excision. This process occurs with a high degree of accuracy to generate the functional transcriptome of a cell.

Key words Exon, Intron, Splicing, Pre-mRNA, Transcript, Splice site, Intronless, Spliceosome

1 Introduction

Most eukaryotic protein-coding genes are transcribed into precursor-messenger RNAs (pre-mRNAs) in which the two terminal untranslated regions and the protein-coding regions, together called exons, are interrupted by noncoding intervening segments called introns. This pre-mRNA must undergo an RNA processing reaction called splicing that removes introns and ligates the exons to generate a mature translatable mRNA. Splicing occurs in two transesterification reactions that are catalyzed by a large *r*ibo*n*ucleo*p*rotein (RNP) complex called the spliceosome. All of the introns must be excised from a gene transcript prior to the export of the mature mRNA from the nucleus to the cytoplasm for translation.

In 1993, Phil Sharp and Richard Roberts were awarded the Nobel Prize in Physiology and Medicine for their discovery of introns in adenovirus transcripts [1, 2]. A historical perspective of research that led to the discovery of introns and the spliceosome is described in earlier reviews and a very recent one [3–9]. In addition to the spliceosomal introns found in transcripts of nuclear protein-coding genes, a variety of organisms contain other intron types including Group I and Group II introns that are removed by different mechanisms [10]. This chapter describes features of the eukaryotic spliceosomal pre-mRNA introns and the splicing reaction chemistry.

Klemens J. Hertel (ed.), *Spliceosomal Pre-mRNA Splicing: Methods and Protocols*, Methods in Molecular Biology, vol. 1126, DOI 10.1007/978-1-62703-980-2_1,

2 Types of Introns

Introns are present in pre-mRNAs of nearly all eukaryotic organisms, but their numbers vary amongst different species. In budding yeast, *Saccharomyces cerevisiae* of the ~6,000 genes, only 283 contain a total of 298 introns [11–15]. In the majority of cases, only one intron is found per gene, and average intron size is about 100–400 nucleotides. In fission yeast, *Schizosaccharomyces pombe*, 4,730 introns are distributed in 43 % of the 4,824 genes with majority containing only one intron [16, 17]. By contrast, of the 20,000–25,000 human protein-coding genes, only 700 are intronless [18–20]. On an average, there are eight introns and nine exons per human gene. About 80 % of the human exons are <200 nucleotides in length. Intron size on the other hand is more variable with an average length of ~3,000 nucleotides and extending to more than 11,000 nucleotides in ~10 %. Online resources for human and yeast introns and exons and for human intronless genes are listed in Table 1.

Specific sequence elements at the exon–intron boundaries and within an intron determine its recognition and removal by the spliceosomal components. Essential elements include the 5′ splice site and the 3′ splice site, which are preceded by the branch point sequence and the polypyrimidine tract (Fig. 1). Metazoan pre-mRNAs contain two types of introns that are distinguished by their characteristic splice site sequences and their excision by different spliceosomal complexes [21]. The majority of metazoan introns are of the major class or U2-type that contain the canonical GT–AG intron boundaries. In most U2 introns, the polypyrimidine tract and the branch point sequence are within 50 nucleotides upstream of the 3′ splice site. These introns have relatively degenerate splicing signals and are processed by the U2-dependent spliceosome.

Table 1
Online resources for human and yeast introns and exons

Database name	URL	Reference
Yeast Intron Database	http://intron.ucsc.edu/yeast4.1	[14]
Saccharomyces Genome Database	http://www.yeastgenome.org/	[15]
PomBase	http://www.pombase.org/	[17]
HEXEvent	http://hexevent.mmg.uci.edu	[19]
U12 Database	http://genome.crg.es/cgi-bin/u12db/u12db.cgi	[27]
Intronless Gene Database	http://www.bioinfo-cbs.org/igd	[18]

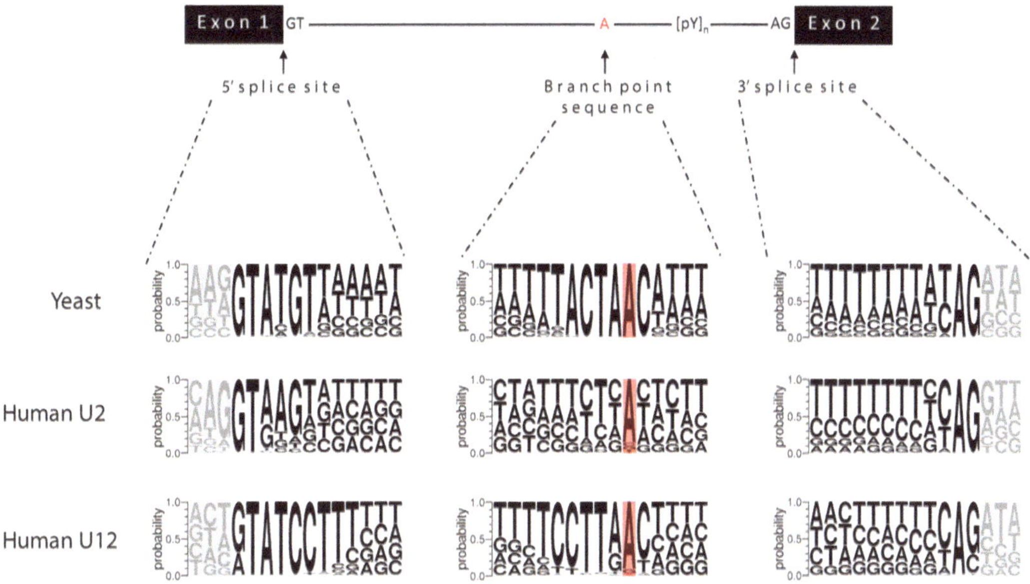

Fig. 1 Splicing elements in *S. cerevisiae* and human introns. Consensus sequences for splicing elements were generated using WebLogo 3 [59]. For budding yeast, *S. cerevisiae*, consensus 5′ and 3′ splice site sequences were generated from 298 intron sequences obtained from Ares Lab Yeast Intron Database 4.1 [14]. For the branch point, 233 introns that contained sequence CTAAC within 50 nucleotides upstream of the 3′ splice site were used for generating a consensus sequence. Human intronic sequences flanking constitutive exons on chromosome 1 were extracted from RefSeq using a constitutive exon dataset from a database of human exon splicing events, HEXEvent [19]. The sequences derived were filtered against non-GT/AG splice sites and against the U12 intron database. This gave 13,324 5′- and 13,341 3′ splice sites, which were used in the consensus analysis. For the human branch point sequences analysis, 59 previously reported experimentally curated sequences were used [31]. All 695 human U12 intron sequences were obtained from a U12 database used for the analysis [27]. *Gray-shaded boxes* represent exonic regions, *red-shaded box* represents the branch point adenosine, and $[pY]_n$ indicates the polypyrimidine tract

The second intron type is the minor class or U12-type, catalyzed by the U12-dependent spliceosome [8, 22]. U12 introns were first identified as rare introns that contain noncanonical AT AC termini [23, 24]. However, subsequent genomic analysis showed that the GT–AG termini are more frequent in minor introns than the AT–AC termini, and a few U2 introns also have AT–AC termini [25]. The U12 introns are present in many animal, plant, and fungal genomes, but are missing in some species including the common model organisms *Caenorhabditis elegans* and *S. cerevisiae* [26]. Amongst the metazoan genomes, the number of U12 introns is highly variable. The human and Arabidopsis genomes have ~695 and ~300 U12 introns, respectively [27, 28]. In Drosophila, only 19 U12 introns have been reported so far [29].

In *S. cerevisiae* and other hemiascomycetous yeasts, all introns are of the U2 type and contain the canonical GT–AG sequences at the termini [30]. The splice site sequences in these introns are highly conserved with the consensus motifs for the 5′ splice site, 3′ splice site, and branch point in *S. cerevisiae* being GTATGT, YAG (Y = pyrimidine), and TACTAAC, respectively (Fig. 1). In the 5′ splice site sequence, variations are tolerated at +4 and +6 positions relative to the exon–intron junction. At the 3′ splice site, TAG and CAG are equally frequent and only seven introns contain an AAG motif [30]. The core CTAAC branch point sequence is present in 95 % of the 298 yeast introns and TACTAAC is present in 83 %. Other low-frequency motifs include GACTAAC, AACTAAC, TACTAAC, and TGCTAAC.

The major U2-type human intron splicing signals are highly degenerate (Fig. 1). The 5′ splice site consensus sequence is AG/GTRAGT (where R is A or G and the "/" denotes the exon–intron junction). The consensus for the branch site and the 3′ splice site sequences are YTNAN (where N is any nucleotide) and YAG, respectively [31, 32]. The polypyrimidine tracts are 15–20 nucleotides long and rich in pyrimidines, especially in uridines. The sequence information in these signals is not sufficient to define the splice sites within the long mammalian introns that contain many cryptic matches to these elements. Initial splice site recognition is promoted by proteins of the serine–arginine (SR) family that bind exonic splicing enhancer elements (ESEs) [33–35]. Through their protein-interacting RS domains, SR proteins facilitate recruitment of spliceosomal components to the 5′ and 3′ splice sites in a process called exon definition [36, 37]. In later steps during spliceosome assembly and activation, the splice sites are paired across introns prior to splicing catalysis [38].

In comparison to the U2-type introns, the 5′ splice site and the branch point sequence of U12-type introns are extended and less variable (Fig. 1). The U12 introns lack a conspicuous polypyrimidine tract between the branch point sequence and the 3′ splice site. The distance between the branch point and the 3′ splice site is short (12–15 nucleotides) and more constrained than that in the U2-type introns [39]. Recognition of the 5′ splice site and the highly conserved branch point primarily determines 3′ splice site identification [39].

The splicing signals in introns are recognized by components of the U2 and U12 spliceosomes, which are large and dynamic complexes that assemble from small nuclear RNPs (snRNPs) and auxiliary proteins. The snRNPs are RNP complexes of spliceosomal snRNAs, Sm proteins, and particle-specific proteins. The U2-dependent spliceosome consists of the five snRNPs U1, U2, U4, U5, and U6 and ~250 proteins [40, 41]. The U12 spliceosome forms from a different complement of snRNPs [42, 43]. The U11, U12, and U4atac/U6atac snRNAs are functional analogs of the U1, U2, and U4/U6 snRNAs [22, 44]. The U5 snRNA is common to both types of spliceosomes. The U2- and U12-dependent spliceosomes share many proteins and also contain distinct sets of proteins.

The spliceosomal complexes assemble onto an intron from sequential binding of their components [22, 41]. For U2 introns the 5′ splice site is initially recognized by the U1 snRNP. At the 3′ end of the intron, the branch point, the polypyrimidine tract, and the 3′ splice site are recognized by the splicing factor 1 (SF1), U2 auxiliary factor 65 (U2AF65), and U2AF35 proteins, respectively. In a subsequent step, U2 snRNP displaces the SF1 protein at the branch point. The minor intron 5′ and 3′ splice sites are recognized by simultaneous binding of the U11–U12 di-snRNPs [45]. Biochemical data indicate a lack of U2AF requirement in U12-intron splicing [46]. Recruitment of the tri-snRNPs, U4/U6-U5 to major introns, and U4atac/U6atac-U5 to minor introns forms the complete spliceosomal complex, which undergoes several structural rearrangements prior to formation of the active spliceosome in which the splicing reaction occurs. During the activation process, the spliceosome undergoes extensive remodeling leading to loss of U1 and U4 snRNPs and formation of the catalytic core from the U2, U5, and U6 snRNAs and the Prp8 protein [47, 48].

3 The Splicing Reaction

The chemistry of the splicing reaction in yeast and mammals was elucidated almost simultaneously in many labs [49–53]. The development of in vitro pre-mRNA splicing systems enabled these studies, which showed that splicing requires ATP and $MgCl_2$ and monovalent cations [54, 55]. Importantly, these in vitro analyses led to the discovery of the unusual configuration of the excised intron as “a circle containing a tail with a branch” and was later termed the lariat [52]. The atypical properties that enabled characterization of the lariat intron include its anomalously slow mobility in polyacrylamide gels, exonuclease resistance, and block to reverse transcription due to the presence of the branch structure.

The chemistry of the splicing reaction is the same for removal of the U2 and U12 introns. Splicing occurs in two isoenergetic transesterification steps (Fig. 2). In the first step, the 2′ hydroxyl group of the conserved branch point adenosine makes a nucleophilic attack at the phosphate of the 5′ splice site and cleaves the phosphodiester bond at the exon–intron junction. This releases the 5′ exon (exon 1 in Fig. 2) with the hydroxyl group at its 3′ end. Concomitantly, a new phosphodiester bond is formed at the branch point. The commonly occurring phosphodiester bond in nucleic acid backbone is between the 3′ hydroxyl and 5′ phosphate (5′→3′) of adjacent nucleotides. The intron branch is formed by a phosphodiester bond between the 5′ phosphate of the 5′ guanosine residue of the intron and the 2′ hydroxyl (5′→2′) of the conserved branch point adenosine. This first transesterification produces the two intermediates of the reaction: the detached exon

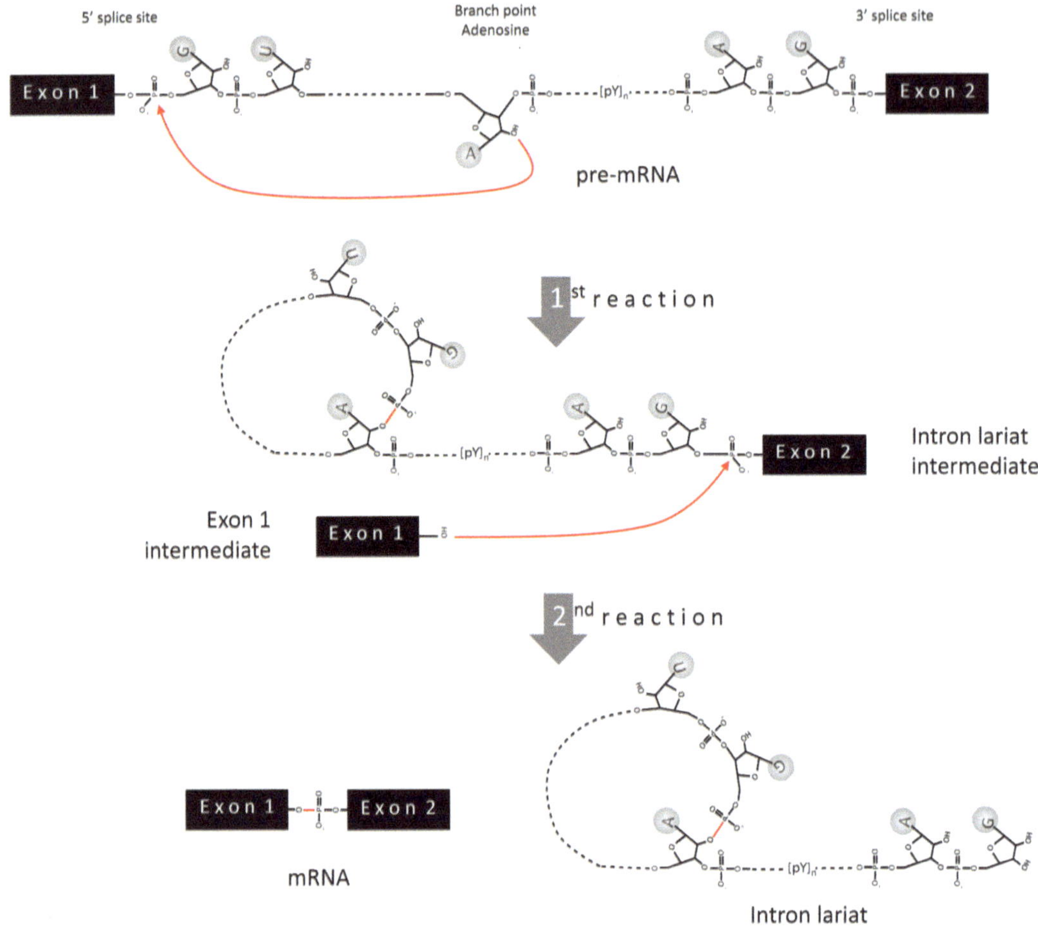

Fig. 2 The splicing reaction. The splicing reaction occurs in two transesterification steps. The first step forms two intermediates, free exon1 and intron–exon2 lariat. The second transesterification ligates the exons and releases the intron lariat. *Black broken line* = intron; *red arrow* = nucleophilic attack; *red line* = newly formed bond; $[pY]_n$ = polypyrimidine tract

1 and the intron–exon 2 fragment in a loop structure called the lariat (Fig. 2). In the second transesterification, a nucleophilic attack by the 3′ hydroxyl group of the detached exon1 at the phosphate of the 3′ splice site results in ligation of the two exons and release of the intron lariat.

The process of spliceosome assembly and the transesterification reactions must occur very precisely on each intron of a primary transcript to form the mature functional mRNA. This involves the selection of authentic splice sites over cryptic sites and is determined by several factors that include splice site strength, the presence of exonic and intronic enhancer sequences, exon–intron length, and RNA secondary structure [37, 38, 56, 57]. Furthermore, the fidelity of the spliceosome assembly process and the catalysis steps is maintained by proofreading of the splicing complex intermediates by the DEAD-box helicases that are associated with the spliceosome [58]. Together, these regulatory mechanisms ensure high fidelity of intron removal and exon ligation during pre-mRNA splicing, which is an essential step in gene expression.

Acknowledgments

The authors are thankful to Beverly Lin and Areum Han for providing the sequences and to Douglas Black for advice and support during the preparation of this manuscript. S.S. is supported by a National Cancer Institute R21 award CA170786.

References

1. Berget SM, Moore C, Sharp PA (1977) Spliced segments at the 5' terminus of adenovirus 2 late mRNA. Proc Natl Acad Sci USA 74(8): 3171–3175
2. Chow LT, Gelinas RE, Broker TR, Roberts RJ (1977) An amazing sequence arrangement at the 5' ends of adenovirus 2 messenger RNA. Cell 12(1):1–8
3. Darnell JE Jr (2013) Reflections on the history of pre-mRNA processing and highlights of current knowledge: a unified picture. RNA. doi:10.1261/rna.038596.113
4. Padgett RA, Grabowski PJ, Konarska MM, Seiler S, Sharp PA (1986) Splicing of messenger RNA precursors. Annu Rev Biochem 55:1119–1150. doi:10.1146/annurev.bi.55.070186.005351
5. Sharp PA (1985) On the origin of RNA splicing and introns. Cell 42(2):397–400
6. Crick F (1979) Split genes and RNA splicing. Science 204(4390):264–271
7. Guthrie C, Patterson B (1988) Spliceosomal snRNAs. Annu Rev Genet 22:387–419. doi:10.1146/annurev.ge.22.120188.002131
8. Tarn WY, Steitz JA (1997) Pre-mRNA splicing: the discovery of a new spliceosome doubles the challenge. Trends Biochem Sci 22(4):132–137
9. Steitz JA, Black DL, Gerke V, Parker KA, Kramer A, Frendewey D, Keller W (1988) Functions of the abundant U-snRNPs. In: Birnstiel ML (ed) Structure and function of major and minor SNURPS. Springer, Heildelberg, pp 115–154
10. Lambowitz AM, Capara MG, Zimmerly S, Perlman PS (1999) Group I and Group II ribozymes as RNPs: clues to the past and guides to the future. In: Gesteland RF, Cech TR, Atkins JF (eds) RNA world, 2nd edn. Cold Spring Harbor Laboratory Press, Cold Spring Harbor, NY, pp 451–485
11. Guldener U, Munsterkotter M, Kastenmuller G, Strack N, van Helden J, Lemer C, Richelles J, Wodak SJ, Garcia-Martinez J, Perez-Ortin JE, Michael H, Kaps A, Talla E, Dujon B, Andre B, Souciet JL, De Montigny J, Bon E, Gaillardin C, Mewes HW (2005) CYGD: the Comprehensive Yeast Genome Database. Nucleic Acids Res 33 (Database issue):D364–D368. doi:10.1093/nar/gki053
12. Spingola M, Grate L, Haussler D, Ares M Jr (1999) Genome-wide bioinformatic and molecular analysis of introns in Saccharomyces cerevisiae. RNA 5(2):221–234
13. Juneau K, Palm C, Miranda M, Davis RW (2007) High-density yeast-tiling array reveals previously undiscovered introns and extensive regulation of meiotic splicing. Proc Natl Acad Sci USA 104(5):1522–1527. doi:10.1073/pnas.0610354104
14. Grate L, Ares M Jr (2002) Searching yeast intron data at Ares lab Web site. Methods Enzymol 350:380–392
15. Hirschman JE, Balakrishnan R, Christie KR, Costanzo MC, Dwight SS, Engel SR, Fisk DG, Hong EL, Livstone MS, Nash R, Park J, Oughtred R, Skrzypek M, Starr B, Theesfeld CL, Williams J, Andrada R, Binkley G, Dong Q, Lane C, Miyasato S, Sethuraman A, Schroeder M, Thanawala MK, Weng S, Dolinski K, Botstein D, Cherry JM (2006) Genome Snapshot: a new resource at the Saccharomyces Genome Database (SGD) presenting an overview of the Saccharomyces cerevisiae genome. Nucleic Acids Res 34 (Database issue):D442–D445. doi:10.1093/nar/gkj117
16. Wood V, Gwilliam R, Rajandream MA, Lyne M, Lyne R, Stewart A, Sgouros J, Peat N,

Hayles J, Baker S, Basham D, Bowman S, Brooks K, Brown D, Brown S, Chillingworth T, Churcher C, Collins M, Connor R, Cronin A, Davis P, Feltwell T, Fraser A, Gentles S, Goble A, Hamlin N, Harris D, Hidalgo J, Hodgson G, Holroyd S, Hornsby T, Howarth S, Huckle EJ, Hunt S, Jagels K, James K, Jones L, Jones M, Leather S, McDonald S, McLean J, Mooney P, Moule S, Mungall K, Murphy L, Niblett D, Odell C, Oliver K, O'Neil S, Pearson D, Quail MA, Rabbinowitsch E, Rutherford K, Rutter S, Saunders D, Seeger K, Sharp S, Skelton J, Simmonds M, Squares R, Squares S, Stevens K, Taylor K, Taylor RG, Tivey A, Walsh S, Warren T, Whitehead S, Woodward J, Volckaert G, Aert R, Robben J, Grymonprez B, Weltjens I, Vanstreels E, Rieger M, Schafer M, Muller-Auer S, Gabel C, Fuchs M, Dusterhoft A, Fritzc C, Holzer E, Moestl D, Hilbert H, Borzym K, Langer I, Beck A, Lehrach H, Reinhardt R, Pohl TM, Eger P, Zimmermann W, Wedler H, Wambutt R, Purnelle B, Goffeau A, Cadieu E, Dreano S, Gloux S, Lelaure V, Mottier S, Galibert F, Aves SJ, Xiang Z, Hunt C, Moore K, Hurst SM, Lucas M, Rochet M, Gaillardin C, Tallada VA, Garzon A, Thode G, Daga RR, Cruzado L, Jimenez J, Sanchez M, del Rey F, Benito J, Dominguez A, Revuelta JL, Moreno S, Armstrong J, Forsburg SL, Cerutti L, Lowe T, McCombie WR, Paulsen I, Potashkin J, Shpakovski GV, Ussery D, Barrell BG, Nurse P (2002) The genome sequence of Schizosaccharomyces pombe. Nature 415(6874):871–880. doi:10.1038/nature724

17. Wood V, Harris MA, McDowall MD, Rutherford K, Vaughan BW, Staines DM, Aslett M, Lock A, Bahler J, Kersey PJ, Oliver SG (2012) PomBase: a comprehensive online resource for fission yeast. Nucleic Acids Res 40 (Database issue):D695–D699. doi:10.1093/nar/gkr853

18. Louhichi A, Fourati A, Rebai A (2011) IGD: a resource for intronless genes in the human genome. Gene 488(1–2):35–40. doi:10.1016/j.gene.2011.08.013

19. Busch A, Hertel KJ (2013) HEXEvent: a database of Human EXon splicing Events. Nucleic Acids Res 41:118–124. doi:10.1093/nar/gks969

20. Lander ES, Linton LM, Birren B, Nusbaum C, Zody MC, Baldwin J, Devon K, Dewar K, Doyle M, FitzHugh W, Funke R, Gage D, Harris K, Heaford A, Howland J, Kann L, Lehoczky J, LeVine R, McEwan P, McKernan K, Meldrim J, Mesirov JP, Miranda C, Morris W, Naylor J, Raymond C, Rosetti M, Santos R, Sheridan A, Sougnez C, Stange-Thomann N, Stojanovic N, Subramanian A, Wyman D, Rogers J, Sulston J, Ainscough R, Beck S, Bentley D, Burton J, Clee C, Carter N, Coulson A, Deadman R, Deloukas P, Dunham A, Dunham I, Durbin R, French L, Grafham D, Gregory S, Hubbard T, Humphray S, Hunt A, Jones M, Lloyd C, McMurray A, Matthews L, Mercer S, Milne S, Mullikin JC, Mungall A, Plumb R, Ross M, Shownkeen R, Sims S, Waterston RH, Wilson RK, Hillier LW, McPherson JD, Marra MA, Mardis ER, Fulton LA, Chinwalla AT, Pepin KH, Gish WR, Chissoe SL, Wendl MC, Delehaunty KD, Miner TL, Delehaunty A, Kramer JB, Cook LL, Fulton RS, Johnson DL, Minx PJ, Clifton SW, Hawkins T, Branscomb E, Predki P, Richardson P, Wenning S, Slezak T, Doggett N, Cheng JF, Olsen A, Lucas S, Elkin C, Uberbacher E, Frazier M, Gibbs RA, Muzny DM, Scherer SE, Bouck JB, Sodergren EJ, Worley KC, Rives CM, Gorrell JH, Metzker ML, Naylor SL, Kucherlapati RS, Nelson DL, Weinstock GM, Sakaki Y, Fujiyama A, Hattori M, Yada T, Toyoda A, Itoh T, Kawagoe C, Watanabe H, Totoki Y, Taylor T, Weissenbach J, Heilig R, Saurin W, Artiguenave F, Brottier P, Bruls T, Pelletier E, Robert C, Wincker P, Smith DR, Doucette-Stamm L, Rubenfield M, Weinstock K, Lee HM, Dubois J, Rosenthal A, Platzer M, Nyakatura G, Taudien S, Rump A, Yang H, Yu J, Wang J, Huang G, Gu J, Hood L, Rowen L, Madan A, Qin S, Davis RW, Federspiel NA, Abola AP, Proctor MJ, Myers RM, Schmutz J, Dickson M, Grimwood J, Cox DR, Olson MV, Kaul R, Shimizu N, Kawasaki K, Minoshima S, Evans GA, Athanasiou M, Schultz R, Roe BA, Chen F, Pan H, Ramser J, Lehrach H, Reinhardt R, McCombie WR, de la Bastide M, Dedhia N, Blocker H, Hornischer K, Nordsiek G, Agarwala R, Aravind L, Bailey JA, Bateman A, Batzoglou S, Birney E, Bork P, Brown DG, Burge CB, Cerutti L, Chen HC, Church D, Clamp M, Copley RR, Doerks T, Eddy SR, Eichler EE, Furey TS, Galagan J, Gilbert JG, Harmon C, Hayashizaki Y, Haussler D, Hermjakob H, Hokamp K, Jang W, Johnson LS, Jones TA, Kasif S, Kaspryzk A, Kennedy S, Kent WJ, Kitts P, Koonin EV, Korf I, Kulp D, Lancet D, Lowe TM, McLysaght A, Mikkelsen T, Moran JV, Mulder N, Pollara VJ, Ponting CP, Schuler G, Schultz J, Slater G, Smit AF, Stupka E, Szustakowski J, Thierry-Mieg D, Thierry-Mieg J, Wagner L, Wallis J, Wheeler R, Williams A, Wolf YI, Wolfe KH, Yang SP,

Yeh RF, Collins F, Guyer MS, Peterson J, Felsenfeld A, Wetterstrand KA, Patrinos A, Morgan MJ, de Jong P, Catanese JJ, Osoegawa K, Shizuya H, Choi S, Chen YJ (2001) Initial sequencing and analysis of the human genome. Nature 409 (6822):860–921. doi:10.1038/35057062

21. Sharp PA, Burge CB (1997) Classification of introns: U2-type or U12-type. Cell 91(7): 875–879
22. Turunen JJ, Niemela EH, Verma B, Frilander MJ (2013) The significant other: splicing by the minor spliceosome. Wiley Interdiscip Rev RNA 4(1):61–76. doi:10.1002/wrna.1141
23. Hall SL, Padgett RA (1994) Conserved sequences in a class of rare eukaryotic nuclear introns with non-consensus splice sites. J Mol Biol 239(3):357–365. doi:10.1006/jmbi.1994.1377
24. Jackson IJ (1991) A reappraisal of non-consensus mRNA splice sites. Nucleic Acids Res 19(14):3795–3798
25. Burge CB, Padgett RA, Sharp PA (1998) Evolutionary fates and origins of U12-type introns. Mol Cell 2(6):773–785
26. Bartschat S, Samuelsson T (2010) U12 type introns were lost at multiple occasions during evolution. BMC Genomics 11:106. doi:10.1186/1471-2164-11-106
27. Alioto TS (2007) U12DB: a database of orthologous U12-type spliceosomal introns. Nucleic Acids Res 35 (Database issue):D110–D115. doi:10.1093/nar/gkl796
28. Sheth N, Roca X, Hastings ML, Roeder T, Krainer AR, Sachidanandam R (2006) Comprehensive splice-site analysis using comparative genomics. Nucleic Acids Res 34(14):3955–3967. doi:10.1093/nar/gkl556
29. Janice J, Pande A, Weiner J, Lin CF, Makalowski W (2012) U12-type spliceosomal introns of Insecta. Int J Biol Sci 8(3):344–352. doi:10.7150/ijbs.3933
30. Bon E, Casaregola S, Blandin G, Llorente B, Neuveglise C, Munsterkotter M, Guldener U, Mewes HW, Van Helden J, Dujon B, Gaillardin C (2003) Molecular evolution of eukaryotic genomes: hemiascomycetous yeast spliceosomal introns. Nucleic Acids Res 31(4):1121–1135
31. Gao K, Masuda A, Matsuura T, Ohno K (2008) Human branch point consensus sequence is yUnAy. Nucleic Acids Res 36(7):2257–2267. doi:10.1093/nar/gkn073
32. Burge BB, Tuschl T, Sharp PA (1999) Splicing of precursors to mRNAs by the spliceosomes. In: Gesteland RF, Cech TR, Atkin JF (eds) RNA world, 2nd edn. Cold Spring Harbor Laboratory Press, Cold Spring Harbor, NY, pp 525–560
33. Shepard PJ, Hertel KJ (2009) The SR protein family. Genome Biol 10(10):242. doi:10.1186/gb-2009-10-10-242
34. Reed R, Maniatis T (1986) A role for exon sequences and splice-site proximity in splice-site selection. Cell 46(5):681–690
35. Valcarcel J, Green MR (1996) The SR protein family: pleiotropic functions in pre-mRNA splicing. Trends Biochem Sci 21(8):296–301
36. Berget SM (1995) Exon recognition in vertebrate splicing. J Biol Chem 270(6):2411–2414
37. Hertel KJ (2008) Combinatorial control of exon recognition. J Biol Chem 283(3):1211–1215. doi:10.1074/jbc.R700035200
38. De Conti L, Baralle M, Buratti E (2013) Exon and intron definition in pre-mRNA splicing. Wiley Interdiscip Rev RNA 4(1):49–60. doi:10.1002/wrna.1140
39. Dietrich RC, Peris MJ, Seyboldt AS, Padgett RA (2001) Role of the 3' splice site in U12-dependent intron splicing. Mol Cell Biol 21(6):1942–1952. doi:10.1128/MCB.21.6.1942-1952.2001
40. Wahl MC, Will CL, Luhrmann R (2009) The spliceosome: design principles of a dynamic RNP machine. Cell 136(4):701–718
41. Will CL, Luhrmann R (2011) Spliceosome structure and function. Cold Spring Harb Perspect Biol 3(7):Pii: a003707. doi:10.1101/cshperspect.a003707
42. Tarn WY, Steitz JA (1996) A novel spliceosome containing U11, U12, and U5 snRNPs excises a minor class (AT-AC) intron in vitro. Cell 84(5):801–811
43. Hall SL, Padgett RA (1996) Requirement of U12 snRNA for in vivo splicing of a minor class of eukaryotic nuclear pre-mRNA introns. Science 271(5256):1716–1718
44. Will CL, Luhrmann R (2005) Splicing of a rare class of introns by the U12-dependent spliceosome. Biol Chem 386(8):713–724. doi:10.1515/BC.2005.084
45. Frilander MJ, Steitz JA (1999) Initial recognition of U12-dependent introns requires both U11/5' splice-site and U12/branchpoint interactions. Genes Dev 13(7):851–863
46. Shen H, Green MR (2007) RS domain-splicing signal interactions in splicing of U12-type and U2-type introns. Nat Struct Mol Biol 14(7):597–603. doi:10.1038/nsmb1263

47. Grainger RJ, Beggs JD (2005) Prp8 protein: at the heart of the spliceosome. RNA 11(5):533–557. doi:10.1261/rna.2220705
48. Valadkhan S (2010) Role of the snRNAs in spliceosomal active site. RNA Biol 7(3):345–353
49. Rodriguez JR, Pikielny CW, Rosbash M (1984) In vivo characterization of yeast mRNA processing intermediates. Cell 39(3 Pt 2):603–610
50. Padgett RA, Konarska MM, Grabowski PJ, Hardy SF, Sharp PA (1984) Lariat RNA's as intermediates and products in the splicing of messenger RNA precursors. Science 225(4665):898–903
51. Ruskin B, Krainer AR, Maniatis T, Green MR (1984) Excision of an intact intron as a novel lariat structure during pre-mRNA splicing in vitro. Cell 38(1):317–331
52. Keller W (1984) The RNA lariat: a new ring to the splicing of mRNA precursors. Cell 39(3 Pt 2):423–425
53. Domdey H, Apostol B, Lin RJ, Newman A, Brody E, Abelson J (1984) Lariat structures are in vivo intermediates in yeast pre-mRNA splicing. Cell 39(3 Pt 2):611–621
54. Hernandez N, Keller W (1983) Splicing of in vitro synthesized messenger RNA precursors in HeLa cell extracts. Cell 35(1): 89–99
55. Hardy SF, Grabowski PJ, Padgett RA, Sharp PA (1984) Cofactor requirements of splicing of purified messenger RNA precursors. Nature 308(5957):375–377
56. Roca X, Krainer AR, Eperon IC (2013) Pick one, but be quick: 5′ splice sites and the problems of too many choices. Genes Dev 27(2):129–144. doi:10.1101/gad.209759.112
57. Ke S, Chasin LA (2010) Intronic motif pairs cooperate across exons to promote pre-mRNA splicing. Genome Biol 11(8):R84. doi:10.1186/gb-2010-11-8-r84
58. Semlow DR, Staley JP (2012) Staying on message: ensuring fidelity in pre-mRNA splicing. Trends Biochem Sci 37(7):263–273. doi:10.1016/j.tibs.2012.04.001
59. Crooks GE, Hon G, Chandonia JM, Brenner SE (2004) WebLogo: a sequence logo generator. Genome Res 14(6):1188–1190. doi:10.1101/gr.849004

Chapter 2

Diversity and Evolution of Spliceosomal Systems

Scott William Roy and Manuel Irimia

Abstract

The intron–exon structures of eukaryotic nuclear genomes exhibit tremendous diversity across different species. The availability of many genomes from diverse eukaryotic species now allows for the reconstruction of the evolutionary history of this diversity. Consideration of spliceosomal systems in comparative context reveals a surprising and very complex portrait: in contrast to many expectations, gene structures in early eukaryotic ancestors were highly complex and "animal or plant-like" in many of their spliceosomal structures; pronounced simplification of gene structures, splicing signals, and spliceosomal machinery has occurred independently in many lineages. In addition, next-generation sequencing of transcripts has revealed that alternative splicing is more common across eukaryotes than previously thought. However, much alternative splicing in diverse eukaryotes appears to play a regulatory role: alternative splicing fulfilling the most famous role for alternative splicing—production of multiple different proteins from a single gene—appears to be much more common in animal species than in nearly any other lineage.

Key words Spliceosomal introns, Evolution, Alternative splicing, Eukaryotes, Convergence

1 Similarities and Differences in the Spliceosomal System Across Species

Chapter 1 summarized the splicing reaction, describing a large number of the key features of the spliceosomal intron splicing machinery (the spliceosome) as well as the target of this machinery—the introns and more broadly the pre-mRNA transcripts themselves. The vast majority of our understanding of these topics comes from decades of study of a relatively small number of model species—in particular *S. cerevisiae*. More recently, genomic and transcriptomic sequencing of diverse species has allowed comparisons of these features between more eukaryotic lineages. These studies have ranged across approaches, topics, species, and conclusions, showing both differences and similarities in a wide variety of spliceosome-related phenomena. Surprisingly, given this diversity, the most important points of these studies may be largely summarized in two clear concepts: *(1) the spliceosomal system is ancestral, specific, and (nearly) universal to eukaryotes; and (2) the*

Klemens J. Hertel (ed.), *Spliceosomal Pre-mRNA Splicing: Methods and Protocols*, Methods in Molecular Biology, vol. 1126, DOI 10.1007/978-1-62703-980-2_2,

spliceosomal system shows phylogenetically complex patterns across eukaryotes, indicating recurrent transformation in diverse eukaryotes. We devote the next two sections to these two observations.

1.1 The Spliceosomal System is Ancestral, Specific, and (Nearly) Universal to Eukaryotes

Every fully sequenced nuclear genome from a eukaryotic organism contains both spliceosomal introns and recognizable spliceosomal components [1] (although *see* [2] for the one reported possible exception and [3] for the one known qualified exception). Moreover, the core features that define introns are also (nearly) completely conserved [4, 5]. The vast majority of known introns in every studied species begin with a donor site showing complete or partial complementarity to a standard U1 RNA sequence, in particular a 5′ "GT" dinucleotide, and nearly all introns in all studied species end with a 3′ terminal "AG" (e.g., Fig. 1). Available evidence suggests that the structure of the branchpoint sequence is also conserved across nearly all species: a region base pairing with the U2 RNA, with a "looped out" adenosine residue that performs the first nucleophilic attack. Also widespread across studied species is the polypyrimidine tract located somewhere within the 3′ end of the intron, although more diversity is found for this signal [5]. These observations about different species' intronic sequences interleave with observations about the core spliceosomal RNA components: U1–U6 snRNAs have been found across a wide variety of eukaryotes [6], with generally well-conserved RNA secondary structures and strict conservation of regions involved in base pairing between different snRNAs as well as between snRNAs and corresponding regions of pre-mRNA transcripts. Thus, all available evidence points to a highly conserved core spliceosomal reaction present in a wide variety of studied eukaryotes. Since the organisms known to share these features include representatives of all major known eukaryotic groups (or kingdoms), this implies that the spliceosome and spliceosomal introns were present in the eukaryotic ancestor and that the spliceosomal system has been retained in all or nearly all species through eukaryotic evolution.

On the other hand, no sequenced prokaryotic organism contains spliceosomal introns or any recognizable component of a spliceosome, indicating that the spliceosomal system is specific to eukaryotes. Interpretation of this second finding has been more contentious. The simplest interpretation is that the spliceosomal system, including a recognizably modern core splicing machinery and intron sequence characteristics, arose in the last common ancestor of eukaryotes (the modern "Introns-Late" hypothesis [7]). This interpretation mirrors findings that many cellular structures and processes are ancestral and specific to eukaryotes, suggesting a general interpretation that the lineage leading to the last ancestor of eukaryotes experienced an unmatched degree of fundamental cell and molecular structural innovation, including the rise of the spliceosomal system. While many authors have concluded that this hypothesis is by far the more likely alternative, this

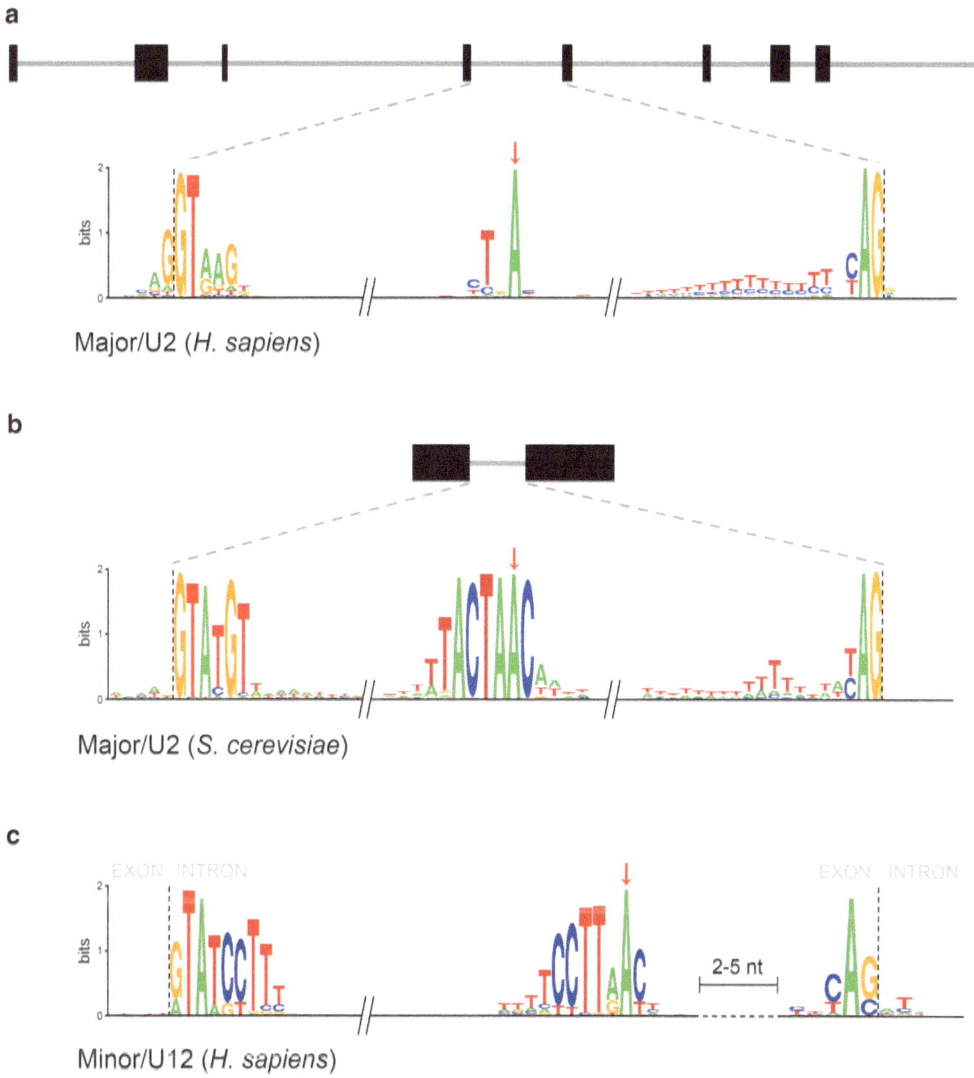

a

Major/U2 (*H. sapiens*)

b

Major/U2 (*S. cerevisiae*)

c

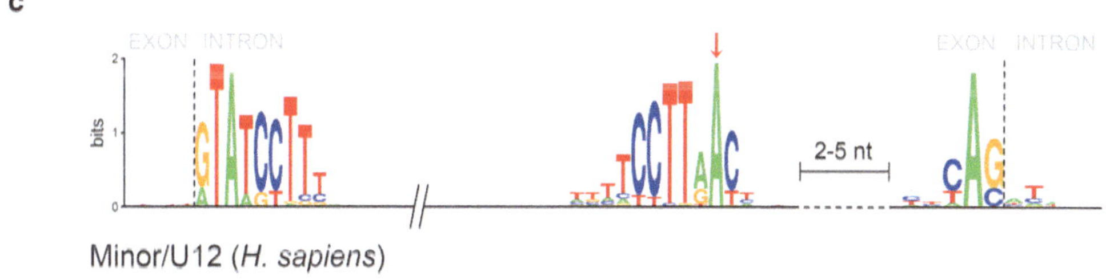

Minor/U12 (*H. sapiens*)

Fig. 1 Intron–exon structures and sequences of U2 (major) and U12 (minor) spliceosomal introns. (**a**) Human genes have frequent and long introns (*lines*) and correspondingly short exons (*boxes*). Human U2-type introns (accounting for >99 % of all human introns) have relatively little sequence homogeneity across intron sequences at the 5′ splice site (*left*), branchpoint (*center*), and 3′ splice site (*right*). (**b**) Introns in the model yeast *S. cerevisiae* are rarer and shorter, and exons longer, with much higher levels of homogeneity at core splice sites. (**c**) In contrast to U2 introns in most species, rare U12 introns show high levels of sequence homogeneity even in species where U2 introns show little homogeneity

perspective has failed to win over a variety of researchers who continue to favor the hypothesis that a system with at least some similarities to the modern spliceosomal system (for instance, high intron density) is even much older than early eukaryotes. Supporters of this "Introns-Early" perspective posit that introns were common in the ancestors of eukaryotes and prokaryotes and have been secondarily lost in both bacteria and archaea [8, 9].

1.2 Phylogenetically Complex Patterns of the Spliceosomal System in Eukaryotes

In stark contrast to this general conservation of the core splicing reaction and its associated machinery, early indications showed that many other aspects of the intron–exon structures of eukaryotic genomes are highly variable across species. Perhaps most striking is the difference in intron numbers. Intron number varies by many orders of magnitude per genome ([10]; Figs. 1 and 2). Whereas human genic transcripts are interrupted by an average of ~8.5 introns, *S. cerevisiae* genes contain only 0.05 introns on average, and extensive next-generation RNA sequencing of the protistan parasite *Trypanosoma brucei* has continually confirmed only two introns in this species' genome [11, 12]. The simplest explanation for these differences would be that intron number had been low in ancestral eukaryotes, with a single massive expansion leading to high intron numbers in one subset of eukaryotes (or alternatively, a single instance of massive loss from an intron-rich eukaryotic ancestor). In this case, we would expect to see high intron numbers to be characteristic of a group of related organisms: for example, in the case of massive expansion in a single event, all intron-rich species would be related. Instead, a very complex pattern is observed, with neither intron-rich nor intron-poor species forming a coherent phylogenetic group (Fig. 2). Very intron-poor organisms (say, with <0.1 introns per gene on average; blue in Fig. 2) are found in diverse eukaryotic groups whose most recent common ancestor is the last common ancestor of all eukaryotes. The same is true of intron-rich species: species with intron densities of at least a few introns per gene are found in disparate groups [4]. This pattern alone implies many different episodes of dramatic genomic change between states in which genomes are alternately nearly intronless or riddled with introns.

Intron length is also highly variable, with intron length distributions ranging widely across species. Median intron lengths range from 19 nts in the nucleomorph (a "mini" green algal nucleus) of the chlorarachniophyte protist *Bigelowiella natans* up to some 2 kb in humans (Fig. 2). Other aspects of the intron length distribution are very different across species as well—whereas the introns in the *B. natans* NM are nearly all within a few nucleotides in length (18–21 nts), human intron lengths are highly diverse, ranging from a few dozen to nearly one million nts. Moreover, intron length distributions can vary between closely related lineages. For example, the introns of tapeworms are sharply distributed around two main lengths (36 and 73 nt), whereas the related animal parasite *Schistosoma* shows only introns of 36 nt [13], implying either gain or loss or transformation of the 73 nt intron type across these species' history. Indeed, intron length distributions may differ significantly even between different classes of introns within a single genome, as recently reported for mammalian introns with different GC content [14].

Different organisms also show striking differences in their sequence characteristics. Particularly clear differences exist in the

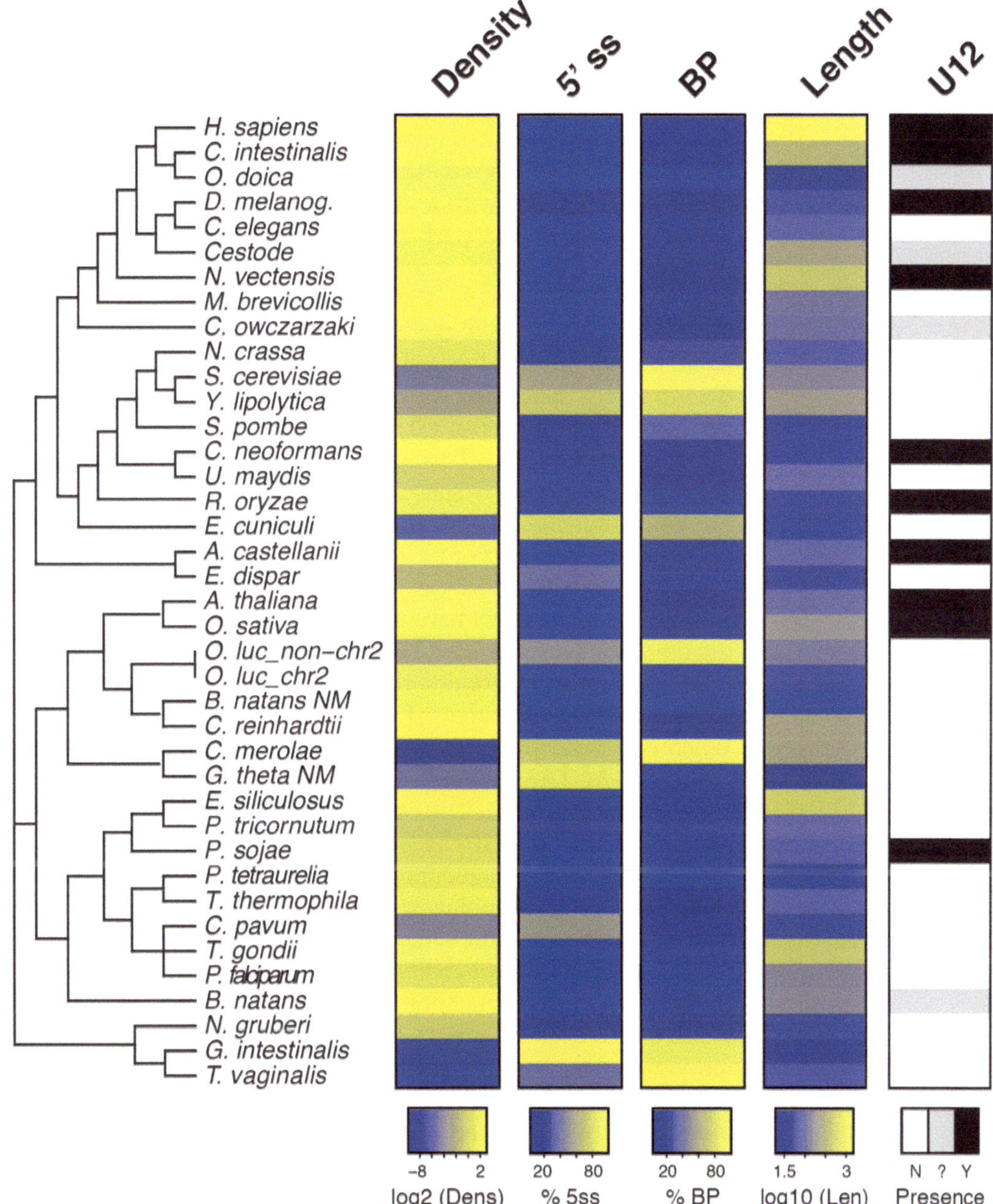

Fig. 2 Diversity of intron–exon structures across eukaryotes. Depicted are as follows: (1) intron density, in number of introns per gene; (2) the probability that two random introns have the same 5′ splice site beyond the canonical GT (in positions 3–6); (3) the fraction of introns exhibiting the exact same seven nucleotide branchpoint motif; (4) median intron length; and (5) presence/absence of minor/U12-type introns and associated splicing machinery

degree of "regularity" of core sequence motifs across introns within a species. For example, whereas nearly all introns in all species maintain significant complementarity between the 5′ splice site sequence and the U1 snRNA, this is accomplished in very different ways. In the model baker's yeast *S. cerevisiae*, this complementarity is packed into a strongly conserved hexamer region at the very beginning of the intron: some three-quarters of S. cerevisiae introns share the same tetramer sequence downstream of the canonical GT (i.e., positions +3 to +6, GTATGT), and nearly all remaining introns have a motif with a single nucleotide difference from this sequence (Fig. 1a). In stark contrast, exonic regions immediately upstream of the intron sequence (e.g., −3 to −1) do not show much preferential complementarity to the U1 sequence: base pairing is largely restricted to the beginning of the intron. On the other hand, human introns' base pairing to the U1 is less concentrated in the intronic 5′ splice site, with most introns having intron-U1 base pairs spread out across an extended region spanning both sides of the 5′ splice site. This flexibility of base pairing is reflected in a great diversity of core 5′ splice site sequences (Fig. 1b). One simple way of quantifying this diversity is to calculate the probability that two random introns from a species will have the same extended splice site sequence (positions +3 to +6). For instance, two random *S. cerevisiae* introns will have the same 5′ splice site nearly 58 % of the time, compared to 5.5 % of the time for human introns (Fig. 2).

Comparative genomics reveals similarly pronounced differences for other features of the core spliceosomal sequences. Whereas *S. cerevisiae* uses a highly regular extended branchpoint sequence (ACTAAC, where A is the branchpoint A) with exact complementarity to the corresponding U2 region, human branchpoint sequences are extremely diverse, to the extent that different sites can be used as branchpoints in a single intron [15]. Among characterized branchpoint sequences, the probability that two human introns share the same branchpoint motif is <1 %, whereas for *S. cerevisiae* the probability is 94 % (Fig. 2; [16]). Regularity of the position of the branchpoint relative to the 3′ end of the intron is also qualitatively different across species: the probability that two random introns have the exact same branchpoint position is <2 % in humans [16] (and is even low, 2 %, in *S. cerevisiae*), but is 67 % in the yeast species *Yarrowia lipolytica* (93 % of introns have the branchpoint A 6 nts (80 %) or 7 nts (13 %) upstream of the 3′ splice site.) As with intron number, species with regular and heterogeneous splicing signals are entwined on the evolutionary tree (Fig. 2; [4, 5]).

Species also show important differences in mechanisms and patterns of splicing. For example, while some components of the spliceosome—most notably the core snRNAs—are (nearly) universally conserved across species, other splicing factors show very different patterns. For instance, a new splicing factor involved in regulating the alternative splicing (AS) of a large number of genes in *Drosophila* was shown to have arisen in *Drosophila* ancestors by duplication of an ancestral factor and functional divergence [17].

This divergence included acquisition of new RNA sequence binding preferences and new biological functions (regulation of AS of dozens of genes in the testes). In other cases, proteins that are evolutionarily old may have acquired new splicing functions (i.e., non-splicing factors have become splicing factors) in specific lineages. One potentially interesting case may involve the splicing factor Nova. Nova is an important AS factor in metazoans [18–20], but Nova plant homologs may be involved in defense mechanisms against RNA viruses [21]. However more data on Nova and other deeply splicing factors in diverse eukaryotic lineages are necessary to confidently reconstruct the evolutionary history of the functions of auxiliary splicing factors.

2 Reconstructing the Evolutionary History of Spliceosomal Systems

Understanding the origins of the diversity of spliceosomal systems not only is interesting in its own right but is an indispensable starting point in understanding the evolution of key splicing innovations in specific lineages (for instance, alternative splicing in animals, see below), since the evolutionary history constrains hypotheses about the possible sets of evolutionary steps leading to these innovations. Therefore, we turn next to results of reconstructions of the evolutionary history of spliceosomal systems.

2.1 The Evolution of the Spliceosome(s)

Crucial to understanding the evolution of spliceosomal systems is understanding the history of the components of the spliceosome. A variety of comparative studies have confirmed that the majority of central and secondary spliceosomal proteins appear to date to the last common ancestor of all eukaryotes [1], completing the portrait of ancestral eukaryotes as having contained a recognizably modern spliceosomal system with a complex spliceosome splicing a large number of introns through a recognition system likely utilizing a diversity of intronic and exonic signals [22]. However, the spliceosomal machinery also appears to have undergone various elaborations in different lineages. In particular, animals and plants appear to have experienced an increase in the number of SR proteins (a family of splicing proteins with diverse core and auxiliary roles in splicing) and other accessory proteins by processes that are likely to have involved both duplication of SR proteins and evolution of new splicing roles for ancestral non-spliceosomal proteins [23, 24]. On the other hand, other lineages have seemingly lost some of the ancestral spliceosomal components, usually in association with massive intron loss. For instance, several human spliceosomal proteins seem to have no ortholog in the *S. cerevisiae* spliceosome [25].

Another question concerns the relative prevalence of intron definition and exon definition. While ultimately detailed molecular experiments are necessary to determine the mechanism of splicing of a given intron in a given species, the fact that the two different mechanisms tend to lead to different types of splicing variation in transcripts allows us to make educated guesses. Because in exon definition a spliceosome assembles across the length of an exon, failure of the spliceosome to assemble tends to lead to failure to "splice in" that exon, yielding exclusion of an exon in a transcript (called "exon skipping"). On the other hand, failure of a spliceosome to assemble across the length of an intron, in intron definition, tends to lead to failure to "splice out" that intron, leading to intron inclusion. These expected differences apply not only to splicing "errors" (nonfunctional splicing variants) but also to functional AS, since regulation of functional splicing generally occurs through modulation of spliceosomal assembly. Thus the relative incidence of exon skipping and intron retention in a species can yield insights into whether the species splices using exon definition, intron definition, or both mechanisms.

The largest many-species survey of splicing to date mapped available EST data from 42 species to their corresponding genomes to identify splicing variation [26]. They found that for the vast majority of species, levels of splicing variation were far lower than is found in characterized animals. They also found that the mode of splicing variation in most groups of organisms differed from that in animals: whereas animals use extensive exon skipping, nearly all nonanimal species studied had a higher incidence of intron retention. More recent studies of individual species have complicated the issue in plants, which appear to exhibit relatively frequent (and functional) exon skipping [27, 28]; however, the general pattern has held: the major mode of splicing variation in most species is intron retention. These results suggest that the vast majority of eukaryotic lineages primarily splice by intron definition and thus that intron definition is the ancestral mode of intron recognition, with exon definition arising during the evolution of animals (and perhaps, independently, in other lineages [29, 30]).

2.1.1 Notes on the U12 Spliceosomal System

Given the central focus of the book, we have focused on the "major" or "U2" spliceosome and its associated introns. U2 introns make up the vast majority of introns (typically >99 %) in all studied species. However, in some species there also exists a second separate spliceosome which is responsible for splicing of a small subset of introns. This second system (both machinery and associated introns) is referred to as the "U12" or "minor" system, after one of the four separate snRNAs that form the core of the U12 spliceosome. Termed U11, U12, U4atac, and U6atac, these components roughly correspond respectively to the U1, U2, U4, and U6 snRNAs of the major spliceosome (also called the U2 spliceosome). The U5 snRNA is involved in both spliceosomal systems. Spliceosomal proteins show a more complex pattern, with some proteins showing specificity for either the U2 or U12 spliceosome and others being associated with both systems. Splicing signals of the U12 system broadly correspond to those in the U2 system, with important and intriguing differences. Relative to U2 introns, U12 introns show more flexibility at core splice sites (with both GT…AG and AT…AC boundaries observed) but less flexibility at extended 5′ splice site and branchpoint signals (Fig. 1c; [33]). U12 branchpoints also show more conserved and more 3′ proximal positions (Fig. 1c), the latter of which is likely related to the general lack of a 3′ polypyrimidine tract. The evolutionary origins and functional importance of this remarkable "dual" spliceosomal system remain matters of debate.

Comparative genomics has revealed the broad contours of the evolutionary history of the U12 system. First, the U12 spliceosomal system (both U12-specific components and U12 introns) is found in a variety of very distantly related eukaryotic lineages, in a pattern that strongly suggests presence of a U12 system in the ancestor of all eukaryotes [6, 31]. Second, comparison of orthologous genes has revealed a large number of apparent cases of U12-to-U2 conversions, but few cases of U2-to-U12 conversion [32, 33]. Perhaps relatedly, whereas the U2 spliceosomal system has shown remarkable resilience across species (with no clear case of complete loss of the U2 system known), the U12 system appears to have been lost completely dozens of times independently through eukaryotic evolution, with ancestral U12 introns being either deleted from genomes or converted into U2 introns (Fig. 2) [6].

2.2 The Evolution of Spliceosomal Introns

In this section we will discuss various studies that have reconstructed the evolution of the three major intron features outlined above: intron density, intron sequence, and intron length. Before we proceed, however, it is worthwhile to clearly distinguish between two aspects of an intron: intron position and intron sequence. "Intron sequence" refers to the specific sequence of nucleotides of a specific intron (i.e., the region removed from RNA transcripts). "Intron position" is defined with reference to the final pre-mRNA transcript sequence—that is, the position of the junction between two flanking exons following intron removal (Fig. 3). In many lineages, these two traits of an intron show very different, even opposed, modes of evolution. Consistent with their removal from transcripts and subsequent degradation, most intron sequences evolve quickly, primarily by classic "micro" mutations (base pair substitutions and small indels or transposable element insertion and deletions). A change in intron position, by contrast, involves either gain or loss of an entire intron (and thus gain/loss of an intron position [34]) or intron sliding (a poorly understood and debated mutation or series of mutations leading to movement of an intron along the sequence of a gene [35, 36]). In some lineages, such intron loss and gain mutations are quite rare (see

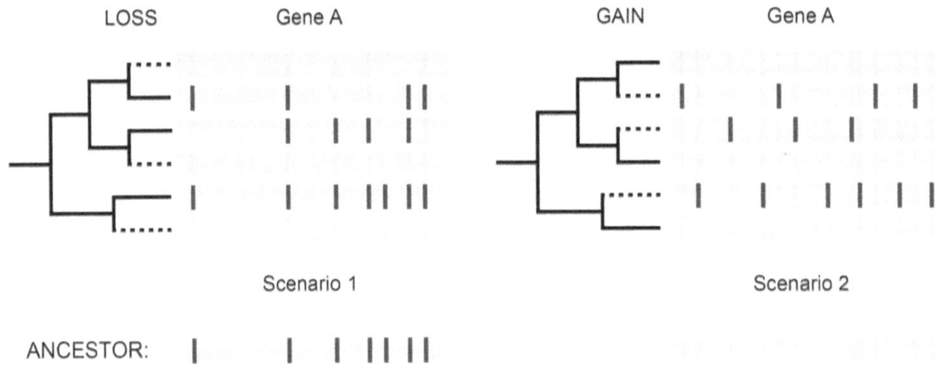

Fig. 3 Intron position comparisons reveal ancestral intron density. Illustrations are given for the cases in which (1) intron positions are shared across species, revealing the presence of introns in the ancestor (*Scenario 1*), or (2) intron positions are largely different across species, revealing that modern introns have been inserted since the common ancestor of the species (*Scenario 2*). In each case, the *gray boxes* represent aligned coding sequence (i.e., after intron removal), with the *blue vertical lines* representing intron positions (i.e., the position of the intronic sequence before removal). In the accompanying phylogenies, *dotted lines* represent lineages undergoing pronounced change, whether primarily intron *loss* (*on the left*) or intron *gain* (*on the right*)

below): in this case intron sequences generally evolve quickly, while intron positions evolve very slowly.

2.2.1 Intron Density

In the simplest case, the dramatic differences in intron–exon structures observed across all species (Fig. 2) could be explained by a single process—either intron loss (deletion) or gain (creation)—acting through eukaryotic evolution. It became clear relatively early on that the situation was not so simple. Study of two duplicated insulin genes in rat showed that one copy had lost an intron [37], while restriction of some introns in the triose-phosphate isomerase gene to one or a few related species provided strong evidence for intron gain [38]. With both processes demonstrated, debate turned to distinguishing the two processes' relative roles and importance in evolution and to reconstruct intron density in ancestral genes.

The most common comparative approach to infer intron gain/loss and reconstruct ancestral states is relatively straightforward (Fig. 3). If an ancestor of two modern organisms had few introns, and the introns in each organism have been created since their divergence, we might expect that the intron positions in these two species—that is, the positions at which the introns interrupt the coding sequence—would have little or no correspondence above random chance (Fig. 3, right). By contrast, if the ancestor had a large number of introns, and if these introns have not been lost, we would expect to find introns in the same position—that is, they would interrupt the coding portion of genes at corresponding (homologous) positions (Fig. 3, left). Closely following on the availability of the first full and partial genome sequences, a few studies sought to compare intron positions across species to probe intron loss and gain dynamics. By comparing intron positions in 1,560 pairs of homologous genes in humans and mouse, we found nearly complete intron correspondence (>99 % of human introns were matched by an intron at the exact same position in mouse), indicating that both intron loss and gain can be very slow in some lineages [34]. At a much deeper level, genomic sequencing of a handful of genes from jakobid protists showed that intron positions in these deeply diverged organisms showed surprising correspondence to intron positions in homologs from very distantly related eukaryotes, with half found at the exact homologous position in the gene [39]. An eight-species study also showed a high percentage of exact intron position correspondence over long evolutionary distances, with, for instance, a quarter of intron positions corresponding between humans and *Arabidopsis* [40].

While these studies would seem to indicate that many modern introns are very old, another possibility is that these coinciding intron positions in different species are just that: coincidences, with introns being inserted into identical (homologous) positions multiple times independently. However, direct tests from a set of "natural biological" experiments, in which introns are known to have been independently inserted into homologous genes in different organisms, found few correspondences [41–43]. These observations suggest that a large fraction of the observed coincident positions reflect true ancestral introns that have been retained in modern species, indicating that early eukaryotic ancestors were relatively intron rich (i.e., at the least, genes in early eukaryotic ancestors had one or a few introns per gene).

In the past few years, a series of statistical models of increasing sophistication (taking into account the possibility of convergent intron insertion and differences in rates of loss and gain across sites and across lineages), as well as ever-expanding comparative genomic databases, have been used to estimate ancestral intron densities [44–51]. Nearly all of these studies have estimated that intron densities in early eukaryotic ancestors were high by modern standards, falling within the range of modern animal species [52, 53]. Additional studies of intron loss and gain across different groups of organisms have further clarified the evolutionary history, leading to a general picture that most eukaryotic lineages experience very few intron gains (and generally more intron loss, ranging from slightly and dramatically more [54–57]). However, a growing number of exceptional lineages have been reported, in which intron gain is an active and ongoing process, potentially "replenishing" relatively intron-poor organisms with a large number of new introns [58–61].

2.2.2 Intron Structures: Splicing Sequence Motifs

As mentioned above, eukaryotic organisms differ considerably in their splicing motifs, ranging from the highly homogeneous 5′ splice site and branchpoint site sequences and branchpoint positions found in the yeast *Yarrowia lipolytica* to the heterogeneous structures characterizing human intron sequences. Notably, as discussed in more detail elsewhere in this book, these differences seem to involve a greater reliance on auxiliary splicing signals (generally lying in proximal regions of introns and exons) by species with heterogeneous core splicing signals. For instance, in humans, the boundaries of exons (i.e., exonic regions near intron–exon boundaries) are enriched in certain sequence motifs, which affect splicing by serving as "exonic splicing enhancers" (ESEs) by binding spliceosomal proteins and promoting splicing at the neighboring splice site [62]. By contrast, in species such as *S. cerevisiae*, ESEs are thought to not play a major role in splicing—intron recognition signals are concentrated in the core intronic splicing motifs.

What is the history of these recognition systems and splicing motifs? Initially it was often assumed that the "simpler" system of *S. cerevisiae* was ancestral and that increased complexity of mechanism arose in animals [63]. Widespread genomic evidence allowed for the possibility to test this notion. We studied full-genome intron complements from 50 diverse eukaryotic species to reconstruct the evolution of intron sequences and recognition [4]. First, we examined 5′ splice signals. We found that 5′ splice sites are heterogeneous in most species and that cases such as *S. cerevisiae* represent exceptions. For nearly all species studied, the probability that two random introns use the same hexamer splice site was <5 % (Fig. 2). However, there were a few clear exceptions, with several distantly related species showing a much higher level of homogeneity. Viewed on the evolutionary tree, these exceptional lineages fall within much larger phylogenetic groups of species with more typical splice signals. This phylogenetic pattern suggests that ancestral splice site sequences were heterogeneous and that the several species or groups of species with homogeneous splice sites evolved independently.

Even more unexpectedly, scrutiny of the specific lineages that have acquired homogeneous signals revealed that they were exactly the same lineages known to have very low modern intron densities (<0.1 introns per gene, blue in Fig. 2), with no known exceptions. Together these patterns indicate that early eukaryotic ancestral genes were roughly "animal-like" in their intron–exon structures, with high intron densities and heterogeneous 5′ splice sites, and that at several times through evolution, different lineages have experienced massive intron loss tightly coupled to the evolution of homogeneous 5′ splice site signals.

We and others also studied 3′ intron sequences [5, 64]. First, we studied branchpoint motifs. Because branchpoints in some species can be so diverse as to be difficult to identify computationally

[15, 65], we used a different metric: the fraction of introns that exhibited the same branchpoint-like sequence motif (i.e., a motif with the potential to base pair with the U2 snRNA with a protruding A nucleotide). For most organisms, we found no single dominating branchpoint motif, indicating heterogeneous branchpoint sequences (Fig. 2). However, again, a small subset of organisms including *S. cerevisiae* exhibited homogeneous branchpoints, with a majority of introns having the same clear branchpoint-like sequence [5]. This subset of organisms proved to be a subset of the studied intron-poor species. Thus low intron density appears to be closely associated with, but not sufficient for, the evolution of homogeneous branchpoint signals.

Finally, we studied the stretch of intronic nucleotides just upstream of the 3′ splice site. Again, for most species we found no clear motif preference (with the exception of a weak polypyrimidine tract). However a few species showed a clear preferred extended 3′ splice site, which was found to represent a branchpoint motif falling at a regular distance from the 3′ terminus—that is, the branchpoint is "anchored" to the 3′ end of the intron at a highly constrained distance [5]. These species proved to be a subset of species that have homogeneous branchpoint motifs. In total, then, these studies may be summarized as follows: all intron-poor lineages have homogeneous 5′ splice sites, a subset of which have homogeneous branchpoints, a subset of which have homogeneous 3′ splice sites owing to anchoring of the homogeneous branchpoint at a specific position a few nucleotides upstream of the 3′ terminus.

This unexpectedly clear pattern is still not well understood. The most obvious hypothesis would be that these changes in the recognition signals are associated with changes in the spliceosome. This hypothesis initially defied direct testing until a natural experiment presented itself, in the form of the sequenced genomes of multiple species from an evolutionarily old group of related algae. Each species' genome showed striking differentiation in intron density across genomic regions: in contrast to genes in most of the genome, which have very few introns (~0.1 per gene), the genes on one chromosome have much higher intron densities (around two introns per gene) [66]. Scrutiny of the genome sequence revealed a single set of core spliceosomal components [5], indicating that there is no evidence that entirely separate spliceosomes are responsible for splicing in the two genomic regions: thus if changes in the spliceosome are responsible for (or closely associated with) changes in splice signals, we would expect introns in both regions of the genome to show similar levels of splice signal homogeneity. Instead, the genomic regions show clear differentiation along the exact lines expected from the across-species comparisons: introns in the intron-rich region of the genome show very heterogeneous splice signals and no recognizable branchpoints, while introns in the intron-poor majority of the genome have homogeneous 5′ splice sites and branchpoint sequences [5]. The differences in intron number and splice motif homogeneity are found across distantly related species likely spanning many millions of years of evolution; thus, this association is long-lived, not transient.

Another issue involves the evolution of ESEs, which are abundant in animal genomes but absent or nearly absent from *S. cerevisiae*. ESEs were initially recognized at the genome-wide level by identifying sequence motifs that were overrepresented in the portions of exons near intron–exon boundaries relative to more distant portions of exons, and overrepresented near intronic splice sites that were "weak" (i.e., had low predicted binding to spliceosomal uRNAs), and which were subsequently confirmed by in vitro and in vivo studies to affect splicing [67, 68]. To test whether a similar signal existed in diverse other eukaryotes, Warnecke and coauthors [67] sought motifs that were overrepresented near exon–intron boundaries relative to interior regions of exons. They found putative ESE motifs in most studied intron-rich eukaryotes, but no evidence for ESEs in studied intron-poor species. This again suggested that the animal-like state (considerable reliance on ESEs for splicing) was ancestral to eukaryotes and that the spliceosomal systems in intron-poor lineages such as *S. cerevisiae* have been altered through evolution.

In total, then, comparative studies of intronic and exonic sequences over long evolutionary distances within eukaryotes support a model in which ancestral eukaryotes had "animal-like" intron–exon structures, with frequent introns spliced by use of a combination of diffuse motifs including frequent ESEs and heterogeneous core splicing motifs. Over the course of evolution, many lineages have changed significantly, shedding the vast majority of their introns, evolving homogeneous core splicing motifs, and significantly decreasing dependence on auxiliary splicing motifs such as ESEs.

2.2.3 Intron Length

The third feature of introns that shows striking diversity is intron length. Introns show a wide variety of lengths both within and between organisms, with lengths spanning multiple orders of magnitude. Studies across many eukaryotic organisms, particularly whole genome sequencing projects, have shown that the vast majority of species have relatively short introns, often with a peak around 60 nucleotides. While it is difficult to directly reconstruct intron length over long evolutionary distances, as introns appear to readily expand and contract along with genome size [69–71], this clear preference for generally short intron length across eukaryotes suggests that it represents the ancestral condition (although it has been suggested that the most ancestral introns, presumably evolved from self-splicing group II introns, may have been much longer, perhaps around 2,000 nts [53]).

Against this backdrop of generally short introns, several lineages show very different patterns. On the one hand, many different lineages from very different groups (animals [72, 73], relatives of green algae [74], and ciliates [75]) have evolved very short introns with median lengths around 20 nts. The clearest exception at the other end of the spectrum is some animals, particularly mammals [76], in which many species have median intron lengths ranging from a couple hundred to a couple thousand nucleotides. It seems likely that there are other lineages with generally long introns yet to be discovered, particularly given that (1) the correspondence between intron and genome size suggests that organisms with long introns would tend to have large genomes; (2) genome sequencing efforts tend to be biased specifically against organisms with large genomes, because of technical difficulties of sequencing and annotation.

3 Diversity and Evolution of Alternative Splicing

Up to this point, we have focused on differences in the genomic structures and in the splicing machinery and intron recognition mechanisms. We now briefly turn to the ways that these structures are used to generate transcriptional diversity by differential splicing of transcripts of the same gene, that is, alternative splicing (AS). The types, mechanisms, and functions of AS will be discussed extensively in Chapters 4 and 5, so here we confine our discussion to AS in the broader context of intron and genome evolution.

The most well-known function of AS is to generate multiple proteins with distinct functional properties from a single gene. However, decades of research have made clear that other forms of splicing diversity in which some transcript variants do not encode proteins are very common. Many genes in animals harbor alternatively spliced "poison exons" whose inclusion in transcripts leads to disruption of the protein-coding sequence [77]. Many of these transcripts are rapidly degraded by the nonsense-mediated decay (NMD) machinery; the fates of others remain obscure, however, the lack of an extended protein-coding region suggests these transcripts are unlikely to encode proteins. Such nonprotein coding variation is usually referred to "unproductive" AS, in contrast to "productive" or multi-protein AS [78]. It is important to point out that very clear evidence exists for functional roles for many of these cases of unproductive splicing: much unproductive splicing is evolutionarily conserved and/or regulated across environmental conditions, development, life cycles, or tissue or cell types [77, 79]. However, it is also likely that nonfunctional splicing errors that lead to transcript diversity with no function also occur (even if it is the case that confidently classifying a given AS event as either nonfunctional variation or functional nonproductive AS can be technically different). Thus in the following we distinguish between three types of AS: productive, unproductive, and nonfunctional.

AS is an extremely important and active process in animals, with the vast majority of multi-exon genes undergoing AS in diverse animal species (e.g., an estimated 95 % in humans [80, 81] and 60 % in fruit fly [82]). Animal AS uses a wide variety of mechanisms including single exon skipping, coordinated splicing of groups of exons, mutually exclusive splicing of pairs (or sets) of exons, alternative 5′ and 3′ splice sites, and intron retention [83]. AS is involved in a wide array of biological processes from sex determination to development to negative autoregulation and generates both productive and unproductive transcripts (*see* Chapters 4 and 5 for further examples).

Initial studies of nonanimal eukaryotes found a dearth of animal-like productive AS. In comparison to the thousands of cases of productive AS uncovered by transcriptomic studies in animals, for a long time no productive AS was known in *S. cerevisiae*, and cases in other species were only few and far between. Both reason and evidence suggest that AS would be facilitated by a variety of features of animals' intron–exon structures: (1) Large numbers of introns provide many opportunities for AS. (2) Heterogeneous intron boundaries, with associated differences in the strength of base pairing with the spliceosomal RNAs, allow for the possibility of regions for which recognition by the spliceosome might be "borderline"—leading to non-constitutive splicing of these regions. (3) Utilization of a variety of heterogeneous splicing signals—exonic and intronic splicing regulators, in addition to core splicing signals—allows for the possibility of regulating local splicing by regulation of the splicing factors that bind subsets of these signals. (4) Long introns increase opportunities for novel alternative exon creation [84–86] and are associated with AS in vertebrates [76].

The fact that these features each differ considerably between AS-rich animals and the model organism for splicing, *S. cerevisiae*, initially suggested that a wholesale remodeling of gene structures had occurred in animals roughly coincident with a rise of ubiquitous AS. However, as discussed above, genomic-era studies have shown that the story is quite different from this: many of the features associated with AS in animals—frequent introns, heterogeneous splicing boundaries, introns with lengths exceeding "minimal" intron lengths, and utilization of auxiliary splicing signals—are not specific to animals, but are in fact quite common in modern eukaryotes as well as characteristic of eukaryotic ancestors [22]. Thus, the hypothesis that widespread productive AS in animals is "due" to these features, a hypothesis still commonly invoked in passing in publications, is strongly rejected, since these features are common in organisms with little or no productive AS.

Furthermore, more recently, transcriptomic studies have opened up questions about the incidence of AS in diverse eukaryotic organisms. Initially it was thought by some authors that AS was absent or very rare in unicellular species [63]. However, genomic and transcriptomic data has greatly changed that picture. Perhaps the clearest case involves splicing of ribosomal protein-coding genes in *S. cerevisiae* [87, 88]. Introns in *S. cerevisiae* are massively overrepresented in ribosomal protein-coding genes, with half of the introns in the genome packed into only a few percent of the genes. A series of studies have shown that many ribosomal protein-coding gene (RPG) introns are regulated in response to environmental changes to produce either spliced protein-coding or unspliced sterile transcripts. This apparent regulatory role for RPG introns suggests that overrepresentation of introns in RPGs reflects selection favoring retention and/or creation of specifically these introns. This would in turn imply that at least half of introns in *S. cerevisiae* have been retained through evolution due to functional AS.

Other studies have begun to suggest that AS plays important roles in a wide variety of eukaryotes. Transcriptomic studies have found between several dozen and several hundred apparent cases of AS in the genomes of nearly all species studied to date, including diverse fungi [89–91], plants [27, 92–94], apicomplexans [95], cryptophytes [96], green algae [97], ciliates [98], and amoebozoa [99] (although studies of two other protists have drawn the opposite conclusion [100]). Nearly all of these studies have found a preponderance of intron retentions, with far smaller numbers of exon skipping events (and often intermediate numbers of alternative splice sites), even in plants [101]. These observations suggest that intron retention has predominated through eukaryotic history in diverse organisms. The one clear exception described so far is the chlorarachniophyte *Bigelowiella natans* [96], which shows striking levels of both intron retention and exon skipping, the latter only comparable to AS levels in the human cortex, which exhibits the highest levels of AS described so far [102].

In total, then, genomic and transcriptomic data have painted a very different picture of the history of AS (productive and otherwise) in animals. Features of animal intron–exon structures (long and frequent introns with diverse splicing signals) are not closely associated with animal-type AS, and AS is far from exclusive to animals, being found across phylogenetically and biologically diverse eukaryotic organisms. The one remaining feature of animal genomes that may still be rare in other organisms is exon definition. Therefore, it has been suggested that the evolution of exon definition, together with the specific expansion of SR proteins and other splicing factors, may be behind the transition from intron retention to exon skipping at the origin of animals [29].

4 Summary

A comparative perspective on spliceosomal systems of diverse eukaryotes paints a surprising portrait: ancestral eukaryotic genes were riddled with introns characterized by heterogeneous splice signals, requiring two distinct complex spliceosomes for intron removal and quite possibly involving some level of functional regulatory alternative splicing, likely dominated by intron retention. Since that time, different lineages have experienced very different evolutionary trajectories ranging from nearly complete intron loss to intron length expansion and episodic intron creation. The one feature of animal gene structures that remains as clearly exceptional is the widespread production of multiple proteins from one gene, although recent findings in *B. natans* suggest that animals may not be entirely alone in this characteristic.

References

1. Collins L, Penny D (2005) Complex spliceosomal organization ancestral to extant eukaryotes. Mol Biol Evol 22:1053–1066
2. Andersson JO, Sjögren AM, Horner DS et al (2007) A genomic survey of the fish parasite Spironucleus salmonicida indicates genomic plasticity among diplomonads and significant lateral gene transfer in eukaryote genome evolution. BMC Genomics 8:51
3. Lane CE, van den Heuvel K, Kozera C et al (2007) Nucleomorph genome of Hemiselmis andersenii reveals complete intron loss and compaction as a driver of protein structure and function. Proc Natl Acad Sci USA 104:19908–19913
4. Irimia M, Penny D, Roy SW (2007) Co-evolution of genomic intron number and splice sites. Trends Genet 23:321–325
5. Irimia M, Roy SW (2008) Evolutionary convergence on highly-conserved 3′ intron structures in intron-poor eukaryotes and insights into the ancestral eukaryotic genome. PLoS Genet 4:e1000148
6. Dávila LM, Rosenblad MA, Samuelsson T (2008) Computational screen for spliceosomal RNA genes aids in defining the phylogenetic distribution of major and minor spliceosomal components. Nucleic Acids Res 36:3001–3010
7. Koonin EV (2006) The origin of introns and their role in eukaryogenesis: a compromise solution to the introns-early versus introns-late debate? Biol Direct 1:22
8. Vibranovski M, Sakabe N, Oliveira R et al (2005) Signs of ancient and modern exon-shuffling are correlated to the distribution of ancient and modern domains along proteins. J Mol Evol 61:341–350
9. Penny D, Hoeppner MP, Poole AM et al (2009) An overview of the introns-first theory. J Mol Evol 69:527–540
10. Logsdon J (1998) The recent origins of spliceosomal introns revisited. Curr Opin Genet Dev 8:637–648
11. Siegel TN, Hekstra DR, Wang X et al (2010) Genome-wide analysis of mRNA abundance in two life-cycle stages of Trypanosoma brucei and identification of splicing and polyadenylation sites. Nucleic Acids Res 38: 4946–4957
12. Kolev NG, Franklin JB, Carmi S et al (2010) The transcriptome of the human pathogen Trypanosoma brucei at single-nucleotide resolution. PLoS Pathog 6:e1001090
13. Tsai IJ, Zarowiecki M, Holroyd N et al (2013) The genomes of four tapeworm species reveal adaptations to parasitism. Nature 496(7443):57–63
14. Amit M, Donyo M, Hollander D et al (2012) Differential GC content between exons and introns establishes distinct strategies of splice-site recognition. Cell Rep 1:543–556
15. Kol G, Lev-Maor G, Ast G (2005) Human-mouse comparative analysis reveals that branch-site plasticity contributes to splicing regulation. Hum Mol Genet 14:1559–1568
16. Gao K, Masuda A, Matsuura T et al (2008) Human branch point consensus sequence is yUnAy. Nucleic Acids Res 36:2257–2267
17. Taliaferro JM, Alvarez N, Green RE et al (2011) Evolution of a tissue-specific splicing network. Genes Dev 25:608–620

18. Brooks AN, Yang L, Duff MO et al (2011) Conservation of an RNA regulatory map between Drosophila and mammals. Genome Res 21:193–202
19. Irimia M, Denuc A, Burguera D et al (2011) Stepwise assembly of the nova-regulated alternative splicing network in the vertebrate brain. Proc Natl Acad Sci USA 108:5319–5324
20. Jensen KB, Dredge BK, Stefani G et al (2000) Nova-1 regulates neuron-specific alternative splicing and is essential for neuronal viability. Neuron 25:359–371
21. Fujisaki K, Ishikawa M (2008) Identification of an Arabidopsis thaliana protein that binds to tomato mosaic virus genomic RNA and inhibits its multiplication. Virology 380:402–411
22. Roy SW, Irimia M (2009) Splicing in the eukaryotic ancestor: form, function and dysfunction. Trends Ecol Evol 24:447–455
23. Barbosa-Morais NL, Carmo-Fonseca M, Aparicio S (2006) Systematic genome-wide annotation of spliceosomal proteins reveals differential gene family expansion. Genome Res 16:66–77
24. Reddy AS, Shad AG (2011) Plant serine/arginine-rich proteins: roles in precursor messenger RNA splicing, plant development, and stress responses. Wiley Interdiscip Rev RNA 2:875–889
25. Plass M, Agirre E, Reyes D et al (2008) Co-evolution of the branch site and SR proteins in eukaryotes. Trends Genet 24:590–594
26. McGuire A, Pearson M, Neafsey D et al (2008) Cross-kingdom patterns of alternative splicing and splice recognition. Genome Biol 9:R50
27. Marquez Y, Brown JW, Simpson C et al (2012) Transcriptome survey reveals increased complexity of the alternative splicing landscape in Arabidopsis. Genome Res 22:1184–1195
28. Carvalho RF, Feijão CV, Duque P (2012) On the physiological significance of alternative splicing events in higher plants. Protoplasma 250(3):639–650
29. Keren H, Lev-Maor G, Ast G (2010) Alternative splicing and evolution: diversification, exon definition and function. Nat Rev Genet 11:345–355
30. Ram O, Ast G (2007) SR proteins: a foot on the exon before the transition from intron to exon definition. Trends Genet 23:5–7
31. Russell AG, Charette JM, Spencer DF et al (2006) An early evolutionary origin for the minor spliceosome. Nature 443:863–866
32. Burge CB, Padgett RA, Sharp PA (1998) Evolutionary fates and origins of U12-type introns. Mol Cell 2:773–785
33. Alioto TS (2007) U12DB: a database of orthologous U12-type spliceosomal introns. Nucleic Acids Res 35:D110–D115
34. Roy SW, Fedorov A, Gilbert W (2003) Large-scale comparison of intron positions in mammalian genes shows intron loss but no gain. Proc Natl Acad Sci USA 100:7158–7162
35. Tarrío R, Ayala FJ, Rodríguez-Trelles F (2008) Alternative splicing: a missing piece in the puzzle of intron gain. Proc Natl Acad Sci USA 105:7223–7228
36. Rogozin IB, Lyons-Weiler J, Koonin EV (2000) Intron sliding in conserved gene families. Trends Genet 16:430–432
37. Perler F, Efstratiadis A, Lomedico P et al (1980) The evolution of genes: the chicken preproinsulin gene. Cell 20:555–566
38. Logsdon J Jr, Tyshenko M, Dixon C et al (1995) Seven newly discovered intron positions in the triose-phosphate isomerase gene: evidence for the introns-late theory. Proc Natl Acad Sci USA 92:8507–8511
39. Archibald J, O'Kelly C, Doolittle W (2002) The chaperonin genes of jakobid and jakobid-like flagellates: implications for eukaryotic evolution. Mol Biol Evol 19:422–431
40. Rogozin I, Sverdlov A, Babenko V et al (2005) Analysis of evolution of exon–intron structure of eukaryotic genes. Brief Bioinform 6:118–134
41. Roy SW, Penny D (2007) A very high fraction of unique intron positions in the intron-rich diatom Thalassiosira pseudonana indicates widespread intron gain. Mol Biol Evol 24: 1447–1457
42. Ahmadinejad N, Dagan T, Gruenheit N et al (2010) Evolution of spliceosomal introns following endosymbiotic gene transfer. BMC Evol Biol 10:57
43. Yoshihama M, Nakao A, Nguyen HD et al (2006) Analysis of ribosomal protein gene structures: implications for intron evolution. PLoS Genet 2:e25
44. Roy SW, Gilbert W (2005) Complex early genes. Proc Natl Acad Sci USA 102: 1986–1991
45. Csuros M (2006) On the estimation of intron evolution. PLoS Comput Biol 2:e84
46. Csuros M (2008) Malin: maximum likelihood analysis of intron evolution in eukaryotes. Bioinformatics 24:1538–1539

47. Csurös M (2005). Likely scenarios of intron evolution. In: Third RECOMB Satellite workshop on comparative genomics. Springer LNCS 3678, p 47–60
48. Csurös M, Rogozin IB, Koonin EV (2008) Extremely intron-rich genes in the alveolate ancestors inferred with a flexible maximum-likelihood approach. Mol Biol Evol 25:903–911
49. Nguyen H, Yoshihama M, Kenmochi N (2005) New maximum likelihood estimators for eukaryotic intron evolution. PLoS Comput Biol 1:e79
50. Carmel L, Wolf YI, Rogozin IB et al (2007) Three distinct modes of intron dynamics in the evolution of eukaryotes. Genome Res 17:1034–1044
51. Carmel L, Rogozin IB, Wolf YI et al (2009) A maximum likelihood method for reconstruction of the evolution of eukaryotic gene structure. Methods Mol Biol 541:357–371
52. Rogozin IB, Carmel L, Csuros M et al (2012) Origin and evolution of spliceosomal introns. Biol Direct 7:11
53. Koonin EV (2009) Intron-dominated genomes of early ancestors of eukaryotes. J Hered 100:618–623
54. Roy SW, Irimia M, Penny D (2006) Very little intron gain in Entamoeba histolytica genes laterally transferred from prokaryotes. Mol Biol Evol 23:1824–1827
55. Roy SW, Penny D (2006) Smoke without fire: most reported cases of intron gain in nematodes instead reflect intron losses. Mol Biol Evol 23:2259–2262
56. Stajich JE, Dietrich FS, Roy SW (2007) Comparative genomic analysis of fungal genomes reveals intron-rich ancestors. Genome Biol 8:R223
57. Coulombe-Huntington J, Majewski J (2007) Intron loss and gain in Drosophila. Mol Biol Evol 24:2842–2850
58. Li W, Tucker AE, Sung W et al (2009) Extensive, recent intron gains in Daphnia populations. Science 326:1260–1262
59. Worden AZ, Lee JH, Mock T et al (2009) Green evolution and dynamic adaptations revealed by genomes of the marine picoeukaryotes Micromonas. Science 324:268–272
60. van der Burgt A, Severing E, de Wit PJGM et al (2012) Birth of new spliceosomal introns in fungi by multiplication of introner-like elements. Curr Biol 22(13):1260–1265
61. Roy SW, Irimia M (2012) Genome evolution: where do new introns come from? Curr Biol 22:R529–R531
62. Lim KH, Ferraris L, Filloux ME et al (2011) Using positional distribution to identify splicing elements and predict pre-mRNA processing defects in human genes. Proc Natl Acad Sci USA 108:11093–11098
63. Ast G (2004) How did alternative splicing evolve? Nat Rev Genet 5:773–782
64. Schwartz S, Silva J, Burstein D et al (2008) Large-scale comparative analysis of splicing signals and their corresponding splicing factors in eukaryotes. Genome Res 18:88–103
65. Tolstrup N, Rouze P, Brunak S (1997) A branch point consensus from Arabidopsis found by non-circular analysis allows for better prediction of acceptor sites. Nucleic Acids Res 25:3159–3163
66. Vaulot D, Lepère C, Toulza E et al (2012) Metagenomes of the picoalga Bathycoccus from the Chile coastal upwelling. PLoS One 7:e39648
67. Warnecke T, Parmley JL, Hurst LD (2008) Finding exonic islands in a sea of non-coding sequence: splicing related constraints on protein composition and evolution are common in intron-rich genomes. Genome Biol 9:R29
68. Fairbrother WG, Yeh R-F, Sharp PA et al (2002) Predictive identification of exonic splicing enhancers in human genes. Science 297:1007–1013
69. McLysaght A, Enright AJ, Skrabanek L et al (2000) Estimation of synteny conservation and genome compaction between pufferfish (Fugu) and human. Yeast 17:22–36
70. Deutsch M, Long M (1999) Intron–exon structures of eukaryotic model organisms. Nucleic Acids Res 27:3219–3228
71. Moriyama EN, Petrov DA, Hartl DL (1998) Genome size and intron size in Drosophila. Mol Biol Evol 15:770–773
72. Aruga J, Odaka YS, Kamiya A et al (2007) Dicyema Pax6 and Zic: tool-kit genes in a highly simplified bilaterian. BMC Evol Biol 7:201
73. Ogino K, Tsuneki K, Furuya H (2010) Unique genome of dicyemid mesozoan: highly shortened spliceosomal introns in conservative exon/intron structure. Gene 449:70–76
74. Gilson PR, Su V, Slamovits CH et al (2006) Complete nucleotide sequence of the chlorarachniophyte nucleomorph: nature's smallest nucleus. Proc Natl Acad Sci 103:9566–9571
75. Russell CB, Fraga D, Hinrichsen RD (1994) Extremely short 20–33 nucleotide introns are the standard length in Paramecium tetraurelia. Nucleic Acids Res 22:1221–1225

76. Gelfman S, Burstein D, Penn O et al (2012) Changes in exon–intron structure during vertebrate evolution affect the splicing pattern of exons. Genome Res 22:35–50
77. Lewis BP, Green RE, Brenner SE (2003) Evidence for the widespread coupling of alternative splicing and nonsense-mediated mRNA decay in humans. Proc Natl Acad Sci USA 100:189–192
78. Lareau LF, Brooks AN, Soergel DAW et al (2007) The coupling of alternative splicing and nonsense mediated mRNA decay. In: Blencowe BJ, Graveley BR (eds) Alternative splicing in the postgenomic era. Landes Bioscience and Springer Science&Business Media, Austin, TX, pp 190–211
79. Lareau LF, Inada M, Green RE et al (2007) Unproductive splicing of SR genes associated with highly conserved and ultraconserved DNA elements. Nature 446:926–929
80. Wang ET, Sandberg R, Luo S et al (2008) Alternative isoform regulation in human tissue transcriptomes. Nature 456:470–476
81. Pan Q, Shai O, Lee LJ et al (2008) Deep surveying of alternative splicing complexity in the human transcriptome by high-throughput sequencing. Nat Genet 40:1413–1415
82. Graveley BR, Brooks AN, Carlson JW et al (2011) The developmental transcriptome of Drosophila melanogaster. Nature 471:473–479
83. Irimia M, Blencowe BJ (2012) Alternative splicing: decoding an expansive regulatory layer. Curr Opin Cell Biol 24:323–332
84. Irimia M, Rukov JL, Penny D et al (2008) Widespread evolutionary conservation of alternatively spliced exons in Caenorhabditis. Mol Biol Evol 25:375–382
85. Irimia M, Rukov JL, Roy SW et al (2009) Quantitative regulation of alternative splicing in evolution and development. Bioessays 31:40–50
86. Roy M, Kim N, Xing Y et al (2008) The effect of intron length on exon creation ratios during the evolution of mammalian genomes. RNA 14:2261–2273
87. Pleiss JA, Whitworth GB, Bergkessel M et al (2007) Rapid, transcript-specific changes in splicing in response to environmental stress. Mol Cell 27:928–937
88. Parenteau J, Durand M, Morin G et al (2011) Introns within ribosomal protein genes regulate the production and function of yeast ribosomes. Cell 147:320–331
89. Yin Y, Yu G, Chen Y et al (2012) Genome-wide transcriptome and proteome analysis on different developmental stages of Cordyceps militaris. PLoS One 7:e51853
90. Zhao C, Waalwijk C, de Wit PJ et al (2013) RNA-Seq analysis reveals new gene models and alternative splicing in the fungal pathogen Fusarium graminearum. BMC Genomics 14:21
91. Wang B, Guo G, Wang C et al (2010) Survey of the transcriptome of Aspergillus oryzae via massively parallel mRNA sequencing. Nucleic Acids Res 38:5075–5087
92. Campbell MA, Haas BJ, Hamilton JP et al (2006) Comprehensive analysis of alternative splicing in rice and comparative analyses with Arabidopsis. BMC Genomics 7:327
93. Iida K, Seki M, Sakurai T et al (2004) Genome-wide analysis of alternative pre-mRNA splicing in Arabidopsis thaliana based on full-length cDNA sequences. Nucleic Acids Res 32:5096–5103
94. Ner-Gaon H, Halachmi R, Savaldi-Goldstein S et al (2004) Intron retention is a major phenomenon in alternative splicing in Arabidopsis. Plant J 39:877–885
95. Sorber K, Dimon MT, DeRisi JL (2011) RNA-Seq analysis of splicing in Plasmodium falciparum uncovers new splice junctions, alternative splicing and splicing of antisense transcripts. Nucleic Acids Res 39:3820–3835
96. Curtis BA, Tanifuji G, Burki F et al (2012) Algal genomes reveal evolutionary mosaicism and the fate of nucleomorphs. Nature 492: 59–65
97. Labadorf A, Link A, Rogers MF et al (2010) Genome-wide analysis of alternative splicing in Chlamydomonas reinhardtii. BMC Genomics 11:114
98. Xiong J, Lu X, Zhou Z et al (2012) Transcriptome analysis of the model protozoan, Tetrahymena thermophila, using Deep RNA sequencing. PLoS One 7:e30630
99. Glöckner G, Golderer G, Werner-Felmayer G et al (2008) A first glimpse at the transcriptome of Physarum polycephalum. BMC Genomics 9:6
100. Jaillon O, Bouhouche K, Gout J-F et al (2008) Translational control of intron splicing in eukaryotes. Nature 451:359–362
101. Wang B-B, Brendel V (2006) Molecular characterization and phylogeny of U2AF35 homologs in plants. Plant Physiol 140: 624–636
102. Barbosa-Morais NL, Irimia M, Pan Q et al (2012) The evolutionary landscape of alternative splicing in vertebrate species. Science 338:1587–1593

Chapter 3

Mechanisms of Spliceosomal Assembly

Ni-ting Chiou and Kristen W. Lynch

Abstract

Pre-mRNA splicing is a key step for generating mature protein-coding mRNA. An RNA–protein complex known as the spliceosome carries out the chemistry of pre-mRNA splicing. However, several pre-spliceosomal intermediates are assembled on the pre-mRNA before the formation of the catalytically activated spliceosome. The progression to the activated spliceosome involves a cascade of the rearrangement events of the RNA–RNA, RNA–protein, and protein–protein interactions within the pre-spliceosomal intermediates. These rearrangements generate multiple combinatorial interactions of the spliceosome with the substrate, which enhances the accuracy of the splice site selection. Each rearrangement also represents a step at which splicing can potentially be subjected to regulation. The aim of this chapter is to provide an overview of the components of the spliceosome and their rearrangements along the spliceosome assembly pathway.

Key words Spliceosome assembly, Ribonucleoproteins, snRNP, Prp19, NTC, Splicing

1 Introduction

In 1977, Phillip Sharp and his colleagues first provided evidence for the presence of the introns in nascent transcripts [1]. In 1985, the spliceosome, the ribonucleoprotein (RNP) machine which catalyzes pre-mRNA splicing, was identified [2]. The components of the spliceosome contain five small nuclear RNAs (snRNAs) and hundreds of proteins. In contrast to other RNPs, the catalytically active spliceosome are not preassembled before they bind to the pre-mRNA substrate. Instead, the components of the spliceosome interact with the substrates in a stepwise way to assemble a series of pre-spliceosomal intermediates, which leads to the formation of active site of the spliceosome. Since the 1990s, these pre-spliceosomal intermediates have been trapped and analyzed in vitro using a variety of approaches [3]. The biochemical characterizations of these intermediate complexes, combined with yeast genetic functional studies of the individual spliceosomal components, have revealed much insight into the spliceosome assembly pathway.

Klemens J. Hertel (ed.), *Spliceosomal Pre-mRNA Splicing: Methods and Protocols*, Methods in Molecular Biology, vol. 1126, DOI 10.1007/978-1-62703-980-2_3, © Springer Science+Business Media, LLC 2014

Recently, with the advance of the mass spectrometry and electric microscopy techniques, the components and structures of the various pre-spliceosomal complexes have been even more precisely mapped [4]. Importantly, this detailed knowledge of spliceosome assembly forms the basis of our understanding of the mechanisms that govern splicing fidelity, alternative splicing, and the regulation of alternative splicing, all of which are the key steps of gene expression and are discussed in Chapters 4–6.

2 Basic Spliceosome Assembly Pathway

2.1 Stepwise Assembly of the snRNPs Across an Intron

Nuclear pre-mRNA splicing involves two transesterification steps to remove the intron from the pre-mRNA to generate mature protein-coding mRNA. The three reactive regions on the pre-mRNA are the 5′ splice site (5′SS), the 3′ splice site (3′SS), and the branch point site (BPS), which are all defined by short consensus sequence. In addition to three regions, metazoan introns contain a conserved polypyrimidine tract (PPT) between the 3′ss and the BPS (Fig. 1a; *see* also Chapter 1).

The spliceosome, the largest RNP machine in nucleus, recognizes and positions these reactive regions to catalyze pre-mRNA splicing (for a recent extensive review, *see* ref. 5). The main building blocks of spliceosome are small nuclear ribonucleoproteins (snRNPs). There are five spliceosomal snRNPs: U1, U2, U4, U5, and U6 snRNP. Each snRNP contains a single snRNA and at least seven protein subunits. The snRNPs and additional non-snRNP-associated proteins (such as SF1, U2AF, and the Prp19 complex (NTC); *see* Subheading 3.2 below) are assembled on the pre-mRNA substrate in a stepwise way to form the pre-spliceosomal E, A, B, and, finally, the catalytic-spliceosomal C complex (Fig. 1b). During the stepwise assembly processes, multiple combinatorial interactions are generated between the spliceosomal components and the reactive regions of the substrate. Although these reactive regions have very limited conservation (*see* Fig. 1a), the multiple interactions provide the spliceosome multiple opportunities to double-check the fidelity of interactions, thereby increasing the accuracy of site selection.

To begin the assembly of the spliceosome, U1 snRNP engages with the 5′SS, while SF1 binds to the BPS, in an ATP-independent manner to form the pre-spliceosomal E complex (Fig. 1b; E). In metazoan systems, the 65 and 35 subunits of the U2AF protein heterodimer also bind to the PPT and 3′ ss respectively during this ATP-independent step to further promote correct identification of the 3′ end of the intron. In the presence of ATP, several rearrangements of the snRNPs then occur to progress assembly from the E to A, B, and C complexes. The first rearrangement is that U2 snRNP displaces SF1 from the BPS to form the A complex (Fig. 1b; A).

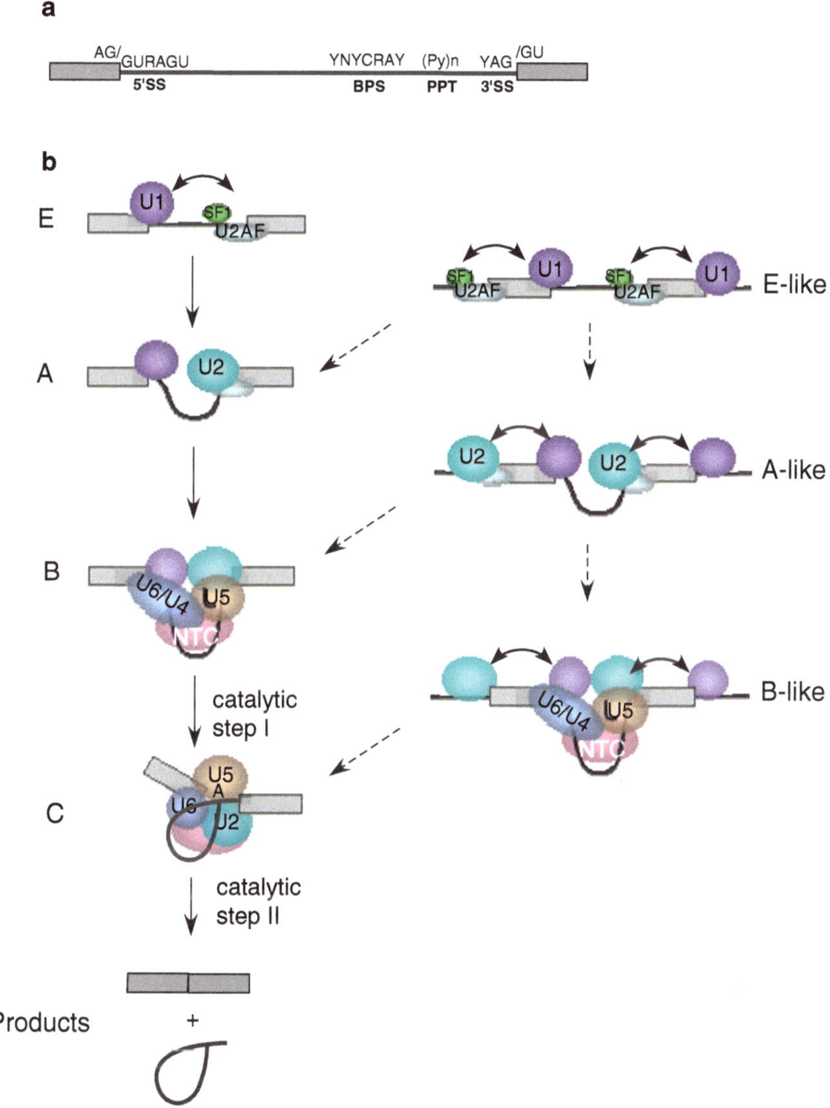

Fig. 1 The pathways of the spliceosome assembly during the pre-mRNA splicing. (**a**) The consensus nucleotide sequences of metazoan pre-mRNAs. Here, two exons (*boxes*) are separated by an intron (*line*). The consensus sequence at the 5′ splice site (5′SS), branch point sequence (BPS), polypyrimidine tract (PPT), and 3′ splice site (3′SS) are indicated above the *line*. In these sequences, R stands for either G or A; Y stands for either U or C. The A within the BPS forms the branch point of the intron lariat produced by splicing. (**b**) The stepwise assembly of U1, U2, U4/U6, U5 snRNP, and NTC on the consensus sequences in the removal of an intron from a pre-mRNA is depicted. *Left*, the cross-intron assembly. *Right*, corresponding exon-defined version of each step

At this point, the preassembled U4/U6.U5 tri-snRNP and NTC are recruited to form the B complex (Fig. 1b; B). The U1 and U4 snRNP are then released followed by the association of the U6 snRNP with the 5′SS and with the U2 snRNA. These rearrangement events promote the first catalytic step to occur, i.e., cleavage of the 5′SS with concurrent formation of a covalent bond between the first nucleotide of the intron and an A residue at the BPS to results in the C complex formation (Fig. 1b; C). In the C complex, the second catalytic step proceeds to excise the lariat intron and join 5′ and 3′ exons to generate mature mRNA (Fig. 1b; products).

2.2 The Exon-Definition Complex

The combinatorial interactions described above are built across the introns. However, the average lengths for exons and introns of human protein-coding genes are, respectively, 145 and 3,364 nucleotides [6]. Since the exons are significantly shorter than introns, it is expected that initially identifying exons during the spliceosome assembly would help the splicing components to be deposited across the introns more precisely, and hence avoid the use of the cryptic splice sites. Thus, it is envisioned that the cross-exon interactions of the snRNPs occur first or simultaneously with the cross-intron interactions in each stage of the assembly (Fig. 1b, right column). Indeed, the U1 and U2 snRNP-containing exon-defined A complexes have been observed for several exons (Fig. 1b; A-like) [7, 8]. In addition, some exon-definition complexes have been shown to contain the tri-snRNP, and the exon-bound tri-snRNP can directly interact with the upstream 5′SS to assemble the B complex across the intron (Fig. 1b; B-like) [9]. However, despite the characterization of some exon-defined complexes, it is still unknown exactly what interactions of the RNA and protein components are involved in building the exon-defined and the connection or conversion of exon-defined to intron-defined complexes.

3 The Rearrangements of the Spliceosome During the Assembly Processes

3.1 The Dynamics of the RNA–RNA Interactions

Much of the structural rearrangements during spliceosome assembly and establishment of its active site involves the remodeling of the base-pairing interactions among five snRNAs and three reactive regions of the pre-mRNA [10, 11]. In the pre-spliceosomal complexes, such as the A and B complex, the base-pairing interactions with the 5′ SS and BPS involve the 5′ end of U1 snRNA and the internal region of U2 snRNA, respectively (Fig. 2a; purple and cyan box of U1 and U2). For the tri-snRNP in the pre-spliceosomal complex, U4 and U6 snRNA are held firmly by base-pairing interactions, and U5 snRNA is associated through RNA–protein interactions (Fig. 2a; tri-snRNP). The U4 and U6 base-pairing interactions inactivate the catalytically important regions of U6 to prevent from cleaving pre-mRNA prematurely.

During the integration of the tri-snRNP into the spliceosome, U1 is displaced from 5′SS, and U4/U6 base-pairing interactions are taken apart. This rearrangement frees U6 snRNA and allows it to form two new base-pairing interactions. One of the interactions involves the ACAGA motif of U6 snRNA engaging in 5′SS interaction, and the other involves the region downstream of ACAGA

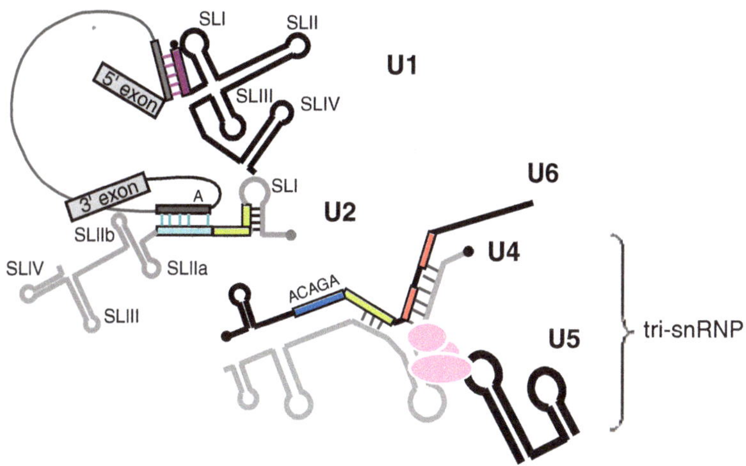

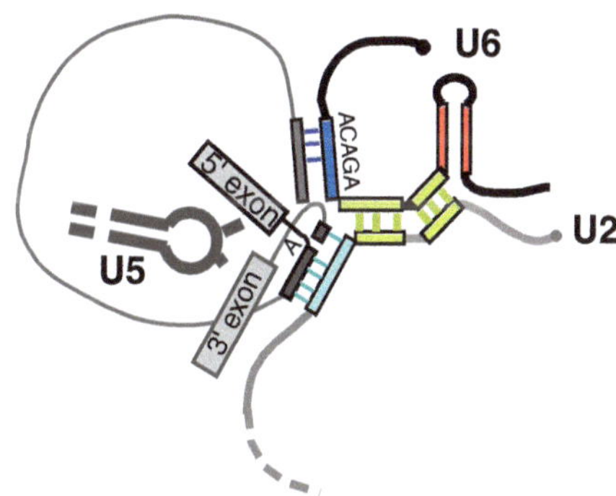

Fig. 2 The rearrangements of RNA–RNA interaction networks during the transition to the catalytic-spliceosomal complex. (**a**) The secondary structures of the five human U snRNAs and the base-pairing interactions of the snRNAs–pre-mRNA within the pre-spliceosomal complex. The stem-loops of the U1 and U2 snRNAs are numbered, and the proteins that associate with these stem-loops are listed in Table 1. The regions of U1 snRNA and U2 snRNA base pair with the 5′SS or BPS are highlighted in *purple* and *cyan*. The tri-snRNP shown here has not integrated into the spliceosome. Within the tri-snRNP, U6 and U4 snRNA are held through base-pairing interactions, while the U5 snRNA and U4/U6 di-snRNA are held by the proteins shown as the *pink circles*. (**b**) During the transition to the catalytic-spliceosomal C complex, U1 and U4 snRNA are released, while U2, U5, and U6 snRNA form the new base-pairing interactions. For simplicity, only the base-pairing interactions of snRNA–snRNA or snRNA–pre-mRNA are shown, but the secondary structures of the snRNAs are not depicted. The regions of the U6 snRNA that are engaged in the base-pairing interactions with the 5′SS and U2 snRNA are highlighted in *purple* and *yellow*, respectively. The stem-loop region of U5 snRNA makes a few contacts with the 5′ and 3′-exon

motif base pairing with the U2 snRNA (Fig. 2b; blue and yellow box of U6). It is interesting that the region of U2 snRNA that base pairs with U6 snRNA is immediately proximal to the region interacting with the pre-mRNA (Fig. 2b; yellow box of U2). Thus, these two newly formed interactions bring the splice sites together to allow the first catalytic step to occur. Following first catalytic step, the stem-loop region of U5 snRNA contacts the nucleotides of the 5′ and 3′exon to bring the two exons into proximity for the second catalytic step.

The RNA base-pairing interactions illustrated above involves from three to eight base pairs that are typically not fully complementary. Thus, proteins certainly play an important role in creating and stabilizing these RNA–RNA interactions. In turn, these RNA–RNA interactions also influence the protein–RNA and protein–protein interactions.

3.2 The Changes of Protein–Protein or Protein–RNA Interactions

The human spliceosomal complexes contain ~45 distinct snRNP-associated proteins and ~170 spliceosome-associated factor [12, 13]. During the splicing processes, proteins enter and leave the spliceosomes from one stage to the next. Thus, the number of the total spliceosomal proteins varies among different pre-spliceosomal intermediates. In general, each of the spliceosomal A, B, and C complex contains ~125 proteins or less (in the case of the A complex) [14–16]. Table 1 lists some of these proteins that have the well-known functions.

During the progression from the A to B complex, ~35 tri-snRNP proteins and ~25 non-snRNP proteins are recruited (*see* Table 1). The major part of these non-snRNP proteins is the Prp19 complex (NTC). The human Prp19 complex is comprised of seven distinct subunits with four copies of prp19 protein. This complex is thus similar to the size of the snRNPs, but unlike the snRNPs, the NTC contains no RNAs [17]. The B complex then transition to the C complex, involving the release not only of the U1 and U4 RNAs but also the protein components of the U1 and U4 snRNP. Although U6 snRNA is not released, most of its protein components also fall off in the C complex. It is possible that the non-snRNP proteins in the C complex, such as the NTC, form new interactions with U6 snRNA to promote its interaction with pre-mRNA in creating the active splice sites of the spliceosome. Moreover, there are ~30 non-snRNP proteins which are recruited during C complex to promote the catalytic site formation (Table 1).

Importantly, many of the proteins that are recruited during the assembly steps are RNA-dependent ATPases/helicases, which are required for the various RNA rearrangements [11]. For example, Brr2, the U5 snRNP component, is involved in unwinding U6/U4 duplex. Subsequently, Prp28, also a U5 snRNP component, mediates the transfer of the 5′SS from the 5′ end of U1 snRNA to the ACAGAG motif of U6 snRNA (Fig. 2). Prp16 and

Table 1
The representative protein components of human U snRNPs, NTC, and spliceosomal factors

U snRNP/NTC/ spliceosomal factors	Representative proteins	Present in the complex A	B	C	Functions/interactions/modifications
A complex factors (~10)	RBM5	+			Block the conversion from a cross-exon to a cross-intron complex
U1 snRNP (~14)	Sm (7)	+	+		Bind to the sm site of U1 snRNA
	U170K	+	+		Bind to the SLI of U1 snRNA
	U1A	+	+		Bind to the SLII of U1 snRNA
	U1C	+	+		Mediate the base-pairing interactions between U1 snRNA and 5′SS
U2 (~17)	Sm (7)	+	+	+	Bind to the sm site of U2 snRNA
	SF3a (3)	+	+	+	Bind to SLI and SLIIb of U2 snRNA
	SF3b (7)	+	+	+	Mediate the base-pairing interactions between U2 and BPS
U2-related (~10)	U2AF35	+	+		Bind to AG nucleotide at 3′SS
	U2AF65	+	+		Bind to PPT
	SPF30	+	+		Bridges an interactions between U2AF35 and Prp3
	Prp5/DDX46	+	+		Bridges a U1 and U2 snRNP interaction network
U5 (~14)	Sm (7)		+	+	Bind to sm site of U5 snRNA
	hSnu114		+	+	GTPase; promote Brr2 helicase activity
	hBrr2		+	+	RNA helicase; unwinding U4/U6 hairpin
	hPrp8		+	+	Bind to both of 5′ and 3′ exon during the catalytic step II
	hPrp6		+	+	Phosphorylated during B complex formation
	hPrp28		+	+	RNA helicase; exchange of U1 for U6 snRNP at 5′SS
U5-related (~11)	hPrp38		+	+	Promote U4/U6 snRNA dissociation
U4 (~12)	Sm (7)		+		Bind to sm site of U4 snRNA
	hPrp31		+		Phosphorylated during B complex formation
	hPrp4		+		Phosphorylate Prp31 and Prp6
	hPrp3		+		Ubiquitinated by NTC
	hPrp24		+		Facilitates the association of U4 and U6 snRNPs
U6 (7)	Lsm2-8 (7)		+		Bind to the Lsm site of U6 snRNA
NTC (7)	Prp19		+	+	Stabilize the association of U5 and U6 with the spliceosome after U4 is dissociated
	CDC5L		+	+	
	SPF27		+	+	
	PRLG1		+	+	
NTC-related (~12)	RBM22		+	+	Promote the catalytic conformation
B complex factors (~8)	UBL5		+		Unknown
C complex factors (~37)	Prp22			+	RNA helicase; required for catalytic step II
	Prp16			+	RNA helicase; required for both catalytic steps
	Slu7			+	Mediate 3′ splice site choice

The functions, interaction, or modifications of the representative proteins are compiled from several sources [5, 11, 23–26]. The association of these representative proteins with the spliceosomal complexes is based on the review paper [5]. Numbers indicate the total number of individual proteins in a particular group

Prp22, the non-snRNP proteins, are recruited during the transition to the C complex and required for the second transesterification step [18, 19].

4 Conclusion and Future Perspectives

The mechanism of spliceosome assembly provides an extraordinary model in illustrating how RNAs and proteins cooperate as they work together to recognize the reactive regions of the pre-mRNA and catalyze its splicing. The other fact that makes the spliceosome an important RNP machine is that 90–95 % of human genome is alternatively spliced and at least 10 % of human genetic disease arises from the mutations either in the splice sites or in the splicing regulatory sequences [20–22]. This indicates that the assembly of the spliceosome is highly regulated in the cell and sensitive to the minor changes of the pre-mRNA sequences. Thus, the mechanism of spliceosome assembly not only is critical for understanding the principles that govern alternative splicing but also brings new opportunities to the possible treatment of human genetic diseases.

References

1. Berget SM, Moore C, Sharp PA (1977) Spliced segments at the 5′ terminus of adenovirus 2 late mRNA. Proc Natl Acad Sci U S A 74:3171–3175
2. Brody E, Abelson J (1985) The "spliceosome": yeast pre-messenger RNA associates with a 40S complex in a splicing-dependent reaction. Science 228:963–967
3. Jurica MS, Moore MJ (2002) Capturing splicing complexes to study structure and mechanism. Methods 28:336–345
4. Luhrmann R, Stark H (2009) Structural mapping of spliceosomes by electron microscopy. Curr Opin Struct Biol 19:96–102
5. Wahl MC, Will CL, Luhrmann R (2009) The spliceosome: design principles of a dynamic RNP machine. Cell 136:701–718
6. Consortium IHGS (2004) Finishing the euchromatic sequence of the human genome. Nature 431:931–945
7. House AE, Lynch KW (2006) An exonic splicing silencer represses spliceosome assembly after ATP-dependent exon recognition. Nat Struct Mol Biol 13:937–944
8. Sharma S, Kohlstaedt LA, Damianov A, Rio DC, Black DL (2008) Polypyrimidine tract binding protein controls the transition from exon definition to an intron defined spliceosome. Nat Struct Mol Biol 15:183–191
9. Schneider M, Will CL, Anokhina M, Tazi J, Urlaub H et al (2010) Exon definition complexes contain the tri-snRNP and can be directly converted into B-like precatalytic splicing complexes. Mol Cell 38:223–235
10. Nilsen TW (1994) RNA-RNA interactions in the spliceosome: unraveling the ties that bind. Cell 78:1–4
11. Staley JP, Guthrie C (1998) Mechanical devices of the spliceosome: motors, clocks, springs, and things. Cell 92:315–326
12. Zhou Z, Licklider LJ, Gygi SP, Reed R (2002) Comprehensive proteomic analysis of the human spliceosome. Nature 419:182–185
13. Rappsilber J, Ryder U, Lamond AI, Mann M (2002) Large-scale proteomic analysis of the human spliceosome. Genome Res 12:1231–1245
14. Behzadnia N, Golas MM, Hartmuth K, Sander B, Kastner B et al (2007) Composition and three-dimensional EM structure of double affinity-purified, human prespliceosomal A complexes. EMBO J 26:1737–1748
15. Deckert J, Hartmuth K, Boehringer D, Behzadnia N, Will CL et al (2006) Protein composition and electron microscopy structure of affinity-purified human spliceosomal B complexes isolated under physiological conditions. Mol Cell Biol 26:5528–5543
16. Jurica MS, Licklider LJ, Gygi SR, Grigorieff N, Moore MJ (2002) Purification and characterization of native spliceosomes suitable for three-dimensional structural analysis. RNA 8:426–439

17. Grote M, Wolf E, Will CL, Lemm I, Agafonov DE et al (2010) Molecular architecture of the human Prp19/CDC5L complex. Mol Cell Biol 30:2105–2119
18. Schwer B (2008) A conformational rearrangement in the spliceosome sets the stage for Prp22-dependent mRNA release. Mol Cell 30:743–754
19. Schwer B, Guthrie C (1992) A conformational rearrangement in the spliceosome is dependent on PRP16 and ATP hydrolysis. EMBO J 11:5033–5039
20. Pan Q, Shai O, Lee LJ, Frey BJ, Blencowe BJ (2008) Deep surveying of alternative splicing complexity in the human transcriptome by high-throughput sequencing. Nat Genet 40: 1413–1415
21. Cooper TA, Wan L, Dreyfuss G (2009) RNA and disease. Cell 136:777–793
22. Wang ET, Sandberg R, Luo S, Khrebtukova I, Zhang L et al (2008) Alternative isoform regulation in human tissue transcriptomes. Nature 456:470–476
23. Schneider M, Hsiao HH, Will CL, Giet R, Urlaub H et al (2010) Human PRP4 kinase is required for stable tri-snRNP association during spliceosomal B complex formation. Nat Struct Mol Biol 17:216–221
24. Pomeranz Krummel DA, Oubridge C, Leung AK, Li J, Nagai K (2009) Crystal structure of human spliceosomal U1 snRNP at 5.5 A resolution. Nature 458:475–480
25. Dybkov O, Will CL, Deckert J, Behzadnia N, Hartmuth K et al (2006) U2 snRNA-protein contacts in purified human 17S U2 snRNPs and in spliceosomal A and B complexes. Mol Cell Biol 26:2803–2816
26. Song EJ, Werner SL, Neubauer J, Stegmeier F, Aspden J et al (2010) The Prp19 complex and the Usp4Sart3 deubiquitinating enzyme control reversible ubiquitination at the spliceosome. Genes Dev 24:1434–1447

Chapter 4

Alternative Pre-mRNA Splicing

Stacey D. Wagner and J. Andrew Berglund

Abstract

Alternative pre-mRNA splicing is an integral part of gene regulation in eukaryotes. Here we provide a basic overview of the various types of alternative splicing, as well as the functional role, highlighting how alternative splicing varies across phylogeny. Regulated alternative splicing can affect protein function and ultimately impact biological outcomes. We examine the possibility that portions of alternatively spliced transcripts are the result of stochastic processes rather than regulated. We discuss the implications of misregulated alternative splicing and explore of the role of alternative splicing in human disease.

Key words Alternative splicing, Myotonic dystrophy (DM), Neurological disease paraneoplastic opsoclonus-myoclonus ataxia (POMA), Frasier's syndrome, Stem cell pluripotency

1 What Is Alternative Splicing?

During splicing constitutive exons are recognized and ligated together. Within a particular gene, constitutive exons are always included in the mature mRNA. Alternative splicing within a gene can create different versions of an mRNA, called isoforms. The individual splicing decisions within a gene are referred to as events. Most genes contain both constitutive exons and alternatively spliced events. It has been suggested that one of the primary purposes of alternative splicing is to expand the proteome coding potential of genes. Significantly, alternative splicing coordinates the expression of the appropriate version of an mRNA in a spatial-temporal manner. Splicing decisions are influenced by tissue and development specific *trans*-regulatory splicing factors.

1.1 The Many Ways to Splice Alternatively

There are several types of alternative splicing events including cassette exon, mutually exclusive exon, alternate 5′ splice sites (alternate donor site), alternate 3′ splice site (alternative acceptor site), intron retention, mutually exclusive 5′ untranslated regions (UTRs), and mutually exclusive 3′ UTRs (Fig. 1). Cassette exon events are a commonly illustrated textbook example of alternative

Klemens J. Hertel (ed.), *Spliceosomal Pre-mRNA Splicing: Methods and Protocols*, Methods in Molecular Biology, vol. 1126, DOI 10.1007/978-1-62703-980-2_4,

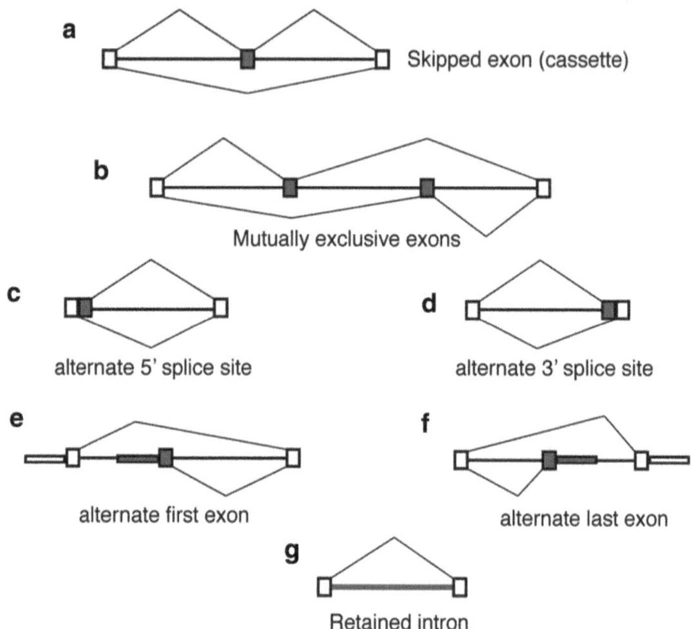

Fig. 1 The different types of alternative splicing are depicted: (**a**) exon skipping (cassette exon), (**b**) mutually exclusive exons, (**c**) alternative 5′ splice sites, (**d**) alternative 3′ splice sites, (**e**) alternate first exon, (**f**) alternate last exon, and (**g**) retained intron

splicing and result from the skipping of the alternative exon (Fig. 1a). Mutually exclusive alternative splicing occurs when a pre-mRNA containing adjacent exons includes only one or the other exon, but not both exons in the same mRNA (Fig. 1b). Alternative 5′ splice sites arise when competing 5′ splice sites are available. An example is shown in Fig. 1c in which an upstream 5′ splice site is selected as the donor splice site, thus truncating the exon at its 3′ end compared to if the downstream 5′ splice site was chosen. An analogous situation arises when different 3′ splice sites are available as the possible accepter site (Fig. 1d). Alternative promoters that alter transcription start sites result in mutually exclusive first exons as shown in Fig. 1e. Similarly, mutually exclusive last exons arise from regulation of alternative polyadenylation sites (Fig. 1f). Intron retention, or lack of splicing, is a common event in many organisms (Fig. 1g). This type of alternative splicing is more commonly observed with short introns, in single-celled eukaryotes and in plants.

1.2 A Phylogenetic Perspective of Alternative Splicing

While genome size may not correlate with organismal complexity, it is clear that alternative splicing is more prevalent in higher organisms. Large-scale profiling experiments have made considerable progress on the understanding of alternative splicing within different species. The proportion of genes that are alternatively spliced and the types of alternative splicing detected vary across species. Interestingly, species within the same kingdom share the same predominant type of alternative splicing [1]. For example, the predominance of intron retention versus cassette exon can be predictive of which kingdom a particular organism belongs. Animals are unique in that they utilize cassette exon splicing much more frequently than fungi and protists, which predominantly use intron retention. Cassette exon alternative splicing is more frequent in plants than protists and fungi, placing plants between fungi and animals in how often cassette exon splicing occurs within their mRNAs.

Organisms within the fungi kingdom differ in how many splice variants are expressed and appear to group by unicellular or multicellular. Three alternative events were detected in the single-celled fission yeast *Schizosaccharomyces pombe,* while 1,091 events were detected in the filamentous fungus *Cryptococcus neoformans* [1]. Not only were differences detected in the amount of alternative splicing but also in the type. In *C. neoformans,* 18 cassette exons were detected, while only intron retention was detected in *S. pombe.* Another example from this kingdom is the filamentous fungus *Aspergillus oryzae* which also undergoes more alternative splicing than the single cell *S. pombe* with 1,375 alternative events (9 % of *A. oryzae* genes) [2]. All types of alternative splicing were observed in *A. oryzae* except mutually exclusive exons. As with the other fungi, intron retention accounted for most of the alternative splicing observed, representing approximately 92 % of the events.

Similar amounts of alternative events were detected in various protists (including *Phytophthora infestans* and *Tetrahymena thermophila*) [1]. Intron retention was also the most common type of alternative splicing, and cassette exons were rarely detected. Contrasting fungi and protists, plants undergo a higher proportion of cassette exon alternative splicing, but intron retention was also the predominant type of alternative splicing in many plant species including *Arabidopsis thaliana* and *Physcomitrella patens* [1, 3, 4]. At least 42 % of genes that contain introns are alternatively spliced in *A. thaliana* [3]. Many intron retention events were affected by abiotic stress, a majority of which contain premature termination codons (PTCs) and undergo either NMD (nonsense-mediated mRNA decay) or RUST (regulated and unproductive splicing and translation) [1, 5, 6]. The coupling of NMD and RUST to splicing is discussed in Subheading 2.1.

The ratio of cassette exon to intron retention (CE/(CE + IR)) events is dramatically increased in animals compared with other kingdoms. Looking at the most extreme examples, in the fungi *S. cerevisiae* and *P. patens,* ratios are 0 and 0.32, respectively, while the animals *Schistosoma mansoni* and *Branchiostoma floridae* are 0.28 and 0.95, respectively [1]. Alternative splicing generally appears to be more prevalent in animals compared to the other kingdoms.

An estimated 25 % of *Caenorhabditis elegans* genes undergo alternative splicing with an average of 2.2 isoforms produced per gene [7]. At least 574 alternative events are developmentally regulated. Contrasting fungi, protists, and plants, every alternative splicing category is equally distributed except for the rare mutually exclusive events [8]. Alternative splicing is important to the worm's development with, on average, 280 genes switching isoforms between developmental stages [9].

Alternative splicing occurs even more frequently in the model organism *Drosophila melanogaster*. Of the expressed multi-exon genes, 60.7 % of them contain one or more alternative splicing events [10]. Alternative splicing in fly development and sex determination has been well documented (reviewed in refs. 11, 12). 66 % of the known alternative splicing events undergo significant change during fly development with 119 genes regulated in the sex determination pathway [10].

Most mammalian genes have been estimated to express between 2 and 20 isoforms [13]. In vertebrates, multi-exon genes produce an average of 6.3 isoforms [14]. In humans, it has been estimated that most genes (95 %) are alternatively spliced [15, 16]. The expression patterns of isoforms vary by tissue [17]. The recent Encode project evaluated the transcriptome of 15 different human cell lines using deep sequencing [18]. They observed many isoforms of the same gene expressed at once, with the most observed being 12 isoforms per gene. They also observed that isoforms were not expressed at equal levels; one dominant isoform was typically present at least 30 % of the time for a given condition. They also found that as the number of isoforms increased for a gene, so did the likelihood that more than one dominant isoform was expressed.

2 Significance of Alternative Splicing: Is It All Regulated?

With 95 % of the human genes undergoing alternative splicing, one might ask if all of these splicing events are functional. Let us consider a pre-mRNA that generates two isoforms, a major isoform that is spliced 99.8 % of the time and a minor isoform that is spliced 0.2 % of the time. Would one conclude that the 0.2 % is relevant or that it is due to stochastic noise? How highly expressed does an isoform need to be considered "important?"

Studies have investigated the "noise" of splicing and determined that 2 % of the transcripts from a gene are mis-spliced [19]. These isoforms map to previously unannotated splice junctions, indicating that these events may not be functionally relevant. Often conservation of alternative splicing is used as an indication if alternative events are functionally relevant (*see* ref. 20 for a study on events conserved between mouse and human). Indeed, Pickrell et al. noted that low-abundance isoforms did not appear to be under selective pressure relative to annotated alternative events in placental mammals. In concordance with this study, Sorek et al. compared cassette exon events conserved between mouse and human to events that are not conserved and concluded that a large portion (75 %) of the detected alternative splicing is not conserved and likely not functional [21].

Two recent studies investigating the evolution of alternative splicing in vertebrates found most alternative splicing to be more related to the species than to the organ type [22, 23]. This result argues that many alternative events have recently evolved and that conservation may not be the most informative parameter used to determine if an alternative splicing event is important. Additionally, the authors observed that differences in alternative splicing contribute more significantly to phenotypic variation than gene expression levels [22, 23].

2.1 Alternative Splicing Modifies Protein Function

While many isoforms generated by alternative splicing may be attributed to noise, the most conclusive evidence for functionality of alternative splicing is demonstrated at the protein and phenotype level. Many alternative splicing events may not be under selective pressure and most are functionally inconsequential [24]. Another limitation for determining the widespread significance of the effect of alternative splicing on protein function is that experiments addressing these questions can be time consuming and difficult to interpret due to the complex cause/effect relationship between gene expression and phenotype. Animal models that are easy to manipulate have been instrumental in addressing these questions. Also, high-throughput methods investigating the expression of protein isoforms may help determine which mRNAs are translated into stable protein, as studies addressing this question have been pursued in *D. melanogaster* [25].

Even though the importance of alternative events has yet to be characterized globally, numerous studies have demonstrated the importance of alternative splicing on the function of specific proteins. Isoform switching can impact a protein's enzymatic activity, localization, stability/expression, molecular interactions, and structure (reviewed in ref. 26). These changes can influence a vast array of biological processes and outcomes including transcription, alternative splicing, cell motility, differentiation, ion channel function, proliferation, angiogenesis, neuronal plasticity, and apoptosis.

Changing a single alternative event can impact downstream signaling and biological fate. For example, alternative splicing can serve as a switch for developmental processes. The forkhead transcription factor Foxp1 plays a pivotal role in human embryonic stem cell pluripotency and reprogramming [27]. A mutually exclusive splicing event in Foxp1 creates a unique isoform in human embryonic stem cells, while the canonical version is present in other cell types. This splice variant has different DNA binding specificity and regulates a different set of genes, specifically those involved in pluripotency, while the canonical version is involved in differentiation. This event was shown to function analogously in mouse and appears to be conserved in vertebrates [28].

Like Foxp1, the role of alternative splicing on the fibroblast growth factor 8 (FGF8) gene in development has been studied in multiple organisms (reviewed in ref. 29). Two of the conserved isoforms are produced using mutually exclusive 3′ splice sites. The resulting proteins only differ by 11 amino acids, but only one isoform induces mesoderm differentiation in *Xenopus laevis* embryos [30]. In the mouse, the different isoforms also have distinct functions in mouse development before and during gastrulation [31] (reviewed in refs. 29, 32).

Alternative splicing often alters the expression level of a protein through the NMD pathway and RUST. As discussed earlier, RUST and NMD appear to be regulated by external cues and play an important role for *A. thaliana* to cope with stress caused by drought, heat, cold, and salt changes [1, 5]. Additionally, over one-third of alternative events in mouse and human introduce a premature stop codon [33]. There is debate in the field as to how much of this process is regulated. In one study, it was concluded that most alternative splicing is not coupled to NMD because of a lack of selective pressure on most of the NMD events [34]. Another study that investigated alternative splicing events conserved between mouse and human found 21 % (192 of 900 single-exon skipping cases) subject to NMD [35]. Furthermore, the authors noted that 25 % of NMD events are exon inclusion events and regulating the inclusion of these alternative exons creates a switch to downregulate the message.

One example of an alternative splicing/NMD switch to upregulate a message is the SSAT gene (spermine/spermidine acetyltransferase) (additional examples of functionally important coupling of alternative splicing can be found in the following review [36]). Degradation of the SSAT message is regulated by the SSAT enzyme substrates (i.e., polyamines). When polyamine concentrations are low, an alternative exon containing premature stop codons is included causing degradation of the transcript through NMD [37]. When polyamine concentrations are high and the enzyme is needed, the isoforms switch so that the PTC-containing exon is excluded and the message is translated to produce functional enzyme.

2.2 Alternative Splicing Gone Wrong

Pre-mRNAs can be alternatively spliced in different ways within an organism as well as between species with the result impacting protein function in biological systems and leading to dramatic differences in phenotype. However, when mis-splicing of important genes occur, dysfunction of the biological system can result in disease. Alternative splicing has been shown to play a role in many diseases, including cancer, muscular dystrophies, developmental, and neurological diseases [38–43].

For some diseases, it is clear that a mutation within a single pre-mRNA cis-regulatory sequence motif can alter the ratio of isoforms created through alternative splicing. A point mutation in the Wilm's tumor gene (WT1) is an example in which a mutation eliminates alternative splicing at one of two adjacent competing 5′ splice sites leading to the loss of one WT1 isoform. The WT1 isoform that is lost lacks three amino acids (KTS) and has been linked to Frasier's syndrome, a developmental disease that severely affects gonadal development [44, 45]. Normally the two isoforms of WT1 (+/-KTS) are found in a ratio of 60/40. The presence of the KTS tripeptide alters the DNA binding and transcriptional activity of WT1 [46, 47]. Several reviews describe WT1 function and its role in cancer and developmental diseases [48–50].

Contrasting WT1, in some diseases, many alternative splicing events are misregulated, some due to a change in a master trans-regulator. Entire alternative splicing programs can change due to certain cues or processes (*see* ref. 51 for a review on the effects of alternative splicing on signaling pathways). Diseases where misregulation of splicing contributes a significant component to the disease are called spliceopathies. Myotonic dystrophy (DM) is a disease in which many alternative splicing changes are linked to causing symptoms of the disease (reviewed in refs. 52, 53). DM is caused by the expression of an expanded CUG or CCUG repeat RNA that sequesters a family of RNA binding proteins (muscleblind proteins) that regulate alternative splicing (reviewed in refs. 54–56). In addition to the sequestration of the muscleblind proteins, the levels of another family of splicing factors, CELF proteins, are increased in DM [57]. Several studies have demonstrated that many alternative splicing events are altered in DM due to the sequestration of muscleblind proteins and the increased levels of CELF proteins [57–62]. A well-characterized alteration of splicing that leads to a disease symptom is the chloride channel, voltage-sensitive 1 (CLCN1), an important transmembrane protein in skeletal muscle [63]. The change in splicing leads to the production of isoforms in which the mRNA is degraded due to the inclusion of a premature stop codon. The levels of total and functional chloride channel are reduced in DM, leading to a decrease in chloride conduction that results in myotonia. The other characterized alternative splicing events that are affected in DM are mostly cassette exon events, and many have been correlated with aberrant function of the resulting proteins causing heart, muscle, and cognitive defects.

In the neurological disease paraneoplastic opsoclonus-myoclonus ataxia (POMA), antibodies target the Nova family of alternative splicing factors leading to loss of Nova function. This loss of function leads to many changes in alternative splicing specific within the brain. The improper alternative splicing of these genes is proposed to lead to protein isoforms that no longer function properly in synaptic function (reviewed in ref. 64). Nova regulates the inclusion/exclusion of many alternative exons, and loss of this regulation has serious consequences. For example, in a double mouse knockout of Nova1 and Nova2, the loss of an alternative exon in agrin is linked to lack of proper motor neuron synapse formation [65]. Studies on Nova have led to many breakthroughs in the use of high-throughput methods to identify large numbers of regulated alternative splicing events and binding sites in pre-mRNAs for RNA binding proteins [66, 67].

With 95 % of human genes undergoing alternative splicing, it is not surprising that alternative splicing is affected in so many diseases. For many of these observations, it remains to be determined if the changes in alternative splicing are causative or if the changes in alternative splicing are downstream events.

Acknowledgment

Research in the Berglund laboratory is supported by NIH (AR059833) and the Myotonic Dystrophy Foundation.

References

1. McGuire AM, Pearson MD, Neafsey DE et al (2008) Cross-kingdom patterns of alternative splicing and splice recognition. Genome Biol 9:R50
2. Wang B, Guo G, Wang C et al (2010) Survey of the transcriptome of *Aspergillus oryzae* via massively parallel mRNA sequencing. Nucleic Acids Res 38:5075–5087
3. Filichkin SA, Priest HD, Givan SA et al (2010) Genome-wide mapping of alternative splicing in *Arabidopsis thaliana*. Genome Res 20:45–58
4. Wang B-B, Brendel V (2006) Genomewide comparative analysis of alternative splicing in plants. Proc Natl Acad Sci U S A 103: 7175–7180
5. Kalyna M, Simpson CG, Syed NH et al (2012) Alternative splicing and nonsense-mediated decay modulate expression of important regulatory genes in Arabidopsis. Nucleic Acids Res 40:2454–2469
6. Gan X, Stegle O, Behr J et al (2011) Multiple reference genomes and transcriptomes for *Arabidopsis thaliana*. Nature 477:419–423
7. Ramani AK, Calarco JA, Pan Q et al (2011) Genome-wide analysis of alternative splicing in *Caenorhabditis elegans*. Genome Res 21: 342–348
8. Yook K, Harris TW, Bieri T et al (2012) WormBase 2012: more genomes, more data, new website. Nucleic Acids Res 40:735–741
9. Gerstein MB, Lu ZJ, Van Nostrand EL et al (2010) Integrative analysis of the *Caenorhabditis elegans* genome by the modENCODE project. Science 330:1775–1787
10. Graveley BR, Brooks AN, Carlson JW et al (2011) The developmental transcriptome of *Drosophila melanogaster*. Nature 471:473–479
11. Venables JP, Tazi J, Juge F (2012) Regulated functional alternative splicing in Drosophila. Nucleic Acids Res 40:1–10
12. Salz HK (2011) Sex determination in insects: a binary decision based on alternative splicing. Curr Opin Genet Dev 21:395–400
13. Katz Y, Wang ET, Airoldi EM et al (2010) Analysis and design of RNA sequencing experiments for identifying isoform regulation. Nat Methods 7:1009–1015
14. Frankish A, Mudge JM, Thomas M et al (2012) The importance of identifying alternative splicing in vertebrate genome annotation. Database 2012:bas014. doi:10.1093/database/bas014
15. Pan Q, Shai O, Lee JL et al (2008) Deep surveying of alternative splicing complexity in the human transcriptome by high-throughput sequencing. Nat Genet 40:1413–1415

16. Wang ET, Sandberg R, Luo S et al (2008) Alternative isoform regulation in human tissue transcriptomes. Nature 456:470–476
17. Yeo G, Holste D, Kreiman G et al (2004) Variation in alternative splicing across human tissues. Genome Biol 5:R74
18. Djebali S, Davis CA, Merkel A et al (2012) Landscape of transcription in human cells. Nature 489:101–108
19. Pickrell JK, Pai AA, Gilad Y et al (2010) Noisy splicing drives mRNA isoform diversity in human cells. PLoS Genet 6:e1001236
20. Sugnet CW, Kent WJ, Ares M Jr et al (2004) Transcriptome and genome conservation of alternative splicing events in humans and mice. Pac Symp Biocomput 9:66–77
21. Sorek R, Shamir R, Ast G (2004) How prevalent is functional alternative splicing in the human genome? Trends Genet 20:68–71
22. Merkin J, Russell C, Chen P et al (2012) Evolutionary dynamics of gene and isoform regulation in mammalian tissues. Science 338:1593–1599
23. Barbosa-Morais NL, Irimia M, Pan Q et al (2012) The evolutionary landscape of alternative splicing in vertebrate species. Science 338:1587–1593
24. Tress ML, Martelli PL, Frankish A et al (2007) The implications of alternative splicing in the ENCODE protein complement. Proc Natl Acad Sci U S A 104:5495–5500
25. Tress ML, Bodenmiller B, Aebersold R et al (2008) Proteomics studies confirm the presence of alternative protein isoforms on a large scale. Genome Biol 9:R162
26. Kelemen O, Convertini P, Zhang Z et al (2012) Function of alternative splicing. Gene 514:1–30
27. Gabut M, Samavarchi-Tehrani P, Wang X et al (2011) An alternative splicing switch regulates embryonic stem cell pluripotency and reprogramming. Cell 147:132–146
28. Salomonis N, Schlieve CR, Pereira L et al (2010) Alternative splicing regulates mouse embryonic stem cell pluripotency and differentiation. Proc Natl Acad Sci U S A 107:10514–10519
29. Sunmonu NA, Li K, Li JYH (2011) Numerous isoforms of Fgf8 reflect its multiple roles in the developing brain. J Cell Physiol 226: 1722–1726
30. Fletcher RB, Baker JC, Harland RM (2006) FGF8 spliceforms mediate early mesoderm and posterior neural tissue formation in Xenopus. Development 133:1703–1714
31. Guo Q, Li JYH (2007) Distinct functions of the major Fgf8 spliceform, Fgf8b, before and during mouse gastrulation. Development 134:2251–2260
32. Itoh N (2007) The Fgf families in humans, mice, and zebrafish: their evolutional processes and roles in development, metabolism, and disease. Biol Pharm Bull 30:1819–1825
33. Lewis BP, Green RE, Brenner SE (2002) Evidence for the widespread coupling of alternative splicing and nonsense-mediated mRNA decay in humans. Proc Natl Acad Sci U S A 100:189–192
34. Pan Q, Saltzman AL, Kim YK et al (2006) Quantitative microarray profiling provides evidence against widespread coupling of alternative splicing with nonsense-mediated mRNA decay to control gene expression. Genes Dev 20:153–158
35. Baek D, Green P (2005) Sequence conservation, relative isoform frequencies, and nonsense-mediated decay in evolutionarily conserved alternative splicing. Proc Natl Acad Sci U S A 102:12813–12818
36. Lareau LF, Brooks AN, Soergel D et al (2007) The coupling of alternative splicing and nonsense mediated mRNA decay. In: Blencowe B, Graveley B (eds) Alternative splicing in the postgenomic era. Landes Biosciences, Austin, TX, pp 190–211
37. Hyvönen MT, Uimari A, Keinänen TA et al (2006) Polyamine-regulated unproductive splicing and translation of spermidine/spermine N1-acetyltransferase. RNA 12:1569–1582
38. Wang G-S, Cooper TA (2007) Splicing in disease: disruption of the splicing code and the decoding machinery. Nat Rev Genet 8:749–761
39. Venables JP (2004) Aberrant and alternative splicing in cancer. Cancer Res 64:7647–7654
40. Biamonti G, Bonomi S, Gallo S et al (2012) Making alternative splicing decisions during epithelial-to-mesenchymal transition (EMT). Cell Mol Life Sci 69:2515–2526
41. Singh RK, Cooper TA (2012) Pre-mRNA splicing in disease and therapeutics. Trends Mol Med 18:472–482
42. Mills JD, Janitz M (2012) Alternative splicing of mRNA in the molecular pathology of neurodegenerative diseases. Neurobiol Aging 33:11–24
43. Poulos MG, Batra R, Charizanis K (2011) Developments in RNA splicing and disease. Cold Spring Harb Perspect Biol 3:a000778
44. Klamt B, Koziell A, Poulat F et al (1998) Frasier syndrome is caused by defective alternative splicing of WT1 leading to an altered ratio of WT1 +/-KTS splice isoforms. Hum Mol Genet 7:709–714
45. Hammes A, Guo JK, Lutsch G et al (2001) Two splice variants of the Wilms' tumor 1 gene have distinct functions during sex determination and nephron formation. Cell 106:319–329
46. Lee SB, Huang K, Palmer R et al (1999) The Wilms tumor suppressor WT1 encodes a tran-

scriptional activator of amphiregulin. Cell 98: 663–673
47. Reynolds PA, Smolen GA, Palmer RE et al (2003) Identification of a DNA-binding site and transcriptional target for the EWS-WT1(+KTS) oncoprotein. Genes Dev 17:2094–2107
48. Morrison AA, Viney RL, Ladomery MR (2008) The post-transcriptional roles of WT1, a multifunctional zinc-finger protein. Biochim Biophys Acta 1785:55–62
49. Huff V (2011) Wilms' tumours: about tumour suppressor genes, an oncogene and a chameleon gene. Nat Rev Cancer 11:111–121
50. Chau Y-Y, Hastie ND (2012) The role of Wt1 in regulating mesenchyme in cancer, development, and tissue homeostasis. Trends Genet 28:515–524
51. Shin C, Manley JL (2004) Cell signalling and the control of pre-mRNA splicing. Nat Rev Mol Cell Biol 5:727–738
52. Day JW, Ranum LPW (2005) RNA pathogenesis of the myotonic dystrophies. Neuromuscul Disord 15:5–16
53. Osborne RJ, Thornton CA (2006) RNA-dominant diseases. Hum Mol Genet 15: 162–169
54. Ranum LPW, Day JW (2004) Myotonic dystrophy: RNA pathogenesis comes into focus. Am J Hum Genet 74:793–804
55. Cooper TA, Wan L, Dreyfuss G (2009) RNA and disease. Cell 136:777–793
56. Mahadevan MS (2011) Myotonic muscular dystrophy, RNA toxicity, and the brain: trouble making the connection? Cell Stem Cell 8:349–350
57. Wang G-S, Kearney DL, De Biasi M et al (2007) Elevation of RNA-binding protein CUGBP1 is an early event in an inducible heart-specific mouse model of myotonic dystrophy. J Clin Invest 117:2802–2811
58. Du H, Cline MS, Osborne RJ et al (2010) Aberrant alternative splicing and extracellular matrix gene expression in mouse models of myotonic dystrophy. Nat Struct Mol Biol 17:187–193
59. Wang ET, Cody NAL, Jog S et al (2012) Transcriptome-wide regulation of pre-mRNA splicing and mRNA localization by muscleblind proteins. Cell 150:710–724
60. Paul S, Dansithong W, Kim D et al (2006) Interaction of muscleblind, CUG-BP1 and hnRNP H proteins in DM1-associated aberrant IR splicing. EMBO J 25: 4271–4283
61. Koshelev M, Sarma S, Price RE et al (2010) Heart-specific overexpression of CUGBP1 reproduces functional and molecular abnormalities of myotonic dystrophy type 1. Hum Mol Genet 19:1066–1075
62. Kino Y, Washizu C, Oma Y et al (2009) MBNL and CELF proteins regulate alternative splicing of the skeletal muscle chloride channel CLCN1. Nucleic Acids Res 37: 6477–6490
63. Mankodi A, Takahashi MP, Jiang H et al (2002) Expanded CUG repeats trigger aberrant splicing of ClC-1 chloride channel pre-mRNA and hyperexcitability of skeletal muscle in myotonic dystrophy. Mol Cell 10:35–44
64. Darnell RB, Posner JB (2003) Paraneoplastic syndromes involving the nervous system. N Engl J Med 349:1543–1554
65. Ruggiu M, Herbst R, Kim N et al (2009) Rescuing Z+ agrin splicing in Nova null mice restores synapse formation and unmasks a physiologic defect in motor neuron firing. Proc Natl Acad Sci U S A 106:3513–3518
66. Ule J, Stefani G, Mele A et al (2006) An RNA map predicting Nova-dependent splicing regulation. Nature 444:580–586
67. Licatalosi DD, Mele A, Fak JJ et al (2008) HITS-CLIP yields genome-wide insights into brain alternative RNA processing. Nature 456:464–469

Chapter 5

Regulation of Alternative Pre-mRNA Splicing

Miguel B. Coelho and Christopher W.J. Smith

Abstract

Alternative splicing plays a prevalent role in generating functionally diversified proteomes from genomes with a more limited repertoire of protein-coding genes. Alternative splicing is frequently regulated with cell type or developmental specificity and in response to signaling pathways, and its mis-regulation can lead to disease. Co-regulated programs of alternative splicing involve interplay between a host of *cis*-acting transcript features and *trans*-acting RNA-binding proteins. Here, we review the current state of understanding of the logic and mechanism of regulated alternative splicing and indicate how this understanding can be exploited to manipulate splicing for therapeutic purposes.

Key words Alternative splicing, Pre-mRNA splicing, Isoforms, Transcriptome, RNA-binding proteins

1 Introduction

As outlined in the preceding chapters, alternative pre-mRNA splicing (AS) is prevalent, has a central role in generating functionally diversified proteomes, and is frequently regulated in a cell-type, developmental, or signal-specific manner. For these reasons, it is not surprising that pathologies arise when splicing or its regulation goes awry [1]. Major efforts have consequently been devoted to understanding the mechanisms responsible for the regulation of alternative splicing. Furthermore, as our understanding increases, efforts have increasingly been made to manipulate splicing for therapeutic ends [2, 3].

Alternative splicing can be understood at a number of levels. First, one can assemble "parts-lists" of regulatory components—transcript features and *trans*-acting factors (proteins or RNAs)—that are able to influence splicing events [4]. This approach can be carried out by molecular dissection of individual alternative splicing events. However, it has been transformed by the availability of global methods for transcriptome profiling (splice-sensitive arrays, high-throughput RT-PCR, and mRNA-Seq) and for defining the in vivo RNA targets of RNA-binding proteins (e.g., CLIP, RIP),

Klemens J. Hertel (ed.), *Spliceosomal Pre-mRNA Splicing: Methods and Protocols*, Methods in Molecular Biology, vol. 1126, DOI 10.1007/978-1-62703-980-2_5, © Springer Science+Business Media, LLC 2014

and by the development of computational techniques to harness, integrate, and exploit the large data sets to define the combinations of features that are associated with particular co-regulated programs of AS [4–8]. Having identified all the influences on a particular alternative splicing event, one can then address the question of how the splicing pattern is switched in particular cell types or in response to signaling pathways. What are the key regulators whose activities are altered, and how are their activities regulated? Finally, one can seek to understand in molecular terms the mechanism by which splicing is regulated; how is splicing complex assembly enhanced or inhibited at particular splice sites? Although insights into mechanism can be provided by global analyses, this level of understanding can only really be achieved by dissection of individual model systems, an approach that has been employed since alternative splicing was first characterized in the 1980s.

It is a truism that AS is regulated by RNA-binding proteins, although the extent to which individual RNA-binding proteins are sufficient to determine cell-specific splicing outcomes is less clear. Indeed many mechanistic investigations of AS in vitro are predicated on the idea that interactions between RNAs and proteins will be sufficient to largely explain regulation of AS. Against this view, an increasing weight of evidence has accumulated to support the view that transcriptional kinetics and delivery of factors by RNA polymerase mean that AS can only be fully understood as a co-transcriptional process; indeed, a process that occurs in a dynamic chromatin context. Early evidence demonstrated the potential for this form of regulation, but there is now a wealth of evidence demonstrating that transcriptional kinetics, sometimes regulated by chromatin modifications, regulates AS in genuine physiological contexts [9]. These topical issues are discussed in Chapters 6 and 7. Here, we focus on the role of transcript features and interacting proteins as the primary agents of regulation of AS.

2 Defining the Regulatory Parts-Lists

2.1 Cis: Transcript Features

A large number of transcript features are known to affect the efficiency with which splice sites are selected. These include the consensus splice site sequences recognized by core splicing factors, auxiliary sequences, and transcript structural features (Fig. 1).

2.1.1 Splice Site Consensus Sequences

Splice sites are defined by consensus sequences that encompass the nearly invariant GU and AG dinucleotides at the intron termini, and the branch point, which is usually an adenosine. The 5′ splice site consensus (in mammals, [A/C]AG|GURAGU) is complementary to, and is recognized by, the 5′ end of U1 snRNA. The branch point sequence (YUR<u>A</u>C) is recognized by SF1/BBP in the

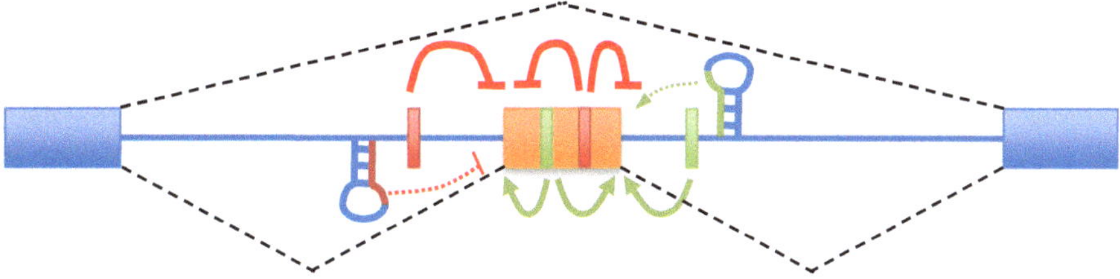

Fig. 1 Transcript features that can influence alternative splicing. Schematic representation of a cassette exon alternative splicing event. Exons shown as *boxes* (*blue* for constitutive and *orange* for alternative) and introns as *thin lines*. Splicing enhancers (*green*) or silencers (*red*) are shown as *thinner rectangles*. These elements bind activator or repressor proteins (not shown) and, like the consensus splice site elements, are usually recognized in single-stranded form. Secondary structure can sequester enhancers or silencer elements and antagonize their activity (indicated by *faint dashed lines*)

E-complex and subsequently by base pairing with U2 snRNA in the A-complex. The 3′ splice site (CAG|G) and polypyrimidine tract preceding it are recognized by the 35 and 65 kDa subunits, respectively, of the U2AF heterodimer in the E-complex. In budding yeast, *S. cerevisiae*, these consensus sequences are highly conserved and nearly invariant, but mammalian splice site consensus sequences are quite degenerate, and the degree of match to the consensus is highly variable [10]. Indeed, taking the GU as invariant, over half of the 16,256 possible sequences have been observed at authentic human 5′ splice sites [11]. To a first approximation, splice site elements that match the consensus well interact more strongly with the cognate binding factors—U1 snRNP, U2 snRNP, U2AF65, or U2AF35—and are functionally stronger; if two or more splice sites are in competition, the site with a better match to consensus tends to win out. However, elevated levels of SRSF1 can promote binding of U1 snRNP binding to strong and weak 5′ splice sites alike, and under these conditions of equal occupancy, more proximal sites are favored (discussed in detail in ref. 11). Global analyses indicate that alternative exons tend to have weaker matches to splice site consensus than constitutive exons [12]. Authentic splice sites with a poor match to the consensus are nevertheless often functional due to the assistance of auxiliary elements.

2.1.2 Auxiliary Elements

For many years it was apparent that, in contrast to budding yeast, mammalian consensus splice site elements contained insufficient information to precisely define authentic splice sites over the numerous nonfunctional intronic sites that resemble consensus sites [10]. The resolution to this “information deficit” was provided by the identification of a plethora of auxiliary sequences commonly known by their location and activity as exon/intron splicing enhancers/silencers, duly abbreviated as ESE, ESS, ISE, and ISS (Fig. 1). These auxiliary elements were identified by a range of approaches, from low-throughput mutagenesis, medium-throughput binding, and functional SELEX for known activator or repressor proteins to global computationally led analyses for motif enrichment in locations consistent with activity as enhancer or silencer [4, 5]. As a result of these combined efforts, it is now apparent that there are huge numbers of auxiliary elements—indeed it has been estimated that more than half of a typical exon is covered by elements with predicted ESE or ESS activity [13]. Collectively, these elements contribute the information "missing" from consensus sites, and accordingly authentic exons are observed to contain a higher ratio of ESE/ESS compared to bulk intron or pseudo-exons [14, 15].

Many auxiliary elements are binding sites for known splicing regulatory proteins, while many "orphan" motifs are probably binding sites for RNA-binding proteins that have either not yet been identified or whose preferred binding motifs have yet to be identified. Auxiliary elements typically have degenerate sequences, presumably reflecting the loose binding specificity of interacting proteins as well as the need for many of these elements to be superimposed on the amino acid encoding content of exons. An exception is the GCAUG motif, which is the tightly defined binding site for RbFOX proteins [16]. Another interesting property of many auxiliary elements, reflecting the properties of the proteins that bind to them, is that they can have opposite activities depending upon their location [17–20]. As well as simple *cis* elements that are binding sites for individual RBPs, there are a number of identified cases of more complex elements resembling splice sites or exons. These typically act in a negative fashion, possibly by acting as "decoy" or nonfunctional splice sites, e.g., [21–23].

2.1.3 Secondary Structure

Like other RNA-mediated processes, splicing can be affected by secondary structure [24]. Indeed, 4 % of conserved alternative splicing events are associated with conserved secondary structure [25]. Most obviously, sequence motifs that are recognized in single-stranded form can be masked by secondary structure [15, 26] (Fig. 1). In this way secondary structure can act negatively (by masking splice sites or enhancers) or positively (by masking silencers). For example, stem-loop structures affect the 5′ splice site of SMN2 exon 7 [27] and Tau exon 10 [28]. In the latter case, intronic mutations that disrupt the structure lead to increased exon inclusion, and the resulting imbalance of Tau isoforms causes the neurodegenerative condition FTDP-17.

A second way in which secondary structure can play an important role is by bringing distantly separated RNA elements into proximity. In the FGFR2 pre-mRNA, secondary structure-mediated juxtaposition of two distantly separated elements is essential for activation of the IIIb exon [29]. Perhaps the most spectacular examples of secondary structure guided AS decisions are in the complex arrays of mutually exclusive exons found in insect genes such as the *Drosophila melanogaster Dscam* gene [30–33]. Here, selection between ~48 mutually exclusive exon variants has been suggested to involve formation of base-paired RNA structures between a common selector element, adjacent to the upstream constitutive element, base-paired to any one of 48 different "docker" elements lying just upstream of each mutually exclusive exon. In these cases, alternative secondary structures provide an elegant solution to the problem of selecting only one exon from among a large number of variants. This type of mechanism, controlled by alternative long-range secondary structures, appears to be prevalent among insect arrays of mutually exclusive exons [32, 33]. Another biologically interesting example employing a more elaborate RNA structure is provided by a thiamine pyrophosphate-binding riboswitch that regulates alternative splicing in *Neurospora crassa* [34]. Nevertheless, while RNA secondary structure clearly influences some splicing events, RNA folding is always in competition with packaging by general RNA-binding proteins (hnRNPs) that assemble onto the nascent pre-mRNA co-transcriptionally and limit the potential for long-range RNA folding [35].

2.1.4 Transcript "Architecture"

The relative location of splice sites and auxiliary elements within a transcript can have profound influences upon splice site selection. In general, closer proximity between 5′ and 3′ splice sites across introns promotes more efficient splice site pairing. All other factors being equal, the proximal of a pair of competing splice sites tends to be selected. However, below a threshold size, primarily dictated by the distance between the 5′ splice site and the branch point, an intron cannot be spliced despite being bounded by functional splice sites. Such an arrangement, with an unspliceable intron, can be used to enforce mutually exclusive selection between exons [36]. There is no such clear-cut upper limit on the size of introns, with many human introns in the size range 10^5–10^6 nt.

In contrast to introns, exons have a preferred size range; human exons have an average size of 140 nucleotides [4], and exons that are significantly shorter or larger than this tend to be spliced inefficiently or employ additional mechanisms that bypass the normal size constraints. The basis for the size constraints of exons was first indicated by the pioneering work in Berget's lab that led to the "exon definition" model. This suggests that splicing complexes initially recognize the exon, rather than the intron, as a unit of assembly, at least for exons flanked by long introns [37]. Expansion of exons above a threshold size of ~300 nt leads to exon skipping. Likewise, very short exons tend to be recognized less efficiently, and alternative cassette exons tend to be shorter than constitutively spliced exons [38]. However, surprisingly, there is no absolute minimum size for an exon; many exons as short as 3 or 6 nt are found, and "zero-length exons"—immediately adjacent 3′ and 5′ sites—have been suggested as a mechanism for recursive splicing of long introns [39]. Nevertheless, very short or very long exons tend to be used less efficiently. The basis for the optimal size of exons is not entirely clear. Splicing factors need to assemble at the consensus splice sites, and very short exons might lead to steric obstruction between splicing factors binding on each side. Within conventional exons, binding of SR proteins to ESEs commonly bridges the interaction between factors bound at the splice sites. It is possible that the ~300 nt threshold might reflect a competition between productive splicing complex assembly and general packaging by hnRNPs. More recently, it has been found that nucleosomes tend to assemble on exonic DNA sequences in the genome, and it has been suggested that the optimal exon size, and even exon definition itself, might be related to nucleosome positioning at the DNA level [40, 41].

The relative positioning of auxiliary elements can also influence their activity. As mentioned above, many auxiliary elements and their cognate binding factors can have opposite activities depending upon their location relative to an exon or splice site. In addition, the distance between an ESE and a consensus splice site can be important. ESE activity declines with distance from a regulated 3′ss [42, 43]. However, a greater separation can also provide for greater inducibility. In the *Drosophila doublesex* gene, a complex ESE located in a 3′ UTR more than 300 nt downstream of a regulated 3′ss is only active in female cells where the female-specific Tra protein binds cooperatively with SR proteins and Tra2 to the enhancer repeats. In the absence of Tra, binding of the SR proteins and Tra2 is weaker and insufficient for activity. However, if the ESE is brought closer to the 3′ss, this weak binding is sufficient for activity, independent of the presence of Tra [43].

2.1.5 RNA Modifications

The effects of posttranscriptional editing and modification of RNA upon splicing are an underexplored area. Adenosine to inosine conversion by ADARs is the commonest form of editing and can affect splicing (as well as translation). Inosine is similar to guanosine in its base-pairing abilities (it can form I–C and I–U base pairs with two hydrogen bonds, and identical geometry to G–C and G–U base pairs), and the Gs at the termini of introns can both be replaced by I [44]. Splice sites can be created by editing, e.g., editing of AA to AI can create a 3′ss [45]. A-to-I editing is necessary for the exonization of some Alu elements [46]. A-to-I editing could affect the activity of auxiliary elements, but this has only been tested by replacing A with G in experimental constructs [46], and it remains possible that ESE binding proteins would not recognize I identically to G. In addition, editing could weaken secondary structure leading to increased accessibility of regulatory elements. Noncoding modifications are more difficult to monitor globally. However, RNA A6 methylation has recently been profiled globally; knockdown of the methylase *METTL3* led to alterations in alternative splicing, and alternatively spliced exons and introns were found to have higher levels of methylation than constitutive exons [47]. Perhaps the most obvious way in which methylation might influence splicing is by affecting the binding of a regulatory factor to its cognate site (similar to the example of CTCF binding to DNA being antagonized by C5 methylation [48]: *see* Chapters 6 and 7).

Thus various base modifications might have effects upon splice site selection. However, these modifications are presumably themselves guided by underlying transcript properties; A-to-I modification is specified by intramolecular base pairing, while A6 methylation occurs within a preferred GACU context.

2.2 Trans-Acting Regulators

Numerous proteins are known to regulate splicing, and these have been joined by a smaller number of *trans*-acting RNAs, e.g., [49, 50]. In many cases, the preferred binding sites of regulatory proteins have been determined, and these correspond to cognate enhancer or silencer motifs. Discussions of splicing regulatory proteins frequently make a distinction between "core" and "regulatory" proteins and among the regulatory proteins between activators and repressors. These distinctions are conceptually helpful; nevertheless, it is becoming clear that many proteins do not fall neatly into a single category. For example, core splicing factors have been shown to influence alternative splicing when their levels are altered [51, 52]. The recent demonstration of expressed variant human U1 snRNAs also supports earlier suggestions that U1 snRNA variants might be responsible for activating suboptimal 5′ splice site, although there is as yet no direct evidence that they play such a role [53]. In addition to the unclear demarcation between core and regulatory factors, many regulatory proteins can either activate or repress depending upon their position of binding relative to a target exon. Perhaps this functional flexibility should not come as a surprise; the founder member of the SR protein family, SRSF1, was discovered simultaneously both as an essential splicing factor [54] and as a positive regulator of alternative splicing [55]. It was later found to repress splicing in some circumstances [56], as well as being involved in numerous other nuclear and cytoplasmic roles [57, 58].

2.2.1 Proteins with RS Domains

A large number of core and regulatory splicing factors contain domains enriched in arginine–serine (RS) dipeptides, usually in combination with additional domains, such as the RNA recognition motif (RRM) RNA-binding domain [16]. Core splicing factors with RS domains include both subunits of U2AF and the U1 70K subunit of U1 snRNP. Perhaps the best-known family of proteins with RS domains is the SR protein family [57, 58], whose members are characterized by an N-terminal RRM domain and a C-terminal RS domain of varying length. Some members of the family also have a second RRM domain, referred to as an RRM homolog or RRMH. By the criterion that they are able to confer splicing activity to an otherwise inactive cytoplasmic S-100 extract, SRSF1 and other members of the SR family can be classed as essential core splicing factors [54, 59]. However, SR proteins can also alter alternative splicing patterns in a concentration-dependent manner [55]. In humans the SR family comprises nine members, and in addition a number of SR-related or SR-like proteins have also been identified. SR proteins affect numerous steps in splicing including recruitment of U1 snRNP to 5′ splice sites, of U2AF to 3′ splice sites, and of the U4/5/6 triple snRNP to assembling spliceosomes (reviewed in refs. 57, 58, 60). While the SR proteins were initially characterized by their functional redundancy in a constitutive splicing assay, it has subsequently become clear that they have many transcript-specific nonredundant roles, particularly in ESE-dependent splicing. Indeed, it is suggested that there is a constant remodeling of the constellation of SR proteins bound to a single mRNA during its life cycle from the nuclear pre-mRNA to a translation-competent cytoplasmic mRNA [61]. The RRM domains of different SR protein family members have distinct RNA-binding preferences, and RNA sequences selected for optimal protein binding act as ESEs for the cognate SR protein [62]. Direct functional selection in the presence of individual SR proteins also reveals distinct ESEs corresponding to different SR proteins [63]. Moreover, global analysis of the RNA-binding sites of different SR proteins indicates a high degree of nonoverlapping sites, in agreement with a nonredundant role [64]. By analogy with the activation domains of transcription factors, the RS domain of SR proteins can be thought of as an "effector" domain in ESE-dependent splicing [43, 65]. The RS domains can be extensively phosphorylated by the SRPK and CLK families of protein kinases. Both phosphorylation and dephosphorylation are required during the spliceosome cycle, and the phosphorylation state also modulates nuclear–cytoplasmic shuttling and protein–protein interactions [57, 58]. However, SRSF1 can also alter alternative splicing of substrates that do not appear to have cognate high-affinity binding sites, and this activity does not require the RS domain.

Despite being best known as activators of splicing, RS domain-containing proteins are also now well known to be able to act as repressors, particularly when binding in introns [56].

Indeed the SR-related protein SRrp38 acts as a general repressor of splicing during mitosis or heat shock, when it is hypo-phosphorylated [66, 67]. In contrast, when its RS domain is phosphorylated, it acts as an activator when bound to high-affinity ESEs [68].

2.2.2 Other RNA-Binding Proteins

Numerous other RNA-binding proteins also regulate splicing. These include members of the hnRNP family [69], which contain RRM or KH RNA-binding domains as well as many other proteins with various RNA-binding domains (Zn finger, Y-box, etc.). The hnRNP proteins tend to be expressed widely in many, but not all, cell types. Some of the hnRNP proteins have undergone gene duplication, and in addition to the more widely expressed protein, paralog proteins show more restricted expression. For example nPTB (nPTB/PTBp2) and hnRNPLL are expressed in differentiated neurons and activated T cells, respectively, while PTB and hnRNPL are more widely expressed [70, 71]. Other non-hnRNP proteins, such as the FOX, CELF, MBNL, nSR100, and NOVA proteins, also show more tissue-restricted expression. Although a number of hnRNP proteins were initially characterized via their repressive activity on splicing, it is now becoming clear that most of the proteins have functional flexibility and can act as either repressors or activators of splicing, depending upon their location of binding [19, 20, 69].

2.3 Maps and Tissue-Specific Codes

The development of global methods for profiling tissue-specific splicing patterns and the binding sites of RNA-binding proteins, combined with computational tools to integrate the various large data sets, has begun to illuminate the details of tissue-specific splicing codes and the contributions of widely expressed and more cell-restricted regulatory factors. Some of the key types of data set that are now available include:

- Global transcriptome data sets at the exon, exon-junction, or single-nucleotide resolution. These were originally generated by splice-sensitive arrays, but mRNA-Seq is now the method of choice. These data sets allow the characterization of large groups of co-regulated events, e.g., cassette exons that are included in neurons or that are differentially regulated upon knockdown of a known regulatory protein. Sets of co-regulated exons can then be analyzed statistically for characteristic features such as exon length, splice site strengths, or enrichment of sequence motifs in particular locations. For cassette exons—the commonest class of ASE—motif enrichments are analyzed within seven transcript regions: the alternative exon itself, the two flanking constitutive exons, and both ends of each of the introns flanking the alternative exon. Enriched motifs are often recognizable as binding sites for known regulatory proteins, particularly in a knockdown/knockout experiment where the binding specificity of the protein is known. More interestingly, motif enrichments can suggest the involvement of particular proteins with tissue-specific splicing programs. Further, by extracting information from the same transcriptomic experiment about the expression levels of the suspected regulatory protein, it is possible to infer whether the protein acts as an activator or repressor. For example, RbFOX-binding sites (GCAUG) downstream of neuron-specific exons act as enhancers in neurons where the cognate protein levels are high, while PTB sites upstream of muscle and neuron-specific exons act as silencers, consistent with the higher expression of PTB in most cells other than neurons and muscle cells [72, 73]. However, in addition to implicating the "usual suspects," or known RBPs in unexpected contexts, these analyses can also reveal "orphan" motifs, indicative of a role for as-yet-unidentified proteins.

- Global data sets of RNAs bound by individual proteins in vivo generated by RIP (RNP immunoprecipitation), which identifies mRNA/pre-mRNA species, or by CLIP (UV cross-linking and immunoprecipitation) which identifies not just the RNA but the location on the RNA where binding occurs [20]. The current generation of CLIP techniques (e.g., iCLIP, PAR-CLIP) provides single-nucleotide resolution of binding. Consensus sequences can be derived from the cross-linking sites, which usually match the consensus sequences identified by in vitro experiments such as SELEX where these have been carried out (see below). Moreover, cellular fractionation before generation of sequencing libraries allows identification of different sets of RNA targets with different biological functions [74].
- Information on the optimal binding sites of RBPs generated by SELEX and related methods [75]. These rely on the availability of recombinant protein and so have mostly been carried out in low-throughput format. However, progress is being made to implement this type of approach in a medium-/high-throughput format [76]. If successful this could help to reduce the number of "orphan" motifs associated with particular regulated programs of splicing.

The preceding approaches have their particular associated strengths and weaknesses. However, these are often complementary and a combination of approaches and integration of different data sets is often particularly powerful [8]. For example, a combination of CLIP with global transcriptome profiling in response to knockdown of the same protein allows the generation of "splicing maps" that reveal how binding of particular proteins can give rise to activation or repression of splicing by binding to specific locations with respect to a regulated exon (reviewed in [20]). Most commonly for proteins without RS domains (e.g., RbFox, NOVA, PTB, MBNL, ESRP), binding of a regulatory protein within or immediately upstream of an exon is associated with exon skipping, while binding on the downstream side is associated with activation. The opposite appears to be true for SR protein-binding sites, which act as enhancers in exons, but silencers in introns [19, 56].

Perhaps the most ambitious use of these large data sets is the attempt to generate computational splicing codes able to predict tissue-specific changes in splicing [7]. The two key inputs for the splicing code are tissue-specific data sets of alternative splicing (splice-sensitive array or mRNA-Seq) and a comprehensive catalog of transcript features. The latter includes the *cis*-acting features known or suspected to influence alternative splicing (sequence motifs, secondary structure, etc.), as discussed above. It also includes features such as conservation, which could not be used by cellular mechanisms (the splicing machinery in the nucleus of a mouse cell cannot take into account whether or not a pre-mRNA sequence it encounters is conserved in human or puffer fish). In a machine-learning approach, a splicing code is assembled using the combinations of transcript features that are best able to predict observed splicing differences between tissues. The first attempt at generating such a code was based on splice-sensitive array data for 3,665 mouse cassette exons across 26 tissue samples [7]. While being unable to predict the actual level of exon inclusion, the code was very successful in being able to predict the direction of change in splicing between different tissue groups (94 % true positive rate). Key observations for the different tissue exon inclusion or exclusion codes were that the most informative features are associated with the regulated exon itself and its immediate flanking introns, and many of these features are interdependent. However, the features associated with the exon tend to be "architectural" rather than sequence motifs. For example, neuron-specific exons tend to be short, non-frameshifting, and not to cause introduction of a premature termination codon when skipped (Fig. 2). The major sites of motif enrichment are in the immediate intron flanks; the neuron-specific inclusion code, for example, has NOVA and RbFOX motifs on the downstream side and PTB/nPTB motifs on both sides of the exon. Pleasingly, many of the code features are consistent with previous detailed experimental dissection of model exons (Figs. 2 and 3), confirming the generality of previous observations, but new regulatory features are also revealed. Work is now in progress to develop similar codes for different species, to examine how the splicing codes evolve [77, 78], and lastly to refine the codes. For example, many functionally important AS events are regulated between different neurons or in response to neuronal activity. Clearly, a single neuronal splicing code cannot account for this level of regulation.

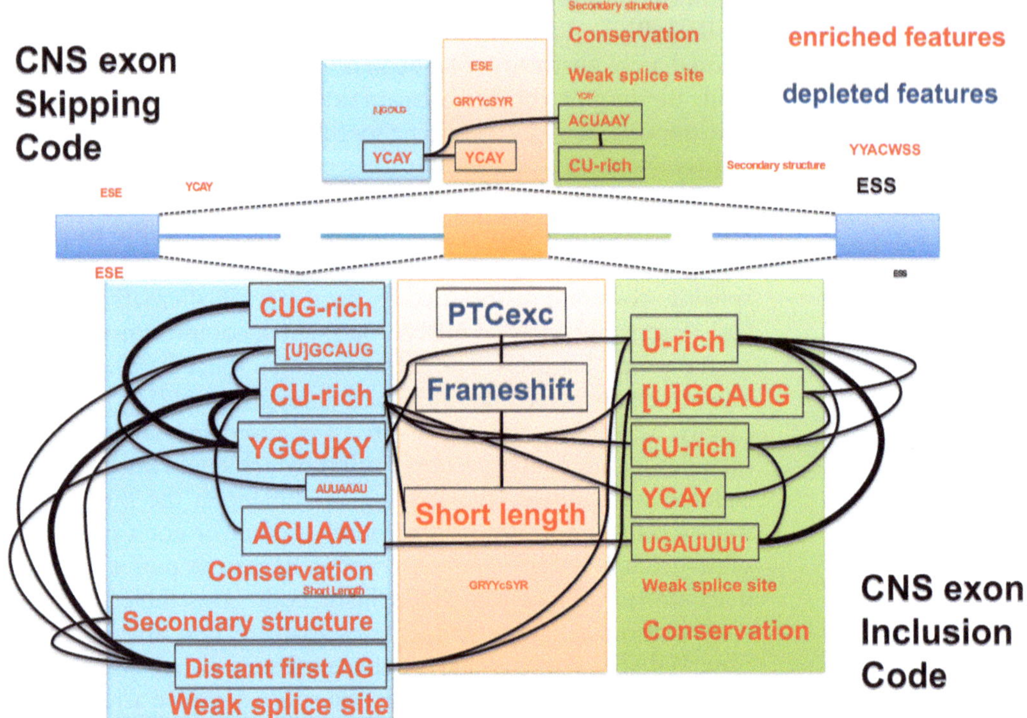

Fig. 2 Neuronal splicing code. The figure is an adaptation from Barash et al. [7] and summarizes the computationally derived "splicing code" for exon skipping (*top*) or exon inclusion (*bottom*) in the central nervous system (CNS). Features are shown for each of the seven transcript regions including the cassette exon, its immediate intronic flanks, the two flanking constitutive exons, and their adjacent intron regions. Transcript features enriched in the code are shown in *red text*, while those in *blue* are depleted features. *Larger text font* represents greater enrichment or depletion of the feature. *Boxed* features show significant co-association with other features; the *black connecting lines* indicate these associations, with the thickness of the connector giving an indication of the significance of the association. Features include sequence motifs (e.g., CU-rich motifs which are binding sites for PTB) as well as architectural features, such as exon size and tendency to cause frameshift when skipped. Note that the majority of sequence features are associated with immediate intronic flanks of the exon. CNS-specific exons tend to be short, not to cause frameshifting when skipped, and are flanked by sequence motifs including binding sites for PTB and nPTB and for RbFOX proteins ([U]GCAUG) downstream. See text for further details

3 Switching Splicing Patterns

Having established the range of influences upon splicing patterns, how can these splicing patterns be altered? The simplest mechanism is regulated expression, or activity of a key regulatory factor. Such regulation could be at the level of transcription or RNA processing (including alternative splicing) of its pre-mRNA, translation or turnover of its mRNA, or posttranslational modification, localization, or turnover of the regulatory protein.

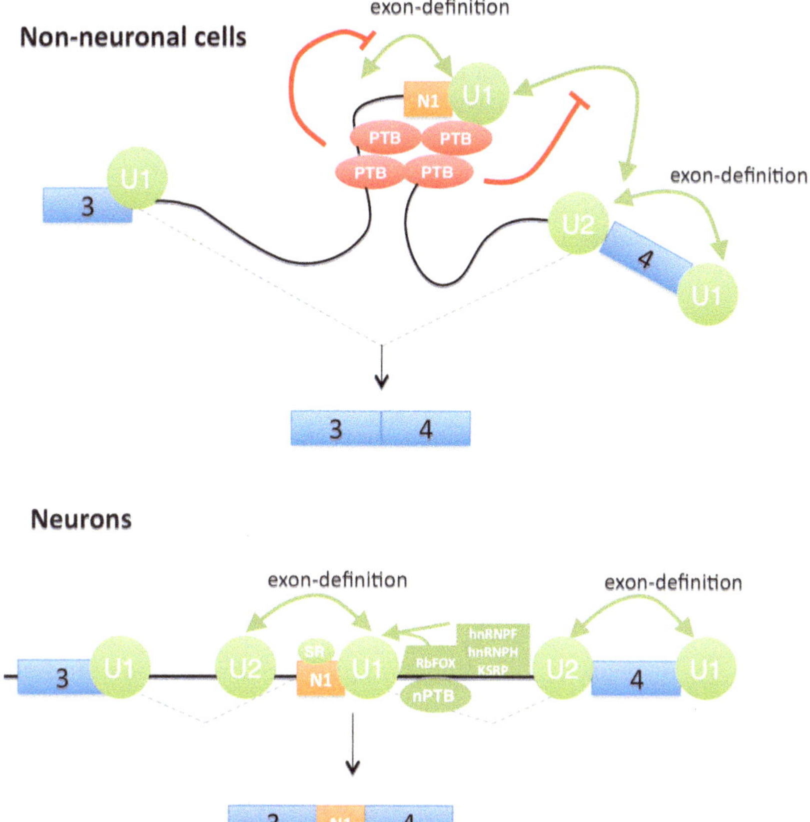

Fig. 3 Neuron-specific alternative splicing of *c-src* N1 exon. The N1 exon of *c-src* is included specifically in neurons. A series of experimental investigations by the Black lab (e.g., [70, 123, 139, 140, 154, 155]) have shown that regulation of N1 splicing involves many of the features subsequently identified in the CNS-specific exon inclusion code (Fig. 2). At 18 nt, the N1 exon is short, which limits its splicing efficiency [154], and non-frameshifting. In nonneuronal cells (*upper panel*), the exon is skipped as the result of PTB binding to a series of flanking CU-rich silencer elements. PTB binding prevents exon definition, but not U1 snRNP binding [123, 139]; indeed PTB directly contacts the stem-loop IV of U1 snRNA in this complex which may prevent productive cross-intron interactions involving U1 snRNP at the N1 5′ splice site [140]. In neurons (*lower panel*) the N1 exon is included. This results in part from the replacement of PTB by the neuronal nPTB paralog [70], which binds to a downstream site. A downstream intron splicing enhancer (the downstream control sequence) binds a complex containing nPTB, hnRNPF and H, and KSRP. RbFox proteins also bind to a GCAUG element promoting exon inclusion [155]

Some the earliest well-characterized cell-specific splicing events were in the pathway determining sexual dimorphism in somatic cells of *Drosophila melanogaster*, where the female-specific proteins sex lethal and transformer are both produced as a result of sex-specific nonproductive splicing and themselves act as splicing regulators [79]. Sex lethal has two RRM domains and acts as a negative regulator of splicing of both its own exon 3 and a 3′ splice site in *Tra*. Transformer has an RS domain, but no RNA-binding domains, and is a key activator of a regulated splice site in the *doublesex* gene.

Characterization of this pathway established the paradigm that the presence or absence of dedicated cell-type-specific proteins can be responsible for determining splicing outcomes.

Early analyses of mammalian AS identified did not provide such simple explanations; a number of regulatory proteins were identified, many of them are members of the SR and hnRNP protein families. In many cases expression of these proteins did not show a simple correlation with regulation of the target splicing events. However, some proteins could be associated with particular cell or tissue-specific splicing events. For example, PTB, which is expressed in numerous cell types, but not in neurons or skeletal muscle cells, was found to repress exons that are neuron or muscle specific [80]. Subsequently, a number of mammalian regulatory proteins with much more tissue-restricted expression have also been identified by a variety of approaches, and some tissue- or cell-type-specific proteins, such as the brain-specific NOVA [81] and nSR100 [82], and the epithelial cell-specific ESRP proteins [83] can play a determining role in cell-specific splicing patterns.

Expression and activity of splicing regulatory proteins can be regulated at levels other than transcription. For example, during neuronal differentiation increased expression of miRNA-124 leads to reduced expression of PTB via target sites in the 3′ UTR [84]. Because PTB represses nPTB expression by inducing a frameshifting exon-skipping event, the reduced PTB levels upon miR124 expression lead to upregulated nPTB [70, 84, 85]. While PTB and nPTB are ~75 % identical at the amino acid level and have similar RNA-binding properties, a subset of ASEs in differentiating neurons are sensitive to the switch between the closely related paralogs [70]. Later during neuronal development, nPTB expression is also reduced leading to a second set of splicing changes of those ASEs that are affected by PTB or nPTB [86, 87]. Many cardiac- and skeletal muscle-specific exons are also repressed by PTB [80]. A set of splicing changes in developing mouse heart are associated with reduced levels of PTB which, in contrast to the micro-RNA inhibition in differentiating neurons, results from developmentally programmed cleavage of PTB by caspases [88]. While the preceding examples involve substantial changes in the levels of splicing regulators, in some cases apparently modest changes in levels can have profound consequences. For example, twofold overexpression of SRSF1 is sufficient to lead to changes in splicing of numerous signaling proteins leading to anchorage-independent growth and cell transformation, leading to the designation of SRSF1 as an oncoprotein [89, 90].

Many changes in alternative splicing respond to altered expression of more than one factor. For instance, during development of cardiac and skeletal muscle, a series of splicing changes occur in response to increased expression of MBNL proteins and decreased nuclear levels of CELF proteins [91], the decrease in CELF expression being mediated by micro-RNAs 23a/b [92]. Members of the MBNL and CELF families generally act antagonistically on target ASEs, so the inverse changes in activities of the two sets of factors reinforce each other. This concerted set of developmentally regulated splicing changes is affected in the CUG triplet expansion disease myotonic dystrophy (DM1) [1]. CUG expansions contain multiple overlapping copies of the optimal MBNL binding motif UGCU [93], and consequently MBNL proteins become sequestered in nuclear foci containing the CUG-repeat RNA [94]. Surprisingly, the CUG expansions not only reduce the effective levels of MBNL proteins, they also lead to elevated levels of the antagonistic CELF proteins. The pathway connecting CUG-expansion RNA and CELF proteins has not been fully elucidated, but it appears to involve inhibition of PKC-phosphorylation-induced cytoplasmic localization of CELF proteins [95]. As a result of the perturbed levels of MBNL and CELF proteins, adult DM muscles express a series of normally embryonic isoforms, a number of which give rise to discrete disease symptoms. The molecular pathology of DM1 involves a toxic gain of function RNA, but it also hints at possible cellular mechanisms of regulation by noncoding RNAs. Indeed, the abundant noncoding MALAT1 RNA binds SR proteins, and knockdown of MALAT1 or overexpression of SR proteins affects the same set of ASEs [96].

4 Signaling to Splicing Regulation

Gene expression is well known to be regulated by signal transduction pathways. However, in contrast to transcription and translation, the number of cases in which each link has been established in the pathway connecting extracellular or cytoplasmic signals and regulation of splicing is still relatively limited [97–99]. Signaling pathways that impact upon splicing typically result in phosphorylation or some other posttranslational modification of an RNA-binding splicing regulator. The consequences for the protein can include altered localization, turnover, or ability to interact with other proteins [99]. T-cell activation in response to antigens, neuronal excitation, and epithelial to mesenchymal transition are good examples of processes involving signal-induced modulation of alternative splicing of target pre-mRNAs.

The transition from naïve to activated T cells occurs upon exposure to an antigen and has been extensively characterized for its changes in phenotype and expression of different transcription factors [100, 101]. Recent findings have uncovered an additional layer of regulation at the level of alternative splicing. The receptor tyrosine phosphatase *CD45* exons 4, 5, and 6 are skipped upon activation of T cells, and this is attributable to two splicing factors, PTB-associated splicing factor (PSF) and hnRNPL-like (hnRNPLL).

HnRNPLL is a paralog of hnRNPL; the two proteins are connected by a cross-regulatory network [71] in a similar manner to PTB and nPTB. HnRNPL is major splicing regulator in T cells and its expression is increased upon activation [102], which partially accounts for the increased skipping; nevertheless, the complete regulation requires active PSF. In resting T cells, PSF is phosphorylated by glycogen synthase kinase-3 (GSK3), and in this form is bound tightly to the thyroid-hormone receptor-associated protein 150 (TRAP150) and is unable to bind to RNA [103]. Upon T-cell activation, GSK3 activity decreases and the resulting unphosphorylated PSF, free of TRAP150, is able to bind to an ESS in exon 4 of the *CD45* gene and promote skipping [103].

SR proteins are also important targets of posttranslational modifications in splicing regulation. They can be extensively phosphorylated in their RS domains, and both phosphorylation and dephosphorylation are important at different steps of splicing. For example, in its hypo-phosphorylated form, the RS domain of SRSF1 interacts with its RRM domain, preventing the RRM from interacting with the RRM of U1 70K protein. Upon phosphorylation of SRSF1, this intramolecular interaction is disrupted [104], allowing interaction between the RRMs of SRSF1 and U1 70K protein, thereby recruiting U1 snRNP. Two main families of kinases, the SRPK and CLK families, can phosphorylate SR proteins. While they both act on the RS domain, the serines they modify are not the same, suggesting that they differentially modify the activity of SR proteins. Regulation of SR protein localization in response to phosphorylation can also influence alternative splicing. For example, in Wilms tumor SRPK1 is overexpressed leading to increased phosphorylation and nuclear localization of SRSF1, which is sufficient to induce a switch in VEGF alternative splicing to favor production of pro-angiogenic isoforms rather than the anti-angiogenic VEGFb isoforms [105]. Autophosphorylation-induced relocalization of kinases can also lead to altered splicing. SRPK is predominantly cytoplasmic while SR proteins are mostly nuclear, but autophosphorylation of SRPK can alter its distribution. The EGF signaling pathway leads to activation of Akt kinase, which binds to SRPK and promotes its autophosphorylation. This in turn leads to an increase in the nuclear translocation of SRPK and consequent SR protein phosphorylation [106]

One of the best characterized RNA-binding proteins regulated by signal transduction is Sam68 (Src associated in mitosis 68 kDa protein), first identified as a target of c-src tyrosine kinase during mitosis [107, 108]. Sam68 is a member of the STAR family (*s*ignal *t*ransduction and *a*ctivation of *R*NA) of proteins, which are characterized by a STAR domain, consisting of a maxi-KH RNA-binding domain flanked by QUA1 and QUA2 elements, as well as other accessory domains [109]. Sam68 can be extensively modified by serine/threonine and tyrosine phosphorylation, arginine methylation, lysine acetylation, and SUMOylation, which variously affect its function, localization, and RNA binding [4, 5, 30, 83, 105, 106]. In the nucleus of T-lymphoma cells, Sam68 is the downstream target of the MAP-kinase pathway, which can be activated in response to phorbol ester stimulation. Activation of this pathway results in Erk1/2-mediated phosphorylation of Sam68, which then binds efficiently to an exonic AAAAUU site in CD44 exon v5 and promotes its inclusion [110]. Regulation by Sam68 of this event is also coordinated with the transcription machinery. As described in Chapter 6, the kinetics of RNA polymerase II can affect the chance of a weak splice site to be recognized by the spliceosome. The chromatin-remodeling protein Brm is well known for its role in facilitating the binding of transcription factors to promoter regions, but also influences alternative splicing by altering the kinetics of RNA polymerase II. Binding of Brm to the region of the *CD44* variable exons results in an accumulation of paused RNA polymerase II, increasing the time available for assembly of spliceosomes around these exons [111]. This event has serious consequences for cell physiology as expression of *CD44* isoforms containing exon v5 correlates with enhanced malignancy and invasiveness of some tumors [112].

Another Sam68-regulated ASE with profound consequences for cell fate is the alternative 5′ splice site choice on exon 2 of the Bcl transcript, leading to either the longer anti-apoptotic isoform Bcl-x(L) or the shorter Bcl-x(s) pro-apoptotic isoform. Genome-wide screens found many regulators of this event including Sam68, which promotes the use of the proximal 5′ splice site resulting in an increase of the pro-apoptotic Bcl-x(s) [113, 114]. The MAPK pathway did not significantly affect this splicing event, but another Sam68 regulator, the src-like FYN kinase, reverted the effect of overexpressing Sam68. FYN kinase phosphorylates tyrosine residues and impairs Sam68 binding to the Bcl-x pre-mRNA and leads to the concentration of Sam68 into discreet nuclear foci [114].

Sam68 is an excellent example of a splicing factor that can be regulated by several kinases with nonredundant outcomes. Two of the six hallmarks of cancer are tissue invasion and apoptosis evasion [115]. These can both result from deregulation of kinase pathways like MAPK and src-like kinases in cancer cells [116–118], which can activate Sam68 to promote CD44 v5 containing isoforms and also protect against apoptosis by preventing Sam68 activation of Bcl-x(s) splicing.

5 Mechanisms of Splicing Regulation

Numerous model systems of alternative splicing have been analyzed biochemically. It is convenient to consider regulation by activation or repression separately, although one must bear in mind

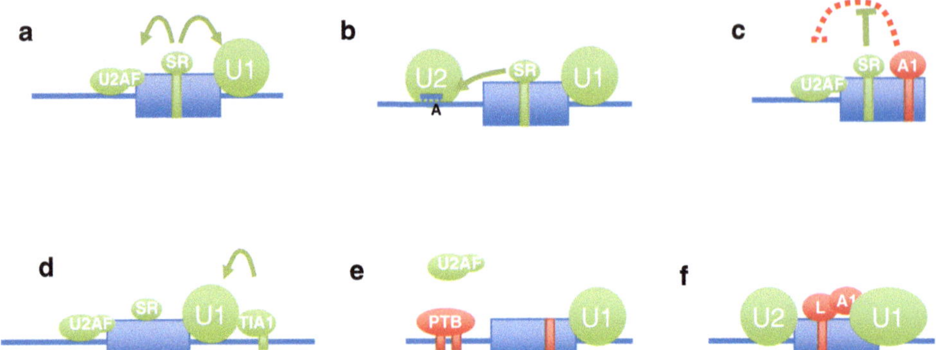

Fig. 4 Mechanisms of splicing regulation. Some of the mechanisms that have been shown to effect alternative splicing have been illustrated schematically. SR proteins can act positively from exon splicing enhancers by promoting recruitment of U2AF or U1 snRNP (**a**), by promoting base pairing of U2 snRNA with the branch point sequence in the A-complex (**b**) [131, 132], or by sterically obstructing the propagative binding of repressive hnRNP A1 from a downstream exon splicing silencer (**c**) [119]. HnRNP proteins or related proteins such as TIA1 can activate splicing from downstream splicing enhancers by promoting recruitment of U1 snRNP via a direct interaction with U1C protein (**d**) [125]. Inhibition of splicing can involve simple steric obstruction of splicing factor binding to splice site elements, e.g., PTB can directly compete with U2AF65 binding to pyrimidine tracts (**e**) [126, 156]. Inhibition can also involve formation of stalled, dead-end, splicing complexes. For example, binding of hnRNPL to an exon splicing silencer can lead to a stalled A-like complex in which the interaction of U1 snRNA with pre-mRNA is hyper-stabilized by extended base-pairing upstream of the 5′ splice site, associated with binding of hnRNP A1 [142]

that activation can occur directly or by antagonism of repressive mechanisms. For example, while splicing activation mediated by the RS domains is the best-known mechanism for ESE-mediated splicing activation by SR proteins (see below), they can also operate in RS domain-independent ways. RRM-mediated binding of SRSF1 to an ESE is able to block propagative binding of hnRNPA1 from a downstream ESS, thereby antagonizing the repressor action of hnRNPA1 [119] (Fig. 4c). This anti-repressor activity of SRFS1 requires its RRMs, but not its RS domain.

Whether the regulatory mechanism involves repression or activation, one of the key questions to address is the step in splicing complex assembly that is affected by regulation. As discussed in Chapter 3, spliceosome assembly occurs in a stepwise fashion [60]. A key point is that the earliest detectable splicing-related complex (the E-complex) is already committed to the splicing pathway (by the criterion that it is resistant to subsequent challenge with an excess of unlabeled self-competitor) [120, 121], all consensus splice site elements are recognized by splicing factors, and finally the complex involves interaction between the two ends of the intron [122]. The key recognition events are the interaction of the 5′ splice site with U1 snRNA by RNA:RNA base pairing and binding of the proteins SF1/BBP to the branch point sequence, U2AF65 to the polypyrimidine tract, U2AF35 to the 3′ splice site, and SR proteins to ESEs, if present. Formation of the E-complex therefore appears to be an ideal stage at which regulatory proteins could intervene to promote or inhibit splicing. Subsequent assembly steps involve displacement of BBP and recognition of the branch point by base pairing with U2 snRNA in the A-complex, and then replacement of U1 snRNA and more limited base pairing of the 5′ splice site with U6 snRNA in the catalytically activated B*-complex [60]. In addition to the splicing-related complexes, all RNAs incubated in nuclear extract form the so-called H-complexes, which in contrast to E-complexes are not essential intermediates on the splicing pathway. The "H" denotes heterogeneous, reflecting the fact that different combinations of proteins associate with different RNAs depending upon sequence. H-complexes are commonly overlooked when considering constitutive splicing mechanisms, but their composition can be important for regulated splicing and influences the ability of the RNA to assemble into productive splicing complexes [123].

Unsurprisingly, many splicing regulators influence formation of productive E-complexes by affecting the binding of U1 snRNP or the U2AF heterodimer. ESE-bound SR proteins can promote the recruitment of U1 snRNP to a weak 5′ splice site [19] or of U2AF to a 3′ splice site with a weak polypyrimidine tract [65] (Fig. 4a). This can be explained by the ability of SR proteins to interact with the U1 70K component of U1 snRNP and with U2AF35 [124]. The non-SR protein TIA1 can activate a weak 5′ splice site from an adjacent downstream ISE by promoting U1 snRNP binding via a direct protein–protein interaction with U1C protein [125] (Fig. 4d). Likewise, repressors can interfere with the same recognition events, in the simplest scenario by binding to overlapping sites. For example, both SXL and PTB can compete directly with U2AF65 binding at specific pyrimidine tracts due to their more restricted preference for subsets of pyrimidine tracts compared to the more flexible requirements of U2AF65 [126] (Fig. 4e). Repression can also occur without affecting initial splicing factor recruitment. SR proteins binding downstream of a 5′ splice site, or hnRNP proteins upstream, repressed splicing without affecting U1 snRNP recruitment [19], suggesting that these repressors affect subsequent productive pairing of the affected 5′ splice site with a 3′ splice site. By contrast, 5′ splice site activation by binding of SR proteins upstream or of hnRNPs downstream was accompanied by increased U1 recruitment.

Despite the clear role of regulation at the earliest steps of spliceosome assembly, it is also evident that regulation can occur at later steps as well. For example, selection between alternative 3′ splice sites, alternative 5′ splice sites, and cassette exon inclusion vs. skipping is not committed within E-complexes, even though the complexes are committed to splicing [122]. However, by the stage of the A-complex pairing of splice sites is committed [127, 128].

Single-molecule analysis also shows that the E-to-A-complex transition is accompanied by the removal of surplus U1 snRNPs, reflecting the commitment to splice site pairing [129]. In addition, elements identified as silencers of competing 5′ splice site had no effect on the kinetics of splicing in the absence of the competing site [130]. This is again consistent with the silencer affecting commitment to splice site pairing, a step that appears to be non-rate limiting in the absence of the competing site. RS domains artificially tethered to an ESE location in a 3′ exon contact the pre-mRNA at the branch site in A-complexes [131, 132]. This suggests that they may play a role in stabilizing the short intramolecular duplex between the branch site and U2 snRNA which first forms in the A-complex, concomitant with commitment to splice site pairing [127, 128] (Fig. 4b). The commitment to splice site pairing at the A-complex stage is consistent with the fact that this accompanies the first of a series of ATP-hydrolysis-dependent transitions that may not be readily reversible [60]. However, single-molecule analysis of yeast splicing shows that the complex assembly steps involving recruitment of U1 snRNP, U2 snRNP, U4/5/6 snRNP, and the PRP19 complex are all reversible [133], and even the catalytic steps of splicing can be reversed [134]. This suggests that some regulators of splicing could intervene at relatively late stages of splicing complex assembly, with re-pairing of splice sites after reversal of some assembly steps. Indeed, investigations of *Drosophila sex-lethal* autoregulation showed that Sxl protein can repress splicing after the first catalytic step [135].

Another important point to consider is that much of the preceding discussion is framed around the splicing complex assembly pathway ($E \rightarrow A \rightarrow B \rightarrow B^{act} \rightarrow C$) determined for pre-mRNA substrates with a single intron, sometimes with competing splice sites at one end. However, internal cassette exons are the commonest type of alternative splicing event, and for most of these, it is likely that the exon definition model applies. The pathway of complex assembly via initially exon-defined complexes followed by exon juxtaposition and formation of cross-intron complexes [37] is much less well defined than for single-intron substrates. However, complexes assembled on a single exon flanked by functional splice sites can contain not just U1 and U2 snRNPs but also U4/5/6 [136]. This suggests that the formation of cross-intron complexes (i.e., splice site pairing) can involve preformed B-like complexes rather than A-like complexes as originally suggested. This in turn suggests that regulation of exon selection could occur in what were previously viewed as later complexes.

A number of studies have indeed indicated that regulatory proteins can act at later steps of assembly after initial exon definition. The protein RBM5 represses *FAS* exon 6 splicing after exon definition [137], and an ESS in *CD45* exon 4 also blocks progression of complex assembly after formation of an ATP-dependent exon complex containing both U1 and U2 snRNPs [138]. These cases demonstrate the importance of considering exon-defined complexes to explore mechanisms of regulation. Investigations of repression of the N1 exon of *c-src* show that it is also important to consider the ability of the flanking constitutive exons to assemble exon definition complexes. PTB inhibits the N1 exon of *c-src* by binding to sites in both flanking introns (Fig. 3). PTB does not prevent U1 snRNP from binding to the N1 5′ splice site, but it does prevent productive cross-intron interactions between this U1 snRNP and the downstream exon 4 while not preventing cross-intron interactions between the 5′ splice site of exon 3 and the 3′ splice site of exon 4 [123]. Experiments using transcripts in which exon 4 lacks its downstream 5′ splice site indicated that association of U2AF with the 3′ splice site of exon 4 was inhibited by PTB. In contrast, similar experiments in which exon 4 had an intact 5′ splice site showed that a functional A-like complex assembled across exon 4, but that this complex was unable to form a subsequent cross-intron splicing complex with the U1 snRNP at the N1 exon [139]. These experiments suggested that PTB blocks the ability of U1 snRNP at the N1 5′ splice site from making productive cross-intron interactions. The basis of this inhibition appears to be via a direct interaction between the N-terminal RRM domains of PTB and the stem-loop IV of U1 snRNA bound at the 5′ splice site [140], which is accompanied by an extension of the U1 snRNP footprint around the 5′ splice site. Indeed, there is evidence from other systems that repression of splicing can be associated by more extended and/or hyper-stabilized U1 snRNP interaction with pre-mRNA [11]. For example, PTB inhibition of *FAS* exon 6 was accompanied by stronger association of U1 snRNP with the 5′ splice site of exon 6 [141]. Moreover, the repression of *CD45* exon 4 by hnRNPL involves extended base pairing of U1 snRNA with sequences in the exon RNA. The hyper-stabilized U1 snRNA inhibits the exchange between U1 and U6 snRNA and so prevents stable integration of the U4/5/6 tri-snRNP and the Prp19 complex [142] (Fig. 4f). Thus, a commonly emerging theme is that exon skipping can be promoted either by interference with early E-like exon definition complexes or by formation of nonproductive complexes sometimes involving hyper-stabilized interactions of snRNPs in the complex [142, 143].

6 Building on the Knowledge of Regulated Splicing: Therapeutic Modulation

As outlined above, a large amount of knowledge has been gained about the transcript features and cellular proteins that influence splicing patterns, the ways in which splicing patterns can be altered by differing availability of active splicing regulators, and finally the ways in which regulators influence spliceosome assembly. One of the surprising lessons has been the huge variety of sequence motifs that have splicing enhancer or silencing activities. As well as illuminating our understanding of alternative splicing, this has also enriched our view of the effects of disease-associated mutations and other genomic sequence variants and has also informed approaches to splicing-based therapies.

Current estimates from the Human Genome Mutation Database (HGMD) predict around 15 % of mutations leading to genetic diseases are located within splice sites, but more than a third of disease-causing SNPs have the potential to disrupt splicing due to the effects of disrupting auxiliary elements [144]. In particular, sequence variants within exons, the consequences of which have typically been interpreted solely in terms of the effects upon the codon in which they occur, have the potential to cause exon skipping if they affect an ESE. Exon skipping can have far more drastic consequences for protein function than alteration of a single amino acid, leading instead to deletion of a segment of the protein or, more drastically, to frameshifting, premature termination, and a C-terminally truncated protein or nonsense-mediated decay of the aberrant mRNA.

Knowledge of the complete panoply of splicing regulatory elements has also informed new approaches to splicing-based therapies. Antisense molecules, either oligonucleotide based (AONs) or delivered by vectors, have the potential to be used for very specific interference with expression. When used to manipulate splicing, these approaches can be harnessed to give rise to increased expression. For example, Duchenne muscular dystrophy (DMD) is frequently caused by splicing mutations in the dystrophin gene that lead to exon skipping. When the exon skipping occurs in regions encoding the N- or C-terminal regions or if it causes a frameshift in the central dystrophin-repeat encoding region, DMD results. In contrast, the much milder Becker's muscular dystrophy results from in-frame exon skipping in the central repeat region. This produces a shorter version of dystrophin that provides partial function [145]. Based upon this observation antisense therapeutic strategies have been developed for DMD in which additional exon-skipping events are induced in the central region of dystrophin, leading to restoration of the open reading frame allowing production of a slightly truncated, but functional, dystrophin [11, 155]. In a dog model of DMD, such treatments show demonstrable relief of clinical symptoms [146].

Inducing exon skipping is conceptually straightforward. However, in the case of spinal muscular atrophy, antisense approaches have been used successfully to induce exon inclusion by targeting intronic splicing silencers. SMA is caused by loss of function of the *SMN1* gene. In humans there is a duplicated SMN2 gene, but due to small number of sequence variations between *SMN1* and 2, exon 7 of *SMN2* is mainly skipped leading to a lack of functional protein [147]. Splicing-based therapies aim to promote *SMN2* exon 7 splicing, thereby compensating for the loss of *SMN1* [148]. Various strategies have been tested including ANOs that target an exon 7 sequence that varies between SMN1 and 2 and that acts as an ESS in SMN2, or bifunctional molecules containing an antisense-targeting domain linked to an effector ESE domain [149]. Peptide nucleic acid antisense-targeting domains linked directly to an arginine–serine peptide to directly provide ESE function without the need to recruit proteins have also been used [150]. However, the most successful approach used an ANO tiling strategy and ultimately identified potent ISS targets downstream of exon 7 of the SMN2 gene [151]. Trials in mouse models of SMA have provided very encouraging long-term improvements in response to administration of the ANOs by direct injection into cerebrospinal fluid [152, 153]. These and other splicing-based therapeutics provide a compelling demonstration of how knowledge of a fundamental biological process built up over the years can lead to unanticipated applications.

Acknowledgments

Work in the authors' lab is supported by grants from the Wellcome Trust (092900) and the Biotechnology and Biological Sciences Research Council (BB/H004203/1 and BB/J0014567/1).

References

1. Cooper TA, Wan L, Dreyfuss G (2009) RNA and disease. Cell 136:777–793
2. Muntoni F, Wood MJ (2011) Targeting RNA to treat neuromuscular disease. Nat Rev Drug Discov 10:621–637
3. Singh RK, Cooper TA (2012) Pre-mRNA splicing in disease and therapeutics. Trends Mol Med 18:472–482
4. Wang Z, Burge CB (2008) Splicing regulation: from a parts list of regulatory elements to an integrated splicing code. RNA 14:802–813
5. Blencowe BJ (2006) Alternative splicing: new insights from global analyses. Cell 126:37–47
6. Hallegger M, Llorian M, Smith CW (2010) Alternative splicing: global insights. FEBS J 277:856–866
7. Barash Y, Calarco JA, Gao W et al (2010) Deciphering the splicing code. Nature 465: 53–59
8. Zhang C, Frias MA, Mele A et al (2010) Integrative modeling defines the Nova splicing-regulatory network and its combinatorial controls. Science 329:439–443
9. Luco RF, Allo M, Schor IE et al (2012) Epigenetics in alternative pre-mRNA splicing. Cell 144:16–26
10. Burge C, Tuschl T, Sharp P (1999) Splicing of precursors to mRNAs by spliceosomes. In: Gestetland R, Cech T, Atkins J (eds) The RNA world. Cold Spring Harbor Laboratory Press, Cold Spring Harbor, NY, pp 525–560
11. Roca X, Krainer AR, Eperon IC (2013) Pick one, but be quick: 5′ splice sites and the problems of too many choices. Genes Dev 27:129–144
12. Keren H, Lev-Maor G, Ast G (2010) Alternative splicing and evolution: diversification, exon definition and function. Nat Rev Genet 11:345–355
13. Chasin LA (2007) Searching for splicing motifs. Adv Exp Med Biol 623:85–106
14. Zhang XH, Chasin LA (2004) Computational definition of sequence motifs governing constitutive exon splicing. Genes Dev 18:1241–1250
15. Ke S, Shang S, Kalachikov SM et al (2011) Quantitative evaluation of all hexamers as exonic splicing elements. Genome Res 21:1360–1374

16. Clery A, Blatter M, Allain FH (2008) RNA recognition motifs: boring? Not quite. Curr Opin Struct Biol 18:290–298
17. Goren A, Ram O, Amit M et al (2006) Comparative analysis identifies exonic splicing regulatory sequences: the complex definition of enhancers and silencers. Mol Cell 22:769–781
18. Wang Y, Ma M, Xiao X et al (2012) Intronic splicing enhancers, cognate splicing factors and context-dependent regulation rules. Nat Struct Mol Biol 19:1044–1052
19. Erkelenz S, Mueller WF, Evans MS et al (2013) Position-dependent splicing activation and repression by SR and hnRNP proteins rely on common mechanisms. RNA 19:96–102
20. Witten JT, Ule J (2011) Understanding splicing regulation through RNA splicing maps. Trends Genet 27:89–97
21. Siebel CW, Fresco LD, Rio DC (1992) The mechanism of somatic inhibition of Drosophila P-element pre-mRNA splicing: multiprotein complexes at an exon pseudo-5′ splice site control U1 snRNP binding. Genes Dev 6:1386–1401
22. Cote J, Dupuis S, Jiang Z et al (2001) Caspase-2 pre-mRNA alternative splicing: identification of an intronic element containing a decoy 3′ acceptor site. Proc Natl Acad Sci USA 98:938–943
23. Pagani F, Buratti E, Stuani C et al (2002) A new type of mutation causes a splicing defect in ATM. Nat Genet 30:426–429
24. Buratti E, Baralle FE (2004) Influence of RNA secondary structure on the pre-mRNA splicing process. Mol Cell Biol 24:10505–10514
25. Shepard PJ, Hertel KJ (2008) Conserved RNA secondary structures promote alternative splicing. RNA 14:1463–1469
26. Hiller M, Zhang Z, Backofen R et al (2007) Pre-mRNA secondary structures influence exon recognition. PLoS Genet 3:e204
27. Singh NN, Singh RN, Androphy EJ (2007) Modulating role of RNA structure in alternative splicing of a critical exon in the spinal muscular atrophy genes. Nucleic Acids Res 35:371–389
28. Grover A, Houlden H, Baker M et al (1999) 5′ splice site mutations in tau associated with the inherited dementia FTDP-17 affect a stem-loop structure that regulates alternative splicing of exon 10. J Biol Chem 274:15134–15143
29. Baraniak AP, Lasda EL, Wagner EJ et al (2003) A stem structure in fibroblast growth factor receptor 2 transcripts mediates cell-type-specific splicing by approximating intronic control elements. Mol Cell Biol 23: 9327–9337
30. Graveley BR (2005) Mutually exclusive splicing of the insect Dscam pre-mRNA directed by competing intronic RNA secondary structures. Cell 123:65–73
31. Olson S, Blanchette M, Park J et al (2007) A regulator of Dscam mutually exclusive splicing fidelity. Nat Struct Mol Biol 14:1134–1140
32. Wang X, Li G, Yang Y et al (2012) An RNA architectural locus control region involved in Dscam mutually exclusive splicing. Nat Commun 3:1255
33. Yang Y, Zhan L, Zhang W et al (2011) RNA secondary structure in mutually exclusive splicing. Nat Struct Mol Biol 18:159–168
34. Cheah MT, Wachter A, Sudarsan N et al (2007) Control of alternative RNA splicing and gene expression by eukaryotic riboswitches. Nature 447:497–500
35. Eperon LP, Graham IR, Griffiths AD et al (1988) Effects of RNA secondary structure on alternative splicing of pre-mRNA: is folding limited to a region behind the transcribing RNA polymerase? Cell 54:393–401
36. Smith CW, Nadal-Ginard B (1989) Mutually exclusive splicing of alpha-tropomyosin exons enforced by an unusual lariat branch point location: implications for constitutive splicing. Cell 56:749–758
37. Berget SM (1995) Exon recognition in vertebrate splicing. J Biol Chem 270:2411–2414
38. Sorek R, Shamir R, Ast G (2004) How prevalent is functional alternative splicing in the human genome? Trends Genet 20:68–71
39. Burnette JM, Miyamoto-Sato E, Schaub MA et al (2005) Subdivision of large introns in Drosophila by recursive splicing at nonexonic elements. Genetics 170:661–674
40. Schwartz S, Meshorer E, Ast G (2009) Chromatin organization marks exon-intron structure. Nat Struct Mol Biol 16:990–995
41. Tilgner H, Nikolaou C, Althammer S et al (2009) Nucleosome positioning as a determinant of exon recognition. Nat Struct Mol Biol 16:996–1001
42. Lavigueur A, La Branche H, Kornblihtt AR et al (1993) A splicing enhancer in the human fibronectin alternate ED1 exon interacts with SR proteins and stimulates U2 snRNP binding. Genes Dev 7:2405–2417
43. Graveley BR, Hertel KJ, Maniatis T (1998) A systematic analysis of the factors that determine the strength of pre-mRNA splicing enhancers. EMBO J 17:6747–6756

44. Scadden ADJ, Smith CWJ (1995) Interactions between the terminal bases of mammalian introns are retained in inosine-containing pre-mRNAs. EMBO J 14:3236–3246
45. Rueter SM, Dawson TR, Emeson RB (1999) Regulation of alternative splicing by RNA editing. Nature 399:75–80
46. Lev-Maor G, Sorek R, Levanon EY et al (2007) RNA-editing-mediated exon evolution. Genome Biol 8:R29
47. Dominissini D, Moshitch-Moshkovitz S, Schwartz S et al (2012) Topology of the human and mouse m6A RNA methylomes revealed by m6A-seq. Nature 485:201–206
48. Shukla S, Kavak E, Gregory M et al (2011) CTCF-promoted RNA polymerase II pausing links DNA methylation to splicing. Nature 479:74–79
49. Kishore S, Khanna A, Zhang Z et al (2010) The snoRNA MBII-52 (SNORD 115) is processed into smaller RNAs and regulates alternative splicing. Hum Mol Genet 19: 1153–1164
50. Kishore S, Stamm S (2006) The snoRNA HBII-52 regulates alternative splicing of the serotonin receptor 2C. Science 311:230–232
51. Park JW, Parisky K, Celotto AM et al (2004) Identification of alternative splicing regulators by RNA interference in Drosophila. Proc Natl Acad Sci USA 101:15974–15979
52. Saltzman AL, Pan Q, Blencowe BJ (2011) Regulation of alternative splicing by the core spliceosomal machinery. Genes Dev 25: 373–384
53. O'Reilly D, Dienstbier M, Cowley SA et al (2012) Differentially expressed, variant U1 snRNAs regulate gene expression in human cells. Genome Res 23:281–291
54. Krainer AR, Mayeda A, Kozak D et al (1991) Functional expression of cloned human splicing factor SF2: homology to RNA-binding proteins, U1 70K, and Drosophila splicing regulators. Cell 66:383–394
55. Ge H, Manley JL (1990) A protein factor, ASF, controls cell-specific alternative splicing of SV40 early pre-mRNA in vitro. Cell 62:25–34
56. Kanopka A, Muhlemann O, Akusjarvi G (1996) Inhibition by SR proteins of splicing of a regulated adenovirus pre-mRNA. Nature 381:535–538
57. Long JC, Caceres JF (2009) The SR protein family of splicing factors: master regulators of gene expression. Biochem J 417:15–27
58. Shepard PJ, Hertel KJ (2009) The SR protein family. Genome Biol 10:242
59. Zahler AM, Lane WS, Stolk JA et al (1992) SR proteins: a conserved family of pre-mRNA splicing factors. Genes Dev 6:837–847
60. Wahl MC, Will CL, Luhrmann R (2009) The spliceosome: design principles of a dynamic RNP machine. Cell 136:701–718
61. Sapra AK, Ankö M-L, Grishina I et al (2009) SR protein family members display diverse activities in the formation of nascent and mature mRNPs in vivo. Mol cell 34: 179–190
62. Tacke R, Manley JL (1995) The human splicing factors ASF/SF2 and SC35 possess distinct, functionally significant RNA binding specificities. EMBO J 14:3540–3551
63. Liu HX, Zhang M, Krainer AR (1998) Identification of functional exonic splicing enhancer motifs recognized by individual SR proteins. Genes Dev 12:1998–2012
64. Änkö M-L, Müller-McNicoll M, Brandl H et al (2012) The RNA-binding landscapes of two SR proteins reveal unique functions and binding to diverse RNA classes. Genome Biol 13:R17
65. Graveley BR, Hertel KJ, Maniatis T (2001) The role of U2AF35 and U2AF65 in enhancer-dependent splicing. RNA 7:806–818
66. Shin C, Feng Y, Manley JL (2004) Dephosphorylated SRp38 acts as a splicing repressor in response to heat shock. Nature 427:553–558
67. Shin C, Manley JL (2002) The SR protein SRp38 represses splicing in M phase cells. Cell 111:407–417
68. Feng Y, Chen M, Manley JL (2008) Phosphorylation switches the general splicing repressor SRp38 to a sequence-specific activator. Nat Struct Mol Biol 15:1040–1048
69. Huelga SC, Vu AQ, Arnold JD et al (2012) Integrative genome-wide analysis reveals cooperative regulation of alternative splicing by hnRNP proteins. Cell Rep 1:167–178
70. Boutz PL, Stoilov P, Li Q et al (2007) A post-transcriptional regulatory switch in polypyrimidine tract-binding proteins reprograms alternative splicing in developing neurons. Genes Dev 21:1636–1652
71. Rossbach O, Hung LH, Schreiner S et al (2009) Auto- and cross-regulation of the hnRNP L proteins by alternative splicing. Mol Cell Biol 29:1442–1451
72. Castle JC, Zhang C, Shah JK et al (2008) Expression of 24,426 human alternative splicing events and predicted *cis* regulation in 48 tissues and cell lines. Nat Genet 40: 1416–1425
73. Wang ET, Sandberg R, Luo S et al (2008) Alternative isoform regulation in human tissue transcriptomes. Nature 456:470–476
74. Wang ET, Cody NA, Jog S et al (2012) Transcriptome-wide regulation of pre-mRNA splicing and mRNA localization by muscleblind proteins. Cell 150:710–724

75. Tuerk C, Gold L (1990) Systematic evolution of ligands by exponential enrichment: RNA ligands to bacteriophage T4 DNA polymerase. Science 249:505–510
76. Ray D, Kazan H, Chan ET et al (2009) Rapid and systematic analysis of the RNA recognition specificities of RNA-binding proteins. Nat Biotechnol 27:667–670
77. Barbosa-Morais NL, Irimia M, Pan Q et al (2012) The evolutionary landscape of alternative splicing in vertebrate species. Science 338:1587–1593
78. Merkin J, Russell C, Chen P, Burge CB (2012) Evolutionary dynamics of gene and isoform regulation in Mammalian tissues. Science 338:1593–1599
79. Lopez AJ (1998) Alternative splicing of pre-mRNA: developmental consequences and mechanisms of regulation. Annu Rev Genet 32:279–305
80. Llorian M, Schwartz S, Clark TA et al (2010) Position-dependent alternative splicing activity revealed by global profiling of alternative splicing events regulated by PTB. Nat Struct Mol Biol 17:1114–1123
81. Buckanovich RJ, Posner JB, Darnell RB (1993) Nova, the paraneoplastic Ri antigen, is homologous to an RNA-binding protein and is specifically expressed in the developing motor system. Neuron 11:657–672
82. Calarco JA, Superina S, O'Hanlon D et al (2009) Regulation of vertebrate nervous system alternative splicing and development by an SR-related protein. Cell 138:898–910
83. Warzecha CC, Sato TK, Nabet B et al (2009) ESRP1 and ESRP2 are epithelial cell-type-specific regulators of FGFR2 splicing. Mol Cell 33:591–601
84. Makeyev EV, Zhang J, Carrasco MA et al (2007) The MicroRNA miR-124 promotes neuronal differentiation by triggering brain-specific alternative pre-mRNA splicing. Mol Cell 27:435–448
85. Spellman R, Llorian M, Smith CW (2007) Crossregulation and functional redundancy between the splicing regulator PTB and its paralogs nPTB and ROD1. Mol Cell 27:420–434
86. Zheng S, Gray EE, Chawla G et al (2012) PSD-95 is post-transcriptionally repressed during early neural development by PTBP1 and PTBP2. Nat Neurosci 15:381–388, S1
87. Licatalosi DD, Yano M, Fak JJ et al (2012) Ptbp2 represses adult-specific splicing to regulate the generation of neuronal precursors in the embryonic brain. Genes Dev 26:1626–1642
88. Ye J, Llorian M, Cardona M et al (2013) A pathway involving HDAC5, cFLIP and caspases regulates expression of the splicing regulator Polypyrimidine Tract Binding Protein in the heart. J Cell Sci 126:1682–1691
89. Anczukow O, Rosenberg AZ, Akerman M et al (2012) The splicing factor SRSF1 regulates apoptosis and proliferation to promote mammary epithelial cell transformation. Nat Struct Mol Biol 19:220–228
90. Karni R, de Stanchina E, Lowe SW et al (2007) The gene encoding the splicing factor SF2/ASF is a proto-oncogene. Nat Struct Mol Biol 14:185–193
91. Kalsotra A, Xiao X, Ward AJ et al (2008) A postnatal switch of CELF and MBNL proteins reprograms alternative splicing in the developing heart. Proc Natl Acad Sci USA 105:20333–20338
92. Kalsotra A, Wang K, Li PF et al (2010) MicroRNAs coordinate an alternative splicing network during mouse postnatal heart development. Genes Dev 24:653–658
93. Goers ES, Purcell J, Voelker RB et al (2010) MBNL1 binds GC motifs embedded in pyrimidines to regulate alternative splicing. Nucleic Acids Res 38:2467–2484
94. Miller JW, Urbinati CR, Teng-Umnuay P et al (2000) Recruitment of human muscleblind proteins to (CUG)(n) expansions associated with myotonic dystrophy. EMBO J 19: 4439–4448
95. Kuyumcu-Martinez NM, Wang GS, Cooper TA (2007) Increased steady-state levels of CUGBP1 in myotonic dystrophy 1 are due to PKC-mediated hyperphosphorylation. Mol Cell 28:68–78
96. Tripathi V, Ellis JD, Shen Z et al (2010) The nuclear-retained noncoding RNA MALAT1 regulates alternative splicing by modulating SR splicing factor phosphorylation. Mol Cell 39:925–938
97. Shin C, Manley JL (2004) Cell signalling and the control of pre-mRNA splicing. Nat Rev Mol Cell Biol 5:727–738
98. Lynch KW (2007) Regulation of alternative splicing by signal transduction pathways. Adv Exp Med Biol 623:161–174
99. Heyd F, Lynch KW (2011) Degrade, move, regroup: signaling control of splicing proteins. Trends Biochem Sci 36:397–404
100. Ellisen LW, Palmer RE, Maki RG et al (2001) Cascades of transcriptional induction during human lymphocyte activation. Eur J Cell Biol 80:321–328
101. Teague TK, Hildeman D, Kedl RM et al (1999) Activation changes the spectrum but not the diversity of genes expressed by T cells. Proc Natl Acad Sci USA 96:12691–12696
102. Oberdoerffer S, Moita LF, Neems D et al (2008) Regulation of CD45 alternative

splicing by heterogeneous ribonucleoprotein, hnRNPLL. Science 321:686–691

103. Heyd F, Lynch KW (2010) Phosphorylation-dependent regulation of PSF by GSK3 controls CD45 alternative splicing. Mol Cell 40:126–137
104. Cho S, Hoang A, Sinha R et al (2011) Interaction between the RNA binding domains of Ser-Arg splicing factor 1 and U1-70K snRNP protein determines early spliceosome assembly. Proc Natl Acad Sci USA 108:8233–8238
105. Amin EM, Oltean S, Hua J et al (2011) WT1 mutants reveal SRPK1 to be a downstream angiogenesis target by altering VEGF splicing. Cancer Cell 20:768–780
106. Zhou Z, Qiu J, Liu W et al (2012) The Akt-SRPK-SR axis constitutes a major pathway in transducing EGF signaling to regulate alternative splicing in the nucleus. Mol Cell 47:422–433
107. Fumagalli S, Totty NF, Hsuan JJ et al (1994) A target for Src in mitosis. Nature 368:871–874
108. Taylor SJ, Shalloway D (1994) An RNA-binding protein associated with Src through its SH2 and SH3 domains in mitosis. Nature 368:867–871
109. Vernet C, Artzt K (1997) STAR, a gene family involved in signal transduction and activation of RNA. Trends Genet 13:479–484
110. Matter N, Herrlich P, König H (2002) Signal-dependent regulation of splicing via phosphorylation of Sam68. Nature 420:691–695
111. Batsché E, Yaniv M, Muchardt C (2006) The human SWI/SNF subunit Brm is a regulator of alternative splicing. Nat Struct Mol Biol 13:22–29
112. Cheng C, Sharp PA (2006) Regulation of CD44 alternative splicing by SRm160 and its potential role in tumor cell invasion. Mol Cell Biol 26:362–370
113. Moore MJ, Wang Q, Kennedy CJ et al (2010) An alternative splicing network links cell-cycle control to apoptosis. Cell 142:625–636
114. Paronetto MP, Achsel T, Massiello A et al (2007) The RNA-binding protein Sam68 modulates the alternative splicing of Bcl-x. J Cell Biol 176:929–939
115. Hanahan D, Weinberg RA (2000) The hallmarks of cancer. Cell 100:57–70
116. Dhillon AS, Hagan S, Rath O et al (2007) MAP kinase signalling pathways in cancer. Oncogene 26:3279–3290
117. Irby RB, Yeatman TJ (2000) Role of Src expression and activation in human cancer. Oncogene 19:5636–5642
118. Paronetto MP, Venables JP, Elliott DJ et al (2003) Tr-kit promotes the formation of a multimolecular complex composed by Fyn, PLCgamma1 and Sam68. Oncogene 22:8707–8715
119. Zhu J, Mayeda A, Krainer AR (2001) Exon identity established through differential antagonism between exonic splicing silencer-bound hnRNP A1 and enhancer-bound SR proteins. Mol Cell 8:1351–1361
120. Jamison SF, Crow A, Garcia-Blanco MA (1992) The spliceosome assembly pathway in mammalian extracts. Mol Cell Biol 12: 4279–4287
121. Michaud S, Reed R (1991) An ATP-independent complex commits pre-mRNA to the mammalian spliceosome assembly pathway. Genes Dev 5:2534–2546
122. Michaud S, Reed R (1993) A functional association between the 5′ and 3′ splice site is established in the earliest prespliceosome complex (E) in mammals. Genes Dev 7:1008–1020
123. Sharma S, Falick AM, Black DL (2005) Polypyrimidine tract binding protein blocks the 5′ splice site-dependent assembly of U2AF and the prespliceosomal E complex. Mol Cell 19:485–496
124. Wu JY, Maniatis T (1993) Specific interactions between proteins implicated in splice site selection and regulated alternative splicing. Cell 75:1061–1070
125. Forch P, Puig O, Martinez C et al (2002) The splicing regulator TIA-1 interacts with U1-C to promote U1 snRNP recruitment to 5′ splice sites. EMBO J 21:6882–6892
126. Singh R, Valcarcel J, Green MR (1995) Distinct binding specificities and functions of higher eukaryotic polypyrimidine tract-binding proteins. Science 268:1173–1176
127. Lim SR, Hertel KJ (2004) Commitment to splice site pairing coincides with A complex formation. Mol Cell 15:477–483
128. Kotlajich MV, Crabb TL, Hertel KJ (2009) Spliceosome assembly pathways for different types of alternative splicing converge during commitment to splice site pairing in the A complex. Mol Cell Biol 29:1072–1082
129. Hodson MJ, Hudson AJ, Cherny D et al (2012) The transition in spliceosome assembly from complex E to complex A purges surplus U1 snRNPs from alternative splice sites. Nucleic Acids Res 40:6850–6862
130. Yu Y, Maroney PA, Denker JA et al (2008) Dynamic regulation of alternative splicing by silencers that modulate 5′ splice site competition. Cell 135:1224–1236
131. Shen H, Green MR (2004) A pathway of sequential arginine-serine-rich domain-

splicing signal interactions during mammalian spliceosome assembly. Mol Cell 16:363–373
132. Shen H, Green MR (2006) RS domains contact splicing signals and promote splicing by a common mechanism in yeast through humans. Genes Dev 20:1755–1765
133. Hoskins AA, Friedman LJ, Gallagher SS et al (2011) Ordered and dynamic assembly of single spliceosomes. Science 331:1289–1295
134. Tseng CK, Cheng SC (2008) Both catalytic steps of nuclear pre-mRNA splicing are reversible. Science 320:1782–1784
135. Lallena MJ, Chalmers KJ, Llamazares S et al (2002) Splicing regulation at the second catalytic step by Sex-lethal involves 3′ splice site recognition by SPF45. Cell 109:285–296
136. Schneider M, Will CL, Anokhina M et al (2010) Exon definition complexes contain the tri-snRNP and can be directly converted into B-like precatalytic splicing complexes. Mol Cell 38:223–235
137. Bonnal S, Martinez C, Forch P et al (2008) RBM5/Luca-15/H37 regulates Fas alternative splice site pairing after exon definition. Mol Cell 32:81–95
138. House AE, Lynch KW (2006) An exonic splicing silencer represses spliceosome assembly after ATP-dependent exon recognition. Nat Struct Mol Biol 13:937–944
139. Sharma S, Kohlstaedt LA, Damianov A et al (2008) Polypyrimidine tract binding protein controls the transition from exon definition to an intron defined spliceosome. Nat Struct Mol Biol 15:183–191
140. Sharma S, Maris C, Allain FH et al (2011) U1 snRNA directly interacts with polypyrimidine tract-binding protein during splicing repression. Mol Cell 41:579–588
141. Izquierdo JM, Majos N, Bonnal S et al (2005) Regulation of Fas alternative splicing by antagonistic effects of TIA-1 and PTB on exon definition. Mol Cell 19:475–484
142. Chiou N-T, Shankarling G, Lynch KW (2013) HnRNP L and HnRNP A1 induce extended U1 snRNA interactions with an exon to repress spliceosome assembly. Mol Cell 49:972–982
143. Motta-Mena LB, Heyd F, Lynch KW (2010) Context-dependent regulatory mechanism of the splicing factor hnRNP L. Mol Cell 37: 223–234
144. Lim KH, Ferraris L, Filloux ME et al (2011) Using positional distribution to identify splicing elements and predict pre-mRNA processing defects in human genes. Proc Natl Acad Sci USA 108:11093–11098
145. Muntoni F, Torelli S, Ferlini A (2003) Dystrophin and mutations: one gene, several proteins, multiple phenotypes. Lancet Neurol 2:731–740
146. Yokota T, Lu Q-L, Partridge T et al (2009) Efficacy of systemic morpholino exon-skipping in Duchenne dystrophy dogs. Ann Neurol 65:667–676
147. Lorson CL, Hahnen E, Androphy EJ et al (1999) A single nucleotide in the SMN gene regulates splicing and is responsible for spinal muscular atrophy. Proc Natl Acad Sci USA 96:6307–6311
148. Nlend Nlend R, Meyer K, Schümperli D (2010) Repair of pre-mRNA splicing: prospects for a therapy for spinal muscular atrophy. RNA Biol 7:430–440
149. Skordis LA, Dunckley MG, Yue B et al (2003) Bifunctional antisense oligonucleotides provide a *trans*-acting splicing enhancer that stimulates SMN2 gene expression in patient fibroblasts. Proc Natl Acad Sci USA 100:4114–4119
150. Cartegni L, Krainer AR (2003) Correction of disease-associated exon skipping by synthetic exon-specific activators. Nat Struct Biol 10:120–125
151. Singh NK, Singh NN, Androphy EJ et al (2006) Splicing of a critical exon of human Survival Motor Neuron is regulated by a unique silencer element located in the last intron. Mol Cell Biol 26:1333–1346
152. Hua Y, Sahashi K, Hung G et al (2010) Antisense correction of SMN2 splicing in the CNS rescues necrosis in a type III SMA mouse model. Genes Dev 24: 1634–1644
153. Passini MA, Bu J, Richards AM et al (2011) Antisense oligonucleotides delivered to the mouse CNS ameliorate symptoms of severe spinal muscular atrophy. Sci Transl Med 3: 72ra18
154. Black DL (1991) Does steric interference between splice sites block the splicing of a short c-src neuron-specific exon in non-neuronal cells? Genes Dev 5:389–402
155. Underwood JG, Boutz PL, Dougherty JD et al (2005) Homologues of the Caenorhabditis elegans Fox-1 protein are neuronal splicing regulators in mammals. Mol Cell Biol 25:10005–10016
156. Matlin AJ, Southby J, Gooding C et al (2007) Repression of alpha-actinin SM exon splicing by assisted binding of PTB to the polypyrimidine tract. RNA 13:1214–1223

Chapter 6

Introduction to Cotranscriptional RNA Splicing

Evan C. Merkhofer, Peter Hu, and Tracy L. Johnson

Abstract

The discovery that many intron-containing genes can be cotranscriptionally spliced has led to an increased understanding of how splicing and transcription are intricately intertwined. Cotranscriptional splicing has been demonstrated in a number of different organisms and has been shown to play roles in coordinating both constitutive and alternative splicing. The nature of cotranscriptional splicing suggests that changes in transcription can dramatically affect splicing, and new evidence suggests that splicing can, in turn, influence transcription. In this chapter, we discuss the mechanisms and consequences of cotranscriptional splicing and introduce some of the tools used to measure this process.

Key words Splicing, Cotranscriptional, RNA, Spliceosome, Transcription, Intron, RNA polymerase II

1 Early Indications of Cotranscriptional Splicing

The last decade has seen a rapid evolution of our understanding of the process of pre-messenger RNA splicing. While elegant genetics and biochemistry have provided a "parts list" of the components of the core splicing machinery and important insights into the well-conserved functions of these proteins and RNAs, it has also become clear that in vivo assembly of the spliceosome and at least some splicing catalysis occur cotranscriptionally, while the elongating RNA polymerase is still actively engaged with a chromatin template (Fig. 1). Furthermore, the cotranscriptional nature of splicing has important functional and regulatory implications.

Remarkably, within a decade of the discovery of split genes, elegant studies by Ann Beyer and Yvonne Osheim using electron microscopy to examine the highly transcribed *Drosophila* chorion genes showed spliceosomes associated at splice junctions of nascent transcripts [1]. These EM results nicely complemented Beyer's earlier studies of nascent transcripts in dipteran embryos, which revealed formation of looped structures (lariats) on the nascent transcripts prior to their release [2], and similar phenomena were later reported on amphibian oocyte lampbrush chromosomes [3]. While these

Klemens J. Hertel (ed.), *Spliceosomal Pre-mRNA Splicing: Methods and Protocols*, Methods in Molecular Biology, vol. 1126, DOI 10.1007/978-1-62703-980-2_6, © Springer Science+Business Media, LLC 2014

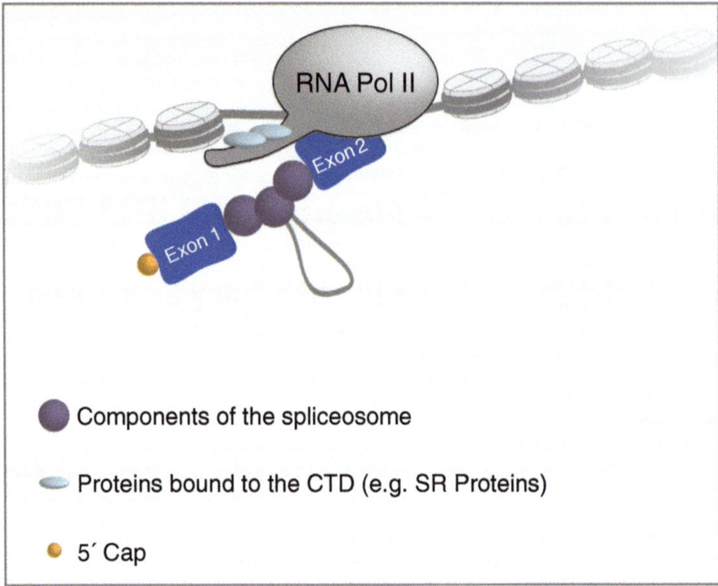

Fig. 1 Coupling between pre-mRNA splicing and transcription. Components of the splicing machinery localize to the nascent RNA while transcription is occurring. SR proteins (*blue ovals*) facilitate cross talk between the CTD tail of RNA polymerase II and the splicing machinery. Cross talk also occurs between the splicing machinery and modified histones. Other RNA processing events, such as 5′ capping, also occur cotranscriptionally

studies showed evidence of cotranscriptional splicing, it is important to point out that they did not address whether the transcription and splicing machineries are functionally coupled. Nonetheless, the ground was set early for studies to explore both the extent of cotranscriptional splicing and the mechanism by which it occurs.

2 Evidence of Widespread Cotranscriptional Splicing

These early studies suggest that both spliceosome assembly and catalysis of splicing can occur in a cotranscriptional manner. Assembly of the spliceosome has been shown to occur in a highly ordered and stepwise fashion in vitro (Chapter 1), and the same is true of spliceosome assembly that occurs on nascent transcripts [4]. In fact, chromatin immunoprecipitation (ChIP) experiments of experimentally tractable genes were the first to demonstrate that the stepwise assembly of the spliceosome in cotranscriptional splicing is akin to how the spliceosome is understood to assemble in in vitro experiments in yeast [5, 6] and in metazoans [7]. While ChIP experiments detect interactions between proteins and DNA, since nascent RNPs lie adjacent to the DNA axis [8], protein—nucleic acid interactions can illustrate cotranscriptional spliceosome assembly, and this method has become a useful proxy to study protein-RNA interactions in cotranscriptional splicing [9]. These analyses have begun to address the specific requirements for proper cotranscriptional spliceosome assembly. For example, studies utilizing ChIP to analyze spliceosome assembly on cotranscriptionally spliced genes have revealed that a histone acetyltransferase regulates the association of components of the U2 snRNP to nascent RNAs in *S. cerevisiae* [10].

While assembly of the spliceosome during transcription is a key aspect of cotranscriptional splicing, a key question is whether splicing catalysis occurs cotranscriptionally, before the termination of transcription [11]. The results showing that spliceosome assembly, from the early steps involving the U1 and U2 snRNPs to the later steps involving the U4/U6-U5 tri-snRNP, occurs on nascent RNA support the notion of cotranscriptional splicing catalysis. However, there is still some question about the proportion of introns that are removed cotranscriptionally. Early studies utilizing chromatin immunoprecipitation methods posited that although spliceosome assembly could begin cotranscriptionally, most genes in *Saccharomyces cerevisiae* are spliced posttranscriptionally [12]. The rationale was that yeast genes are too short for cotranscriptional catalysis, since the polymerase would be expected to terminate transcription before exon ligation could occur. However, more recent studies applying global analysis of nascent RNA support the argument that most intron-containing genes in *S. cerevisiae* are indeed spliced cotranscriptionally. Although terminal exons are short, these data show evidence that the polymerase pauses at the terminal exon, effectively allowing time to splice (this phenomenon is explained in greater detail below) [13, 14] (Fig. 2a). Our current understanding of the breadth of cotranscriptional splicing in other organisms continues to evolve, though it has been reported that the majority of intron-containing genes are at least partially spliced cotranscriptionally in *Drosophila* [15], as well as in human tissues and cell lines [16–20], and these transcripts remain associated with chromatin until fully spliced [21]. Nonetheless, it will be important to understand the potential biological significance of posttranscriptional splicing when it occurs.

3 Constitutive and Alternative Cotranscriptional Splicing and the Transcriptional Machinery

It has become clear that cotranscriptional splicing is spatially and temporally linked to transcription, and a key player in coordinating transcription with splicing is the RNA polymerase itself. RNA polymerase II, the polymerase responsible for transcribing intron-containing pre-mRNAs, is distinguished from the other eukaryotic RNA polymerases by the presence of a C-terminal "tail" made up

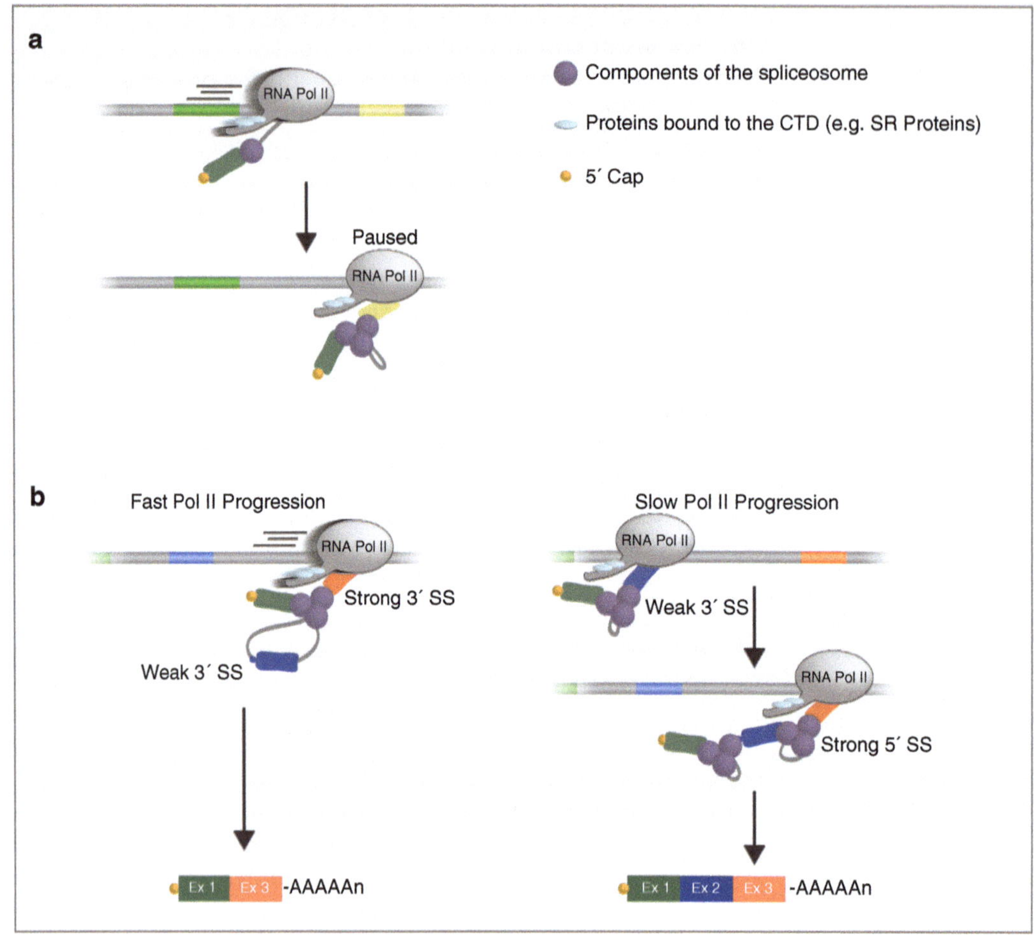

Fig. 2 Kinetic Model of Cotranscriptional Splicing. (**a**) In *Saccharomyces cerevisiae*, RNA Pol II pauses at the terminal exon and/or 3′ SS to facilitate cotranscriptional splicing. (**b**) RNA Pol II rate of elongation modulates cotranscriptional alternative splicing. Fast elongation rate of transcription (*left*) favors skipping of exons with "weak" upstream 3′ splice sites (*blue* exon). Slower Pol II elongation rates (*right*) favor inclusion of exons with weak 3′ SS sites. Constitutive exons (containing strong 3′ SS) are included independently of RNA Pol II elongation rate (not shown)

of numerous heptad repeats (YSPTSPS), the number of which roughly correlates with organism complexity. For example, the CTD consists of 26 repeats in the yeast *Saccharomyces cerevisiae* and 52 repeats in humans. Posttranslational modifications on the CTD tail play a key role in the regulation of Pol II activity, and modifications of the CTD, particularly phosphorylation, help couple transcription and numerous RNA processing events (reviewed in refs. 22–24). Serines 2 and 5 of the CTD have been identified as major phosphorylated residues [25, 26]. Serine 5 is phosphorylated by the basal transcription factor TFIIH at the initiation of transcription [27]. Subsequent to initiation, promoter clearance and transcriptional elongation occur, during which serine 2 of the CTD is phosphorylated by the transcriptional elongation factor P-TEFb. This shift from serine 5 to serine 2 phosphorylation of the CTD during transcriptional elongation may play an important role in the regulation of cotranscriptional splicing [28, 29].

Early studies investigating the role of the CTD in pre-mRNA splicing proposed that the CTD interacts directly with RNA splicing proteins to recruit them to the nascent transcript [30], as truncation or mutation of the CTD of RNAP II leads to changes in splicing in vitro and in vivo [31, 32]. For example, the splicing protein U2AF65 interacts directly with the phosphorylated CTD [33] to promote its association with the pre-mRNA [33, 34]. There is also evidence of interactions between the CTD and serine-arginine (SR) proteins required for constitutive and alternative splicing in metazoans [35] (Fig. 1), although the precise consequence3s of these interactions are still poorly understood. Interestingly, the SR protein SRSF3 negatively regulates the inclusion of the EDI exon of the *fibronectin* gene in a manner that is dependent on the presence of the CTD of RNA Pol II [36]. A number of other members of the SR family of proteins have been shown to functionally interact with the RNA Pol II CTD to affect pre-mRNA splicing [37].

As further evidence of an important role for the CTD in cotranscriptional splicing, in vitro transcription/splicing systems show that the presence of the CTD enhances the rate of splicing, as in vitro T7 RNA polymerase-transcribed RNAs are spliced less efficiently than those transcribed with the CTD-containing RNAP II [38, 39], and this requires CTD phosphorylation [32]. Posttranslational modifications of the CTD may mediate physical interactions between the elongating RNAP II and the splicing machinery by creating a binding platform for splicing factors (bound directly to the CTD or indirectly with other CTD associated proteins), that can then be transferred to the nascent RNA [40].

These interactions between the CTD and splicing proteins represent examples of a "recruitment model" of cotranscriptional splicing, which posits that there are physical contacts between the transcriptional and splicing machineries, and perturbing these alters cotranscriptional splicing. In addition to interactions between the CTD and splicing proteins, interactions between spliceosomal snRNP complexes and transcription elongation factors are likely to be very important for the coupling of these two processes [41, 42], as are interactions between chromatin marks (or proteins associated with chromatin marks) and the spliceosome, which will be explored further in a subsequent chapter.

Posttranslational modifications on the CTD, as well as other factors that affect the rate of elongation of transcription, influence splice site recognition, spliceosome assembly, and splicing patterns [43–45] through kinetic coupling of these two processes. This model has been termed the "kinetic model" of cotranscriptional splicing. For example, as described above, changes in Pol II elongation, specifically Pol II pausing, couple splicing with transcriptional elongation in *Saccharomyces cerevisiae* [13, 14] (Fig. 2a). Thus, genes that are predicted to be spliced posttranscriptionally are in fact spliced cotranscriptionally. While it is unclear what the precise mechanism of pausing is or to what extent this phenomenon occurs in other species, this terminal exon pausing represents a functional coupling between transcription and splicing.

Some of the most compelling evidence of the importance of the rate of Pol II elongation on splicing outcomes comes from analyses of splicing of RNAs containing alternative 3′ splice sites [46], as changes in the kinetics of RNA Pol II elongation can markedly affect splice site selection in alternatively spliced genes (Fig. 2b). Studies in both yeast and mammalian cells expressing constructs in which an intron-containing gene contains a strong 3′ SS downstream of a weak 3' SS support this model [46]. A decrease in the elongation rate of the polymerase or pausing by Pol II favors the inclusion of exons possessing the weak 3′ splice site, whereas Pol II with a normal elongation rate, or without pausing during elongation, favors the exclusion of these exons [43, 47, 48]. Furthermore, an exon containing a suboptimal 3′ SS that is normally not utilized in a reporter minigene is included in the transcribed mRNA in *Drosophila* cells expressing a mutant form of Pol II which transcribes at a slower elongation rate [49]. One particularly intriguing possible mechanism is that nucleosomes, which can form a natural barrier to the transcribing polymerase [50, 51], may alter polymerase elongation rates to facilitate inclusion of weak splice sites. Consistent with this, exons flanked by weak splice sites are more enriched with nucleosomes compared with those containing strong splice sites, and exon inclusion levels correlate with nucleosome occupancy [52, 53]. It is important to note that the elongation rate may also influence the ability of splicing regulators (both positive and negative) to bind to sequences in the nascent RNA, which could also affect exon inclusion and skipping.

In addition to transcription elongation, cotranscriptional alternative splicing can be influenced by promoters and transcriptional activators or repressors. Promoter-swapping experiments indicate that changes in the structure of these sequences result in a change in alternative splice site selection [54]. There is evidence that this promoter-driven effect on splice site selection may occur through interactions between the transcription and splicing machineries modulated by transcriptional activators such as PGC-1 [55]. Recently, the mediator complex has also been implicated in cross talk between transcription and alternative splicing through its ability to link transcriptional activators or repressors that interact with splicing silencers or enhancers (such as hnRNPs and SR proteins) with transcription factors associated with core promoters [56]. These data show that while elongation influences splicing outcomes, early transcriptional events can also affect splicing.

While many of the studies mentioned above appear to support either the "recruitment" or the "kinetic" model of cotranscriptional splicing, these mechanisms are by no means mutually exclusive. It is becoming increasingly clear that both the recruitment of splicing and splicing-associated factors by the transcriptional machinery, as well as the kinetics of the transcription machinery, play critical roles in regulating pre-mRNA splicing.

4 Cotranscriptional Splicing and Its Effects on Transcription

An obvious implication of the close spatial and temporal proximity of the splicing and transcription machineries is that the relationship between transcription and splicing could work both ways, namely, that splicing and splicing factors could also influence transcription. Indeed there is a growing body of evidence indicating that this is the case.

Some of the earliest indications of this came from work as far back as the late 1980s and early 1990s in which it was shown that the presence of an intron increases expression of mouse transgenes [57–59]. In the subsequent years there have been a number of important discoveries that have shed light on the mechanisms by which introns can exert a positive effect on transcription. One of the first was the striking observation that interactions between U snRNPs and the transcription elongation factor TAT-SF1 stimulated RNA polymerase II elongation. More specifically, the authors proposed that stimulation of Pol II elongation was the result of TAT-SF1 interaction with the *p*ositive *t*ranscription *e*longation *f*actor *b* (P-TEFb), which phosphorylates the CTD of RNA polymerase II [42]. This study hinted at a central role for P-TEFb in mediating communication between components of the splicing machinery and the RNA polymerase—a role supported by subsequent studies.

As previously described, SR proteins associate cotranscriptionally with the RNA polymerase during active transcription. In vivo depletion of either of the SR proteins SRSF1 or SRSF2 decreases nascent RNA production, with dramatic effects on transcription elongation seen upon SRSF2 depletion [29]. SRSF2 co-IPs with both P-TEFb and TAT-SF1, and its depletion correlates with defective P-TEFb recruitment. Moreover, in these cells, Pol II accumulates in the body of genes and Ser-2 CTD phosphorylation is abrogated, indicative of defective transcription elongation. One intriguing model is that SR proteins such as SRSF2 dynamically associate with Pol II and enhance elongation by stimulating P-TEFb, and at emerging splice sites, the SR proteins disembark and bind to the appropriate RNA signals. Interestingly, SR proteins have also been shown to bind directly to histones [60], raising the possibility that SR protein binding to histones may affect the state of the chromatin and, in turn, affect transcription. Moreover, SR proteins' interactions with chromatin may facilitate their roles in splicing; since it has been shown that nucleosomes are enriched in exons (discussed elsewhere in this issue), the ability of SR proteins to bind to histones may facilitate their association in exonic RNA sequences. It should be noted that SR proteins are also found associated with intronless genes [61, 62], so the presence of an intron may not be a prerequisite for SR protein effects on transcription. Nonetheless, SR proteins appear to play a central role in mediating the bidirectional relationship between transcription and splicing.

In addition to SR proteins, other proteins involved in RNA processing in general and RNA splicing in particular may affect CTD phosphorylation. In fact, the cap-binding complex, which binds to the 5′ cap structure of pre-mRNAs and has long been known to interact with the core splicing machinery to affect spliceosome assembly, also interacts with P-TEFb. Moreover, the CBC is required for P-TEFb-dependent alternative splicing [63]. This example of the CBC again illustrates the strong bidirectional relationship between splicing and transcription: RNA processing factors can affect transcription, which in turn affects RNA processing. The effect of the CBC on RNA Pol II CTD phosphorylation appears to be conserved, as the yeast cap-binding complex interacts with the yeast ortholog of P-TEFb and stimulates transcription elongation [64].

While there has been a great deal of focus on the effect of splicing factors on transcription elongation, it has also been established that introns can affect early steps of transcription. A functional 5′ SS enhances pre-initiation complex (PIC) formation and stimulates recruitment of general transcription initiation factors [65]. The precise mechanism by which the 5′ SS influences early transcription complex formation is not yet clear; nonetheless, it is likely to involve protein and/or RNA interactions at the 5′ SS. Intriguingly, the U1 snRNA has been shown to associate with TFIIH and regulate transcriptional initiation in a reconstituted transcription system, and promoter proximal 5′ SS recognition by U1 snRNA stimulates TFIIH dependent reinitiation of transcription [66]. Consistent with this, removal of promoter proximal splice signals from a mammalian gene leads to a significant reduction in nascent transcription [67].

While the topic is discussed in more detail elsewhere, it is clear that splicing can influence transcription through its effects on chromatin. The Hu proteins are a family of mammalian RNA binding proteins that act as splicing regulators. Hu proteins are recruited to their RNA-binding sites and interact with histone deacetylase 2 (HDAC2) to inhibit its activity, alter histone acetylation, and, as a consequence, alter RNA polymerase elongation [68]. Several recent studies in mammalian cells demonstrate that histone H3K36me3, a mark of active transcription, is directly influenced by splicing [69, 70]. Moreover, histone H3K4me3 and H3K9ac, both marks of active transcription, are enriched at the first 5′ SS. Removal of endogenous introns or inhibition of splicing using the splicing inhibitor spliceostatin leads to a reduction in the overall H3K4me3 signal [71]. The next exciting challenge will be to determine whether specific components of the splicing machinery interact with the histone-modifying machinery to direct effects on chromatin and, if so, to identify these factors and their modes of action.

Finally, as described above and shown in Fig. 2, elegant yeast studies demonstrate polymerase pausing around the 3′ SS and/or in the 3′ exon, suggesting a model in which the splicing-induced polymerase pausing provides a checkpoint to allow time for splice site recognition and splicing catalysis. It is possible that components of the spliceosome involved in splicing events near the 3′ splice site feedback on the polymerase to induce pausing—either through changes in the chromatin, changes to the RNA polymerase itself (e.g., through CTD phosphorylation), or interactions with components of the transcription elongation machinery. Ongoing studies are aimed at understanding how splicing and/or specific splicing factors provide this feedback to the polymerase to affect pausing.

5 Cotranscriptional Splicing in Disease and Development

Given the requirement of precise expression of genes for normal cellular processes, it is not surprising that dysregulation of gene expression due to defects in pre-mRNA splicing can result in disease. In fact, it is likely that at least 30 % of mutations that cause disease do so by disrupting splicing, through *cis*-acting or *trans*-acting mechanisms [72–74]. Since a significant amount, if not most, of pre-mRNA splicing in humans occurs cotranscriptionally, it would be expected that defects in cotranscriptional splicing would lead to disease as well as defects in development. Consistent with this, genes that are highly cotranscriptionally alternatively spliced in the fetal brain have also been implicated in critical neurodevelopmental processes, suggesting that dysregulation of cotranscriptionally alternatively spliced genes may impair neural development [16]. Mutations in the autoimmune regulator (AIRE) transcription factor lead to autoimmune-polyendocrinopathy-candidiasis-ectodermal dystrophy (APECED), likely due to a decrease in cotranscriptional splicing of the AIRE target genes [75]. Some of the genes that undergo dysregulated cotranscriptional alternative splicing are currently being investigated as targets for antisense oligonucleotide (AON) therapy. For example, antisense oligonucleotides, which induce exon skipping in the *DMD* gene, have shown promise as therapeutic intervention in Duchenne muscular dystrophy [76]. Further analysis to better understand which genes are cotranscriptionally spliced and the mechanisms of cotranscriptional splicing will likely uncover many genes whose dysregulation leads to disease or defects in development.

6 Tools of the Trade: Studying Cotranscriptional Splicing

Understanding cotranscriptional spliceosome assembly and splicing catalysis has evolved as more varied tools have been applied to study these processes. As described above, early EM studies provided direct visual evidence of cotranscriptional splicing, and more recent studies provide insights into the ordered nature of cotranscriptional spliceosome assembly, the extent of cotranscriptional splicing, and the roles played by specific transcription proteins as well as chromatin in directing cotranscriptional splicing.

Our grasp of spliceosome assembly has been largely informed by in vitro studies of the formation of splicing complexes using non-denaturing gel systems. These studies portrayed a picture of a spliceosome that assembled in an ordered, stepwise manner onto the pre-messenger RNA. It was an exciting surprise when it was shown using chromatin immunoprecipitation studies in yeast that, in vivo, the spliceosome followed a similar, stepwise pattern of assembly [5, 6, 9]. This approach allows inference of the kinetics of spliceosome assembly, based on the assumption that distance travelled by RNA Pol II is equivalent to time, as specific snRNPs localize to specific regions of the transcribed gene (e.g., U1 snRNP localizes to the 5′ SS as measured by ChIP) [77]. This is still an indirect measure of cotranscriptional splicing, and therefore, multiple caveats, such as accessibility of epitopes, should be considered when ChIP is used as a tool to measure cotranscriptional splicing [78]. Nonetheless, this approach has proven to be a powerful tool for measuring chromatin-associated RNA-binding proteins such as snRNPs [10, 78].

Advances in live-cell imaging have allowed for in vivo investigation of cotranscriptional splicing that was previously not feasible. For example, photobleaching experiments measuring mobility and distribution of spliceosomal proteins, as well as direct, real-time imaging of fluorescently tagged snRNP components, have provided significantly more insight into the kinetics of cotranscriptional spliceosome assembly in human cells [79–81].

The development of in vitro transcription-splicing coupled systems to study cotranscriptional splicing has also led to a further increase in our understanding of the interactions between the transcription machinery and the nascent transcript, as well as the effect that cotranscriptional splicing has on pre-mRNA stability and splicing efficiency [38, 82, 83]. However, a major drawback of these in vitro systems has been their inability to recapitulate the chromatin setting of cotranscriptionally spliced genes. However, as the technological challenges of in vitro splicing from in vitro assembled chromatin templates are addressed, this assay will certainly yield important mechanistic insights.

As next-generation deep sequencing of genomes and transcriptomes has become less costly, the role of these technologies in examining cotranscriptional splicing has also increased. The sequencing of nascent RNAs using high-density tiling arrays has previously shown that splicing catalysis occurs cotranscriptionally in *S. cerevisiae* [13]. More recently, studies utilizing RNA-Seq of nascent and chromatin-associated RNAs have revealed widespread cotranscriptional splicing in *Drosophila* and human cells [15–17]. The majority of these studies support widespread cotranscriptional splicing across species; however, there are reports suggesting otherwise [20]. The disparity between these observations may be due to cell-type differences and the conditions to which the cells are exposed. Furthermore, different methods used to calculate the frequency of cotranscriptional splicing may also result in a disparity between studies, particularly when assessing cotranscriptional splicing on an intron-to-intron versus entire gene basis (see also ref. 84). Therefore, it is necessary to use a technique such as RT-qPCR to validate high-throughput cotranscriptional splicing results. Even newer methods of deep sequencing and the availability of high-quality databases of transcriptomes will likely provide even further insight into the extent of cotranscriptional splicing across species.

Though newer high-throughput technologies such as RNA-seq have increased our knowledge of the breadth of cotranscriptional splicing, traditional methods such as classical yeast genetics still play a critical role in determining the underlying mechanisms of cotranscriptional splicing, particularly in *S. cerevisiae*. For example, genetic analyses have been instrumental in showing interactions between splicing factors and other cellular machineries, such as histone-modifying machinery and mRNA export factors [78, 85, 86]. The combined use of these tools will lead to a heightened understanding of the mechanisms and spectrum of cotranscriptional splicing.

Acknowledgments

We would like to thank members of the Johnson Lab for critical reading of the manuscript and apologize to any colleagues whose work is not referenced due to unintentional oversight or space constraints. Funding was provided by the National Institutes of General Medical Sciences (GM085474), the National Science Foundation (MCB-1051921), and an IRACDA fellowship to E.C.M. (K12 GM068524).

References

1. Osheim YN, Miller OL Jr et al (1985) RNP particles at splice junction sequences on Drosophila chorion transcripts. Cell 43(1): 143–151
2. Beyer AL, Bouton AH, Miller OL Jr (1981) Correlation of hnRNP structure and nascent transcript cleavage. Cell 26(2 Pt 2):155–165
3. Wu ZA, Murphy C, Callan HG et al (1991) Small nuclear ribonucleoproteins and heterogeneous nuclear ribonucleoproteins in the amphibian germinal vesicle: loops, spheres, and snurposomes. J Cell Biol 113(3):465–483
4. Perales R, Bentley D (2009) "Cotranscriptionality": the transcription elongation com-

plex as a nexus for nuclear transactions. Mol Cell 36(2):178–191

5. Gornemann J, Kotovic KM, Hujer K et al (2005) Cotranscriptional spliceosome assembly occurs in a stepwise fashion and requires the cap binding complex. Mol Cell 19(1):53–63
6. Lacadie SA, Rosbash M (2005) Cotranscriptional spliceosome assembly dynamics and the role of U1 snRNA:5′ss base pairing in yeast. Mol Cell 19(1):65–75
7. Listerman I, Sapra AK, Neugebauer KM (2006) Cotranscriptional coupling of splicing factor recruitment and precursor messenger RNA splicing in mammalian cells. Nat Struct Mol Biol 13(9):815–822
8. Wetterberg I, Zhao J, Masich S et al (2001) In situ transcription and splicing in the Balbiani ring 3 gene. EMBO J 20(10):2564–2574
9. Kotovic KM, Lockshon D, Boric L et al (2003) Cotranscriptional recruitment of the U1 snRNP to intron-containing genes in yeast. Mol Cell Biol 23(16):5768–5779
10. Gunderson FQ, Johnson TL (2009) Acetylation by the transcriptional coactivator Gcn5 plays a novel role in co-transcriptional spliceosome assembly. PLoS Genet 5(10):e1000682
11. Carrillo Oesterreich F, Bieberstein N, Neugebauer KM (2011) Pause locally, splice globally. Trends Cell Biol 21(6):328–335
12. Tardiff DF, Lacadie SA, Rosbash M (2006) A genome-wide analysis indicates that yeast pre-mRNA splicing is predominantly posttranscriptional. Mol Cell 24(6):917–929
13. Carrillo Oesterreich F, Preibisch S, Neugebauer KM (2010) Global analysis of nascent RNA reveals transcriptional pausing in terminal exons. Mol Cell 40(4):571–581
14. Alexander RD, Innocente SA, Barrass JD et al (2010) Splicing-dependent RNA polymerase pausing in yeast. Mol Cell 40(4):582–593
15. Khodor YL, Rodriguez J, Abruzzi KC et al (2011) Nascent-seq indicates widespread cotranscriptional pre-mRNA splicing in Drosophila. Genes Dev 25(23):2502–2512
16. Ameur A, Zaghlool A, Halvardson J et al (2011) Total RNA sequencing reveals nascent transcription and widespread co-transcriptional splicing in the human brain. Nat Struct Mol Biol 18(12):1435–1440
17. Tilgner H, Knowles DG, Johnson R et al (2012) Deep sequencing of subcellular RNA fractions shows splicing to be predominantly co-transcriptional in the human genome but inefficient for lncRNAs. Genome Res 22(9):1616–1625
18. Girard C, Will CL, Peng J et al (2012) Post-transcriptional spliceosomes are retained in nuclear speckles until splicing completion. Nat Commun 3:994
19. Windhager L, Bonfert T, Burger K et al (2012) Ultrashort and progressive 4sU-tagging reveals key characteristics of RNA processing at nucleotide resolution. Genome Res 22(10): 2031–2042
20. Bhatt DM, Pandya-Jones A, Tong AJ et al (2012) Transcript dynamics of proinflammatory genes revealed by sequence analysis of subcellular RNA fractions. Cell 150(2):279–290
21. Pandya-Jones A, Bhatt DM, Lin CH et al (2013) Splicing kinetics and transcript release from the chromatin compartment limit the rate of Lipid A-induced gene expression. RNA 19(6):811–827
22. Hsin JP, Manley JL (2012) The RNA polymerase II CTD coordinates transcription and RNA processing. Genes Dev 26(19):2119–2137
23. de Almeida SF, Carmo-Fonseca M (2008) The CTD role in cotranscriptional RNA processing and surveillance. FEBS lett 582(14): 1971–1976
24. Pandit S, Wang D, Fu XD (2008) Functional integration of transcriptional and RNA processing machineries. Curr Opin Cell Biol 20(3):260–265
25. Corden JL (1990) Tails of RNA polymerase II. Trends Biochem Sci 15(10):383–387
26. West ML, Corden JL (1995) Construction and analysis of yeast RNA polymerase II CTD deletion and substitution mutations. Genetics 140(4):1223–1233
27. Buratowski S (2009) Progression through the RNA polymerase II CTD cycle. Mol Cell 36(4):541–546
28. Barboric M, Lenasi T, Chen H et al (2009) 7SK snRNP/P-TEFb couples transcription elongation with alternative splicing and is essential for vertebrate development. Proc Natl Acad Sci USA 106(19):7798–7803
29. Lin S, Coutinho-Mansfield G, Wang D et al (2008) The splicing factor SC35 has an active role in transcriptional elongation. Nat Struct Mol Biol 15(8):819–826
30. Mortillaro MJ, Blencowe BJ, Wei X et al (1996) A hyperphosphorylated form of the large subunit of RNA polymerase II is associated with splicing complexes and the nuclear matrix. Proc Natl Acad Sci USA 93(16):8253–8257
31. McCracken S, Fong N, Yankulov K et al (1997) The C-terminal domain of RNA polymerase II couples mRNA processing to transcription. Nature 385(6614):357–361
32. Hirose Y, Tacke R, Manley JL (1999) Phosphorylated RNA polymerase II stimulates pre-mRNA splicing. Genes Dev 13(10): 1234–1239
33. David CJ, Boyne AR, Millhouse SR et al (2011) The RNA polymerase II C-terminal

domain promotes splicing activation through recruitment of a U2AF65-Prp19 complex. Genes Dev 25(9):972–983
34. Gu B, Eick D, Bensaude O (2012) CTD serine-2 plays a critical role in splicing and termination factor recruitment to RNA polymerase II in vivo. Nucleic Acids Res 41(3):1591–1603
35. Yuryev A, Patturajan M, Litingtung Y et al (1996) The C-terminal domain of the largest subunit of RNA polymerase II interacts with a novel set of serine/arginine-rich proteins. Proc Natl Acad Sci USA 93(14):6975–6980
36. de la Mata M, Kornblihtt AR (2006) RNA polymerase II C-terminal domain mediates regulation of alternative splicing by SRp20. Nat Struct Mol Biol 13(11):973–980
37. Das R, Yu J, Zhang Z et al (2007) SR proteins function in coupling RNAP II transcription to pre-mRNA splicing. Mol Cell 26(6):867–881
38. Ghosh S, Garcia-Blanco MA (2000) Coupled in vitro synthesis and splicing of RNA polymerase II transcripts. RNA 6(9):1325–1334
39. Das R, Dufu K, Romney B et al (2006) Functional coupling of RNAP II transcription to spliceosome assembly. Genes Dev 20(9): 1100–1109
40. Abruzzi KC, Lacadie S, Rosbash M (2004) Biochemical analysis of TREX complex recruitment to intronless and intron-containing yeast genes. EMBO J 23(13):2620–2631
41. Neugebauer KM (2002) On the importance of being co-transcriptional. J Cell Sci 115 (Pt 20):3865–3871
42. Fong YW, Zhou Q (2001) Stimulatory effect of splicing factors on transcriptional elongation. Nature 414(6866):929–933
43. de la Mata M, Alonso CR, Kadener S et al (2003) A slow RNA polymerase II affects alternative splicing in vivo. Mol Cell 12(2): 525–532
44. Munoz MJ, Perez Santangelo MS, Paronetto MP et al (2009) DNA damage regulates alternative splicing through inhibition of RNA polymerase II elongation. Cell 137(4):708–720
45. Ip JY, Schmidt D, Pan Q et al (2011) Global impact of RNA polymerase II elongation inhibition on alternative splicing regulation. Genome Res 21(3):390–401
46. Dujardin G, Lafaille C, Petrillo E et al (2012) Transcriptional elongation and alternative splicing. Biochimica et Biophysica Acta 1829(1):134–140
47. Kornblihtt AR, de la Mata M, Fededa JP et al (2004) Multiple links between transcription and splicing. RNA 10(10):1489–1498
48. Howe KJ, Kane CM, Ares M Jr (2003) Perturbation of transcription elongation influences the fidelity of internal exon inclusion in Saccharomyces cerevisiae. RNA 9(8):993–1006
49. Chen Y, Chafin D, Price DH et al (1996) Drosophila RNA polymerase II mutants that affect transcription elongation. J Biol Chem 271(11):5993–5999
50. Hodges C, Bintu L, Lubkowska L et al (2009) Nucleosomal fluctuations govern the transcription dynamics of RNA polymerase II. Science 325(5940):626–628
51. Churchman LS, Weissman JS (2011) Nascent transcript sequencing visualizes transcription at nucleotide resolution. Nature 469(7330): 368–373
52. Schwartz S, Meshorer E, Ast G (2009) Chromatin organization marks exon-intron structure. Nat Struct Mol Biol 16(9): 990–995
53. Tilgner H, Nikolaou C, Althammer S et al (2009) Nucleosome positioning as a determinant of exon recognition. Nat Struct Mol Biol 16(9):996–1001
54. Cramer P, Pesce CG, Baralle FE et al (1997) Functional association between promoter structure and transcript alternative splicing. Proc Natl Acad Sci USA 94(21):11456–11460
55. Monsalve M, Wu Z, Adelmant G et al (2000) Direct coupling of transcription and mRNA processing through the thermogenic coactivator PGC-1. Mol Cell 6(2):307–316
56. Huang Y, Li W, Yao X et al (2012) Mediator complex regulates alternative mRNA processing via the MED23 subunit. Mol Cell 45(4):459–469
57. Brinster RL, Allen JM, Behringer RR et al (1988) Introns increase transcriptional efficiency in transgenic mice. Proc Natl Acad Sci USA 85(3):836–840
58. Choi T, Huang M, Gorman C et al (1991) A generic intron increases gene expression in transgenic mice. Mol Cell Biol 11(6): 3070–3074
59. Palmiter RD, Sandgren EP, Avarbock MR et al (1991) Heterologous introns can enhance expression of transgenes in mice. Proc Natl Acad Sci USA 88(2):478–482
60. Loomis RJ, Naoe Y, Parker JB et al (2009) Chromatin binding of SRp20 and ASF/SF2 and dissociation from mitotic chromosomes is modulated by histone H3 serine 10 phosphorylation. Mol Cell 33(4):450–461
61. Huang Y, Steitz JA (2001) Splicing factors SRp20 and 9G8 promote the nucleocytoplasmic export of mRNA. Mol Cell 7(4):899–905
62. Pozzoli U, Riva L, Menozzi G et al (2004) Over-representation of exonic splicing enhancers in human intronless genes suggests multiple functions in mRNA processing. Biochem Biophys Res Commun 322(2):470–476

63. Lenasi T, Peterlin BM, Barboric M (2011) Cap-binding protein complex links pre-mRNA capping to transcription elongation and alternative splicing through positive transcription elongation factor b (P-TEFb). J Biol Chem 286(26):22758–22768
64. Hossain MA, Chung C, Pradhan SK et al (2013) The yeast cap binding complex modulates transcription factor recruitment and establishes proper histone H3K36 trimethylation during active transcription. Mol Cell Biol 33(4):785–799
65. Damgaard CK, Kahns S, Lykke-Andersen S et al (2008) A 5′ splice site enhances the recruitment of basal transcription initiation factors in vivo. Mol Cell 29(2):271–278
66. Kwek KY, Murphy S, Furger A et al (2002) U1 snRNA associates with TFIIH and regulates transcriptional initiation. Nat Struct Biol 9(11):800–805
67. Furger A, O'Sullivan JM, Binnie A et al (2002) Promoter proximal splice sites enhance transcription. Genes Dev 16(21):2792–2799
68. Zhou HL, Hinman MN, Barron VA et al (2011) Hu proteins regulate alternative splicing by inducing localized histone hyperacetylation in an RNA-dependent manner. Proc Natl Acad Sci USA 108(36):E627–E635
69. Kim S, Kim H, Fong N et al (2011) Pre-mRNA splicing is a determinant of histone H3K36 methylation. Proc Natl Acad Sci USA 108(33):13564–13569
70. de Almeida SF, Grosso AR, Koch F et al (2011) Splicing enhances recruitment of methyltransferase HYPB/Setd2 and methylation of histone H3 Lys36. Nat Struct Mol Biol 18(9):977–983
71. Bieberstein NI, Carrillo Oesterreich F, Straube K et al (2012) First exon length controls active chromatin signatures and transcription. Cell Rep 2(1):62–68
72. Lopez-Bigas N, Audit B, Ouzounis C et al (2005) Are splicing mutations the most frequent cause of hereditary disease? FEBS lett 579(9):1900–1903
73. Cooper TA, Wan L, Dreyfuss G (2009) RNA and disease. Cell 136(4):777–793
74. Faustino NA, Cooper TA (2003) Pre-mRNA splicing and human disease. Genes Dev 17(4): 419–437
75. Zumer K, Plemenitas A, Saksela K et al (2011) Patient mutation in AIRE disrupts P-TEFb binding and target gene transcription. Nucleic Acids Res 39(18):7908–7919
76. Pramono ZA, Wee KB, Wang JL et al (2012) A prospective study in the rational design of efficient antisense oligonucleotides for exon skipping in the DMD gene. Hum Gene Ther 23(7):781–790
77. Nilsen TW (2005) Spliceosome assembly in yeast: one ChIP at a time? Nat Struct Mol Biol 12(7):571–573
78. Gunderson FQ, Merkhofer EC, Johnson TL (2011) Dynamic histone acetylation is critical for cotranscriptional spliceosome assembly and spliceosomal rearrangements. Proc Natl Acad Sci USA 108(5):2004–2009
79. Schmidt U, Basyuk E, Robert MC et al (2011) Real-time imaging of cotranscriptional splicing reveals a kinetic model that reduces noise: implications for alternative splicing regulation. J Cell Biol 193(5):819–829
80. Huranova M, Ivani I, Benda A et al (2010) The differential interaction of snRNPs with pre-mRNA reveals splicing kinetics in living cells. J Cell Biol 191(1):75–86
81. Rino J, Carvalho T, Braga J et al (2007) A stochastic view of spliceosome assembly and recycling in the nucleus. PLoS Comput Biol 3(10):2019–2031
82. Yu Y, Das R, Folco EG et al (2010) A model in vitro system for co-transcriptional splicing. Nucleic Acids Res 38(21):7570–7578
83. Hicks MJ, Yang CR, Kotlajich MV et al (2006) Linking splicing to Pol II transcription stabilizes pre-mRNAs and influences splicing patterns. PLoS Biol 4(6):e147
84. Brugiolo M, Herzel L, Neugebauer KM (2013) Counting on co-transcriptional splicing. F1000Prime Rep 5:9
85. Wilmes GM, Bergkessel M, Bandyopadhyay S et al (2008) A genetic interaction map of RNA-processing factors reveals links between Sem1/Dss1-containing complexes and mRNA export and splicing. Mol Cell 32(5):735–746
86. Moehle EA, Ryan CJ, Krogan NJ et al (2012) The yeast SR-like protein Npl3 links chromatin modification to mRNA processing. PLoS Genet 8(11):e1003101

Chapter 7

Chromatin and Splicing

Nazmul Haque and Shalini Oberdoerffer

Abstract

In the past several years, the relationship between chromatin structure and mRNA processing has been the source of significant investigation across diverse disciplines. Central to these efforts was an unanticipated nonrandom distribution of chromatin marks across transcribed regions of protein-coding genes. In addition to the presence of specific histone modifications at the 5′ and 3′ ends of genes, exonic DNA was demonstrated to present a distinct chromatin landscape relative to intronic DNA. As splicing in higher eukaryotes predominantly occurs co-transcriptionally, these studies raised the possibility that chromatin modifications may aid the spliceosome in the detection of exons amidst vast stretches of noncoding intronic sequences. Recent investigations have supported a direct role for chromatin in splicing regulation and have suggested an intriguing role for splicing in the establishment of chromatin modifications. Here we will summarize an accumulating body of data that begins to reveal extensive coupling between chromatin structure and pre-mRNA splicing.

Key words Alternative splicing, Chromatin, RNA polymerase II, Transcription, Epigenetics

1 Introduction

Unlike the genes of lower eukaryotes, in which protein-coding sequences are typically uninterrupted, genes of higher metazoans are characterized by a large number of coding exons separated by long stretches of noncoding introns. As genes are transcribed into mRNA, introns are excised by the megadalton spliceosome complex, which recognizes short consensus sequences at intron–exon boundaries [1]. While introns were initially dubbed as "junk DNA," it is increasingly evident that exon–intron architecture serves as a critical platform for transcriptome diversification via alternative pre-mRNA splicing. Cassette exons thus represent an important aspect of proteome complexity in higher organisms, and current estimates indicate that greater than 90 % of human genes engage in alternative splicing [2, 3]. However, the evolutionary drive for transcriptome expansion has posed the spliceosome with an increasingly difficult task as intron lengths have increased and splice site strengths have

Klemens J. Hertel (ed.), *Spliceosomal Pre-mRNA Splicing: Methods and Protocols*, Methods in Molecular Biology, vol. 1126, DOI 10.1007/978-1-62703-980-2_7, © Springer Science+Business Media, LLC 2014

weakened [4]. Alternative pre-mRNA splicing adds an additional layer of complexity in that splice site recognition must be diversified in a context-dependent manner. To accomplish regulated transcript production within a multivariable framework, pre-mRNA splicing is coordinated at multiple levels. In addition to regulation via RNA-binding protein recognition of *cis*-elements encoded within pre-mRNA [5], the rate of RNA polymerase II (pol II) transcription elongation and chromatin structure contribute to splice site recognition [6]. Rather than operating independently, these processes are highly integrated as a result of co-transcriptional pre-mRNA splicing [7, 8]. The basic mechanisms of alternative splicing regulation via RNA-binding proteins and evidence for co-transcriptional splicing are discussed elsewhere in this volume. Here, we will focus directly on the accumulating evidence in support of a role for chromatin structure in splicing regulation.

2 Chromatin and Co-transcriptional Splicing

A central tenet to the relationship between chromatin structure and alternative splicing is that the majority of splicing in higher eukaryotes occurs co-transcriptionally, while the nascent message is still tethered to the template DNA. This allows for several layers of coupling between the transcription and splicing machineries. Initial efforts to address coupling focused on the potential association between RNA-binding proteins and the pol II carboxy-terminal domain (CTD), such that the factors could be efficiently transferred to the nascent transcript during the process of transcription [9]. Combinatorial association of these factors could in principle influence splicing decisions [5]. Co-transcriptionality further allows for kinetic regulation of splicing decisions. In work pioneered in the Kornblihtt group, it was shown that the rate of transcription elongation impacts splicing decisions such that weak exons are more likely to be excluded from spliced mRNA in response to a rapid elongation rate [10]. While the kinetic model has now been validated in a variety of systems, the physiological barriers to pol II elongation remained comparatively elusive until recently. Genome-wide profiling of chromatin modifications revealed the transcribed DNA template itself as a potential modulator of elongation rate or other aspects of splicing regulation. Intragenic DNA presents a distinct chromatin landscape relative to intergenic DNA, and more importantly to this discussion, exonic DNA presents unique features relative to intronic DNA [11]. These observations raised the possibility that the chromatin structure of transcribed genes may aid the spliceosome in the process of exon definition. In this section, we expand on these themes and examine the various evidences and mechanisms for chromatin-directed splicing.

2.1 Evidence for Co-transcriptional Splicing

While pre-mRNA splicing was initially envisioned as a distinct cellular process that occurred subsequent to the completion of transcription, evidence for co-transcriptional splicing quickly mounted. In a widely cited landmark study, electron micrographs of chromosomal spreads from Drosophila embryos provided a visual demonstration of spliceosome assembly on nascent mRNA, while the RNA was still tethered to the template DNA [12, 13]. Further, indirect evidence of functional cross talk between transcription and pre-mRNA splicing came from studies wherein protein-coding genes were placed downstream of RNA polymerase I or RNA polymerase III promoters [14–17]. These studies indicated that pol I and pol III promoters led to efficient mRNA transcription, but the resulting RNAs were poorly spliced, highlighting an obligatory functional connection between pol II and the splicing machinery [14–17]. Similar results were obtained following deletion of the pol II CTD [18]. The CTD consists of multiple repeats (52 in human) of evolutionary conserved heptapeptide (YSPTSPS) sequences that are dynamically phosphorylated at the different stages of transcription [19]. Transient transfection of CTD-deleted pol II completely abrogated splicing of a β-globin reporter [18], suggesting an important role for the CTD in co-transcriptional splicing. This notion was further strengthened with the development of fluorescent microscopy. Misteli et al. used live cell imaging to demonstrate the relocalization of a fluorescently labeled splicing factor from nuclear speckles to sites of new transcription initiated from a β-tropomyosin minigene [20]. Furthermore, they and others have showed that such splicing factor mobilization is dependent on transcription through RNA polymerase II with an intact CTD [21–24]. More recently, several kinetic studies have shown that the vast majority of splicing in higher eukaryotes occurs co-transcriptionally, in a general 5′–3′ order. Altogether, these studies provide a rationale framework for coupling between elements at the transcribed DNA template and the splicing machinery.

2.2 Kinetic Regulation of Splicing

Almost immediately preceding the formal demonstration of co-transcriptional splicing, Kornblihtt and others revealed an intriguing new connection between these two processes that would ultimately pave the way for a new era in the alternative splicing field. In what would come to be known as the “kinetic” hypothesis, the rate of pol II elongation was implicated in alternative splicing decisions. As a first hint of things to come, it was demonstrated that promoter identity influences splicing decisions. Swapping promoters in minigene constructs resulted in altered splicing of weak exons [25], and recruitment of the splicing factor SF2/ASF (now SRSF1) to enhancers was shown to be promoter dependent [26]. These promoter-related splicing effects were ultimately attributed to elongation rate [27]. In support of a kinetic connection, the Smith group further found that inserting a binding site for the zinc-finger protein MAZ downstream of a weak exon led to pol II pausing and increased inclusion of the exon in spliced mRNA [28]. It was thus proposed that alternative splicing decisions are influenced by a temporal window of opportunity. Subsequent work fully established this notion and ultimately revealed a connection to chromatin (discussed below). Focusing on the fibronectin gene, inhibiting pol II elongation rate through use of a chemical inhibitor, 5,6-dichloro-1-beta-D ribofuranosylbenzimidazole (DRB), increased inclusion of the weak EDI exon, whereas increasing elongation rate through treatment with the histone deacetylase inhibitor trichostatin A (TSA) decreased exon inclusion [71]. A similar increase in EDI inclusion was seen in response to transcription with α-amanitin-resistant pol II with reduced elongation rate [10]. The global relevance of kinetic coupling between transcription and pre-mRNA splicing was later solidified through several genome-wide studies. Consistent throughout these studies, a slow elongation rate was associated with weak exon inclusion. For example, inhibition of pol II elongation with DRB or camptothecin in activated human T cells favored inclusion of weak exons in a subset of genes that were enriched in RNA processing and apoptosis pathways [29]. Similarly, hyperphosphorylation of the pol II CTD by ultraviolet irradiation (UV) and consequent inhibition of elongation rate induced alternative splicing of many genes involved in the DNA damage response [30].

Various models have been put forth to address the physiological signals for variable pol II elongation. For example, polymerase itself may be altered due to differential CTD phosphorylation or interaction with accessory factors [31] (Chapter 6). Alternatively, recent evidence suggests that the chromatin template of transcribed genes may provide polymerase with signals that locally regulate elongation rate, or provide direct information to the spliceosome regarding the location of exons. We expand on both these modes of chromatin-regulated splicing below.

2.3 Chromatin-Mediated Regulation of Splicing

Building on the accumulating evidence in support of co-transcriptional mRNA processing [32, 33], the first hints of a role for chromatin in splicing decisions date back to the early 1990s. Beckmann and Trifonov unexpectedly discovered that the average distance between the 3′ and 5′ splice sites flanking an exon followed a periodic pattern that was very close to the length covered by a single nucleosome [34]. Based on these results, they rationalized that placement of nucleosomes according to exon–intron boundaries may reflect an unanticipated role for chromatin structure in pre-mRNA splicing [34]. At the same time, an involvement for chromatin was also suspected when integrated copies of the adenovirus genome at distinct genomic locations in the same nuclei yielded different splicing outcomes [35]. Given that the viral genomes and cellular contexts were identical, the authors speculated that the chromatin structure at the integration site was responsible for the distinct splicing patterns, possibly through influencing the pol II elongation rate [35]. While these ideas gained momentum in the following years, it wasn't until the advent of genome-wide sequencing data showing distinct chromatin patterns at exonic relative to intronic sequence that chromatin was solidified as a genuine contributor to splicing regulation. Genome-wide studies revealed that exons show a higher rate of nucleosome occupancy, specific histone modifications, and elevated DNA methylation relative to introns, raising the possibility that chromatin may aid the spliceosome in the process of exon definition. We discuss each of these associations in turn below.

2.3.1 Nucleosomes

A fundamental aspect of gene regulation in eukaryotes is the packaging of DNA into higher-order structures. In addition to maintaining genome integrity, this allows for the control of gene expression through the adoption of either transcriptionally inaccessible heterochromatin or relatively open euchromatin [36]. The basic building block of chromatin is the nucleosome, which is composed of approximately 146 base pairs of DNA wound around an octamer of histone proteins [37]. Canonical nucleosomes include two copies each of H2A, H2B, H3, and H4 histones [37]. Nucleosomes are inherent barriers to pol II elongation, as has been effectively demonstrated in vitro [38–40]. In order to accomplish efficient elongation in vivo, a variety of proteins cooperate to disassemble nucleosomes in front of elongating pol II and reestablish them in its wake [11, 41, 42]. Nucleosome turnover is a critical aspect of transcription fidelity, as nucleosome depletion facilitates unwanted cryptic transcription [43–47]. Given these very basic roles for nucleosomes in transcription, it was surprising to find a nonrandom intragenic positioning pattern across the genome: nucleosome occupancy is elevated at exons relative to introns, irrespective of gene expression status [48–50]. Considering the in vitro data demonstrating that nucleosomes are barriers to pol II elongation, these observations suggested that nucleosomes promote exon definition by facilitating transient pol II pausing and consequent spliceosome assembly. While not formally demonstrated due to difficulties in depleting nucleosomes without untoward effects on gene expression, several studies support this premise. For example, exons with weak splice sites show higher nucleosome density compared to constitutive exons, and pseudoexons with strong splice sites are nucleosome depleted [48]. Furthermore, the average size of a mammalian exon (145 base pairs) is similar to the length of DNA wrapped within a single nucleosome [48, 51]. Additionally, a recent study suggested a role for variant histone incorporation in splicing regulation. Depletion of the mammalian-specific H2A variant, H2A.Bbd, which is preferentially found in transcribed sequences, led to a global decrease in splicing efficiency [52]. As several proteins involved in RNA processing as well as major splicing components were coprecipitated with H2A.Bbd [52–55], these studies hint at a direct role for nucleosomes in splicing regulation. This notion is further reinforced by evidence implicating posttranslational modification of histone tails in splicing regulation, as described below.

An additional twist in the interplay between nucleosomes and splicing is related to nucleotide content. Exonic sequences tend to be GC-rich as compared to introns, which inherently favors nucleosome deposition at exons [56, 57]. Nucleotide bias has been implicated in alternative splicing regulation [58], and a recent study identified a role for the DBIRD complex (composed of deleted in breast cancer 1 [DBC1] and ZNF236) in the exclusion of AT-rich exons. Through a yet undefined mechanism, DBIRD interacts with pol II and facilitates efficient passage through AT-rich sequence. In the absence of DBIRD, polymerase stalls at the weak exons and leads to increased inclusion [59]. This study highlights an additional emerging theme: splicing, chromatin, and transcription are highly intertwined.

2.3.2 Modification of Histone Tails

The advent of high-throughput deep sequencing following chromatin immunoprecipitation (ChIP-seq) has allowed for the dissection of the intragenic epigenome across diverse species, including human, mice, *C. elegans*, and *Drosophila* [50, 51, 60, 61]. Several histone marks were found to be particularly prevalent on exonic sequences, including trimethylation of H3 lysine 36 (H3K36me3) [47, 62–65], dimethylation of H3 lysine 27 (H3K27me2) [49, 64], and monomethylation of H3 lysine 79 (H3K79me1) and H2B lysine 5 (H2BK5me1) [62, 63]. In contrast, H3K9 methylation is depleted at exons [66]. Intriguingly, exonic enrichment is not evenly distributed across gene bodies. For example, H3K36me3 levels rise into gene bodies, whereas H3K4me3 is found near transcription start sites [60]. H3K36me3, in particular, has received significant attention as it is exclusively found at actively transcribed genes, suggesting an important role in pre-mRNA processing [60]. Acknowledging that histone modification measurements must be adjusted for the overall increase in nucleosome content at exonic sequences, H3K36me3 enrichment at exons persists after nucleosome correction [64]. While the genome-wide associations have largely focused on histone methylation, additional modifications to histone tails have the potential to influence splicing. For example, a recent study in yeast reported elevated monoubiquitylation of H2B lysine 123 (H2BK123ub1) in introns of transcribed genes, and disruption of H2BK123ub1 altered the distribution of H3K6me3

[67], suggesting functional antagonism in exon–intron definition through specific posttranslational modifications of histone tails.

A direct role for histone modifications in exon definition is supported by a variety of studies involving modulation of specific histone posttranslational marks. As a result, two non-exclusive potential mechanisms by which chromatin can influence alternative splicing decisions have been proposed: (1) local alteration of pol II elongation rate and (2) site-specific recruitment of RNA-binding proteins through interaction with chromatin-binding proteins. Examples of both modes of regulation exist and are largely intertwined as described below.

Histone Modifications in Kinetic Regulation

The strongest evidence in favor of a role for chromatin in kinetic regulation of splicing comes from studies in which exogenous stimuli led to global or local changes in chromatin structure and associated changes in splicing. In general, acetylation of histone tails is associated with an open chromatin context and efficient pol II processivity [68]. Several studies from the Kornblihtt group have shown that tipping the "accessibility" balance in either direction alters exon inclusion of both model genes and genome wide. For example, membrane depolarization of neuronal cells led to increased H3K9 acetylation (H3K9ac) specifically within the vicinity of exon 18 of the NCAM gene and promoted exclusion of the exon from spliced mRNA [69]. This exon appears to be particularly susceptible to kinetic regulation, as evidenced through use of a mutant polymerase with reduced elongation rate. In addition, globally increasing acetylation with the histone deacetylase inhibitor trichostatin A (TSA) decreased exon 18 inclusion [70]. A similar decrease in the fibronectin EDI exon was seen following TSA treatment [71]. In contrast, the opposing, repressive modification, H3K9 methylation, has been shown to mediate exon inclusion in a number of systems. The Muchardt group defined a role for H3K9me3 in alternative splicing of *CD44* pre-mRNA and was further able to uncover a potential physiological link to polymerase pausing. In response to phorbol 12-myristate 13-acetate (PMA) treatment, *CD44* transcripts show increased inclusion of nine tandem alternative exons, which is associated with increased pol II occupancy and a local increase in H3K9me3 detection [72, 73]. Stimulation also resulted in increased detection of the H3K9me3-interacting chromodomain protein HP1γ at the alternative exons. Remarkably, RNAi-mediated depletion of HP1γ abrogated both pol II accumulation and variant exon inclusion [74]. This study suggested that H3K9me3-associated HP1γ somehow bridged the processes of transcription and splicing. Indeed, it has been reported that HP1γ is enriched at hundreds of active genes and promotes co-transcriptional splicing through recruitment of the spliceosomal protein U1-70K and SR family protein SRSF1 [75, 76].

Additional evidence supporting a role for H3K9 methylation in the kinetic regulation of splicing comes from studies involving exogenous introduction of siRNAs directed against intragenic sequence. Analogous to Argonaute protein-dependent transcriptional gene silencing (TGS), wherein siRNA directed against promoter DNA triggers gene silencing through local heterochromatin formation [77–80], extension of TGS into intragenic sequence mediates chromatin changes that locally modulate pol II elongation without affecting overall gene expression. For example, exogenous siRNAs targeted against an intronic sequence proximal to the fibronectin EDI exon in human cells promoted an AGO1-dependent local increase in the repressive chromatin marks H3K9me2 and H3K27me3 and increased exon inclusion [81]. The authors further demonstrated that the altered chromatin structure resulted in HP1α recruitment to the region, suggesting a similar bridging effect as described for HP1γ above. A recent study in the *CD44* model system also directly implicated the Argonaute proteins AGO1 and AGO2 in bridging interactions. AGO1 and AGO2 interact with splicing factors and are recruited to the *CD44* variant exons in a Dicer and HP1γ dependent–manner, culminating in increased exon inclusion through reduced pol II processivity [82]. These complex associations were reverberated in several studies from the Kennedy laboratory examining the Argonaute-related nuclear RNAi defective (NRDE) protein-dependent intragenic TGS pathway in *C. elegans*. Both exogenous and endogenous siRNA-associated recruitment of NRDE factors were found to promote accumulation of H3K9me3 and inhibited pol II elongation in *C. elegans* [83, 84]. Furthermore, endogenous siRNAs directed NRDE-1 to interact with both chromatin and pre-mRNA, thereby revealing a conserved role for Argonaute proteins in connecting these nuclear processes [84, 85].

As is evident throughout these studies, post-transcriptional modification of H3K9 seems to have a central role in the kinetic regulation of splicing. It is worth noting that these observations are somewhat at odds with the intergenic role of H3K9. Trimethylation of H3K9 is a classic feature of heterochromatin formation and is associated with repeat elements and otherwise silenced areas of the genome [86]. However, intragenic H3K9me3 is not strictly associated with transcriptional repression [87]. These studies suggest that H3K9 may mediate context-dependent effects on transcription and/or splicing. It will certainly be interesting to examine the role of this modification in splicing in greater detail in the coming years.

Histone Modifications in Adaptor Function

While several aspects of bridging from chromatin to RNA were highlighted in the discussion of kinetic regulation above, it is unlikely that all alternative exons are strictly under kinetic regulation. Indeed, studies focused on the role of H3K36 and H3K4 methylation in splicing reveal a more general role for chromatin modifications as adaptors for RNA-binding protein recruitment. Interestingly, as shown for H3K36me3, the same chromatin modification can be associated with distinct splicing outcomes dependent on the interacting factors. For example, the Misteli laboratory has shown that H3K36me3 is associated with exclusion of a subset of PTB-dependent exons. PTB is recruited to these exons through interaction with the H3K36me3-binding protein, MRG15 [88]. As a proof of principle, depletion of H3K36me3 levels through RNAi against the K36 methyltransferase, Setd2, increased the inclusion level of these exons [88]. In contrast, the Bickmore laboratory has shown that H3K36me3 is associated with inclusion of a subset of exons due to Psip1-/Ledgf-dependent recruitment of the splicing factor SRSF1. In a strikingly similar mechanism, Psip1 interacts with H3K36me3 and recruits the positive acting splicing factor to a subset of exons, and SRSF1 localization and splicing are altered in Psip1 null cells [89]. These contrasting studies illustrate the clear involvement of additional context-dependent factors that remain to be identified.

Adaptor function has also been demonstrated for H3K4me3, which is enriched at the 5′ ends of active genes. Biochemical purification identified CHD1 as an H3K4me3-interacting protein and CHD1 was also found to interact with the spliceosomal proteins U2 snRNP [90]. Depletion of either CHD1 or H3K4me3 through RNAi reduced U2 association with chromatin and reduced pre-mRNA splicing efficiency [90]. Similarly, the histone 3 acetyltransferase, GCN5, promotes co-transcriptional U2 snRNP recruitment [91], hinting at a possible adaptor function. Altogether, the sum of these studies demonstrates a clear role for chromatin structure in constitutive and alternative splicing regulation, both through kinetic and adaptor mechanisms.

2.4 DNA Methylation

As introduced above, DNA methylation also shows a nonrandom intragenic distribution pattern: methylation levels are significantly enhanced at exonic relative to intronic sequences. However, unlike promoter DNA methylation, which is associated with gene silencing [92], methylation within gene bodies does not have a clear relationship to gene expression levels [93]. Instead, genome-wide methylome analyses in lower eukaryotes foreshadowed a potential role for DNA methylation in splicing regulation. Comparative methylome analyses from the Jacobsen and Zilberman laboratories showed that the acquisition of DNA methylation predates the divergence of plant and animal lineages, and revealed conserved enrichment at exonic sequences relative to introns [94, 95]. In comparing genetically identical honeybee castes, the Maleszka laboratory further showed that differences in queen versus worker bee methylome patterns correlate with changes in alternative splicing patterns. Strikingly, they also found that the low level of DNA methylation is seemingly restricted to exons and is depleted at intronless genes [96]. Subsequent work in additional insect model systems has confirmed an association between intragenic methylation, alternative splicing, and phenotypic diversity [97–99]. While mammalian methylomes are comparatively complex and widespread intergenic methylation is found, these intragenic features are highly conserved [100, 101]. High-resolution bisulfite sequencing of the human genome validated the enrichment of exonic methylation and revealed sharp transitions at exon–intron junctions [102]. A reanalysis of several genome-wide human methylome and RNA-seq datasets established that methylation levels correlate with alternative splicing in human cells. Included exons showed an overall higher level of DNA methylation relative to excluded exons, suggesting a direct role for DNA methylation in exon definition. These associations persisted even after correcting for increased nucleosome and GC content at exons relative to introns [93].

The conserved association between exonic DNA methylation and alternative splicing across diverse taxa suggests regulated mechanisms for the establishment and removal of methylation patterns. While the mechanisms by which DNA methyltransferases (DNMTs) are targeted to exons at distinct stages in development remain unknown, recent studies have begun to reveal a basis for variable DNA methylation. 5-methylcytosine (5-mC) can be converted to 5-hydroxymethylcytosine (5-hmC) through the activity of the TET family of proteins. 5-hmC can stably persist in the genome, or can be further converted to unmethylated cytosine through additional oxidation. Notably, bisulfite sequencing is unable to distinguish 5-mC from 5-hmC, and recent studies aimed at specifically deciphering the 5-hmC methylome have found an overlapping distribution pattern: 5-hmC is also enriched at exons relative to introns [103, 104]. Furthermore, 5-hmC levels were also shown to undergo sharp transitions at exon–intron boundaries in the brain, and alternative exons showed an overall lower level of 5-hmC relative to constitutive exons. Curiously, non-neural tissues showed more 5-mC at exon–intron boundaries [105]. These studies suggest that tissue-specific changes in the ratio of 5-mC to 5-hmC may represent a novel mode of alternative splicing regulation.

While the accumulation of these genome-wide data over the last several years strongly suggested a fundamental role for DNA methylation in exon definition, potential mechanisms had remained more elusive. Given that the majority of splicing occurs co-transcriptionally, possibilities included direct impact of DNA methylation on splicing through kinetic regulation or indirect regulation through variable interaction with auxiliary factors. We recently provided evidence for the latter possibility. Through our analysis of alternative splicing of *CD45* pre-mRNA, we determined that inclusion of variable exon 5 is mediated by reciprocal binding of the zinc-finger protein, CTCF, and 5-methylcytosine. The binding of CTCF to *CD45* DNA acts as a transient barrier to pol II elongation, which kinetically favors spliceosome assembly at the weak splice sites. In contrast, DNA methylation acts to evict CTCF and thereby abolishes pol II pausing and exon 5 inclusion. CTCF-ChIP-seq and RNA-seq in CTCF-depleted cells verified CTCF to be a global regulator of alternative splicing [106]. This study provided the first mechanistic link between DNA methylation and alternative splicing. A similar effect was recently shown for the zinc-finger protein VEZF1: binding of VEZF1 to DNA promotes pol II pausing and results in alternative splicing of a subset of genes [107]. Interestingly, like CTCF, VEZF1 interaction with DNA protects against DNA methylation [108]. In addition, VEZF1 interacts with MRG15, which was previously implicated to be a chromatin-binding adaptor between H3K36me3 and PTB as described above [88, 107]. Together, these studies suggest a basic role for intragenic binding of zinc-finger proteins in kinetically regulated splicing. We predict that many additional examples of DNA-binding regulators of splicing will be revealed in the coming years.

It should further be noted that we found *CD45* exon 5 methylation to be developmentally regulated. Naïve peripheral lymphocytes show enhanced CTCF binding, whereas mature lymphocytes show enhanced exon 5 methylation. Thus, mechanisms certainly exist that promote exon 5 methylation in a stage-specific manner. Support for active remodeling of intragenic methylation can also be found in the honeybee genome. As mentioned above, genetically identical queen and worker honeybee show distinct methylation profiles. Remarkably, RNAi-mediated depletion of the de novo methyltransferase DNMT3 generated adult bees with queen-like characteristics, and queen-specific isoforms have been defined that are distinguished by unique overlapping exonic DNA methylation [96, 109]. Given that DNA methylation is not strictly associated with exon inclusion or exclusion [110], it is possible that intragenic DNA methylation plays a fundamental role in developmentally regulated alternative splicing through association with a complex network of methyl-resistant and methyl-sensitive DNA-binding proteins.

3 Splicing Reciprocally Modulates Chromatin Structure

While the role of chromatin structure in splicing is now well established, an emerging area of coupling is recent evidence showing that splicing can reciprocally influence chromatin modifications. Independent publications from the Carmo-Fonseca and Bentley laboratories indicated that H3K36me3 deposition over gene bodies is splicing dependent [111, 112]. The Bentley laboratory showed that mutation of 3′ splice sites in the upstream introns of a β-globin reporter resulted in repositioning of H3K36me3 from the 5′ to the 3′ end of the reporter. Globally inhibiting splicing with spliceostatin A also resulted in H3K36me3 redistribution further 3′ into genes [111]. The Carmo-Fonseca group further showed that splicing promotes recruitment of the H3K36 methyltransferase HYPB/Setd2 to gene bodies. Inhibition of splicing globally reduced H3K36me3, whereas activating splicing of a model gene had the opposite effect. Furthermore, intronless genes show lower H3K6me3 levels irrespective of expression status [112], implicating splicing rather than general transcription in Setd2 recruitment. The notion that pre-mRNA processing can influence histone modification is further strengthened by the demonstration that the RNA-binding Hu proteins can modulate histone acetylation. The binding of Hu proteins to target sites on mRNA flanking alternative exons led to local inhibition of histone deacetylase 2 (HDAC2) activity and increased histone acetylation and exon exclusion [113]. Altogether, these studies suggest that spliceosomal components actively connect histone-modifying enzymes to transcription through yet unknown mechanisms.

4 Concluding Remarks

As highlighted throughout this chapter, the last several years have revealed an extensive network of coupling between pre-mRNA splicing and chromatin. Studies of model genes and genome-wide analyses have revealed exonic epigenetic signatures. Some of these signatures, such as nucleosomes [48–50], are found independent of transcription status, whereas others, such as H3K36me3, are transcription and splicing dependent [111, 112]. Chromatin modifications have been shown to influence spliceosome assembly through kinetic regulation of pol II elongation and through recruitment of splicing factors to their required sites of action [11]. An area that will be particularly interesting to follow in the coming years is how these changes in chromatin structure are modulated during development. While the current studies have focused on the chromatin template, pol II, and the spliceosome, in reality, a vast network of remodelers is required to effect chromatin changes. For example, how are constitutively expressed histone- and DNA-modifying enzymes targeted to intragenic sequences at specific stages in development? In addition, do additional chromatin-associated factors that are critical for transcription, such as histone chaperones, have a role in splicing regulation? Certainly, our current understanding of chromatin-mediated mRNA splicing field is hazy at best, and future research in this area is likely to be full of surprises.

References

1. Black DL (2003) Mechanisms of alternative pre-messenger RNA splicing. Annu Rev Biochem 72:291–336. doi:10.1146/annurev.biochem.72.121801.161720, 121801.161720[pii]
2. Wang ET, Sandberg R, Luo S et al (2008) Alternative isoform regulation in human tissue transcriptomes. Nature 456(7221):470–476. doi:nature07509, [pii] 10.1038/nature07509

3. Pan Q, Shai O, Lee LJ et al (2008) Deep surveying of alternative splicing complexity in the human transcriptome by high-throughput sequencing. Nat Genet 40(12):1413–1415. doi:ng.259, [pii] 10.1038/ng.259
4. Busch A, Hertel KJ (2011) Evolution of SR protein and hnRNP splicing regulatory factors. Wiley Interdiscip Rev RNA. doi:10.1002/wrna.100
5. Matlin AJ, Clark F, Smith CW (2005) Understanding alternative splicing: towards a cellular code. Nat Rev Mol Cell Biol 6(5): 386–398
6. Allo M, Schor IE, Munoz MJ et al (2010) Chromatin and alternative splicing. Cold Spring Harb Symp Quant Biol 75:103–111. doi:sqb.2010.75.023, [pii] 10.1101/sqb.2010.75.023
7. Oesterreich FC, Bieberstein N, Neugebauer KM (2011) Pause locally, splice globally. Trends Cell Biol 21(6):328–335. doi:S0962-8924(11)00035-3, [pii] 10.1016/j.tcb.2011.03.002
8. Han J, Xiong J, Wang D et al (2011) Pre-mRNA splicing: where and when in the nucleus. Trends Cell Biol 21(6):336–343. doi:S0962-8924(11)00036-5, [pii] 10.1016/j.tcb.2011.03.003
9. Egloff S, Murphy S (2008) Cracking the RNA polymerase II CTD code. Trends Genet 24(6):280–288. doi:S0168-9525(08)00128-5, [pii] 10.1016/j.tig.2008.03.008
10. de la Mata M, Alonso CR, Kadener S et al (2003) A slow RNA polymerase II affects alternative splicing in vivo. Mol Cell 12(2): 525–532. doi:S1097276503003101 [pii]
11. Shukla S, Oberdoerffer S (2012) Co-transcriptional regulation of alternative pre-mRNA splicing. Biochim Biophys Acta 1819(7):673–683. doi:S1874-9399(12)00032-6, [pii] 10.1016/j.bbagrm.2012.01.014
12. Beyer AL, Bouton AH, Miller OL Jr (1981) Correlation of hnRNP structure and nascent transcript cleavage. Cell 26(2 Pt 2):155–165
13. Beyer AL, Osheim YN (1988) Splice site selection, rate of splicing, and alternative splicing on nascent transcripts. Genes Dev 2(6):754–765
14. Dower K, Rosbash M (2002) T7 RNA polymerase-directed transcripts are processed in yeast and link 3′ end formation to mRNA nuclear export. RNA 8(5):686–697
15. McCracken S, Rosonina E, Fong N et al (1998) Role of RNA polymerase II carboxy-terminal domain in coordinating transcription with RNA processing. Cold Spring Harb Symp Quant Biol 63:301–309
16. Sisodia SS, Sollner-Webb B, Cleveland DW (1987) Specificity of RNA maturation pathways: RNAs transcribed by RNA polymerase III are not substrates for splicing or polyadenylation. Mol Cell Biol 7(10):3602–3612
17. Smale ST, Tjian R (1985) Transcription of herpes simplex virus tk sequences under the control of wild-type and mutant human RNA polymerase I promoters. Mol Cell Biol 5(2): 352–362
18. McCracken S, Fong N, Yankulov K et al (1997) The C-terminal domain of RNA polymerase II couples mRNA processing to transcription. Nature 385(6614):357–361. doi:10.1038/385357a0
19. Dahmus ME (1996) Reversible phosphorylation of the C-terminal domain of RNA polymerase II. J Biol Chem 271(32):19009–19012
20. Misteli T, Caceres JF, Spector DL (1997) The dynamics of a pre-mRNA splicing factor in living cells. Nature 387(6632):523–527. doi:10.1038/387523a0
21. de la Mata M, Alonso CR, Kadener S et al (2003) A slow RNA polymerase II affects alternative splicing in vivo. Mol Cell 12(2): 525–532
22. de la Mata M, Kornblihtt AR (2006) RNA polymerase II C-terminal domain mediates regulation of alternative splicing by SRp20. Nat Struct Mol Biol 13(11):973–980. doi:10.1038/nsmb1155
23. Listerman I, Sapra AK, Neugebauer KM (2006) Cotranscriptional coupling of splicing factor recruitment and precursor messenger RNA splicing in mammalian cells. Nat Struct Mol Biol 13(9):815–822. doi:10.1038/nsmb1135
24. Misteli T, Spector DL (1999) RNA polymerase II targets pre-mRNA splicing factors to transcription sites in vivo. Mol Cell 3(6): 697–705
25. Cramer P, Pesce CG, Baralle FE et al (1997) Functional association between promoter structure and transcript alternative splicing. Proc Natl Acad Sci USA 94(21):11456–11460
26. Cramer P, Caceres JF, Cazalla D et al (1999) Coupling of transcription with alternative splicing: RNA pol II promoters modulate SF2/ASF and 9G8 effects on an exonic splicing enhancer. Mol Cell 4(2):251–258. doi:S1097-2765(00)80372-X [pii]
27. Kadener S, Fededa JP, Rosbash M et al (2002) Regulation of alternative splicing by a transcriptional enhancer through RNA pol II elongation. Proc Natl Acad Sci USA 99(12):8185–8190. doi:10.1073/pnas.122246099, 99/12/8185 [pii]
28. Roberts GC, Gooding C, Mak HY et al (1998) Co-transcriptional commitment to alternative splice site selection. Nucleic Acids Res 26(24):5568–5572
29. Ip JY, Schmidt D, Pan Q et al (2011) Global impact of RNA polymerase II elongation inhibition on alternative splicing regulation.

Genome Res 21(3):390–401. doi:10.1101/gr.111070.110
30. Munoz MJ, Perez Santangelo MS, Paronetto MP et al (2009) DNA damage regulates alternative splicing through inhibition of RNA polymerase II elongation. Cell 137(4):708–720. doi:10.1016/j.cell.2009.03.010
31. Munoz MJ, de la Mata M, Kornblihtt AR (2010) The carboxy terminal domain of RNA polymerase II and alternative splicing. Trends Biochem Sci 35(9):497–504. doi:10.1016/j.tibs.2010.03.010
32. Yue Z, Maldonado E, Pillutla R et al (1997) Mammalian capping enzyme complements mutant Saccharomyces cerevisiae lacking mRNA guanylyltransferase and selectively binds the elongating form of RNA polymerase II. Proc Natl Acad Sci USA 94(24):12898–12903
33. Hirose Y, Manley JL (1998) RNA polymerase II is an essential mRNA polyadenylation factor. Nature 395(6697):93–96. doi:10.1038/25786
34. Beckmann JS, Trifonov EN (1991) Splice junctions follow a 205-base ladder. Proc Natl Acad Sci USA 88(6):2380–2383
35. Adami G, Babiss LE (1991) DNA template effect on RNA splicing: two copies of the same gene in the same nucleus are processed differently. EMBO J 10(11):3457–3465
36. Kouzarides T (2007) Chromatin modifications and their function. Cell 128(4):693–705. doi:S0092-8674(07)00184-5, [pii] 10.1016/j.cell.2007.02.005
37. Luger K, Mader AW, Richmond RK et al (1997) Crystal structure of the nucleosome core particle at 2.8 A resolution. Nature 389(6648):251–260. doi:10.1038/38444
38. Knezetic JA, Luse DS (1986) The presence of nucleosomes on a DNA template prevents initiation by RNA polymerase II in vitro. Cell 45(1):95–104
39. Izban MG, Luse DS (1991) Transcription on nucleosomal templates by RNA polymerase II in vitro: inhibition of elongation with enhancement of sequence-specific pausing. Genes Dev 5(4):683–696
40. Bondarenko VA, Steele LM, Ujvari A et al (2006) Nucleosomes can form a polar barrier to transcript elongation by RNA polymerase II. Mol Cell 24(3):469–479. doi:10.1016/j.molcel.2006.09.009
41. Kristjuhan A, Svejstrup JQ (2004) Evidence for distinct mechanisms facilitating transcript elongation through chromatin in vivo. EMBO J 23(21):4243–4252. doi:10.1038/sj.emboj.7600433, 7600433 [pii]
42. Schwabish MA, Struhl K (2004) Evidence for eviction and rapid deposition of histones upon transcriptional elongation by RNA polymerase II. Mol Cell Biol 24(23):10111–10117. doi:24/23/10111, [pii] 10.1128/MCB.24.23.10111-10117.2004
43. Mason PB, Struhl K (2003) The FACT complex travels with elongating RNA polymerase II and is important for the fidelity of transcriptional initiation in vivo. Mol Cell Biol 23(22):8323–8333
44. Schwabish MA, Struhl K (2006) Asf1 mediates histone eviction and deposition during elongation by RNA polymerase II. Mol Cell 22(3):415–422. doi:S1097-2765(06)00180-8, [pii] 10.1016/j.molcel.2006.03.014
45. Kaplan CD, Laprade L, Winston F (2003) Transcription elongation factors repress transcription initiation from cryptic sites. Science 301(5636):1096–1099. doi:10.1126/science.1087374, 301/5636/1096 [pii]
46. Carrozza MJ, Li B, Florens L, Suganuma T et al (2005) Histone H3 methylation by Set2 directs deacetylation of coding regions by Rpd3S to suppress spurious intragenic transcription. Cell 123(4):581–592. doi:S0092-8674(05)01156-6, [pii] 10.1016/j.cell.2005.10.023
47. Kolasinska-Zwierz P, Down T, Latorre I et al (2009) Differential chromatin marking of introns and expressed exons by H3K36me3. Nat Genet 41(3):376–381. doi:ng.322, [pii] 10.1038/ng.322
48. Tilgner H, Nikolaou C, Althammer S et al (2009) Nucleosome positioning as a determinant of exon recognition. Nat Struct Mol Biol 16(9):996–1001. doi:10.1038/nsmb.1658
49. Andersson R, Enroth S, Rada-Iglesias A et al (2009) Nucleosomes are well positioned in exons and carry characteristic histone modifications. Genome Res 19(10):1732–1741. doi:10.1101/gr.092353.109
50. Spies N, Nielsen CB, Padgett RA et al (2009) Biased chromatin signatures around polyadenylation sites and exons. Mol Cell 36(2):245–254. doi:10.1016/j.molcel.2009.10.008
51. Schwartz S, Meshorer E, Ast G (2009) Chromatin organization marks exon-intron structure. Nat Struct Mol Biol 16(9):990–995. doi:10.1038/nsmb.1659
52. Tolstorukov MY, Goldman JA, Gilbert C et al (2012) Histone variant H2A.Bbd is associated with active transcription and mRNA processing in human cells. Mol Cell 47(4):596–607. doi:10.1016/j.molcel.2012.06.011
53. Corrionero A, Minana B, Valcarcel J (2011) Reduced fidelity of branch point recognition and alternative splicing induced by the anti-tumor drug spliceostatin A. Genes Dev 25(5):445–459. doi:10.1101/gad.2014311
54. Folco EG, Coil KE, Reed R (2011) The anti-tumor drug E7107 reveals an essential role for SF3b in remodeling U2 snRNP to expose

the branch point-binding region. Genes Dev 25(5):440–444. doi:10.1101/gad.2009411

55. Gozani O, Feld R, Reed R (1996) Evidence that sequence-independent binding of highly conserved U2 snRNP proteins upstream of the branch site is required for assembly of spliceosomal complex A. Genes Dev 10(2):233–243
56. Bernardi G (1993) The vertebrate genome: isochores and evolution. Mol Biol Evol 10(1):186–204
57. Peckham HE, Thurman RE, Fu Y et al (2007) Nucleosome positioning signals in genomic DNA. Genome Res 17(8):1170–1177. doi:10.1101/gr.6101007
58. Amit M, Donyo M, Hollander D et al (2012) Differential GC content between exons and introns establishes distinct strategies of splice-site recognition. Cell Rep 1(5):543–556. doi:10.1016/j.celrep.2012.03.013
59. Close P, East P, Dirac-Svejstrup AB et al (2012) DBIRD complex integrates alternative mRNA splicing with RNA polymerase II transcript elongation. Nature 484(7394):386–389. doi:10.1038/nature10925
60. Kolasinska-Zwierz P, Down T, Latorre I et al (2009) Differential chromatin marking of introns and expressed exons by H3K36me3. Nat Genet 41(3):376–381. doi:10.1038/ng.322
61. Huff JT, Plocik AM, Guthrie C et al (2010) Reciprocal intronic and exonic histone modification regions in humans. Nat Struct Mol Biol 17(12):1495–1499. doi:10.1038/nsmb.1924
62. Andersson R, Enroth S, Rada-Iglesias A et al (2009) Nucleosomes are well positioned in exons and carry characteristic histone modifications. Genome Res 19(10):1732–1741. doi:gr.092353.109, [pii] 10.1101/gr.092353.109
63. Schwartz S, Meshorer E, Ast G (2009) Chromatin organization marks exon-intron structure. Nat Struct Mol Biol 16(9):990–995. doi:nsmb.1659, [pii] 10.1038/nsmb.1659
64. Spies N, Nielsen CB, Padgett RA et al (2009) Biased chromatin signatures around polyadenylation sites and exons. Mol Cell 36(2):245–254. doi:S1097-2765(09)00743-6, [pii] 10.1016/j.molcel.2009.10.008
65. Huff JT, Plocik AM, Guthrie C et al (2010) Reciprocal intronic and exonic histone modification regions in humans. Nat Struct Mol Biol 17(12):1495–1499. doi:nsmb.1924, [pii] 10.1038/nsmb.1924
66. Dhami P, Saffrey P, Bruce AW et al (2010) Complex exon-intron marking by histone modifications is not determined solely by nucleosome distribution. PLoS One 5(8):e12339. doi:10.1371/journal.pone.0012339
67. Shieh GS, Pan CH, Wu JH et al (2011) H2B ubiquitylation is part of chromatin architecture that marks exon-intron structure in budding yeast. BMC Genomics 12:627. doi:10.1186/1471-2164-12-627
68. Hnilicova J, Hozeifi S, Duskova E et al (2011) Histone deacetylase activity modulates alternative splicing. PLoS One 6(2):e16727. doi:10.1371/journal.pone.0016727
69. Schor IE, Rascovan N, Pelisch F et al (2009) Neuronal cell depolarization induces intragenic chromatin modifications affecting NCAM alternative splicing. Proc Natl Acad Sci USA 106(11):4325–4330. doi:10.1073/pnas.0810666106
70. Schor IE, Rascovan N, Pelisch F et al (2009) Neuronal cell depolarization induces intragenic chromatin modifications affecting NCAM alternative splicing. Proc Natl Acad Sci USA 106(11):4325–4330. doi:0810666106, [pii] 10.1073/pnas.0810666106
71. Nogues G, Kadener S, Cramer P et al (2002) Transcriptional activators differ in their abilities to control alternative splicing. J Biol Chem 277(45):43110–43114. doi:10.1074/jbc.M208418200
72. Batsche E, Yaniv M, Muchardt C (2006) The human SWI/SNF subunit Brm is a regulator of alternative splicing. Nat Struct Mol Biol 13(1):22–29. doi:nsmb1030, [pii] 10.1038/nsmb1030
73. Saint-Andre V, Batsche E, Rachez C et al (2011) Histone H3 lysine 9 trimethylation and HP1gamma favor inclusion of alternative exons. Nat Struct Mol Biol 18(3):337–344. doi:nsmb.1995, [pii] 10.1038/nsmb.1995
74. Saint-Andre V, Batsche E, Rachez C et al (2011) Histone H3 lysine 9 trimethylation and HP1gamma favor inclusion of alternative exons. Nat Struct Mol Biol 18(3):337–344. doi:10.1038/nsmb.1995
75. Smallwood A, Hon GC, Jin F et al (2012) CBX3 regulates efficient RNA processing genome-wide. Genome Res 22(8):1426–1436. doi:10.1101/gr.124818.111
76. Greil F, van der Kraan I, Delrow J et al (2003) Distinct HP1 and Su(var)3-9 complexes bind to sets of developmentally coexpressed genes depending on chromosomal location. Genes Dev 17(22):2825–2838. doi:10.1101/gad.281503
77. Janowski BA, Huffman KE, Schwartz JC et al (2006) Involvement of AGO1 and AGO2 in mammalian transcriptional silencing. Nat Struct Mol Biol 13(9):787–792. doi:10.1038/nsmb1140
78. Morris KV, Chan SW, Jacobsen SE et al (2004) Small interfering RNA-induced transcriptional gene silencing in human cells. Science 305(5688):1289–1292. doi:10.1126/science.1101372

79. Kim DH, Villeneuve LM, Morris KV et al (2006) Argonaute-1 directs siRNA-mediated transcriptional gene silencing in human cells. Nat Struct Mol Biol 13(9):793–797. doi:10.1038/nsmb1142
80. Weinberg MS, Villeneuve LM, Ehsani A et al (2006) The antisense strand of small interfering RNAs directs histone methylation and transcriptional gene silencing in human cells. RNA 12(2):256–262. doi:10.1261/rna.2235106
81. Allo M, Buggiano V, Fededa JP et al (2009) Control of alternative splicing through siRNA-mediated transcriptional gene silencing. Nat Struct Mol Biol 16(7):717–724. doi:10.1038/nsmb.1620
82. Ameyar-Zazoua M, Rachez C, Souidi M et al (2012) Argonaute proteins couple chromatin silencing to alternative splicing. Nat Struct Mol Biol. doi:10.1038/nsmb.2373
83. Guang S, Bochner AF, Burkhart KB et al (2010) Small regulatory RNAs inhibit RNA polymerase II during the elongation phase of transcription. Nature 465(7301):1097–1101. doi:10.1038/nature09095
84. Burkhart KB, Guang S, Buckley BA et al (2011) A pre-mRNA-associating factor links endogenous siRNAs to chromatin regulation. PLoS Genet 7(8):e1002249. doi:10.1371/journal.pgen.1002249, PGENETICS-D-11-00476 [pii]
85. Ameyar-Zazoua M, Rachez C, Souidi M et al (2012) Argonaute proteins couple chromatin silencing to alternative splicing. Nat Struct Mol Biol 19(10):998–1004. doi:nsmb.2373, [pii] 10.1038/nsmb.2373
86. Grewal SI, Jia S (2007) Heterochromatin revisited. Nat Rev Genet 8(1):35–46. doi:nrg2008, [pii] 10.1038/nrg2008
87. Hahn MA, Wu X, Li AX et al (2011) Relationship between gene body DNA methylation and intragenic H3K9me3 and H3K36me3 chromatin marks. PLoS One 6(4):e18844. doi:10.1371/journal.pone.0018844
88. Luco RF, Pan Q, Tominaga K et al (2010) Regulation of alternative splicing by histone modifications. Science 327(5968):996–1000. doi:10.1126/science.1184208
89. Pradeepa MM, Sutherland HG, Ule J et al (2012) Psip1/Ledgf p52 binds methylated histone H3K36 and splicing factors and contributes to the regulation of alternative splicing. PLoS Genet 8(5):e1002717. doi:10.1371/journal.pgen.1002717
90. Sims RJ 3rd, Millhouse S, Chen CF et al (2007) Recognition of trimethylated histone H3 lysine 4 facilitates the recruitment of transcription postinitiation factors and pre-mRNA splicing. Mol Cell 28(4):665–676. doi:10.1016/j.molcel.2007.11.010
91. Gunderson FQ, Johnson TL (2009) Acetylation by the transcriptional coactivator Gcn5 plays a novel role in co-transcriptional spliceosome assembly. PLoS Genet 5(10):e1000682. doi:10.1371/journal.pgen.1000682
92. Klose RJ, Bird AP (2006) Genomic DNA methylation: the mark and its mediators. Trends Biochem Sci 31(2):89–97. doi:S0968-0004(05)00352-X, [pii] 10.1016/j.tibs.2005.12.008
93. Choi JK (2010) Contrasting chromatin organization of CpG islands and exons in the human genome. Genome Biol 11(7):R70. doi:gb-2010-11-7-r70, [pii] 10.1186/gb-2010-11-7-r70
94. Feng S, Cokus SJ, Zhang X et al (2010) Conservation and divergence of methylation patterning in plants and animals. Proc Natl Acad Sci USA 107(19):8689–8694. doi:1002720107, [pii] 10.1073/pnas.1002720107
95. Zemach A, McDaniel IE, Silva P et al (2010) Genome-wide evolutionary analysis of eukaryotic DNA methylation. Science 328(5980):916–919. doi:science.1186366, [pii] 10.1126/science.1186366
96. Lyko F, Foret S, Kucharski R et al (2010) The honey bee epigenomes: differential methylation of brain DNA in queens and workers. PLoS Biol 8(11):e1000506. doi:10.1371/journal.pbio.1000506
97. Park J, Peng Z, Zeng J et al (2011) Comparative analyses of DNA methylation and sequence evolution using Nasonia genomes. Mol Biol Evol 28(12):3345–3354. doi:msr168, [pii] 10.1093/molbev/msr168
98. Bonasio R, Li Q, Lian J et al (2012) Genome-wide and caste-specific DNA methylomes of the ants camponotus floridanus and harpegnathos saltator. Curr Biol 22(19):1755–1764. doi:S0960-9822(12)00867-6, [pii] 10.1016/j.cub.2012.07.042
99. Flores KB, Wolschin F, Allen AN et al (2012) Genome-wide association between DNA methylation and alternative splicing in an invertebrate. BMC Genomics 13(1):480. doi:10.1186/1471-2164-13-480
100. Lister R, Pelizzola M, Dowen RH et al (2009) Human DNA methylomes at base resolution show widespread epigenomic differences. Nature 462(7271):315–322. doi:nature08514, [pii] 10.1038/nature08514
101. Hodges E, Smith AD, Kendall J et al (2009) High definition profiling of mammalian DNA methylation by array capture and single molecule bisulfite sequencing. Genome Res 19(9):1593–1605. doi:gr.095190.109, [pii] 10.1101/gr.095190.109
102. Laurent L, Wong E, Li G et al (2010) Dynamic changes in the human methylome

during differentiation. Genome Res 20(3):320–331. doi:gr.101907.109, [pii] 10.1101/gr.101907.109
103. Pastor WA, Pape UJ, Huang Y et al (2011) Genome-wide mapping of 5-hydroxymethylcytosine in embryonic stem cells. Nature 473(7347):394–397. doi:nature10102, [pii] 10.1038/nature10102
104. Ficz G, Branco MR, Seisenberger S et al (2011) Dynamic regulation of 5-hydroxymethylcytosine in mouse ES cells and during differentiation. Nature 473(7347):398–402. doi:nature10008, [pii] 10.1038/nature10008
105. Khare T, Pai S, Koncevicius K et al (2012) 5-hmC in the brain is abundant in synaptic genes and shows differences at the exon-intron boundary. Nat Struct Mol Biol 19(10):1037–1043. doi:10.1038/nsmb.2372
106. Shukla S, Kavak E, Gregory M et al (2011) CTCF-promoted RNA polymerase II pausing links DNA methylation to splicing. Nature 479(7371):74–79. doi:10.1038/nature10442
107. Gowher H, Brick K, Camerini-Otero RD et al (2012) Vezf1 protein binding sites genome-wide are associated with pausing of elongating RNA polymerase II. Proc Natl Acad Sci USA 109(7):2370–2375. doi:10.1073/pnas.1121538109
108. Dickson J, Gowher H, Strogantsev R et al (2010) VEZF1 elements mediate protection from DNA methylation. PLoS Genet 6(1):e1000804. doi:10.1371/journal.pgen.1000804
109. Kucharski R, Maleszka J, Foret S et al (2008) Nutritional control of reproductive status in honeybees via DNA methylation. Science 319(5871):1827–1830. doi:1153069, [pii] 10.1126/science.1153069
110. Oberdoerffer S (2012) A conserved role for intragenic DNA methylation in alternative pre-mRNA splicing. Transcription 3(3):106–109. doi:19816, [pii] 10.4161/trns.19816
111. Kim S, Kim H, Fong N et al (2011) Pre-mRNA splicing is a determinant of histone H3K36 methylation. Proc Natl Acad Sci USA 108(33):13564–13569. doi:10.1073/pnas.1109475108
112. de Almeida SF, Grosso AR, Koch F et al (2011) Splicing enhances recruitment of methyltransferase HYPB/Setd2 and methylation of histone H3 Lys36. Nat Struct Mol Biol 18(9):977–983. doi:10.1038/nsmb.2123
113. Zhou HL, Hinman MN, Barron VA et al (2011) Hu proteins regulate alternative splicing by inducing localized histone hyperacetylation in an RNA-dependent manner. Proc Natl Acad Sci USA 108(36):E627–E635. doi:10.1073/pnas.1103344108

Part II

Methods Chapters

Chapter 8

Preparation of Splicing Competent Nuclear Extracts

Chiu-Ho T. Webb and Klemens J. Hertel

Abstract

Splicing components play an essential role in mediating accurate and efficient splicing. The complexity of the spliceosome and its regulatory networks increase the difficulty of studying the splicing reaction in detail. Nuclear extracts derived from HeLa cells provide all of the obligatory components to carry out intron removal in vitro. This chapter describes the large-scale preparation of nuclear extract from HeLa cells.

Key words Nuclear extracts, In vitro, Splicing competent, mRNA processing, Alternative splicing

1 Introduction

Crude, nuclear, and cytoplasmic extracts can be used to study the regulation of mRNA processing, such as transcription, pre-mRNA splicing, and polyadenylation [1–3]. As such, they can be used to evaluate molecular mechanisms and interactions through the use of immunoassays, mobility shift assays (EMSA), co-immunoprecipitation (Co-IP), and pull-down assays.

Analyzing pre-mRNA splicing cell-free has permitted to characterize the splicing machinery in detail [4]. Whole cell extracts can be used to support splicing in vitro [5]; however, more than 60 % of the reaction volume has to be dedicated to whole cell extracts. Functional nuclear extracts, originally developed to study transcription in a test tube [6], were first applied to in vitro splicing reactions using β-globin minigenes [7]. Several modified nuclear extract methods were reported since then [8–11], all of which contained all the components required for in vitro splicing of short pre-mRNAs synthesized in a separate transcription reaction [4]. Cytoplasmic S-100 extracts, a by-product of nuclear extract preparations, lack serine/arginine (SR)-rich proteins and are therefore unable to support pre-mRNA splicing unless they are supplemented with recombinant SR proteins [12, 13]. HeLa cells are the most commonly used cells for the preparation of nuclear extract. Nevertheless, the following protocol is suitable for extract preparation from other cell lines as well.

Klemens J. Hertel (ed.), *Spliceosomal Pre-mRNA Splicing: Methods and Protocols*, Methods in Molecular Biology, vol. 1126, DOI 10.1007/978-1-62703-980-2_8, © Springer Science+Business Media, LLC 2014

2 Materials

2.1 Cells

Spinner cultured suspension HeLa-S3 cells (National Cell Culture Center) (*see* **Note 1**).

2.2 Reagents (See Note 2)

1. 1 M dithiothreitol (DTT).
2. 100 mM phenylmethanesulfonyl fluoride (PMSF) in isopropanol.
3. 1× phosphate-buffered saline (PBS): 137 mM sodium chloride (NaCl), 2.7 mM potassium chloride (KCl), 8 mM sodium phosphate dibasic (Na_2HPO_4), 1.5 mM monopotassium phosphate (KH_2PO_4), pH 7.4.
4. Hypotonic buffer: 10 mM 4-(2-hydroxyethyl)-1-piperazineethanesulfonic acid (HEPES), pH 7.9, 1.5 mM magnesium acetate (MgOAc), 10 mM potassium acetate (KOAc), 0.5 mM DTT, 0.2 mM PMSF.
5. Low-salt buffer: 20 mM HEPES, pH 7.9, 25 % glycerol, 1.5 mM MgOAc, 0.02 M KOAc, 0.2 mM EDTA, 0.5 mM DTT, 0.2 mM PMSF.
6. High-salt buffer: 20 mM HEPES, pH 7.9, 25 % glycerol, 1.5 mM MgOAc, 1.2 M KOAc, 0.2 mM EDTA, 0.5 mM DTT, 0.2 mM PMSF.
7. Dialysis buffer: 20 mM HEPES, pH 7.9, 20 % glycerol, 100 mM KOAc, 0.2 mM EDTA, 0.2 mM PMSF, 0.5 mM DTT.
8. Dialysis tubing (10 K MWCO, Fisher Scientific).

3 Methods (*See* Note 3)

1. Wash the cell pellet and determine the total cell number by resuspending the cell pellet with 5 times (X) cell pellet volume of PBS.
2. Determine and record the packed cell volume (PCV) by centrifuging the cells at 1,850 × *g* for 10 min and then remove the supernatant.
3. Wash the cells by resuspending cell pellet with 5× PCV of hypotonic buffer and immediately centrifuging the cells at 1,850 × *g* for 10 min. Discard the supernatant (*see* **Note 4**).
4. Swell the cells by adding hypotonic buffer to a final volume of 3× PCV followed by incubating on ice for 10 min (*see* **Note 5**).
5. Check for cell lysis of pre-dounced cells by staining a small aliquot of cells with trypan blue.

6. Lyse cells by douncing 10–20 plunges in Kontes-B (Wheaton) Dounce homogenizer (Pestle B) and pour into new bottles (*see* **Note 6**).
7. Monitor the dounced cell lysis by staining a small aliquot of the cells with trypan blue.
8. Determine and record the packed nuclear volume (PNV) by centrifuging the cells at 3,300 × *g* for 15 min, then remove the supernatant. The supernatant can be saved for cytoplasmic S100 isolation preparation [10].
9. Resuspend the pellet of nuclei by adding 0.5× PNV of low-salt buffer and transfer to glass beaker. Combine the nuclei into one beaker if there are multiple tubes.
10. Release the soluble proteins from the nuclei by adding 0.5× PNV of high-salt buffer drop-by-drop while gently stirring (*see* **Note 7**).
11. Extract the nuclei on ice while stirring for 30 min.
12. Remove the nuclei by centrifuging at 25,000 × *g* for 30 min, and save the supernatant.
13. Desalt the nuclear extract by dialyzing the supernatant in dialysis tubing with more than 50× supernatant volume of dialysis buffer for 2–2.5 h while stirring.
14. Change the dialysis buffer and dialyze for an additional 2–2.5 h.
15. Remove the precipitate by centrifuging at 25,000 × *g* for 30 min, and then save the supernatant.
16. Aliquot the extract into 1 ml fractions and freeze on dry ice (*see* **Note 8**).
17. Store the extracts at −80 °C (*see* **Note 9**).
18. Validate the activity of the nuclear extracts with in vitro splicing (i.e., β-globin, Chapter 11).

4 Notes

1. This extract prep starts with pelleted HeLa cells, which can either be purchased or grown in the lab.
2. All reagents should be prepared with autoclaved Milli-Q or double-distilled water, followed by sterilization with autoclave or filtration with 0.22 μm filter. DTT and PMSF stock solutions should be stored at −20 °C and added to buffers just prior to use. All other reagents should be stored at 4 °C.
3. To prevent the denature of proteins and RNA, all extraction steps should be carried out on ice in cold room with ice-cold reagents and centrifuge at 4 °C with pre-chilled rotors.

4. This step needs to be carried out quickly because the hypotonic buffer swells the cells and could potentially break them. Consequently, proteins could leak out of the cell and be discarded with the supernatant.
5. The previous washing step using hypotonic buffer may already have initiated the swelling of the cells. Thus, the PCV may have increased. Refer only to the initial PCV that was recorded. For example, the PCV determined in **step 2** is 15 ml, yet after **step 3** it has increased to 25 ml. In **step 4** add hypotonic buffer such that the final volume of cells and buffer is 45 ml.
6. Perform the douncing with gentle strokes and the loose B pestle to ensure only the cell membranes, but not the nuclear membranes are disrupted.
7. The "drop-by-drop" action is important because rapidly increasing the salt concentration may lyse the nuclei. The lysate can be homogenized again by douncing if it is chunky.
8. 30 l of HeLa cell culture with a $4–6 \times 10^5$ cells/ml density would yield about 45 ml of nuclear extract. Every milliliter of nuclear extract should support ~130 (of 25 μl scale) splicing reactions at 30 % NE.
9. The freeze/thaw cycles should be limited to 5 times to avoid compromising extract activity. The non-disturbed extracts can be stored up to years at −80 °C without losing activity; however, the half-life at 4 °C is only 12 h [14].

Acknowledgement

Research in the Hertel laboratory is supported by NIH (RO1 GM62287 and R21 CA149548).

References

1. Roca X, Karginov FV (2012) RNA biology in a test tube – an overview of in vitro systems/assays. Wiley Interdiscip Rev RNA 3:509–527
2. Knapp G, Beckmann JS, Johnson PF et al (1978) Transcription and processing of intervening sequences in yeast tRNA genes. Cell 14:221–236
3. Di Giammartino DC, Shi Y, Manley JL (2013) PARP1 represses PAP and inhibits polyadenylation during heat shock. Mol Cell 49:7–17
4. Hicks MJ, Lam BJ, Hertel KJ (2005) Analyzing mechanisms of alternative pre-mRNA splicing using in vitro splicing assays. Methods 37: 306–313
5. Kataoka N, Dreyfuss G (2004) A simple whole cell lysate system for in vitro splicing reveals a stepwise assembly of the exon-exon junction complex. J Biol Chem 279:7009–7013
6. Dignam JD, Lebovitz RM, Roeder RG (1983) Accurate transcription initiation by RNA polymerase II in a soluble extract from isolated mammalian nuclei. Nucleic Acids Res 11:1475–1489
7. Krainer AR, Maniatis T, Ruskin B et al (1984) Normal and mutant human beta-globin pre-

mRNAs are faithfully and efficiently spliced in vitro. Cell 36:993–1005
8. Pugh BF (1995) Preparation of HeLa nuclear extracts. Methods Mol Biol 37:349–357
9. Mayeda A, Krainer AR (1999) Preparation of HeLa cell nuclear and cytosolic S100 extracts for in vitro splicing. Methods Mol Biol 118:309–314
10. Abmayr SM, Yao T, Parmely T et al (2006) Preparation of nuclear and cytoplasmic extracts from mammalian cells. Curr Protoc Mol Biol. Chapter 12, Unit 12 1.
11. Kataoka N, Dreyfuss G (2008) Preparation of efficient splicing extracts from whole cells, nuclei, and cytoplasmic fractions. Methods Mol Biol 488:357–365
12. Krainer AR, Conway GC, Kozak D (1990) Purification and characterization of pre-mRNA splicing factor SF2 from HeLa cells. Genes Dev 4:1158–1171
13. Fu XD, Mayeda A, Maniatis T et al (1992) General splicing factors SF2 and SC35 have equivalent activities in vitro, and both affect alternative 5′ and 3′ splice site selection. Proc Natl Acad Sci USA 89:11224–11228
14. Carey MF, Peterson CL, Smale ST (2009) Dignam and Roeder nuclear extract preparation. Cold Spring Harb Protoc. pdb prot5330.

Chapter 9

Preparation of Yeast Whole Cell Splicing Extract

Elizabeth A. Dunn and Stephen D. Rader

Abstract

Pre-mRNA splicing, the removal of introns from pre-messenger RNA, is an essential step in eukaryotic gene expression. In humans, it has been estimated that 60 % of noninfectious diseases are caused by errors in splicing, making the study of pre-mRNA splicing a high priority from a health perspective. Pre-mRNA splicing is also complicated: the molecular machine that catalyzes the reaction, the spliceosome, is composed of five small nuclear RNAs, and over 100 proteins, making splicing one of the most complex processes in the cell.

An important tool for studying pre-mRNA splicing is the in vitro splicing assay. With an in vitro assay, it is possible to test the function of each splicing component by removing the endogenous version and replacing it (or reconstituting it) with a modified one. This assay relies on the ability to produce an extract—either whole cell or nuclear—that contains all of the activities required to convert pre-mRNA to mRNA. To date, splicing extracts have only been produced from human and *S. cerevisiae* (yeast) cells. We describe a method to produce whole cell extracts from yeast that support splicing with efficiencies up to 90 %. These extracts have been used to reconstitute snRNAs, screen small molecule libraries for splicing inhibitors, and purify a variety of splicing complexes.

Key words Pre-mRNA splicing, Whole cell extract, Liquid nitrogen, Grinding, Reconstitution

1 Introduction

Pre-mRNA splicing was first reported in the late 1970s when several research groups identified regions of genomic sequence that were absent in the corresponding mature mRNA transcripts [1–4]. Over the next several years, these intervening sequences were characterized, and consensus sequences at the splice junctions were identified [5]. It was soon revealed that pre-mRNA splicing was a general phenomenon that was prevalent across eukaryotes and that many eukaryotic genes contained more than one intervening sequence. However, it was not obvious how the correct splice sites were selected since the consensus sequences at the splice sites were short and relatively degenerate. It became clear that in order to understand the mechanism of splicing, it was critical to develop an in vitro splicing system from which splicing intermediates and splicing factors could be isolated, identified, and characterized.

Klemens J. Hertel (ed.), *Spliceosomal Pre-mRNA Splicing: Methods and Protocols*, Methods in Molecular Biology, vol. 1126, DOI 10.1007/978-1-62703-980-2_9, © Springer Science+Business Media, LLC 2014

The first report of an accurate and efficient in vitro pre-mRNA splicing system did not come until 1984, when Krainer et al. successfully spliced approximately 90 % of the in vitro-transcribed, radiolabeled β-globin pre-mRNA transcript that they incubated in a HeLa cell nuclear extract [6]. In this system, the progress of the splicing reaction could easily be followed by separating the splicing intermediates and final products using gel electrophoresis and visualizing them by autoradiography. The following year, Lin et al. published a yeast whole cell in vitro pre-mRNA splicing system that had the added advantage of applying yeast genetics in parallel to study the components and interactions of the spliceosome, the large macromolecular machine responsible for catalyzing splicing [7]. These systems revealed that splicing proceeds via two chemical steps, as well as identifying the basal requirements for pre-mRNA splicing: ATP, Mg^{2+}, and monovalent cations.

A major advantage of an in vitro splicing system is that mutated RNAs can be reconstituted into functional complexes to address very specific questions. In yeast, in vitro reconstitution systems have been developed for four of the five small nuclear RNAs (snRNAs): U2, U4, U5, and U6 [8–11]. In these systems, the endogenous full-length snRNA in the splicing extract is targeted for RNAse H-dependent cleavage using complementary oligonucleotides, resulting in nonfunctional fragmented snRNAs. The depleted snRNA is then replaced with a mutated synthetic or in vitro-transcribed snRNA. While in vivo studies of mutated yeast snRNAs only reveal a defect in cell growth that could be attributable to a defect in splicing, the in vitro assay allows for the direct assessment of the effect of the mutation on splicing and the ability to distinguish first and second step splicing defects. For example, the U6 snRNA mutant, G50A, shows a severe growth defect in vivo, and when studied in vitro, it shows a strong block only in the second step of splicing [12].

Another advantage of reconstitution in splicing extract is that the exogenous RNA can be modified in a variety of ways, including the incorporation of site-specific nucleotide analogs that allow for the covalent attachment of cross-linkable or reactive chemical groups. By changing the length of the cross-linking species at a specific location, one can deduce the proximity of various components of the spliceosome to the modified residue. For example, 4-thiouridine, a structural analog of uridine, will cross-link to anything that is within a covalent bond distance of the reactive thiol group while other cross-linkers such as azidophenacyl have longer reaches and will cross-link over a longer distance [13, 14]. Ryan et al. used a 4-thiouridine- and a 5-iodouridine-based cross-linking strategy to probe for RNAs that are within a covalent bond distance of 26 different residues in U6 snRNA [15]. They used these data to develop a three-dimensional model of the functional spliceosome.

In vitro splicing assays have also been used extensively to study the process of spliceosome assembly and to identify the RNA and protein species found in various splicing complexes blocked at specific steps. Cheng and Abelson [16] and Konarska and Sharp [17, 18] developed these assembly assays using yeast whole cell or HeLa cell nuclear extract, respectively, revealing a striking similarity between yeast and mammalian spliceosome assembly. In both cases, the first splicing complex to form contains U2 snRNA and requires ATP and a 3′ splice site. The second complex contains U2, U4, U5, and U6 snRNAs and requires both a 5′ and a 3′ splice site. Mutations in the 3′ splice site, which allow the first splicing reaction to occur, but block the second reaction, result in accumulation of the second complex [16]. More recently, mass spectrometry has been used to identify proteins present at each step of splicing (reviewed in Will and Luhrmann [19]).

In vitro splicing and assembly assays have been used to screen small molecules for their ability to inhibit the splicing process [20–22]. The discovery of small molecules that block splicing and result in the accumulation of normally transient intermediate splicing complexes should facilitate investigation of the splicing mechanism. Intriguingly, Spliceostatin A exhibits potent antitumor activity, suppressing the growth of various mouse and human tumors and prolonging the life span of affected mice [23].

Despite the development of in vitro pre-mRNA splicing assays over 25 years ago, the splicing mechanism is still very poorly understood. A major reason for this is the asynchronous assembly of the spliceosome and progress of the splicing reactions in whole cell extract. Splicing is a highly dynamic process that involves numerous structural rearrangements within the spliceosome to position the pre-mRNA substrate in an appropriate orientation for each step of the splicing reaction to take place [24]. Consequently, the signals that are observed in whole cell extract are an average of these processes. To overcome this challenge, single molecule fluorescence assays using whole cell splicing extract have recently been developed to monitor in vitro splicing of individual pre-mRNA transcripts as well as assembly of the spliceosome using fluorescently labeled splicing complexes [25, 26]. These studies have already shed some light on the details of the mechanism of splicing by revealing the ordered pathway of spliceosome assembly and the time- and ATP-dependent conformational states of the pre-mRNA during splicing.

Here, we describe a method for preparing yeast whole cell splicing extract from a protease-deficient yeast strain. We also outline the steps required to prepare a radiolabeled pre-mRNA substrate to monitor the splicing reaction. Finally, we describe the procedure for testing the splicing extract for activity and provide the equations that are necessary to determine the splicing efficiency.

2 Materials

Purchase RNase-free plasticware (tubes, tips, etc.). Bake glassware at 250 °C for 2 h to inactivate any RNases. Allow all baked materials to reach room temperature before moving them to 4 °C or −80 °C, as a rapid change in temperature can cause the materials to break. Autoclave all reagents unless otherwise indicated. Filter sterilize indicated buffers through 0.22 μm nitrocellulose (high protein-binding) bottle-top or syringe filters to remove RNases. Chill all centrifuge rotors and tubes prior to use.

2.1 Cell Growth and Harvest

1. Yeast strain BJ2168 (EJ101) (Jones 1991 [27]; available upon request) or strain of your choice.
2. YPD media (4.2 L and one plate): 1 % Bacto™ Yeast Extract, 2 % Bacto™ Peptone, 2 % dextrose. Add 2 % Bacto™ Agar for plates.
3. 4×2.8 L Fernbach flasks.
4. Incubator and shaker (30 °C).
5. Beckman Coulter Avanti HP-20 XPI Centrifuge, JA8.1000 rotor, and 4×1 L centrifuge bottles (or equivalent).
6. AGK buffer: 10 mM HEPES-KOH, pH 7.9, 1.5 mM $MgCl_2$, 200 mM KCl, 0.5 mM DTT (add fresh), 10 % glycerol. Filter sterilize. Store at 4 °C.
7. 4×50 mL conical tubes, 1×15 mL conical tube.
8. Plastic pitcher, autoclaved.
9. 21-gauge needle and 10 mL syringe.
10. Liquid nitrogen.

2.2 Preparation of Yeast Whole Cell Extract

1. Large (~20 cm diameter) mortar and pestle wrapped in foil, baked, and cooled at −80 °C overnight.
2. 250 mL beaker, small metal spatula, and several Pasteur pipets all wrapped in foil, baked, and chilled at 4 °C.
3. Magnetic stir bar rinsed with ethanol and chilled at 4 °C.
4. 5 mL pipets, autoclaved and chilled at 4 °C, or RNase-free disposable pipets.
5. Liquid nitrogen.
6. Beckman Coulter Avanti HP-20 XPI centrifuge, JA25.50 rotor, 2 Oakridge tubes (or equivalent).
7. 8,000–10,0000 MWCO dialysis membrane tubing.
8. Buffer D: 20 mM HEPES-KOH, pH 7.9, 0.2 mM EDTA, 50 mM KCl, 0.5 mM DTT (add fresh), 20 % glycerol. Store at 4 °C.

2.3 Preparation of Radiolabeled In Vitro-Transcribed Actin Pre-mRNA

1. Plasmid containing the T7 promoter followed by a portion of the actin gene (pJPS149; Vijayraghavan et al. 1986 [28]; available upon request).
2. Microcentrifuge and tubes.
3. Restriction enzyme HindIII, buffer, and BSA.
4. Heat block or water bath (37 °C, 65 °C).
5. Gel extraction kit.
6. 0.8 % agarose gel, gel apparatus, and power supply.
7. 1× TBE: 89 mM Tris base, 89 mM boric acid, 2 mM EDTA, pH 8.0.
8. Spectrophotometer (NanoDrop).
9. T7 RNA polymerase and buffer.
10. RNA nucleotides, 100 mM. Make a mixture of ATP, CTP, and UTP 10 mM each in dH_2O. Make a 0.5 mM stock of GTP. Aliquot and store at −80 °C.
11. Superasin RNAse inhibitor
12. α-^{32}P-GTP, 3,000 Ci/mmol, 10 mCi/mL.
13. TE: 10 mM Tris–HCl, pH 7.5, 1 mM EDTA. Filter sterilize.
14. G25 spin column.
15. Scintillation counter.

2.4 Testing the Extract for Splicing Activity

1. Microcentrifuge and tubes chilled at 4 °C.
2. Radiolabeled pre-mRNA transcript.
3. Splicing extract.
4. Splicing buffer components: dH_2O, 25 mM $MgCl_2$, 1 M potassium phosphate, pH 7 (mix 6.15 mL 1 M K_2HPO_4 and 3.85 mL 1 M KH_2PO_4), 30 % PEG 8000, 100 mM ATP. Filter sterilize each component (except ATP) separately, aliquot, and store at −80 °C.
5. Stop solution: 0.30 M NaOAc, 1 mM EDTA, 1 % SDS, 0.034 mg/mL E. coli tRNA (add tRNA when ready to use). Filter sterilize. Store at room temperature.
6. Heat blocks (30 °C, 65 °C)
7. 25:24:1 phenol/chloroform/isoamyl alcohol, pH 6.7 (Sigma).
8. Chloroform.
9. 70 %, 100 % ethanol, −20 °C.
10. 7 M urea loading buffer.
11. Vertical gel apparatus, plates (14.5 × 16.5 cm), spacers (0.75 mm), comb, and power supply.

12. Acrylamide gel components: 40 % (19:1) acrylamide/bis acrylamide, 20× TBE (1.78 M Tris base, 1.78 M boric Acid, 40 mM EDTA, pH 8.0), 10 % APS (in dH_2O, made fresh within the last month), TEMED.
13. Phosphor screen, cassette, and phosphorimager.

2.5 Time Considerations

Cell growth: 4 days from the glycerol stock.

Cell harvest: 1.5 h.

Extract preparation: 7–8 h including dialysis.

Preparation of actin in vitro transcription template: 7–8 h.

Preparation of radiolabeled actin pre-mRNA: 2 h.

Standard splicing assay: 4 h excluding exposure to a phosphorimager screen.

3 Methods

3.1 Cell Growth and Cell Harvest

Splicing extract can be prepared from any yeast strain; however, the presence of proteases can be problematic. The use of a protease-deficient strain results in the most active splicing extract.

1. Grow 4 L of yeast cells in YPD to an OD_{600} of 2.0–2.5 (*see* **Notes 1, 2**).
2. Harvest the cells at 2,200 × *g* for 15 min in a JA 8.1000 rotor (Beckman Coulter Avanti HP-20 XPI Centrifuge). Pour off the YPD and keep the cells on ice from this point on.
3. Resuspend pellets quickly in 50 mL cold dH_2O/2 L cells by vigorous swirling. Spin 12 min at 2,200 × *g* in the JA 8.1000 rotor. Gently pour off the dH_2O (*see* **Note 3**).
4. Wash each 2 L cell pellet with 50 mL cold AGK buffer as in the previous step.
5. Resuspend each 2 L cell pellet in 7.5 mL cold AGK buffer and combine in one 50 mL conical tube (*see* **Note 4**).
6. Using a 10 mL syringe with a 21-gauge needle, drip the cell suspension into a 1 L sterile plastic pitcher containing about 200 mL of liquid nitrogen (*see* **Note 5**). Pour off the liquid nitrogen and collect the cell drops in two 50 mL conical tubes (*see* **Note 6**). Store at −80 °C until ready to prepare extract (*see* **Note 7**).

3.2 Preparation of Yeast Whole Cell Extract

1. In the cold room, pour the frozen cell drops into the prechilled mortar holding about 50 mL of liquid nitrogen (*see* **Note 8**). Allow most of the liquid nitrogen to evaporate and then grind the cells into a very fine powder using a chilled pestle. Make sure the cells stay frozen by adding extra liquid nitrogen about

every 4 min. Do not allow the powder to get shiny and sticky. Grind for 30–40 min (*see* **Note 9**).

2. Scrape the powder into the 250 mL beaker, containing the stir bar, using the metal spatula, and thaw at 4 °C for 1 h. Place the beaker in an ice water bath on a stir plate in the cold room and stir for 30 min (*see* **Note 10**).
3. Transfer the thawed powder to an Oakridge tube using a cold 5 mL pipet. Spin in a JLA 25.50 rotor (Beckman Coulter Avanti HP-20 XPI Centrifuge) at 18,000 ×*g* for 30 min at 4 °C to remove cell debris.
4. Pipet no more than 8.0 mL of the supernatant from the Oakridge tube into a 70.1 Ti centrifuge tube without disturbing the pellet. Also avoid the floating white lipoproteins at the surface (*see* **Note 11**). Spin at 100,000 ×*g* for 1 h in a Beckman XL-70 ultracentrifuge.
5. Use a Pasteur pipet to remove approximately 4–5 mL of the pale yellow aqueous phase from the middle of the supernatant without disturbing the top film or bottom pellet (*see* **Note 11**). Transfer into a 15 mL conical tube on ice.
6. Dialyze twice against 2 L buffer D at 4 °C for 1.5 h each time in 8,000–10,000 molecular weight cutoff dialysis membrane (*see* **Note 12**).
7. Aliquot 75–100 μL of dialyzed extract into chilled microcentrifuge tubes and snap freeze in liquid nitrogen (*see* **Note 13**). Store at −80 °C (*see* **Note 14**).

3.3 Preparation of Radiolabeled In Vitro-Transcribed Actin Pre-mRNA

In order to follow the splicing reaction, radiolabel an appropriate pre-mRNA substrate. Any RNA containing an intron and at least 50 nt of flanking exonic sequences should work. A standard substrate used in yeast splicing assays is a 590 nt segment of actin pre-mRNA downstream of a T7 promoter.

1. Prepare the T7 actin plasmid (pJPS149; Vijayraghavan 1986 [28]) template for run-off in vitro transcription by linearizing 10 μg in a reaction containing 1× restriction enzyme buffer, 0.1 μg/μL BSA, and 20 U HindIII (*see* **Note 15**). Incubate the reaction for 1 h at 37 °C. Run the linearized plasmid on a 0.8 % agarose gel and purify using a gel purification kit, as described by the manufacturer. Quantify the template by reading the A_{260} on a spectrophotometer.
2. In vitro transcribe radiolabeled actin pre-mRNA. Combine, in order, at room temperature: 1 μL 10× T7 RNA polymerase buffer, 0.5 μL 10 mM ATP, CTP, and UTP, 0.5 μL 0.5 mM GTP, 0.5 μL Superasin RNAse inhibitor, 500 ng linearized pJPS149, 2.5 μL α-^{32}P-GTP, 0.5 μL T7 RNA polymerase, and dH_2O to 10 μL (*see* **Notes 16, 17**). Incubate the transcription reaction for 1.5 h at 37 °C.

3. Remove unincorporated nucleotides from the transcription reaction and quantify the efficiency of incorporation of α-^{32}P-GTP. Dilute the reaction to 50 μL with TE, pH 7.5. Count 1 μL of the diluted reaction in a scintillation counter. Prepare a G25 spin column according to the manufacturer's instructions. Apply the rest of the diluted reaction to the column and elute. Count 1 μL of the eluate in a scintillation counter. Calculate the percent incorporation of α-^{32}P-GTP into the actin transcript and determine the cpm/fmol (*see* **Note 18**). Dilute the actin transcript to 4 fmol/μL in TE, pH 7.5.

3.4 Testing the Extract for Splicing Activity

1. Assemble two splicing reactions on ice by combining 1.4 μL dH_2O, 1 μL 25 mM $MgCl_2$, 0.6 μL 1 M KPO_4, pH 7.0, 1 μL 30 % PEG, 4 μL precleared splicing extract (*see* **Note 19**), 1 μL 100 mM ATP (*see* **Note 20**), and 1 μL radiolabeled actin transcript (4 fmol).
2. Immediately add 200 μL stop solution to one tube and place the reaction on ice. This is a zero-minute negative control reaction to ensure that the actin pre-mRNA is intact. Incubate the second reaction at 30 °C for 30 min to allow splicing to occur, and then add 200 μL stop solution.
3. Phenol/chloroform extract the RNA from both reactions. Add 200 μL phenol/chloroform and invert the tubes 5 times. Incubate at 65 °C for 5 min and then spin the tubes in a microcentrifuge for 5 min at maximum speed. Pipet 170 μL of the aqueous (top layer) phase into a clean microcentrifuge tube and back extract by adding 150 μL chloroform. Invert the tubes 5 times and then spin for 3 min at maximum speed. Remove the chloroform (bottom layer) with a P200 pipetter set to 200 μL. Discard.
4. Ethanol precipitate the RNA. Add 800 μL of cold (−20 °C) 100 % ethanol to each tube, invert 5 times, and then spin for 30 min at maximum speed in a microcentrifuge, at 4 °C. Remove the ethanol with a pipet and add 170 μL of cold (−20 °C) 70 % ethanol. Spin for 3–5 min at maximum speed and then remove all of the ethanol with a pipet. Air-dry the pellet for about 5 min and then resuspend it in 8 μL of 7 M urea loading dye.
5. Resolve the splicing reaction products in a 6 % acrylamide (7 M urea) denaturing gel (*see* **Note 21**). Heat the samples for 3 min at 65 °C and load onto the gel. Run at 400 V for 1 h in 1× TBE. Expose the gel to a phosphorimager screen at −80 °C and visualize the autoradiograph.
6. Quantify the pre-mRNA, splicing intermediates, and products by densitometry and determine the splicing efficiency (*see* Fig. 1, **Note 22**).

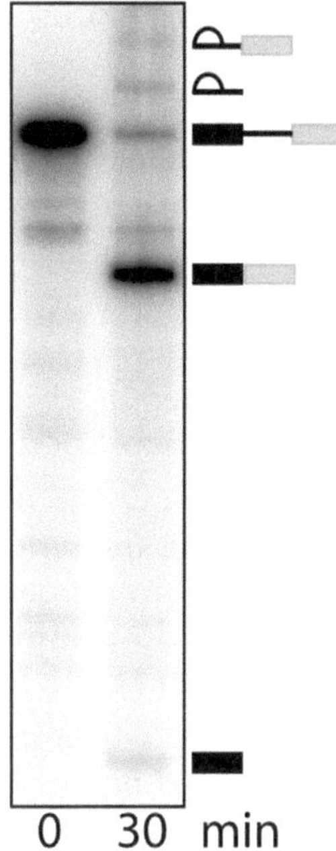

Fig. 1 Testing the activity of splicing extract. Four femtomoles of radiolabeled actin pre-mRNA were incubated in splicing extract and the products of the splicing reaction were separated by electrophoresis in a 6 % polyacrylamide (7 M urea) denaturing gel. The *left lane* is a zero time point showing the location of the pre-mRNA substrate, and the *right lane* shows the splicing products after a 30 min incubation at 30 °C. The identity of the bands is indicated to the right of the gel as follows (*top* to *bottom*): lariat-exon intermediate, lariat intron, pre-mRNA, mRNA, and 5′ exon. Products of the first chemical step of splicing are the free 5′ exon and lariat-exon intermediate, and the final splicing products are mature mRNA and excised lariat intron

4 Notes

1. A minimum of 2 L of culture is recommended in order to prepare a clean extract using this protocol.
2. Streak the yeast strain BJ2168 from a glycerol stock onto a YPD plate and incubate at 30 °C until the colonies have grown to a suitable size (~2 days). Late in the day, inoculate a 2 mL YPD culture with a single colony and grow the culture overnight at 30 °C with shaking (200 rpm). Inoculate 100 mL of YPD with the entire 2 mL culture early the next morning and continue growing. If you do not know the doubling time of your yeast strain, measure the OD_{600} several times during the day to determine it (wait about 3 h before taking measurements to get through the lag phase). The doubling time for BJ2168 is 1.8 h. Later that day, while the culture is still in log

phase, inoculate 4 × 1 L YPD from the 100 mL culture so that the cultures will reach an OD_{600} ~ 2.0–2.5 the next morning. There should be no lag phase as the inoculum is in log phase.

3. Combine 2 pellets so cells are now in two 1 L bottles. Do not transfer to smaller tubes, as the large pellets become much more difficult to resuspend when compacted.
4. If your yeast strain is not protease deficient, add protease inhibitors (e.g., 1 mM AEBSF, 1 mM benzamidine, 1 μg/mL leupeptin, 1 μg/mL aprotinin, or your favorite premade inhibitor cocktail).
5. Remove the needle to fill the syringe with cells. Constantly move the syringe over the beaker while dripping to prevent the drops from aggregating. Be careful not to get the needle too close to the liquid nitrogen, or the cells will freeze in the tip and block it.
6. Poke a hole in the cap of the tube so that the gas from the liquid nitrogen can escape. A nail heated in a Bunsen burner will pierce the cap easily.
7. Cell drops can be stored indefinitely at –80 °C. Alternatively, drip the cell suspension directly into the chilled mortar containing liquid nitrogen and proceed with preparation of the extract.
8. Wear freezer gloves to protect your hands. Grinding is substantially easier if you fashion a holder for the mortar out of a thick piece of Styrofoam (e.g., the lid of a Styrofoam shipping container). Carve out a hollow that snuggly fits the mortar. Line it with foil so that bits of Styrofoam do not get into the extract.
9. Grind slowly at first to prevent the cell drops from popping out of the mortar. Once the cell drops are crushed (5–10 min), grind more vigorously. Grind for about 10 min after the powder is the texture of talcum powder (i.e., very fine—it should feel smooth, not granular, at this stage). Note that excessive grinding will reduce the activity of the extract.
10. Place some ice in the bottom of a 2 L beaker. Add a little water to make a slurry. Add about 1 g of NaCl to lower the temperature of the bath. Nestle the beaker containing the cells into the center, pressing it down to the bottom. The ice should come just part way up the 250 mL beaker so that it does not float.
11. This is the critical step in extract preparation. Be conservative in the amount of supernatant you remove, scrupulously avoiding the pellet and the floating lipids. Although you will end up with less extract, it will be more active and less likely to get hung up in the wells of the splicing gel. Remember: less is more.

12. Buffer D can be reused up to ten times without affecting the quality of the extract. Mark the 2 beakers and always use the same beaker for the first dialysis. Store at 4 °C. Add fresh DTT each time.
13. The protein concentration in the extract should be 15–30 mg/mL, as measured by Bradford.
14. Extract is good for at least a year, probably longer. It can be thawed and refrozen one time without significant loss in activity.
15. We have experienced problems with the pJPS149 plasmid when purified from a glycerol stock of DH5α. Consequently, always freshly transform DH5α cells from the original plasmid stock when more plasmid is needed.
16. The spermidine in the transcription buffer will precipitate nucleic acids at 4 °C or colder, so keep the transcription reagents on ice but assemble the reaction at room temperature.
17. α-^{32}P-UTP can be substituted for the α-^{32}P-GTP. Make up the stocks of ribonucleotides to correspond with the change in the radionucleotide.
18. Calculation for cpm/fmole of actin:

$$\begin{aligned} Total\,moles\,of\,GTP &= Hot\,GTP + Cold\,GTP \\ &= \left(0.835\times10^{-8}\,mmol\right) + (2.5\times10^{-7}\,mmol) \\ &= 2.58\times10^{-7}\,mmol \end{aligned}$$

$$\%\,\text{incorporation of}\,^{32}\text{P} = \frac{\text{Scintillation count before G25}}{\text{Scintillation count after G25}} \times 100$$

$$\text{Total m moles of GTP incorporated} = 2.58\times10^{-7}\,\text{mmol}\times\%\,\text{incorporation}$$

$$\begin{aligned} &\text{Moles of full length actin} \\ &= \text{Total moles of GTP incorporated}\times\left(1\,\text{mol}/1{,}000\,\text{mmol}\right)\times(1\,\text{actin}/118\,\text{Gs}) \end{aligned}$$

$$\text{cpm/fmol} = \frac{\left(\text{Scintillation count after G25}\times 50\text{mL}\right)}{\text{Moles of full length actin}} \times \frac{1\,\text{mol}}{1.0\times10^{15}\,\text{fmol}}$$

$$\text{fmol}/\mu\text{L} = \frac{\text{Scintillation count after G25}}{\text{cpm/fmole}}$$

19. Thaw all splicing reagents on ice. Once the splicing extract has thawed, spin at maximum speed for 5 minutes in a microcentrifuge. Transfer the cleared extract to a fresh tube on ice.

20. The standard splicing assays in the literature contain 2 mM ATP. However, we find that the splicing efficiency is greatly enhanced with 10 mM ATP.
21. Assemble the gel plate, gasket, and spacers. Dissolve 6.3 g of ultrapure urea in about 6 mL of dH_2O. Add 750 μL 20× TBE, 150 μL 10 % APS, 15 μL of TEMED, and dH_2O to 15 mL. Apply the mixture between the gel plates using a disposable 10 mL pipet and then insert the comb. Allow the gel to polymerize for at least 45 min (can be made the day before, wrapped in plastic with a wet paper towel over the comb, and stored at 4 °C). Remove gasket and comb and place in gel box with 1× TBE. Blow out urea and air bubbles from the wells of the gel using a syringe and needle filled with buffer. Make sure there are no air bubbles trapped under the gel. Pre-run the gel at 400 V for 15–30 min. Blow out the wells again before loading samples.
22.

$$\text{Splicing efficiency} = \frac{(\text{mRNA} + \text{lariat intron})}{(\text{Pre-mRNA} + \text{lariat-}3'\text{ exon} + 5'\text{ exon} + \text{lariat intron} + \text{mRNA})} \times 100$$

Acknowledgments

We acknowledge extensive discussions with and advice from Rabiah Mayas, Jon Staley, and Beate Schwer as we optimized this protocol in our lab. We also acknowledge Martha Stark's contributions to developing the method and editing this manuscript. This work was supported by NSERC Discovery Grant 298521 and UNBC Office of Research awards to SDR and an NSERC PGS award to EAD.

References

1. Berget SM, Moore C, Sharp PA (1977) Spliced segments at the 5′ terminus of adenovirus 2 late mRNA. Pro Natl Acad Sci USA 74:3171–3175
2. Chow LT, Gelinas RE, Broker TR et al (1977) An amazing sequence arrangement at the 5′ ends of adenovirus 2 messenger RNA. Cell 12:1–8
3. Kinniburgh AJ, Mertz JE, Ross J (1978) The precursor of mouse beta-globin messenger RNA contains two intervening RNA sequences. Cell 14:681–693
4. Tilghman SM, Tiemeier DC, Seidman JG et al (1978) Intervening sequence of DNA identified in the structural portion of a mouse beta-globin gene. Pro Natl Acad Sci USA 75:725–729
5. Catterall JF, O'Malley BW, Robertson MA et al (1978) Nucleotide sequence homology at 12 intron–exon junctions in the chick ovalbumin gene. Nature 275:510–513
6. Krainer AR, Maniatis T, Ruskin B et al (1984) Normal and mutant human beta-globin pre-mRNAs are faithfully and efficiently spliced in vitro. Cell 36:993–1005
7. Lin RJ, Newman AJ, Cheng SC et al (1985) Yeast mRNA splicing in vitro. J Biol Chem 260:14780–14792

8. McPheeters DS, Fabrizio P, Abelson J (1989) In vitro reconstitution of functional yeast U2 snRNPs. Gene Dev 3:2124–2136
9. Hayduk AJ, Stark MR, Rader SD (2012) In vitro reconstitution of yeast splicing with U4 snRNA reveals multiple roles for the 3′ stem-loop. RNA 18:1075–1090
10. O'Keefe RT, Norman C, Newman AJ (1996) The invariant U5 snRNA loop 1 sequence is dispensable for the first catalytic step of pre-mRNA splicing in yeast. Cell 86:679–689
11. Fabrizio P, McPheeters DS, Abelson J (1989) In vitro assembly of yeast U6 snRNP: a functional assay. Gene Dev 3:2137–2150
12. Madhani HD, Bordonné R, Guthrie C (1990) Multiple roles for U6 snRNA in the splicing pathway. Gene Dev 4:2264–2277
13. Sontheimer EJ (1994) Site-specific RNA crosslinking with 4-thiouridine. Mol Biol Rep 20:35–44
14. Chen JL, Nolan JM, Harris ME et al (1998) Comparative photocross-linking analysis of the tertiary structures of Escherichia coli and Bacillus subtilis RNase P RNAs. EMBO J 17: 1515–1525
15. Ryan DE, Kim CH, Murray JB et al (2004) New tertiary constraints between the RNA components of active yeast spliceosomes: a photo-crosslinking study. RNA 10:1251–1265
16. Cheng SC, Abelson J (1987) Spliceosome assembly in yeast. Gene Dev 1:1014–1027
17. Konarska MM, Sharp PA (1986) Electrophoretic separation of complexes involved in the splicing of precursors to mRNAs. Cell 46:845–855
18. Konarska MM, Sharp PA (1987) Interactions between small nuclear ribonucleoprotein particles in formation of spliceosomes. Cell 49: 763–774
19. Will CL, Lührmann R (2011) Spliceosome structure and function. Cold Spring Harb Perspect Biol 3:1–23
20. Aukema KG, Chohan KK, Plourde GL et al (2009) Small molecule inhibitors of yeast pre-mRNA splicing. ACS Chem Biol 4:759–768
21. Kaida D, Motoyoshi H, Tashiro E et al (2007) Spliceostatin A targets SF3b and inhibits both splicing and nuclear retention of pre-mRNA. Nat Chem Biol 3:576–583
22. O'Brien K, Matlin AJ, Lowell AM et al (2008) The biflavonoid isoginkgetin is a general inhibitor of Pre-mRNA splicing. J Biol Chem 283:33147–33154
23. Nakajima H, Hori Y, Terano H et al (1996) New antitumor substances, FR901463, FR901464 and FR901465. II. Activities against experimental tumors in mice and mechanism of action. J Antibiot 49: 1204–1211
24. Smith DJ, Query CC, Konarska MM (2008) "Nought may endure but mutability": spliceosome dynamics and the regulation of splicing. Mol Cell 30:657–666
25. Crawford DJ, Hoskins AA, Friedman LJ et al (2008) Visualizing the splicing of single pre-mRNA molecules in whole cell extract. RNA 14:170–179
26. Hoskins AA, Friedman LJ, Gallagher SS et al (2011) Ordered and dynamic assembly of single spliceosomes. Science 331:1289–1295
27. Jones EW (1991) Tackling the protease problem in Saccharomyces cerevisiae. Methods Enzymol 194:428–453
28. Vijayraghavan U, Parker R, Tamm J et al (1986) Mutations in conserved intron sequences affect multiple steps in the yeast splicing pathway, particularly assembly of the spliceosome. EMBO J 5:1683–1695

Chapter 10

Efficient Splinted Ligation of Synthetic RNA Using RNA Ligase

Martha R. Stark and Stephen D. Rader

Abstract

RNA ligation allows the creation of large RNA molecules from smaller pieces. This can be useful in a number of contexts: to generate molecules that are larger than can be directly synthesized; to incorporate site-specific changes or RNA modifications within a large RNA in order to facilitate functional and structural studies; to isotopically label segments of large RNAs for NMR structural studies; and to construct libraries of mutant RNAs in which one region is extensively mutagenized or modified. The impediment to widespread use of RNA ligation is the low and variable efficiency of standard ligation strategies, which frequently preclude joining more than two pieces of RNA together.

We describe a method using RNA ligase (Rligation), rather than DNA ligase (Dligation), in a splint-mediated ligation reaction that joins RNA molecules with high efficiency. RNA ligase recognizes single-stranded RNA ends, which are held in proximity to one another by the splint. Monitoring the reaction is easily accomplished by denaturing gel electrophoresis and ethidium bromide staining. Using this technique, it is possible to generate a wide range of modified RNAs from synthetic oligoribonucleotides.

Key words RNA ligation, Rligation, T4 RNA ligase, Oligoribonucleotide, Synthetic RNA, 2′ ACE, RNA modifications, RNA library

1 Introduction

1.1 Applications of RNA Ligation

RNA ligation is an essential method to generate site-specifically mutated, modified, or labeled RNA molecules, as direct synthesis of RNA oligonucleotides is limited to lengths of approximately 100 nucleotides (<50–75 nt for modified oligonucleotides). Chemical synthesis of RNA permits the incorporation of modified residues either internally at the base or sugar phosphate backbone or at either end of the RNA. The only limitation to modification is the availability of the reactive phosphoramidite. Modified residues containing reactive groups can also be incorporated into synthetic RNA to allow for post-synthetic labeling of the RNA when direct incorporation of a label is not possible.

The ligation of several modified RNA oligonucleotides, or modified RNA oligonucleotides with in vitro transcribed RNAs,

Klemens J. Hertel (ed.), *Spliceosomal Pre-mRNA Splicing: Methods and Protocols*, Methods in Molecular Biology, vol. 1126, DOI 10.1007/978-1-62703-980-2_10, © Springer Science+Business Media, LLC 2014

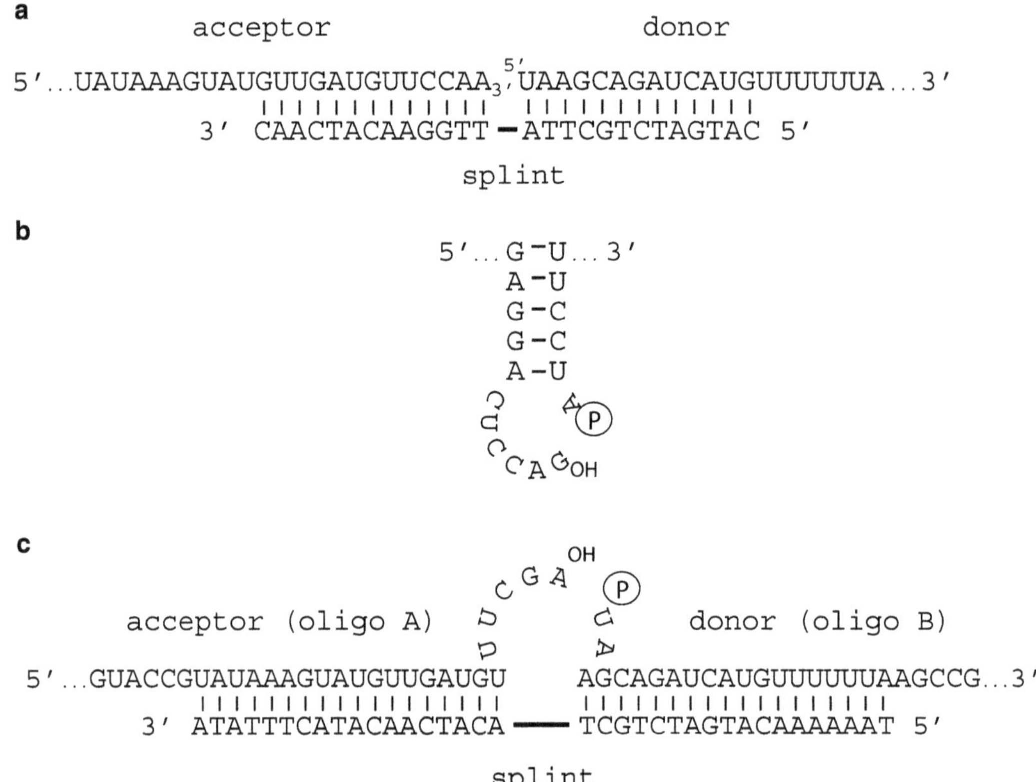

Fig. 1 Substrates and splint designs for ligases. (**a**) Dligation design with splint completely complementary to RNA acceptor and donor. (**b**) The tRNA anticodon loop that is the natural substrate for T4 RNA ligase. (**c**) Rligation design with splint not complementary to ends of RNA acceptor and donor. Circled "P" is the 5′ phosphate, and "OH" is the 3′ hydroxyl that is used in the ligation reaction. 5′ and 3′ ends of each RNA molecule are indicated. The heavy line in the splints simply denotes the continuous DNA backbone (i.e., there are no missing nucleotides)

has led to new insights into the structure and function of large RNAs. For example, to investigate the function of post-transcriptionally modified nucleotides in U2 snRNA in pre-mRNA splicing, Dönmez et al. ligated modified oligonucleotides together to generate full-length U2 with modifications at specific positions [1]. Similarly, RNA ligation has been used to incorporate fluorescent dyes at specific positions of pre-messenger RNA to monitor spliceosome assembly and splicing [2]. Furthermore, the NMR structure of a 100 kDa RNA (310 nt, the internal ribosomal entry site of the hepatitis C virus) was determined after using T4 RNA ligase to generate segmentally ^{15}N-labeled RNA [3].

1.2 Dligation

The most common method for ligating RNA uses T4 DNA ligase on RNA fragments that are held together by a DNA splint (Fig. 1a; [4]). The splint is completely complementary to the ends of the RNA fragments, thereby mimicking the nicked, double-stranded

DNA that is the natural substrate for DNA ligase. The design of a splinted ligation using DNA ligase is relatively straightforward—the ligation junction can be anywhere in the RNA molecule with no sequence restriction. The main requirement is simply the presence of a 5′-monophosphate on the downstream RNA (the donor) and a free 3′-hydroxyl on the upstream RNA (the acceptor, Fig. 1a). Circularization of the donor RNA, often the most important source of unwanted products, is insignificant due to DNA ligase's strict requirement for double-stranded substrate.

Despite the apparent geometric fidelity of this arrangement, RNA turns out to be only a mediocre substrate for DNA ligase, resulting in the need for high concentrations of ligase (often stoichiometric), which can become cost prohibitive when large amounts of product are required. Due to the inefficiency of DNA ligase on an RNA/DNA substrate, long ligation times are required, which increase the chance of RNA degradation. Unproductive hybridization intermediates between RNAs and the DNA splint also contribute to low ligation yields. Ligation efficiency can be increased in some cases by using very long DNA splints to reduce the number of unproductive hybrid complexes formed [5].

1.3 Rligation

To overcome these problems, T4 RNA ligase can be used instead of T4 DNA ligase, but this brings its own set of challenges. The natural substrate for T4 RNA ligase is the anticodon loop of tRNA (Fig. 1b; [6]), but RNA ligase will readily join any single-stranded RNA with a 5′-phosphate and a 3′-hydroxyl. Consequently, 5′-phosphorylated RNA is rapidly circularized by RNA ligase, necessitating a mechanism for conferring specificity to the reaction. The use of a DNA splint can be effective but requires a different design from the splints used with DNA ligase [7, 8].

Since RNA ligase requires single-stranded RNA substrates, the splint must hold the RNA fragments in proximity to one another while allowing the ends the flexibility to reach into the enzyme active site (Fig. 1c). In practice, this is achieved by making the splint complementary to the 3′ end of the phosphate *acceptor* (A) oligonucleotide except at the last 4–8 nucleotides. Similarly, complementarity to the phosphate *donor* (B) oligonucleotide starts with the 3rd nucleotide from the 5′ end. The resulting free ends of 4–8 nucleotides at the 3′ end of A and two nucleotides at the 5′ end of B closely match the 5 and 2 nucleotide single-stranded ends of tRNA prior to ligation.

The presence of the splint is frequently sufficient to sequester the 5′ end of the B oligonucleotide, thereby preventing circularization, but the 2′ ACE protecting groups introduced by 5′-silyl-2′-acetoxy ethyl orthoester solid-phase synthesis chemistry further inhibit side reactions [8]. Chemical protecting groups at the 2′ position can sterically reduce or eliminate reaction at the 3′-OH. By selectively deprotecting only the acceptor RNA fragment, phosphorylating only the donor fragment, and using an appropriate splint, high reaction yields and specificity can be achieved.

When one or more of the RNAs to be ligated are made enzymatically using T7 RNA polymerase, a different approach is necessary to prevent unwanted products. To prevent circularization or concatamerization of the donor molecule, a blocked 3′ end is necessary, either with a 2′,3′-cyclic phosphate, a 3′-phosphate, or a dideoxy residue. On the acceptor molecule, a 5′-OH will prevent the formation of these unwanted products.

1.4 Outline of Method

The RNA ligation described here involves a design stage followed by the actual ligation reactions. First, break points are chosen in the full-length molecule to facilitate incorporation of modifications and to minimize the total number of ligation steps. Second, splint sequences are chosen for each oligonucleotide junction. Finally, the two oligonucleotides and the corresponding splint are mixed in the appropriate molar ratio with RNA ligase. Reaction products are analyzed, and additional ligation steps can be carried out iteratively.

2 Materials

As RNA is highly sensitive to nucleases, and as nucleases are ubiquitous in the environment, it is important to take all possible precautions to avoid nuclease contamination: purchase nuclease-free lab supplies (e.g., pipette tips, microcentrifuge tubes), use ultrapure water to make all buffers and solutions , and filter solutions through high protein-binding nitrocellulose filters.

2.1 Ligation Reaction

1. RNA oligonucleotides: Make a stock solution of the B oligonucleotide by resuspending to a final concentration of 500 μM in water. Check the concentration by measuring the A260 on a spectrophotometer. Do not resuspend the A oligonucleotide in water.
2. DNA (splint) oligonucleotides: Make a stock solution of the splint oligonucleotide by resuspending to a final concentration of 200 μM in water. Check the concentration by measuring the A260 on a spectrophotometer.
3. 2′ ACE deprotection buffer: 100 mM acetate adjusted to pH 3.8 with TEMED.
4. 10× T4 RNA ligase buffer: 50 mM Tris-HCl, pH 7.8, 10 mM $MgCl_2$, 10 mM DTT, 1 mM ATP.
5. T4 RNA ligase.

2.2 Analysis

1. Vertical gel system; glass gel plates, approximately 16 × 14 cm (*see* **Note 1**).
2. 20× Tris–Borate–EDTA (TBE) gel running buffer: 1.8 M Tris base, 1.8 M boric acid, 25 mM EDTA.

3. 7 M urea/12 % acrylamide: Mix 4.5 mL of 40 % (19:1) acrylamide, 750 μL 20× TBE, 6.3 g urea, and 5 mL water in a 50 mL glass beaker. Add a stir bar and stir on a magnetic stir plate until the urea has dissolved completely. Bring the volume to 15 mL with water (*see* **Note 2**).
4. Ammonium persulfate: 10 % solution in water (*see* **Note 3**).
5. *N,N,N′,N′*-tetramethylethylenediamine (TEMED).
6. Urea sample buffer: 4.2 g urea, 500 μL 20× TBE, 20 μL 0.5 M EDTA, 2.5 mg xylene cyanol, 2.5 mg bromophenol blue, water to 10 mL. Filter sterilize.
7. Ethidium Bromide: 10 mg/mL in water.
8. UV gel documentation system.

2.3 Product Purification

1. Formamide.
2. Fluor-coated thin layer chromatography plate.
3. Handheld UV light.
4. Disposable 1.5 mL microcentrifuge tube pestle.
5. DTR gel filtration cartridge (Edge BioSystems).
6. 20 mg/mL glycogen.
7. 3 M NaOAc.
8. 70 and 100 % EtOH.

3 Methods

3.1 Oligonucleotide Design

As RNA secondary structure may interfere with base pairing to the splint, it is helpful to place ligation junctions in single-stranded or loop regions. The use of small loops (left and right arrows, Fig. 2) as ligation junctions has not been extensively examined, but since the natural substrate for RNA ligase is such a loop, it is likely that they would work well. Under favorable conditions, e.g., with an extensive stem stabilizing the arrangement, a splint is not necessary [3, 9]. Large loops and single-stranded regions provide the other likely locations for junctions (Fig. 2, middle arrow, bottom, and top arrow).

Examine the predicted secondary structure of the desired, full-length RNA to identify loops that can serve as the junctions between RNA fragments (e.g., using mfold, [10]). In order to achieve the highest ligation yields possible, one should consider the sequence requirements of T4 RNA ligase when designing a ligation scheme. RNA ligase has been shown to have a slight preference for pyrimidines over purines at the 5′-terminal position of the donor, and ligation efficiency is highest if the last two nucleotides of the acceptor are not uridine—adenosine is best, followed by guanosine and cytidine [7, 11, 12]. Although we have found little or no difference in product yields based on sequence specificity

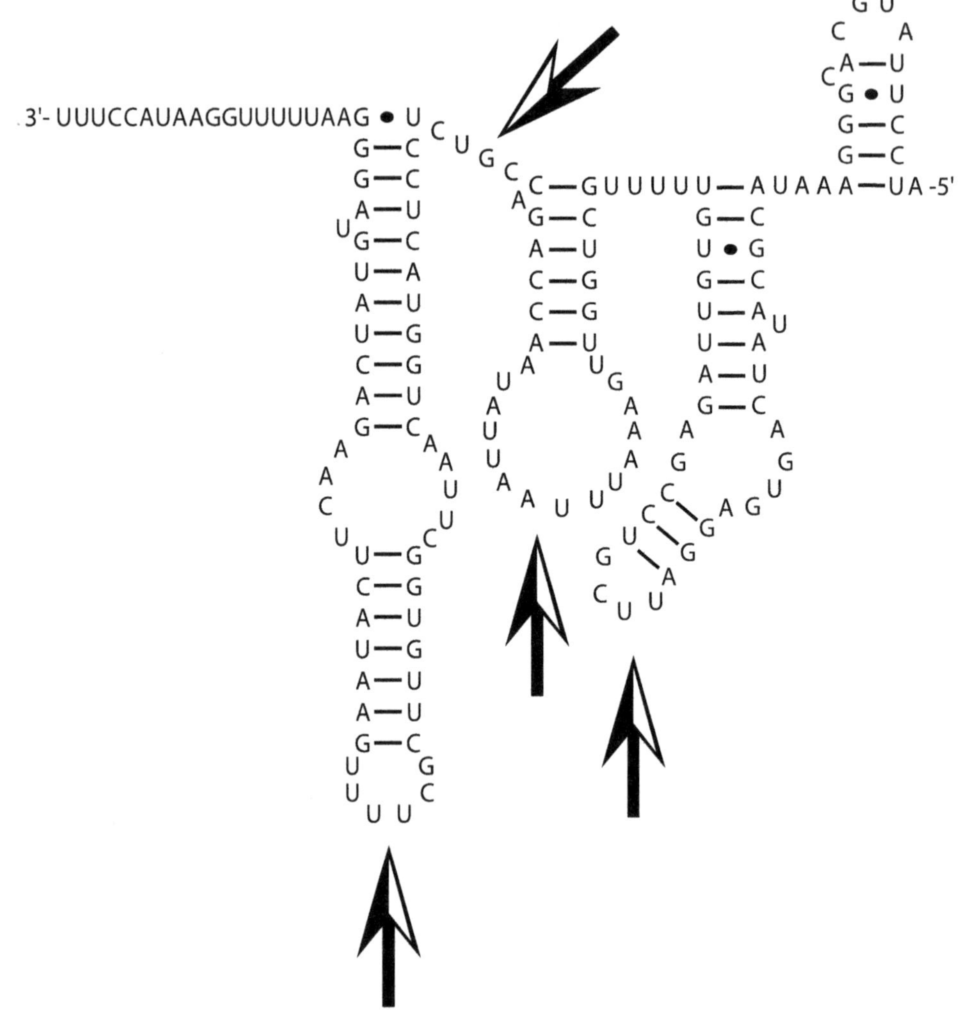

Fig. 2 Example of how to choose optimal ligation junctions for Rligation. *Bottom arrows* indicate loops, and *top arrow* indicates single-stranded region between secondary structure elements

in either the donor or acceptor substrates, it would be prudent to avoid junctions that contain uridines at the acceptor 3′-terminus whenever possible [8]. In addition, although the tRNA substrate of T4 RNA ligase has a single-stranded region of 5 nucleotides on the acceptor and 2 nucleotides on the donor, optimum lengths may vary for other sequences. The length of the donor loop seems to be most sensitive to changes in length, with ligation efficiency decreasing for lengths other than 1 or 2 nucleotides [8].

Current oligonucleotide synthesis limits RNA fragment length to approximately 100 nucleotides, but the limit is generally lower if modified nucleotides are included within the fragment, with the efficiency dropping off dramatically after approximately 50 nt [13]. Design all RNA molecules except the first acceptor (most 5′

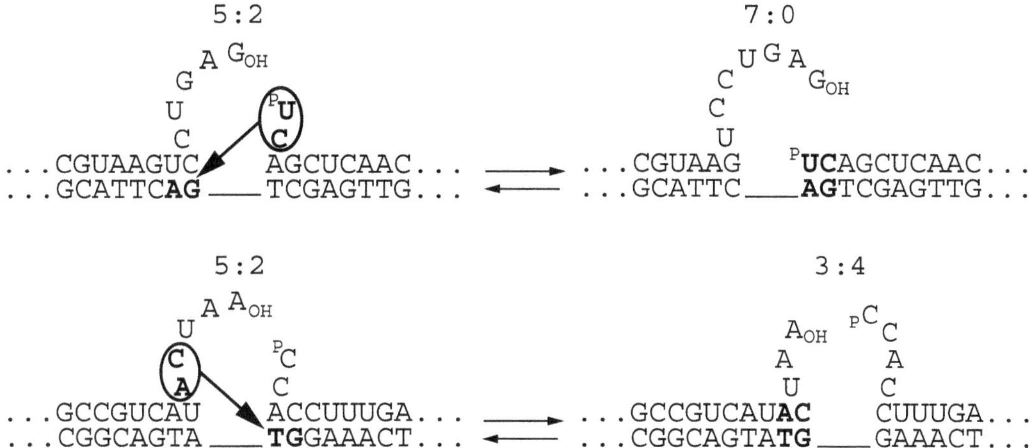

Fig. 3 Example of base-pairing ambiguity between splint and RNA acceptor and donor. In the top example, the 5′ end of the donor can base-pair to the splint, resulting in a 7:0 junction that is a poor substrate for RNA ligase. Similarly, in the bottom example, base pairing of the acceptor strand to the splint can extend further than desired, again resulting in suboptimal ligation efficiency

molecule) with a 5′-phosphate. This guarantees the fidelity of the ligation junction, as only full-length donor molecules will be ligated (any truncated products produced during synthesis are chemically capped and will not contain the necessary phosphate at the 5′ end).

3.2 Splint Design

Design the splint to have approximately equal binding to the A and B oligonucleotides. 18 nucleotides of complementarity work well, but it is important to calculate the T_m for each half to ensure they are comparable. T_ms of at least 40 °C yield good results. Close attention should be paid to ensure that the first two nucleotides of the donor molecule are not complementary to the last two nucleotides in the acceptor-binding region of the splint, which could result in the loss of the single-stranded linker on the donor. Similarly, if the first nucleotides in the acceptor single-stranded loop were complementary to the first nucleotides in the donor-binding region of the splint, then the number of single-stranded nucleotides on the acceptor side of the loop would decrease, while the number on the donor side would increase (Fig. 3).

3.3 Ligation

1. For a two-piece ligation, deprotect oligonucleotide A by resuspending in 400 μL 2′-ACE deprotection buffer. Pipet up and down to completely dissolve the RNA pellet, vortex 10 s, and then spin 10 s. Heat 30 min at 60 °C. Dry down in a SpeedVac at 55 °C. This will take over an hour. Resuspend in water to make a 500 μM stock solution. Determine the concentration by UV spectroscopy.
2. Prepare the polyacrylamide gel: To 15 mL of 12 % acrylamide/7 M urea, add 150 μL 10 % ammonium persulfate and

15 μL TEMED in a fume hood. Swirl to mix. Quickly fill the gel cassette with this acrylamide mixture, being careful to avoid and/or remove any bubbles. Insert the comb, ensuring that the acrylamide reaches all the way to the top of the notched plate. Leave to polymerize at ambient temperature for at least 1 h (*see* **Note 4**).

3. From the concentrated oligonucleotide stocks, make up 20 μM dilutions for the initial small-scale analysis.
4. Anneal the oligonucleotides by mixing them in a 1:1.5:2 molar ratio of A/splint/B. Oligonucleotide concentrations between 1 and 10 μM, depending on the size of the oligonucleotide, are usually sufficient for post-ligation visualization with EtBr (*see* **Note 5**). For example, for a 20 μL ligation reaction, mix 2 μL of 20 μM oligonucleotide A, 3 μL of 20 μM splint, and 4 μL of 20 μM oligonucleotide B in a 0.2 mL PCR tube. Add 2 μL 10× T4 RNA ligase buffer and 8.5 μL of water. Heat at 65 °C for 3 min, followed by 5 min at 25 °C in a thermocycler (*see* **Notes 6–8**).
5. Add 0.5 μL of 10 U/μL T4 RNA ligase. Mix thoroughly by flicking the bottom of the tube (do not vortex), and pulse briefly in a minicentrifuge to ensure the reaction mixture is at the bottom of the tube. Incubate at 37 °C.
6. The reaction can be completed within 5 min, but the appropriate incubation time must be determined empirically for each oligonucleotide/splint combination. From the 20 μL reaction above, remove 5 μL aliquots at 0, 5, 30, and 60 min.

3.4 Analysis

1. Stop the reactions by addition of 100 μL deprotection buffer, heat 30 min at 65 °C, and evaporate to dryness in a SpeedVac (*see* **Notes 9** and **10**).
2. While the sample is drying, pre-run the gel for 15–30 min at 400 V in 1× TBE. Before running, remove the comb and use a needle and syringe to rinse out each well with running buffer. Flush out any bubbles from the bottom of the gel.
3. Resuspend the sample in 10 μL urea sample buffer, and heat for 3 min at 65 °C. While the samples are heating, stop the gel and again rinse out each well. Load heated RNA samples in the prepared polyacrylamide gel.
4. Run the gel at 400 V until the smallest RNA in the sample is about 75 % of the way to the bottom of the gel, as judged by the positions of the bromophenol blue and xylene cyanol dyes—approximately 15 and 40 nt, respectively (see Sambrook [14] for dye migration in denaturing polyacrylamide gels of various percentages).

5. Remove the spacers and then the siliconized glass plate and carefully peel the gel off of the uncoated glass plate into a shallow dish containing 0.5 μg/mL ethidium bromide in water. Shake gently for 15–30 min.
6. Wearing gloves, lift gel out of the staining solution, rinse briefly in water, and place in a gel documentation apparatus equipped with a UV light. Quantitate ligation efficiency based on the amount of the limiting oligonucleotide (A) that has shifted into the A/B product. If efficiency is poor, modify the reaction conditions as outlined in the troubleshooting section.

3.5 Preparation

To produce ligated RNA products on a preparative scale, for subsequent use, scale up the reaction and purify the products as follows:

1. Combine 1–10 nmol of each RNA and DNA oligonucleotide in a minimal reaction volume with RNA ligase buffer (*see* **Note 11**). Denature and anneal as above. Add 5–10 U of RNA ligase/nmol phosphorylated 5′ ends (*see* **Note 12**).
2. Incubate the optimal length of time at 37 °C, as determined above in the small-scale reactions.
3. Stop by addition of an equal volume of formamide (NOT sample buffer with dyes), and then heat 3 min at 65 °C before loading on a pre-run gel, as above. Ensure that the wells are large enough to accommodate the entire sample volume in as few lanes as possible. Load a few μL of urea sample buffer in an empty lane to use as a marker.
4. Run the gel at 400 V until the smallest RNA is 75 % of the way down the gel.
5. Remove the top gel plate and cover the gel with plastic wrap. Flip the gel over and carefully release the gel from the bottom plate using one of the spacers. Fold the plastic wrap over the gel and place on a fluor-coated TLC plate. Using a handheld UV lamp on the short wavelength setting (254 nm), quickly identify the location in the gel of the desired RNA product by looking for the shadow cast by the RNA. Draw a box around the product on the plastic wrap using a permanent marker.
6. Move the wrapped gel onto a clean, scratch-proof surface. Using a sterile scalpel, cut the marked fragment of gel away from the remainder, peel off the plastic wrap, cut into several pieces if necessary, and place in a 1.5 mL microcentrifuge tube(s) (*see* **Note 13**).
7. Crush the gel slice using a disposable pestle. Add 400 μL water and crush some more. Elute the RNA from the gel fragment by heating at 70 °C for 10 min.

8. While heating, spin the DTR cartridge 3 min at 850 × *g*. Discard the flowthrough.
9. Pulse gel solution briefly in a microcentrifuge and then load entire slurry onto the pre-spun DTR cartridge. Spin in microcentrifuge 3 min at 850 × *g* to remove the acrylamide from the eluted RNA.
10. Precipitate the RNA by adding 15 μg glycogen, one tenth volume of 3 M NaOAc, and 2.5 volumes of ice-cold 100 % EtOH. Vortex to mix. Spin 30 min at max speed at 4 °C in microcentrifuge. Aspirate the supernatant, wash the pellet with cold 70 % EtOH, and spin again for 5 min. Aspirate the supernatant and allow the pellet to air-dry for 5 min.
11. If this fragment is to be used in a subsequent ligation step, deprotect as described above, combining all tubes from the gel elution into one (*see* **Note 14**).
12. Resuspend in a small volume of water (i.e., 20 μL), and determine the concentration by UV spectrophotometry.

3.6 Simultaneous Ligations

It is possible to carry out two or more ligation reactions simultaneously, but each combination of reactions must be independently optimized. For a three-way ligation of oligonucleotides A, B, and C, the most probable competing reaction is circularization of the B oligonucleotide, which is the only one that is both phosphorylated and deprotected. Consequently, the splints and other oligonucleotides must be used in excess to sequester the ends of B and minimize its auto-ligation.

1. Order oligonucleotides B and C with 5′ phosphate groups. Deprotect oligonucleotides A and B as described above in ligation **step 1**.
2. A good starting point is to set up the ligation as in **step 4** (small scale), with final concentrations of 2.0 μM oligonucleotide A, 1.5 μM splint 1 (for AB junction), 1.0 μM oligonucleotide B, 1.5 μM splint 2 (for BC junction), and 2.0 μM oligonucleotide C. Anneal and ligate as described above. Ratios of oligonucleotides and splints may need to be modified to obtain better yields of product ABC.
3. To assess the degree to which circularization of B is limiting product yield, carry out a control ligation with only oligonucleotide B. Run this on the gel next to the products of the ABC ligation to ascertain whether a substantial fraction of B in the three-way ligation is circularizing. If so, increase the concentration of the two splints.

3.7 Troubleshooting

When ligation efficiency is not satisfactory:

1. Alter the ratios of RNA oligonucleotides to splint. Other things that might improve product yield include increased ligation time, increase/decrease in ligation temperature, or cycling several times between 65 °C and the ligation temperature. Up to 25 % PEG 8000 can be added to the annealing reaction to increase molecular crowding, and 10–20 % DMSO can be added to disrupt RNA secondary structure. Check the annealed reactions on a native gel to see if the RNA is forming a complex with the splint. If most of the RNA is in the complex, proceed with the ligation.
2. Redesign the ligation junction in the RNA—change the placement of the junction and/or the number of single-stranded nucleotides in one or both sides of the loop. Increased splint length can disrupt secondary structure. Incorporation of the nucleotide purine into the middle of the splint can prevent a stable base-pairing interaction between donor- and acceptor-binding regions of the splint, thereby maintaining the desired single-stranded loop when it is not feasible to change the junction site (*see* Fig. 3).

4 Notes

1. Glass plates separate most reliably when only one of them is siliconized. Clean both plates carefully with dish soap and rinse extensively with deionized water. Allow to dry on a rack. Wipe with 70 % ethanol, apply nonstick solution (e.g., Gel Repel, Aardvark Science) according to the manufacturer's directions to the notched plate only, and buff again with 70 % ethanol. Mark the other side of the plate with a permanent marker. Repeat when notched plate no longer separates cleanly from the gel.
2. Make an acrylamide gel in the 8–15 % range depending on the sizes of the RNAs to be ligated.
3. Ammonium persulfate solution can be stored at 4 °C for up to 1 month.
4. The gel can be stored at 4 °C overnight. Place a wet paper towel over the comb and wrap the gel in plastic wrap prior to storage.
5. Approximately 200 ng of each oligonucleotide gives a strong band when stained with EtBr.
6. The molar ratios of oligonucleotides are designed to maximize incorporation of oligonucleotide A into the final product while minimizing side reactions. The splint is used at a molar excess relative to oligonucleotide A to ensure that all of oligonucleotide A is associated with splint and can therefore participate

in the desired reaction. Similarly, an excess of oligonucleotide B is used to ensure that all of the A/splint complexes can react productively with oligonucleotide B.

7. Annealing in the absence of buffer results in partial deprotection of oligonucleotide B, which increases the rate of side reactions such as circularization.

8. Anneal in the smallest reasonable volume to increase efficiency of RNA/oligonucleotide binding.

9. 2′ ACE-protected RNAs bind ethidium bromide poorly and are consequently difficult or impossible to observe by ethidium bromide staining. Oligonucleotides must therefore be deprotected prior to analysis or, alternatively, detected by another method, e.g., silver stain or Stains-All (Sigma). For an initial, quick screening of the ligation, one can forego deprotection, stopping the ligation instead by adding 10 μL of urea sample buffer, heating 3 min at 65 °C, and placing on ice. Oligonucleotide B will not be visible on the gel, but if the ligation was successful, there will be an obvious decrease in the amount of oligonucleotide A, as well as the appearance of the slower mobility A/B ligated product.

10. If the DNA splint is too close to the same size as one of the RNA reactants or products, it may be necessary to degrade the splint enzymatically prior to visualizing the RNA. To do this, add 1 unit of RNase-free DNase I (Invitrogen) at the end of the ligation reaction, and incubate a further 15 min at 37 °C. Stop the reaction and analyze the products as above. It is useful to run a control sample without DNase to confirm that the DNA oligonucleotide has indeed been degraded.

11. In order to purify the RNA by UV shadowing, at least 0.1 OD_{260} unit of ligated product is required.

12. 1 U of RNA ligase is defined as the amount of enzyme required to convert 1 nmol of 5′-[^{32}P]rA_{16} into a phosphatase-resistant form in 30 min at 37 °C (NEB).

13. Place a maximum of 0.1 g of gel in each tube.

14. If using 2′ ACE-protected oligonucleotides, the ligation reactions proceed most efficiently when performed in a 5′ to 3′ direction. For example, if three oligonucleotides, A, B, and C, are to be ligated, first ligate A and B, deprotect the AB product, and then ligate AB to C.

15. Post-synthetic modification of oligonucleotides prior to ligation may require deprotection. If oligonucleotide A needs to be deprotected before ligation, the overall yield of the ligation may be reduced due to circularization of A. Increasing the ratio of splint/A can decrease the likelihood of this unwanted side reaction (*see* Subheading 3.6).

Acknowledgments

This work was supported by NSERC Discovery Grant 298521 and UNBC Office of Research awards to SDR.

References

1. Dönmez G, Hartmuth K, Lührmann R (2004) Modified nucleotides at the 5′ end of human U2 snRNA are required for spliceosomal E-complex formation. RNA 10:1925–1933
2. Crawford DJ, Hoskins AA, Friedman LJ, Gelles J, Moore MJ (2008) Visualizing the splicing of single pre-mRNA molecules in whole cell extract. RNA 14:170–179
3. Kim I, Lukavsky PJ, Puglisi JD (2002) NMR study of 100 kDa HCV IRES RNA using segmental isotope labeling. J Am Chem Soc 124:9338–9339
4. Moore MJ, Sharp PA (1992) Site-specific modification of pre-mRNA: the 2′-hydroxyl groups at the splice sites. Science 256:992–997
5. Kurschat WC, Müller J, Wombacher R, Helm M (2005) Optimizing splinted ligation of highly structured small RNAs. RNA 11: 1909–1914
6. Amitsur M, Levitz R, Kaufmann G (1987) Bacteriophage T4 anticodon nuclease, polynucleotide kinase and RNA ligase reprocess the host lysine tRNA. EMBO J 6:2499–2503
7. Bain JD, Switzer C (1992) Regioselective ligation of oligoribonucleotides using DNA splints. Nucleic Acids Res 20:4372
8. Stark MR, Pleiss JA, Deras M, Scaringe SA, Rader SD (2006) An RNA ligase-mediated method for the efficient creation of large, synthetic RNAs. RNA 12:2014–2019
9. Höbartner C, Micura R (2004) Chemical synthesis of selenium-modified oligoribonucleotides and their enzymatic ligation leading to an U6 SnRNA stem-loop segment. J Am Chem Soc 126:1141–1149
10. Zuker M (2003) Mfold web server for nucleic acid folding and hybridization prediction. Nucleic Acids Res 31:3406–3415
11. Wittenberg WL, Uhlenbeck OC (1985) Specific replacement of functional groups of uridine-33 in yeast phenylalanine transfer ribonucleic acid. Biochemistry 24:2705–2712
12. Arn EA, Abelson JN (1996) The 2″-5″ RNA ligase of Escherichia coli. Purification, cloning, and genomic disruption. J Biol Chem 271: 31145–31153
13. Scaringe SA (2001) RNA oligonucleotide synthesis via 5″-silyl-2-″orthoester chemistry. Methods 23:206–217
14. Sambrook J, Russell DW (2001) Molecular cloning. Cold Spring Harbor Laboratory Press, Cold Spring Harbor, NY

Chapter 11

In Vitro Assay of Pre-mRNA Splicing in Mammalian Nuclear Extract

Maliheh Movassat, William F. Mueller, and Klemens J. Hertel

Abstract

The in vitro splicing assay is a valuable technique that can be used to study the mechanism and machinery involved in the splicing process. The ability to investigate various aspects of splicing and alternative splicing appears to be endless due to the flexibility of this assay. Here, we describe the tools and techniques necessary to carry out an in vitro splicing assay. Through the use of radiolabeled pre-mRNA and crude nuclear extract, spliced mRNAs can be purified and visualized by autoradiography for downstream analysis.

Key words In vitro splicing, Alternative splicing, Splicing analysis, Pre-mRNA substrate, HeLa cell nuclear extract, In vitro transcription, RNA extraction and purification

1 Introduction

The ability to study biochemical changes associated with pre-mRNA splicing in a cell-free-based assay, also referred to as the in vitro splicing assay, has vastly improved our understanding of this complex, key process of gene expression. Not only has it improved our knowledge of the mechanisms and necessary components involved in splicing, but it has also allowed insights into the regulation of alternative splicing as it is mediated by *cis*-acting elements and *trans*-acting factors.

The ease of use, flexibility, and rapid results provided by an in vitro splicing system allows for tailored investigations into various aspects of the splicing reaction. The major benefit, however, lies with the ability to biochemically manipulate the splicing reaction through utilizing two key components: (1) minigene constructs and (2) mammalian crude nuclear extracts. The use of minigene constructs is a common in vitro technique that employs genomic segments from a gene (introns and exons) that include alternatively spliced regions within flanking genomic regions that are cloned

Klemens J. Hertel (ed.), *Spliceosomal Pre-mRNA Splicing: Methods and Protocols*, Methods in Molecular Biology, vol. 1126, DOI 10.1007/978-1-62703-980-2_11, © Springer Science+Business Media, LLC 2014

downstream of efficient promoters. These minigene constructs allow for identification of specific features that control intron and exon usage as well as the characterization of *cis*-acting elements and *trans*-acting factors that interact and modulate regulatory elements necessary for splicing regulation [1]. Crude nuclear extract is another important component of the in vitro splicing reaction that is usually generated from HeLa cells. Importantly, these nuclear extracts contain the necessary proteins and snRNAs for an efficient splicing reaction (*see* Chapter 8). The advantage associated with in vitro biochemical manipulation allows for insights into various factors and processes. These include, but are not limited to, protein regulatory elements and composition, splice site recognition and selection, the influence of RNA elements and their *trans*-acting factors, the characterization of enhancer and silencer elements, and kinetic insights into the splicing pathway. As with all in vitro-based systems, the assay does come with limitations. The rate of intron removal in vitro is slower than rates determined in vivo [2]. The efficiency of in vitro transcription of pre-mRNA, its purification, and subsequent splicing is restricted by the size of the RNA to be used; RNA should be less than 2,000bp [3]. Because of this, the in vitro splicing assay relies heavily on the use of shorter minigenes that are only a subset of a larger gene. The assay also does not take into account the effects of other events associated with splicing, such as transcription, capping, and polyadenylation.

Methods for in vitro splicing reactions have previously been described [4–7]. In general, these protocols employ the use of radiolabeled pre-mRNAs that are incubated for several hours in nuclear extract supplemented with necessary salts and cofactors. The mRNA is then extracted and purified from the nuclear extract, subjected to denaturation on a polyacrylamide gel, and subsequently dried for visualization by autoradiography via film or phosphor imaging. The pre-mRNA, mRNA, and other intermediates are then identified as bands on the autoradiograph.

2 Materials

All reagents should be high quality, molecular biology grade, and RNase-free. Stock solutions should be stored at 4 °C (unless otherwise indicated). Certain reagents can be substituted for their equivalents from other manufacturers or as otherwise stated. The concentrations of chemicals/reagents listed in the materials are stock concentrations, not final concentrations. Since all steps require working with radioactive isotopes, all necessary precautions must be taken. Carefully follow all hazardous and radioactive waste disposal regulations when disposing of waste materials.

2.1 Splicing Reaction Components

1. Radiolabeled pre-mRNA: generated from an in vitro transcription reaction (*see* **Note 1**).
2. Splicing competent nuclear extract (NE) (*see* Chapter 8).
3. 1 mM adenosine triphosphate (ATP). Store at −20 °C.
4. 0.5 M creatine phosphate (CP). Store at −20 °C.
5. 80 mM magnesium acetate ($Mg(OAc)_2$) (*see* **Note 2**).
6. RNase inhibitor (40 U/μl). Store at −20 °C.
7. 100 mM dithiothreitol (DTT). Store at −20 °C.
8. 1 M potassium acetate (KOAc) (*see* **Note 3**).
9. 0.5 M HEPES buffer, pH 7.9 (*see* **Note 3**).
10. 13 % polyvinyl alcohol (PVA): optional (*see* **Note 4**).
11. Wet ice and dry ice (finely ground or small chunks).
12. Water bath.

2.2 6 % Splicing Gel Components

1. Tris-Borate-EDTA (TBE) buffer: 89 mM Tris Base, 89 mM boric acid, 2 mM EDTA.
2. 7 M urea.
3. 40 % (19:1) acrylamide:bis-acrylamide solution: acrylamide is dissolved in 1× TBE/7 M urea.
4. *N,N,N′,N′*-Tetramethylethylenediamine (TEMED).
5. 10 % Ammonium persulfate (APS).
6. Formamide/EDTA stop dye: formamide with 0.1 % bromophenol blue and 0.1 % xylene cyanol and 2 mM EDTA.
7. Radiolabeled RNA ladder/molecular marker.
8. Electrophoresis glass plates: 8″ × 8″ (two): one glass plate should notch to allow for the addition of a comb.
9. 0.4 mm gel plate spacers (three).
10. 0.4 mm comb (same thickness as the spacers).
11. 1¼″ binder clips (four).
12. Aluminum plate: 8″ × 8″ or longer and precooled (*see* **Note 5**).
13. Silicon Gel Slick® Solution (Lonza Rockland) or equivalent.
14. 70 % ethanol.
15. 30–50 ml syringe, with and without a needle (two).
16. Flat gel loading tips.
17. Putty knife/gel spatula.
18. Vertical gel electrophoresis system.
19. Whatman paper, cut into an 8″ × 8″ square.
20. Plastic wrap (such as Saran™ Wrap).
21. Power pack for an electrophoresis system with a temperature probe.

22. Bio-Rad Gel Dryer or equivalent.
23. Bio-Rad Personal Molecular PhosphorImager System or similar. Film may also be used.

2.3 Splicing Digest and RNA Purification

1. Proteinase K 10 mg/ml.
2. 2× Proteinase K buffer: 20 mM Tris Base, 2 % SDS, 200 mM NaCl, 2 mM EDTA, pH 7.5.
3. 100 % ethanol.
4. Glycogen.
5. Phenol, chloroform, isoamyl alcohol solution (25:24:1 pH 8.0).

3 Methods

Carry out all steps of the reaction on ice unless otherwise stated.

3.1 Splicing Reaction

1. Thaw NE on ice.
2. Thaw ATP, CP, $Mg(OAc)_2$, DTT, HEPES, KOAc, and radio-labeled RNA at room temperature. Once thawed, place them immediately on ice. RNase inhibitor should be kept on ice.
3. Determine the Master Mix reaction volume and reaction size (*see* **Note 6**):
   (a) (# of reactions) + 1 = Master Mix reaction size.
   (b) Reaction volume: 12.5 μl or 25 μl reaction volume total.
4. Mix reagents to a final concentration of 1 mM ATP, 20 mM CP, 3.2 mM $Mg(OAc)_2$, 10 U RNase inhibitor, 1 mM DTT, 10–50 % NE (should be optimized for each extract and substrate used), 72.5 mM KOAc, 12 mM HEPES (*see* **Note 3**), and 3 % PVA (optional). Use sterile water to bring up the Master Mix volume if needed.
5. For each experimental reaction condition: add the appropriate Master Mix volume, 0.01–0.1 nM RNA (~1,000 cpm) (*see* **Note 7**), experimental variant (i.e., protein), and/or sterile water to bring up the volume. Add NE last and pipet carefully to mix (*see* **Note 8**). Keep all reaction tubes on ice. Prepare a time 0 tube as control and immediately place on dry ice after addition of NE (*see* **Note 9**).
6. Incubate all reactions, except the time 0 reaction, at 30 °C water bath for 90 min (*see* **Note 10**).
7. While the splicing reactions are running, prepare the 6 % acrylamide gel.
8. Once the incubation time is complete, immediately place the tubes on dry ice to stop the reactions (*see* **Note 11**).

3.2 Splicing Gel Preparation

1. Prepare a 20 % acrylamide:bis solution: dilute 40 % (19:1) acrylamide:bis-acrylamide solution in 1× TBE with 7 M urea.

2. Prepare a 6 % polyacrylamide mixture from the 20 % acrylamide solution: in a 50 ml conical tube, dilute the 20 % acrylamide solution with the desired amount of 1× TBE/7 M urea buffer to obtain a mixture at the required percentage.
3. Carefully clean the inside face of a siliconized plate (*see* **Note 12**) with 70 % ethanol. Wipe dry with lint-free paper towels.
4. Carefully clean the non-siliconized plate using water and 70 % ethanol. Make sure the gel plates are completely clean, with no small pieces of debris present (*see* **Note 13**). Wipe dry with lint-free paper towels.
5. Place the spacers around the outside edge (bottoms and sides) of a non-siliconized plate. Lay the siliconized notched plate on top and clip the glass plates together using binder clips.
6. Once the gel cassette is ready, add the appropriate amount of 10 % APS and TEMED (*see* **Note 14**) to 20 ml of 6 % acrylamide and mix gently.
7. Using a syringe (without needle), aspirate the acrylamide and gently dispense the mixture between the plates. Once the cassette is filled, lay it flat, place a gel comb with an appropriate well size into the top of the gel, and allow the gel to set at room temperature for approximately 30 min (or until polymerized) (*see* **Note 15**).
8. Pre-run the gel before adding your samples (*see* **Note 16**): clamp the gel cassette onto the vertical gel electrophoresis apparatus, fill the chambers with 1× TBE (*see* **Note 17**), and run the gel at 30 W (100 V), 45 °C, for 15 min.

3.3 Digest

1. Once the last in vitro splicing tube has been placed on dry ice, prepare the Proteinase K digest mix:
   (a) Determine the desired final volume of the Proteinase K Master Mix: reaction volume (μl) × (# of reactions + 1) = Master Mix volume (μl).
   (b) Proteinase K Master Mix: final concentration of 1× Proteinase K buffer, 0.25 mg/ml glycogen, 0.25 mg/ml Proteinase K, and sterile water, for a final volume of 180 μl per reaction.
2. Add 175 μl of Proteinase K Master Mix (*see* **Note 18**) to each reaction tube and incubate at 37 °C for 10–15 min.

3.4 RNA Purification and Precipitation

1. Once the Proteinase K digest has completed, purify the RNA by adding 200 μl of phenol/chloroform, vortex for 30 s, and spin at 16,500 × *g* for 5 min to separate the aqueous and organic layers.
2. To precipitate the RNA: remove the aqueous (top) phase and place into a separate tube (~200 μl). Add 2.5 times the volume of 100 % ice-cold ethanol (for 200 μl of top phase, add 500 μl of ethanol). Incubate at −20 °C or −80 °C for 10–15 min.

3. Centrifuge the tubes at 16,500 × *g* for 10 min at room temperature to pellet.
4. Remove the ethanol supernatant and allow the pellet to air-dry for no more than 5 min (*see* **Note 19**). Resuspend the pellet in a small amount of stop dye within 5–10 min (10 μl or less). Pipet up and down and vortex for 30 s to mix.

3.5 Visualization of Splicing Reaction

1. Load RNA samples onto the pre-run gel. Clamp an aluminum plate to the front glass plate (*see* **Note 20**). Run the gel at 30 W (100 V), 45 °C, for 90 min or until the dye runs off the gel.
2. Remove the gel cassette from the apparatus and dispose of the buffers in appropriate waste containers. Split the plates apart with a putty knife/spatula. The gel should remain attached to the non-siliconized plate.
3. Center the pre-cut Whatman paper on top of the gel and press gently to allow the gel to adhere evenly to the paper. Carefully peel the Whatman paper upward at an angle to allow for the gel to be peeled away from the glass. Cover the gel with plastic wrap, minimizing the presence of any creases (*see* **Note 21**).
4. Dry the gel for 15–20 min using a Bio-Rad Gel Dryer at 80 °C with suction.
5. Expose the gel to film or preferably a phosphor imaging screen (*see* **Note 22**) or similar equipment for the recommended length of time (generally at least 3 h to overnight; *see* **Note 23**).
6. Once the gel has been exposed and imaged, the appearance of spliced product can be used to determine the amount of RNA spliced (% spliced) in each lane (Fig. 1), which can in turn be used to calculate the efficiency of product appearance (*see* Chapter 12 and **Note 24**). Use a suitable computer program to analyze the digital quantitation file (*see* **Note 25**).

4 Notes

1. Generally, for in vitro splicing reactions, DNA is transcribed using T7 polymerase in a reaction containing radiolabeled nucleotides, phosphorus-32 (^{32}P) α-UTP. This reagent is usually in the 0.3–3 nM range, with an incorporation of around 100,000 cpm/μl.
2. Magnesium chloride can also be used; however, chloride has in some cases been shown to inhibit in vitro splicing reactions [8].
3. Potassium chloride or potassium glutamate may also be used although chloride has been shown to inhibit in vitro splicing reactions [8]. KOAc is used in this reaction because the nuclear extract has been prepared in KOAc (*see* Chapter 8). The final volume of KOAc to add to the Master Mix will depend on

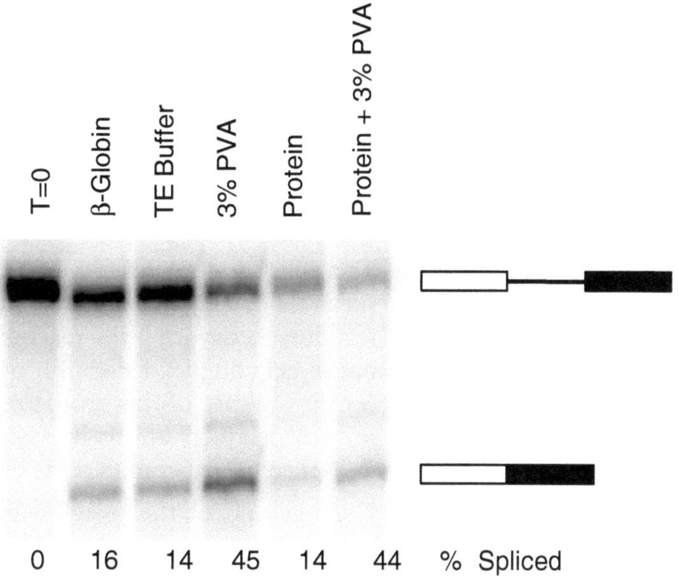

Fig. 1 Autoradiogram of radiolabeled β-globin minigene construct (from *left* to *right*) at time 0, alone, with TE buffer, with 3 % PVA, with 1 μM protein X, with 1 μM protein X with 3 % PVA, run on a 6 % polyacrylamide gel. Analysis of % spliced is performed using Bio-Rad Quantity One (*see* **Note 25**)

how many ions are present in the nuclear extract to begin with. The final volume of HEPES to add will also depend on how many ions are present in the nuclear extract.

4. Addition of PVA is optional but has been shown to potentially increase splicing efficiency in certain reactions [9].

5. Monitor the temperature of the gel using a temperature probe connected to the power pack. It is highly recommended to place a precooled aluminum plate aluminum plate on the front surface the front surface of the gel cassette to keep the cassette cool and prevent it from shattering as well as evenly distribute heat (*see* **Note 19** as well).

6. The extra reaction is to account for pipetting errors. When determining the reaction volume (12.5 μl or 25 μl reaction), consider how many reactions are needed, how much radiolabeled pre-RNA is present, and how radioactive the radiolabeled pre-mRNA is. If the radiolabeled pre-RNA is less than 4,000 cpm/μl, a 12.5 μl reaction may be appropriate with the addition of more pre-mRNA.

7. It is possible to add radiolabeled pre-mRNA to the Master Mix, rather than adding it separately.

8. When adding NE, make sure to prevent any air bubbles from forming. Mix gently by pipetting up and down, and *do not vortex*. Excessive bubbles may reduce splicing efficiency.

9. A time 0 tube should be prepared as a control. Once NE is added to the reaction tube, immediately place the tube on dry ice to prevent the splicing reaction from starting. This time 0 control treatment will be used to adjust for background intensity associated with un-spliced product for all the reactions.
10. The optimal temperature for cleavage at the 5′ splice site is 30 °C [10].
11. The splicing complexes formed on the pre-mRNA will not survive a dry ice freeze/thaw cycle. Therefore, only place the reaction on dry ice if the reaction will not be used to visualize native gel complex formation or for other downstream analyses. In the case of this protocol, only the spliced radiolabeled mRNA products are to be visualized. Therefore, destroying the spliceosomal complexes is not an issue.
12. Coating one of the plates with silicon is not required but highly recommended. A siliconized gel plate allows for easier separation when separating the glass plates. The gel will almost always stick to the uncoated plate, instead of partially sticking to both. Preferably, the notched plate should be siliconized.
13. Both the siliconized and non-siliconized plates should be free of any sort of particles and debris. Make sure to wipe away any debris, as they will form tiny air pockets between the glass plates that will cause leakage when pouring the gel.
14. Altering the amount of APS and TEMED can have different effects on gel polymerization and on how the samples run on the gel [11–13]. Generally a 1:150 dilution of 10 % APS and 1:1,000 dilution of TEMED are used.
15. Avoid the formation of bubbles while making the gel. Hold the clipped gel cassette (with notched plate facing upward) in one hand at a 45° angle, tilted on its corner. Slowly dispense the acrylamide solution. If an air bubble is present, adjust the angle of the cassette to allow the solution to force the air bubble outward. Add the comb immediately before the gel solution has time to harden.
16. Pre-running the gel before adding samples can remove all traces of (APS) and will apply a constant temperature to the gel before use [14].
17. Immediately before loading samples, make sure to flush out the wells with buffer to remove any urea that has leached and deposited into the wells.
18. The spliced RNA solution will be frozen when adding the Proteinase K Master Mix. Pipet the Master Mix up and down slowly in the reaction tube to thaw the spliced RNA. *do not vortex.*
19. Keep track of the orientation of the tubes while centrifuging; the pellet will be very hard to see and sometimes invisible.

If a pellet is not visible, continue to add the stop dye and load samples onto the gel (it is most likely there as the dye will stick to the pellet).

20. As mentioned previously, to prevent the glass plates from cracking and to ensure even conduction of heat, clamp a pre-cooled aluminum plate to the front glass plate using the same binder clips used to hold the gel cassette in place. Make sure the aluminum plate is positioned so that it does not touch any buffer in the lower chamber. Run the gel for an appropriate amount of time; this will differ depending on the splicing products of your reaction and the percent/mix of the gel poured.
21. Make sure there are no creases in the plastic wrap. Remove any extra overhanging plastic wrap using a razor, being careful not to slice the gel. Any extra plastic wrap will bulge and may prevent the gel from being flush with the phosphor imaging screen or film.
22. The PhosphorImager screen is a form of autoradiography that is used to visualize and detect radioactive emission from radiolabeled RNA. Phosphor imaging screens contain BaFBR:Eu^{2+} crystals. When these crystals are exposed to ionizing radiation from radiolabeled RNA, electrons from Eu^{2+} become excited resulting in subsequent oxidation. During screening, the oxidized electrons revert back releasing a photon that can then be detected at certain wavelengths via a photomultiplier system producing a quantitative image [15]. There are many advantages to this method over other methods such as film. These advantages include increased sensitivity over a linear detection range of 5 orders of magnitude, while exposure to film is limited to only 1.5 orders of magnitude, increased exposure time from 10 to 250 times faster than film, easier and faster quantitation of images, and reuse of the phosphor screens indefinitely [16]. Other molecular detection systems similar to the Bio-Rad Molecular Imager are also available.
23. If the radiolabeled pre-mRNA used for the splicing reaction is around 8,000 cpm/μl, 1 h exposure to the PhosphorImager screen or film is sufficient to observe most splicing; however, longer exposures are often needed to see all splicing products or intermediates.
24. Due to the differential rates of decay among some splicing products, not all bands may be suitable for quantification. Depending on the in vitro reaction, lariat formation may be more stable than certain products and can be used as a substitute for calculating % spliced [17]. In addition, certain products may form which will not necessarily be stable in the cell (such as single exons). These RNAs will be degraded in the cell but may persist in an in vitro reaction.

25. % spliced product is obtained by calculating the volume intensity from the digital image of the splicing gel for each band in each lane. Briefly, calculate the sum of the adjusted intensity (taking into account the background of the gel and time 0 reaction) for the spliced and un-spliced band, divide the signal for spliced product by the total signal in the lane, and take the percent:

$$\%\ \text{Spliced} = \frac{\text{Signal from final spliced product}}{\text{Total signal in lane}} \times 100$$

$$(\text{Total signal in lane} = \text{spliced product} + \text{un-spliced product})$$

Acknowledgment

Research in the Hertel laboratory is supported by NIH (RO1 GM62287 and R21 CA149548).

References

1. Cooper TA (2005) Use of minigene systems to dissect alternative splicing elements. Methods 37:331–340
2. Beyer AL, Osheim YN (1988) Splice site selection, rate of splicing, and alternative splicing on nascent transcripts. Genes Dev 2:754–765
3. Hicks MJ, Lam BJ, Hertel KJ (2005) Analyzing mechanisms of alternative pre-mRNA splicing using in vitro splicing assays. Methods 37:306–313
4. Hernandez N, Keller W (1983) Splicing of in vitro synthesized messenger RNA precursors in HeLa cell extracts. Cell 35:89–99
5. Padgett RA, Hardy SF, Sharp PA (1983) Splicing of adenovirus RNA in a cell-free transcription system. Proc Natl Acad Sci USA 80:5230–5234
6. Hardy SF, Grabowski PJ, Padgett RA et al (1984) Cofactor requirements of splicing of purified messenger RNA precursors. Nature 308:375–377
7. Lee KA, Bindereif A, Green MR (1988) A small-scale procedure for preparation of nuclear extracts that support efficient transcription and pre-mRNA splicing. Gene Anal Tech 5:22–31
8. Reichert V, Moore MJ (2000) Better conditions for mammalian in vitro splicing provided by acetate and glutamate as potassium counterions. Nucleic Acids Res 28:416–423
9. Krainer AR, Maniatis T, Ruskin B, Green MR (1984) Normal and mutant human β-globin pre-mRNAs are faithfully and efficiently spliced in vitro. Cell 36:993–1005
10. Furdon PJ, Kole R (1986) Inhibition of splicing but not cleavage at the 5′ splice site by truncating human beta-globin pre-mRNA. Proc Natl Acad Sci USA 83:927–931
11. Dirksen ML, Chrambach A (1972) Studies on the redox state in polyacrylamide gels. Separ Sci 7:747–772
12. Gelfi C, Righetti PG (1981) Polymerization kinetics of polyacrylamide gels. I. Effect of different cross-linkers. Electrophoresis 2: 213–219
13. Righetti PG, Gelfi C, Bosisio AB (1981) Polymerization kinetics of polyacrylamide gels. III. Effect of catalysts. Electrophoresis 2:291–295
14. Rio DC, Ares M, Hannon GJ, Nilsen TW (2010) Polyacrylamide gel electrophoresis of RNA. Cold Spring Harb Protoc 2010:1–6
15. Voytas D, Ke N (1999) Current protocols in molecular biology – detection and quantitation of radiolabeled proteins and DNA in gels and blots. Curr Protoc Mol Biol 48:A.3A.1–A.3A.10
16. Johnston RF, Pickett SC, Barker DL (1990) Autoradiography using storage phosphor technology. Electrophoresis 11:355–360
17. Kotlajich MV, Crabb TL, Hertel KJ (2009) Spliceosome assembly pathways for different types of alternative splicing converge during commitment to splice site pairing in the A complex. Mol Cell Biol 29:1072–1082

Chapter 12

Kinetic Analysis of In Vitro Pre-mRNA Splicing in HeLa Nuclear Extract

William F. Mueller and Klemens J. Hertel

Abstract

Kinetic analysis of in vitro splicing is a valuable technique for understanding splicing regulation. It allows the determination of specific contributions from functional elements for the efficient removal of introns. This chapter will describe the rationale and approach employed to use kinetic analysis to evaluate an in vitro splicing reaction using radiolabeled pre-mRNA incubated in splicing-competent HeLa nuclear extract (NE).

Key words Splicing, Kinetics, In vitro splicing, Splicing rates, Alternative splicing

1 Introduction

In vitro splicing assays have been used to reliably discover new aspects of alternative splicing for many years [1–10]. The ability to manipulate the biochemical system where splicing reactions take place has illuminated the steps of the reaction, the molecular machinery required, and their regulation as splicing occurs. Although cell transfection and subsequent analysis are closer physiologically to a regulated splicing event, they lack the experimental flexibility of the in vitro system. That flexibility allows the study of specific RNA elements, trans-acting factors, and their unique effects that are otherwise difficult to determine in vivo or in cell culture.

There are differences between in vitro and cell culture splicing experiments. The rate of splicing in vitro is much slower than that which occurs in a cell [7, 11]. In vitro splicing occurs without nuclear compartmentalization, allowing the splicing machinery to be decoupled from the transcriptional machinery. While this permits characterization of specific sequence elements influence on splicing, it probably contributes to the decrease in the splicing rate in vitro as spliceosomal recruitment occurs co-transcriptionally in cells [12, 13]. Despite this drawback, the study of *cis-* and *trans-acting* splicing regulatory elements has shown that their actions in cell transfection and in vitro experiments yield parallel outcomes [4, 14].

Klemens J. Hertel (ed.), *Spliceosomal Pre-mRNA Splicing: Methods and Protocols*, Methods in Molecular Biology, vol. 1126, DOI 10.1007/978-1-62703-980-2_12,

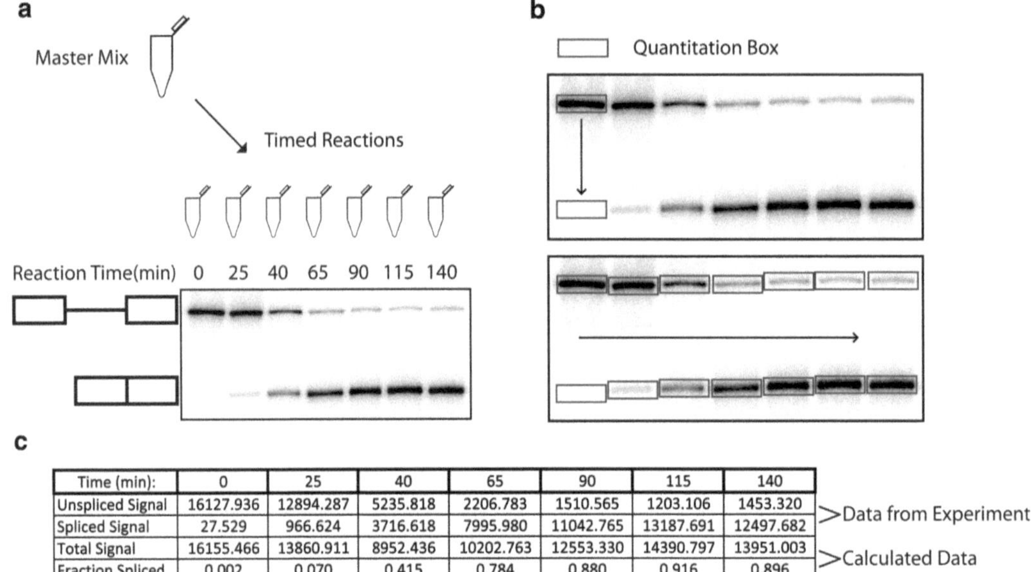

Time (min):	0	25	40	65	90	115	140
Unspliced Signal	16127.936	12894.287	5235.818	2206.783	1510.565	1203.106	1453.320
Spliced Signal	27.529	966.624	3716.618	7995.980	11042.765	13187.691	12497.682
Total Signal	16155.466	13860.911	8952.436	10202.763	12553.330	14390.797	13951.003
Fraction Spliced	0.002	0.070	0.415	0.784	0.880	0.916	0.896

Fig. 1 Analysis and quantitation of the gel scan. (**a**) The timed reactions run out on the gel allow observation of splicing over time. This is observed as the shift in band intensity from the higher pre-mRNA band to the lower spliced RNA band. Cartoon at left depicts the spliced and unspliced RNAs. (**b**) Quantitation boxes should be put around the bands as depicted. (**c**) The values (adjusted for background) for each time point/lane are found in the table. Total signal is the addition of the spliced and unspliced values. Fraction spliced is calculated per time point as the spliced value divided by the total signal. This is then plotted against time and fit with the rate equation to determine a rate constant

The kinetic analysis of in vitro splicing takes advantage of the small amount of pre-mRNA substrate required and the excess of splicing factors contained within the nuclear extract. This allows easy calculation of a rate constant using a pseudo-first-order rate approximation. When following along the time course of a splicing reaction, the first appearance of spliced product can be delayed [7]. This product appearance lag seems to be dependent on the efficiency of intron removal. Reactions that are less efficient or substrates that contain weaker splicing signals typically display longer lags. Once the reaction has proceeded past the lag phase, it enters the linear phase in which it exhibits reliable product appearance until the endpoint of the reaction is reached. That appearance of product can be measured and then fit to the first-order reaction model for the formation of spliced product: $A = C \times (1 - e^{-kt})$ where A is the fraction spliced, C is the fraction spliced at the endpoint of the reaction, k is the apparent rate constant, and t is time from the end of the lag period (*see* Figs. 1 and 2). This equation is a derivative of the standard reactant decay description $A = A_0 \times e^{-kt}$. This rate equation describes an ideal reaction scenario where product formation initiates immediately and reaches 100 % completion. In practice, this is generally not the case so some fraction of the

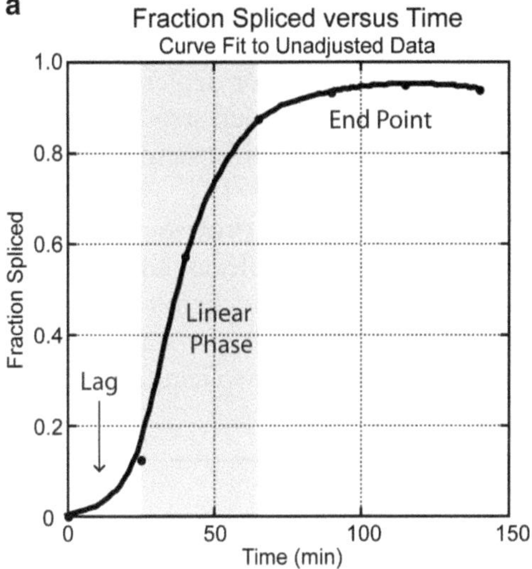

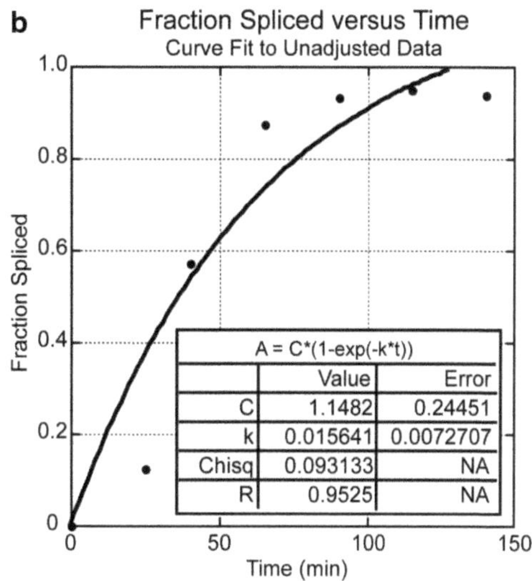

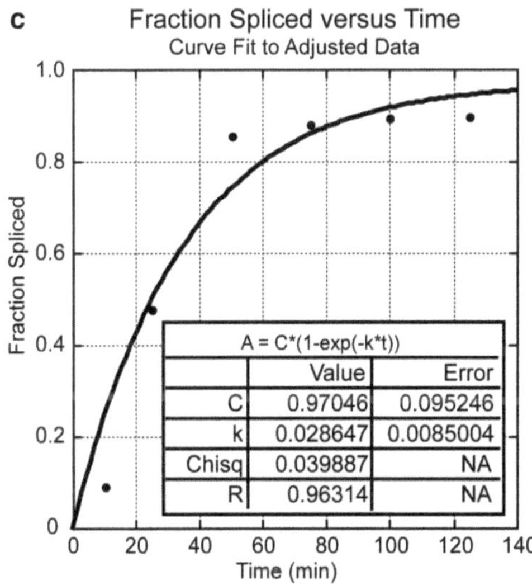

Fig. 2 Plotting and analysis of splicing data. (**a**) The data from Fig. 1c was plotted and fit with a smooth curve to depict the phases of the reaction. There is a slight lag, followed by the linear phase, followed by the endpoint phase. Due to the lack of change between the last three time points, we determine that the endpoint has been reached. (**b**) The data is then fit with a curve following the first-order rate equation. The equation is meant for an ideal reaction and thus gives a final value for *C* that is greater than 1. (**c**) Plotted values after adjusting for time. With the time adjustment, the calculated value for *C* is less than 1 suggesting this value is more accurate. This brings the splicing rate for this reaction to 0.029/min

pre-mRNA supplied may never be used as a splicing substrate. Fluctuations in lag time and overall fractions spliced will differ between NE preparations and pre-mRNA substrates.

The difference in kinetics between experimental treatments of splicing reactions has allowed the discovery of multiple regulatory mechanisms. Here, we describe the approach to carry out a basic kinetic analysis of an in vitro splicing reaction. However, this protocol can be altered to determine other aspects of intron removal. For example, through the addition or depletion of regulatory proteins, one can determine during which step of splicing a specific action is taking place or when a specific factor influences the outcome of splice site recognition [14–17]. Furthermore, slight alterations in the gel type and reaction processing allow the visualization of different spliceosomal complexes and, thus, an analysis of their assembly kinetics [6, 14]. Biochemical tricks can be used to stall spliceosomal assembly at various stages thereby permitting further insights into the kinetics of splicing [6, 18, 19].

2 Materials

The in vitro splicing reaction is described in detail in Chapter 11. To obtain sufficient data point to carry out a kinetic analysis, multiple splicing reactions of identical composition need to be set up. They will be the same as are required to test kinetics of the in vitro splicing reaction.

1. Reagents as in Chapter 11, but sufficient amounts for multiple reactions.
2. Imaging system to quantitate signal from the labeled RNA off of the gel (such as Bio-Rad PhosphorImager).
3. Computer with suitable analysis and graphing software that allows line fitting to an input equation (such as Bio-Rad Quantity One in combination with KaleidaGraph™).

3 Methods

Carry out all steps of the reaction on ice unless otherwise stated.

1. Thaw reagents as in in vitro splicing reaction (Chapter 11).
2. Determine time points needed and the Master Mix (MM) reaction volume: (# of time points + 1) × volume per reaction = MM volume (*see* **Note 1**).
3. Mix reagents as in Chapter 11.
4. Reactions should be aliquoted into separate tubes of equal volume and kept on wet ice. This means that there will be one MM tube and one tube for each time point chosen, for example, if the time course is for 2+ h with 7 time points, the MM tube will be divided into tubes to be incubated for 0, 25, 40, 65, 90, 115, and 140 min. Place the time 0 tube on dry ice immediately after mixing the reaction to prevent the reaction from progressing.

5. Incubate reactions at 30 °C for the desired length of time for each time point. Most in vitro splicing reactions should reach completion within ~2 h.
6. Once reactions have reached their time point, place on dry ice to stop the reaction from progressing, i.e., after 25 min, the 25 min time point tube should be removed from the water bath and placed on dry ice. To freeze the tube and stop the reaction quickly, crush the dry ice into powder so the tube can be immersed.
7. Once the reactions have finished, prepare the samples, and load and run them on the gel as in Chapter 11.
8. Expose the gel to film or preferably a PhosphorImager screen or similar equipment for 1 h to overnight, depending on the radioactivity of the pre-mRNA used (*see* Chapter 11).
9. Once the screen has been exposed, scan the screen for quantitation and subsequent use in the analysis software. The appearance of spliced product on the scan can be observed by a decrease over time in the full length unspliced pre-mRNA and an accompanying increase over time in the correctly sized product band (*see* Fig. 1a).
10. Determine the level of signal for the bands in each lane on the scan pertaining to fully spliced products and the unspliced pre-mRNA. This should be done using suitable quantification software compatible with your imaging system (such as Bio-Rad Quantity One). Using the box quantification tool, a small box containing the largest band should be drawn. A copy of this box should be made and then used to quantify all other desired bands (*see* Fig. 1b and **Note 2**). The signal levels can then be used to calculate the total spliced signal in each lane. Do this by adding all the values determined for each lane/time point together, spliced products, as well as the unspliced band.
11. Determine the fraction spliced for each time point. The fraction spliced is the signal from the spliced product (or products) divided by the total amount of signal within a lane. This value should increase in an efficient splicing reaction as time progresses such that in the final time points there is very little change in the last values (signifying the endpoint has been reached). There is no loading control so comparison between lanes is not useful. Computing this value allows one to observe the changes in splicing that occur without trying to compare between lanes.

    Fraction spliced = Signal from spliced products/total signal from **step 10** (*see* Fig. 1c).
12. The fraction of spliced product and time data should be plotted as the time (*x*-axis) vs. fraction spliced (*y*-axis) for each time point taken. This plot should have at least two parts: a linear

increase to a point (linear phase) that then levels to a plateau or asymptote (endpoint phase; *see* Fig. 2a). Reaching the endpoint of the reaction is important for the proper determination of the rate constant (*see* **Note 3**). There may also be a lag before the linear phase of splicing occurs (*see* **Note 4**). This is due to the competition in the NE for the pre-mRNA by multiple groups of proteins that may impede splicing complex formation.

13. Determine the observed rate of splicing by fitting the data points to an equation that describes first-order rate kinetics (*see* Fig. 2b). This is appropriate because the splicing reaction contains an excess of splicing components and a limiting amount of pre-mRNA (*see* **Note 5**). Using appropriate graphing software, the reaction profile can be fit to the equation $A = C \times (1 - e^{-kt})$, where A is the fraction spliced, C is the fraction spliced at the endpoint of the reaction, k is the apparent rate constant, and t is the time.
14. Make any adjustments to the data to more accurately identify the splicing portion of the reaction. A lag in the splicing of the pre-mRNA can be observed by a period of very little appearance of spliced product for the first ~25 min of the splicing reaction. If there is a lag at the beginning of the in vitro splicing reaction, it may be helpful to adjust the time course by subtracting the length of the lag time from each time point taken. To account for the delay in timing, subtract the amount of time before splicing is observed from all time values. To do this, draw a line along the slope of the linear phase of the splicing reaction—in Fig. 2 the 25–65 min time points. Then use the x-intercept of that line as the lag time and subtract it from each time point. The adjusted profile will more closely follow the actual kinetics of splicing as opposed to including the kinetics of the proteins initial competition for the pre-mRNA (*see* Fig. 2c). Additionally, the fit curve may run to a maximum spliced fraction that is greater than 1. If this is the case, it is most likely due to a need for more time points to more accurately follow the reaction or for a longer reaction time to better determine the endpoint of the reaction (*see* **Notes 3** and **4**).
15. Replot the adjusted data and redetermine the curve fit. Rate constants of different pre-mRNAs or different reaction conditions can then be compared to determine the influence of splicing effectors. Not adjusting the data may result in inaccurate results for values computer using the first-order rate equation. This is usually due to insufficient time points (not accurately following the changes over time or not reaching the endpoint of the reaction) or not accounting for the lag period when splicing is not yet occurring.

4 Notes

1. The number of time points required depends on the resolution required for the rate constant. More data points assure a more accurate rate determination. An initial time course can be run with evenly spaced time points (every 10–15 min.) that will allow to determine the general shape of the reaction analyzed. Following this first attempt, taking more time points during the portions of the reaction in the linear phase and its slow transition into the end phase are recommended. More data in the linear phase is important because this is the area where the most striking changes are observed. More data toward the endpoint is necessary to accurately define maximal splicing levels.
2. Identical volume quantitation areas ensure differences between bands are not due to quantification box volume. A box does not have to be used; other shapes are usable as long as they are all the same around each band. Additionally, make sure to account for background signal either with a setting within the quantification program or by making an extra quantification box around an area where there is no band, giving a value that can then be subtracted from all other bands, removing the background signal. Additionally, make sure the boxes do not overlap.
3. The maximum product formation or fraction of spliced product at the endpoint of the reaction will be different for every pre-mRNA evaluated. Less efficient nuclear extracts and pre-mRNAs with poorer splice sites may have a lower fraction of spliced product at the endpoint of the reaction. This should be verified experimentally by carrying out splicing reactions with extended time points (past 2 h). By carrying out longer experiments, the true endpoint of the reaction can be determined eliminating any error that may occur due to the estimations of the software and curve fitting functions.
4. Determination of the reaction lag time adjustment allows a better fit for the rate equation to the splicing phase of the reaction. Pre-mRNAs that have less binding potential with splicing components have been noted to have longer lag times, suggesting that the lag is occurring due to competition for binding along the pre-mRNA molecule. After this lag, the reaction follows a pseudo-first-order reaction rate profile. This is what is modeled by the rate equation used, the initial linear appearance of product followed by a hyperbolic approach to an asymptotic maximum of product formation.
5. Even though the reaction is second order, based on the concentrations of NE and the pre-mRNA, it can be analyzed as a first-order reaction because the [NE] is in such excess over the [pre-mRNA] that the [NE] does not change over the course of the reaction, i.e., pseudo-first-order reaction conditions

(~2 μg/μL of protein vs. 0.01–0.1 nM RNA in the reaction). This does mean that if you are adding splicing components to the reaction, they should be added to saturating amounts so as to make sure they interact with all radiolabeled pre-mRNAs in the reaction.

Acknowledgement

Research in the Hertel laboratory is supported by NIH (RO1 GM62287 and R21 CA149548).

References

1. Black DL, Chabot B, Steitz JA (1985) U2 as well as U1 small nuclear ribonucleoproteins are involved in premessenger RNA splicing. Cell 42(3):737–750
2. Tarn WY, Steitz JA (1994) SR proteins can compensate for the loss of U1 snRNP functions in vitro. Genes Dev 8(22):2704–2717
3. Tarn WY, Steitz JA (1995) Modulation of 5′ splice site choice in pre-messenger RNA by two distinct steps. Proc Natl Acad Sci USA 92(7):2504–2508
4. Krainer AR, Maniatis T, Ruskin B et al (1984) Normal and mutant human beta-globin pre-mRNAs are faithfully and efficiently spliced in vitro. Cell 36(4):993–1005
5. Ruskin B, Krainer AR, Maniatis T et al (1984) Excision of an intact intron as a novel lariat structure during pre-mRNA splicing in vitro. Cell 38(1):317–331
6. Kotlajich MV, Crabb TL, Hertel KJ (2009) Spliceosome assembly pathways for different types of alternative splicing converge during commitment to splice site pairing in the A complex. Mol Cell Biol 29(4):1072–1082. doi:10.1128/MCB.01071-08, MCB.01071-08 [pii]
7. Hicks MJ, Lam BJ, Hertel KJ (2005) Analyzing mechanisms of alternative pre-mRNA splicing using in vitro splicing assays. Methods 37(4):306–313
8. Zhou Z, Sim J, Griffith J, Reed R (2002) Purification and electron microscopic visualization of functional human spliceosomes. Proc Natl Acad Sci USA 99(19):12203–12207
9. Deckert J, Hartmuth K, Boehringer D et al (2006) Protein composition and electron microscopy structure of affinity-purified human spliceosomal B complexes isolated under physiological conditions. Mol Cell Biol 26(14):5528–5543. doi:10.1128/MCB.00582-06, 26/14/5528 [pii]
10. Hicks MJ, Mueller WF, Shepard PJ et al (2010) Competing upstream 5′ splice sites enhance the rate of proximal splicing. Mol Cell Biol 30(8):1878–1886. doi:10.1128/MCB.01071-09, MCB.01071-09 [pii]
11. Audibert A, Weil D, Dautry F (2002) In vivo kinetics of mRNA splicing and transport in mammalian cells. Mol Cell Biol 22(19):6706–6718
12. Kornblihtt AR, de la Mata M, Fededa JP et al (2004) Multiple links between transcription and splicing. RNA 10(10):1489–1498
13. Hicks MJ, Yang CR, Kotlajich MV et al (2006) Linking splicing to Pol II transcription stabilizes pre-mRNAs and influences splicing patterns. PLoS Biol 4(6):e147
14. Erkelenz S, Mueller WF, Evans MS et al (2013) Position-dependent splicing activation and repression by SR and hnRNP proteins rely on common mechanisms. RNA 19(1):96–102. doi:10.1261/rna.037044.112
15. Lam BJ, Bakshi A, Ekinci FY et al (2003) Enhancer-dependent 5′-splice site control of fruitless pre-mRNA splicing. J Biol Chem 278(25):22740–22747. doi:10.1074/jbc.M301036200, M301036200 [pii]
16. Lam BJ, Hertel KJ (2002) A general role for splicing enhancers in exon definition. RNA 8(10):1233–1241
17. Hertel KJ, Maniatis T (1999) Serine-arginine (SR)-rich splicing factors have an exon-independent function in pre-mRNA splicing. Proc Natl Acad Sci USA 96(6):2651–2655
18. Fox-Walsh KL, Dou Y, Lam BJ et al (2005) The architecture of pre-mRNAs affects mechanisms of splice-site pairing. Proc Natl Acad Sci USA 102(45):16176–16181
19. Lim SR, Hertel KJ (2004) Commitment to splice site pairing coincides with A complex formation. Mol Cell 15(3):477–483

Chapter 13

In Vitro Systems for Coupling RNAP II Transcription to Splicing and Polyadenylation

Eric G. Folco and Robin Reed

Abstract

Studies over the past several years have revealed that steps in gene expression are extensively coupled to one another both physically and functionally. Recently, in vitro systems were developed for understanding the mechanisms involved in coupling transcription by RNA polymerase II to RNA processing. Here we describe an efficient two-way system for coupling transcription to splicing and a robust three-way system for coupling transcription, splicing, and polyadenylation. In these systems a CMV-DNA construct is incubated in HeLa cell nuclear extracts in the presence of ^{32}P-UTP to generate the nascent transcript. Transcription is then stopped by addition of α-amanitin followed by continued incubation to allow RNA processing.

Key words RNAP II, Coupled steps in gene expression, Polyadenylation, Splicing, Transcription

1 Introduction

During gene expression, pre-mRNAs are synthesized in the nucleus by RNAP II and then undergo several processing steps, including capping, splicing, and polyadenylation. These steps are extensively coupled to one another via an extensive network of interactions [1–3]. A number of systems have been developed to investigate the mechanisms for coupling transcription to splicing [4–9], coupling transcription to polyadenylation [10], and coupling transcription to both polyadenylation and splicing [11]. Here we describe methods for two systems that we developed, one for coupling transcription to splicing and one for coupling transcription, splicing, and polyadenylation. In these systems, pre-mRNAs are synthesized by RNAP II in HeLa cell nuclear extracts followed by RNA processing. The method employs nuclear extracts similar to those that were originally optimized for splicing ^{32}P-labeled pre-mRNA synthesized with bacteriophage RNA polymerases [12]. These nuclear extracts are typically prepared in bulk from 10 to 50 L of cells grown in suspension [13] but, for small-scale applications, can also

Klemens J. Hertel (ed.), *Spliceosomal Pre-mRNA Splicing: Methods and Protocols*, Methods in Molecular Biology, vol. 1126, DOI 10.1007/978-1-62703-980-2_13, © Springer Science+Business Media, LLC 2014

be prepared using a few 150 mm plates of HeLa cells grown as adherent monolayers [14]. Preparation of the nuclear extracts was optimized for use in the coupled systems [5]. The DNA template used in the coupled systems is a PCR product containing the CMV promoter fused to a DNA template encoding a splicing substrate. The bovine growth hormone (BGH) polyA signal is also present in the DNA template for the system using polyadenylation.

2 Materials

All solutions are prepared using analytical grade reagents and ultrapure water (Milli-Q water—purified deionized water at a sensitivity of 18 MΩ cm at 25 °C). Storage temperature of each reagent is listed below.

2.1 Preparation of CMV-DNA Constructs

1. Plasmid encoding CMV-Ftz DoF construct containing or lacking BGH polyA signal or encoding constructs of interest. Plasmids should be stored at −20 °C in 1× TE buffer (Tris–HCl, pH 8.0, 1 mM EDTA) at a concentration of 5 ng/μL.
2. Primers for coupled transcription/splicing: Make 500 μL aliquots of Forward primer (5′ tgg agg tcg ctg agt agt gc 3′) and Reverse primer (5′ tag aag gca cag tcg agg 3′) at a final concentration of 1.6 μM. Store at −20 °C.
3. Primers for coupled transcription/splicing/polyadenylation: Make 500 μL aliquots of Forward primer (5′ tgg agg tcg ctg agt agt gc 3′) and Reverse primer (5′ cca cac cct aac tga gac 3′) at a final concentration of 1.6 μM. Store at −20 °C.
4. 10 mM dNTPs. Store at −20 °C.
5. 50 mM $MgSO_4$. Store at −20 °C.
6. Platinum Taq HiFi and 10× HiFi Buffer provided by supplier (Invitrogen). Store at −20 °C.
7. 10× TBE: Combine 432 g Tris-Base, 220 g boric acid, and 37.2 g EDTA. Add water to a final volume of 4 L. Store at room temperature.
8. 10 mg/mL ethidium bromide (10 mg/mL in H_2O).
9. Agarose HS standard/high melt. Store at room temperature.
10. 3 M sodium acetate. Store at room temperature.
11. 100 and 70 % ethanol diluted from 200 proof pure ethanol. Store at room temperature.
12. Phenol/chloroform, pH 7.9. Store at 4 °C.
13. 1 kb DNA ladder. Store at −20 °C.

2.2 Coupled Transcription/Splicing Reaction

1. 12.5 mM ATP. Filter, make 100 μL aliquots, and store at −20 °C.
2. 0.5 M creatine phosphate di-Tris salt (CrPh): Filter, make 100 μL aliquots, and store at −20 °C.
3. 80 mM $MgCl_2$: Filter, make 100 μL aliquots, and store at −20 °C.
4. CMV-DNA template. Make 50 μL aliquots at 200 ng/μL (*see* **Note 1**). Store at −20 °C.
5. [α-^{32}P]-UTP (800 Ci/mmol, 250 μCi). Store at 4 °C.
6. HeLa cell nuclear extract (*see* **Note 2**). Store at −80 °C.
7. α-Amanitin: Dilute to 10 ng/μL with water from 1 mg/mL stock. Store at −20 °C.
8. 2× proteinase K buffer (PK buffer): Mix 20 mL 1 M Tris pH 8.0, 5 mL 0.5 M EDTA, 6 mL 5 M NaCl, and 10 mL 20 % sodium dodecyl sulfate. Add water up to 100 mL. Filter and store at room temperature.
9. Proteinase K (PK). Add water to PK powder to prepare a 10 mg/mL stock. Make 100 μL aliquots. Store at −20 °C.
10. Glycogen, 20 mg/mL. Store at −20 °C.
11. Formamide gel-loading dye: Add 16 mL formamide, 0.4 mL 0.5 M EDTA, 0.8 mL 2.5 % xylene cyanol, and 0.8 mL 2.5 % bromophenol blue. Mix well and make 1 mL aliquots. Store at −20 °C.
12. Phenol:chloroform:isoamyl alcohol, pH 6.6. Store at 4 °C.

2.3 Coupled Transcription/Splicing/Polyadenylation Reaction

1. 12.5 mM ATP. Filter, make 100 μL aliquots, and store at −20 °C.
2. 0.5 M creatine phosphate di-Tris salt (CrPh). Filter, make 100 μL aliquots, and store at −20 °C.
3. 160 mM $MgCl_2$: Filter, make 100 μL aliquots, and store at −20 °C.
4. CMV-DNA template with BGH polyA signal. Make 50 μL aliquoted at 200 ng/μL. Store at −20 °C.
5. 15 % (w/v) polyvinyl alcohol (PVA) dissolved in water. Autoclave to solubilize the PVA. Use low molecular weight PVA. Make 1 mL aliquots and store at −20 °C.
6. 50 μM GTP, CTP, UTP [G, C, U]: Mix 5 μL of each NTP from 10 mM stock solutions with 85 μL water and store at −20 °C.
7. 10 mM UTP. Store at −20 °C.

2.4 Denaturing Polyacrylamide Gel

1. 5 % denaturing gel solution: Mix 215 g urea with 50 mL of 10× TBE and 62.5 mL 40 % acrylamide–bisacrylamide (acrylamide–bisacrylamide solution, 40 % (w/v, 29:1)). Bring volume up to 500 mL with water. Filter and store at 4 °C.
2. *N*,*N*,*N*′,*N*′-Tetramethylethylenediamine (TEMED). Store at 4 °C.
3. 10 % ammonium persulfate (APS). Store at 4 °C.
4. Model V16 polyacrylamide gel electrophoresis system, PROTEAN II xi spacers 0.5 mm, PROTEAN II xi comb (Bio-Rad).
5. Gel-loading tips: Flat orifice, 83 mm × 0.33 mm diameter.
6. Whatman paper 3 MM Chr.
7. Bio-Rad Model 583 gel dryer.
8. Bio-Rad HydroTech vacuum pump.
9. PhosphorImager cassette: Mounted General Purpose, 20 cm × 25 cm, screen and cassette.
10. PhosphorImager: Personal Molecular Imager™ (PMI) System.

3 Methods

Carry out all procedures on ice unless otherwise specified.

3.1 Preparation of the CMV-DNA Template

1. PCR reaction: Mix 2 μL of the CMV-DNA plasmid (CMV-DoF) (*see* **Note 3**), 2 μL 10 mM dNTPs, 2 μL 50 mM $MgSO_4$, 5 μL 10× HiFi Buffer, 12.5 μL of each primer, 13.6 μL of Milli-Q water, and 0.4 μL Platinum Taq HiFi. Start the PCR reaction at 94 °C for 5 min followed by 32 cycles at 94 °C for 30 s, 55 °C for 30 s, and 68 °C for 2 min. The final cycle at 72 °C for 10 min and store the PCR reaction at 15 °C.
2. After the PCR reaction, bring the volume up to 150 μL with water and run a small aliquot (2 μL) on a mini-agarose gel. The PCR product for CMV-DoF should be ~1.5 kb.
3. Purify the DNA template by extracting the PCR reaction with an equal volume of phenol/chloroform (pH 7.9). Transfer the supernatant (aqueous phase) to a new Eppendorf tube.
4. Add 2 μL of glycogen; mix well. Add 1/10 volume 3 M sodium acetate; mix well. Add 3 volumes of 100 % ethanol. Centrifuge at 16,000 × *g* for 15 min to pellet the DNA. Remove the supernatant, without disturbing the pellet. The precipitated DNA should form a translucent pellet at the bottom of the tube. Wash once with 1 mL 70 % ethanol. Air-dry the pellet and dissolve it in 100 μL of water. Estimate the concentration of DNA

by running 2 μL on an agarose gel and comparing the intensity of the band to the known concentrations of bands in the 1 kb DNA ladder (*see* **Notes 4** and **5**).

3.2 Coupled Transcription and Splicing Reaction

1. Preheat two water baths to 30 °C and 37 °C.
2. Prepare a master mix for the total number of reactions you plan to perform. A 1× reaction mixture contains 1 μL CMV-DNA template, 1 μL 12.5 mM ATP, 1 μL 0.5 M CrPh, 1 μL 80 mM $MgCl_2$, 1 μL α-^{32}P-UTP, and 5 μL autoclaved Milli-Q water.
3. Aliquot 10 μL of the master mix per tube.
4. Add 15 μL of nuclear extract and pipet up and down gently to mix (*see* **Note 6**).
5. Incubate the reaction mixtures at 30 °C for 8 min to allow RNAP II transcription (*see* **Note 7**).
6. Add 1 μL of α-amanitin (10 ng/μL) per 25-μL reaction mixture and pipet up and down gently to mix (*see* **Note 8**).
7. Remove a 4 μL aliquot of the reaction at the 8-min time point and transfer it to a microfuge tube containing 100 μL of 2× PK buffer and 91 μL autoclaved Milli-Q water (*see* **Note 9**).
8. Repeat **step 7** for all subsequent time points.
9. Add 5 μL of PK to each sample, mix well, and incubate at 37 °C for 15 min.
10. Add 200 μL of phenol/chloroform pH 6.6 to each sample and mix well by pipetting up and down. Centrifuge for 10 min at 16,000 ×*g*.
11. Transfer 175 μL of the aqueous phase to a new tube containing 2 μL of glycogen. Mix by pipetting up and down.
12. Add 500 μL of 100 % ethanol and mix well by pipetting up and down.
13. Spin at 16,000 ×*g* for 15 min. The RNA pellet should look white/translucent.
14. Carefully remove the supernatant using a pipet.
15. Quick spin at 16,000 ×*g*.
16. Remove the rest of the supernatant using a P200 pipetman (*see* **Note 10**).
17. Resuspend the pellet in 15 μL formamide dye and mix carefully by pipetting.
18. Prepare a gel by combining 15 mL of 5 % denaturing gel solution with 15 μL TEMED and 150 μL 10 % APS. Wait 10 min until the gel is polymerized.
19. Place samples in hot water (75–90 °C) for 10 min.
20. Pre-run the denaturing gel at 20 mAmps for 10 min.

21. Briefly centrifuge the samples and load 7.5 μL of each on the pre-run denaturing gel. Run the gel at 20 mAmps, constant current for 30–45 min (*see* **Note 11**).
22. Transfer the gel to Whatman paper and dry the gel on a gel dryer for 30 min at 80 °C.
23. Place the gel in a PhosphorImager cassette, expose 1–12 h, and scan the gel using a PhosphorImager.
24. See representative results in Fig. 1.

3.3 Coupled Transcription/Splicing/Polyadenylation Reaction

1. Prepare the CMV-DNA template containing the BGH polyA signal by following the steps outlined in Subheading 3.1. Use the Reverse primer specifically designed for the coupled transcription/splicing/polyadenylation system.
2. Prepare a master mix for the total number of reactions you plan to perform. A 1× reaction mixture contains 1 μL CMV-DNA

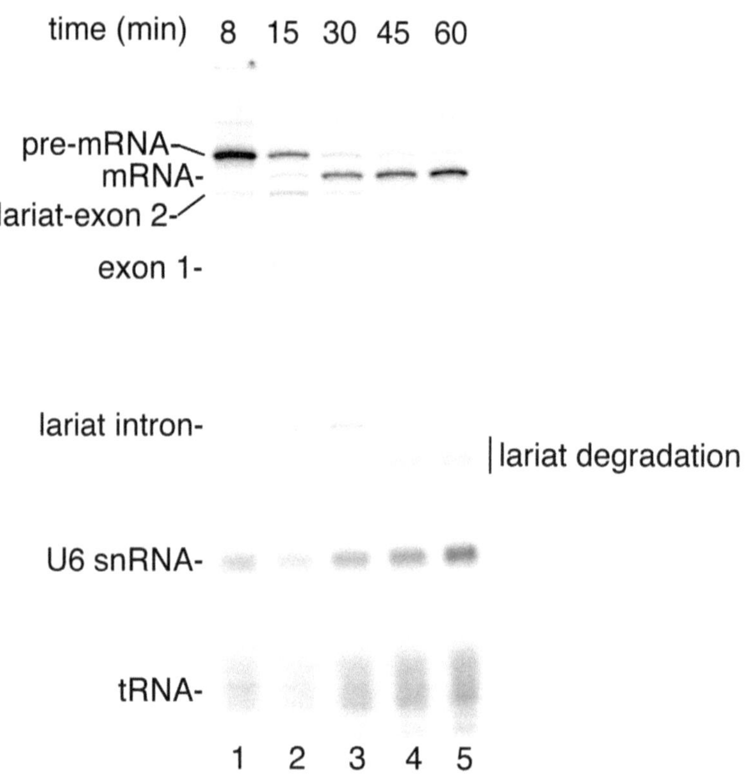

Fig. 1 Coupled RNAP II transcription and pre-mRNA splicing in vitro. ^{32}P-UTP and the CMV-DoF DNA template were incubated under transcription/splicing conditions for 8 min. α-Amanitin was added after the 8-min time point and incubation was continued for the indicated times. Pre-mRNA and the splicing intermediates are indicated. The endogenous U6 snRNA and tRNA in the extract are labeled by the ^{32}P-UTP [15]

template, 0.5 μL 160 mM $MgCl_2$, and 4.1 μL 15 % PVA. Add 1 μL autoclaved Milli-Q water.

3. Aliquot 10 μL of master mix per tube.
4. Add 15 μL of nuclear extract to each reaction mixture and pipet up and down gently to mix (*see* **Note 6**).
5. Incubate the tubes at 30 °C for 20 min (*see* **Note 12**).
6. Add 1 μL 0.5 M CrPh, 2 μL 12.5 mM ATP, 2 μL α-^{32}P-UTP, and 0.5 μL 50 μM [G, C, U]. Mix well.
7. Incubate the tubes at 30 °C for 2–5 min (*see* **Note 13**).
8. Add cold UTP to each sample to a final concentration of 2 mM (*see* **Note 14**).
9. Incubate again at 30 °C for 5–8 min.
10. Add 1 μL of α-amanitin (10 ng/μL) per 25-μL reaction mixture, pipet up and down to mix (*see* **Note 8**).
11. Follow all of the steps from Subheading 3.2, **steps 7–24**.
12. *See* Fig. 2 for representative results.

4 Notes

1. The amount of CMV-DNA template should be titrated to obtain optimal RNAP II transcription efficiency. For our 1.5 kb CMV-DoF DNA, 200 ng/25 μL coupled reaction is optimal.
2. The main difference between the nuclear extract used for the coupled systems and uncoupled systems is the omission of the spin after the dialysis at the end of the standard Dignam protocol.
3. In some preparations of nuclear extract, RNAP II transcribes end to end in a promoter-independent manner. If you are using your own DNA template, use the smallest possible PCR fragment that contains your sequence of interest to avoid large end-to-end transcription products.
4. RNAP II can initiate at nicks in DNA. To avoid nicked DNA, store aliquots at −20 °C in 1× TE, and dilute to 200 ng/μL before use.
5. PCR products should not be purified using mini-columns because these templates are not transcribed well.
6. Time points are taken in 4 μL aliquots from 25 μL reaction mixtures. Typically, 3–5 time points are taken. If more time points are needed, it is best to set up one larger reaction mixture and divide it into 25 μL aliquots for incubation rather than incubating a large-volume reaction.
7. For our CMV-DoF DNA template, time points ranging from 5 to 60 min are used and should be optimized for different DNA templates.

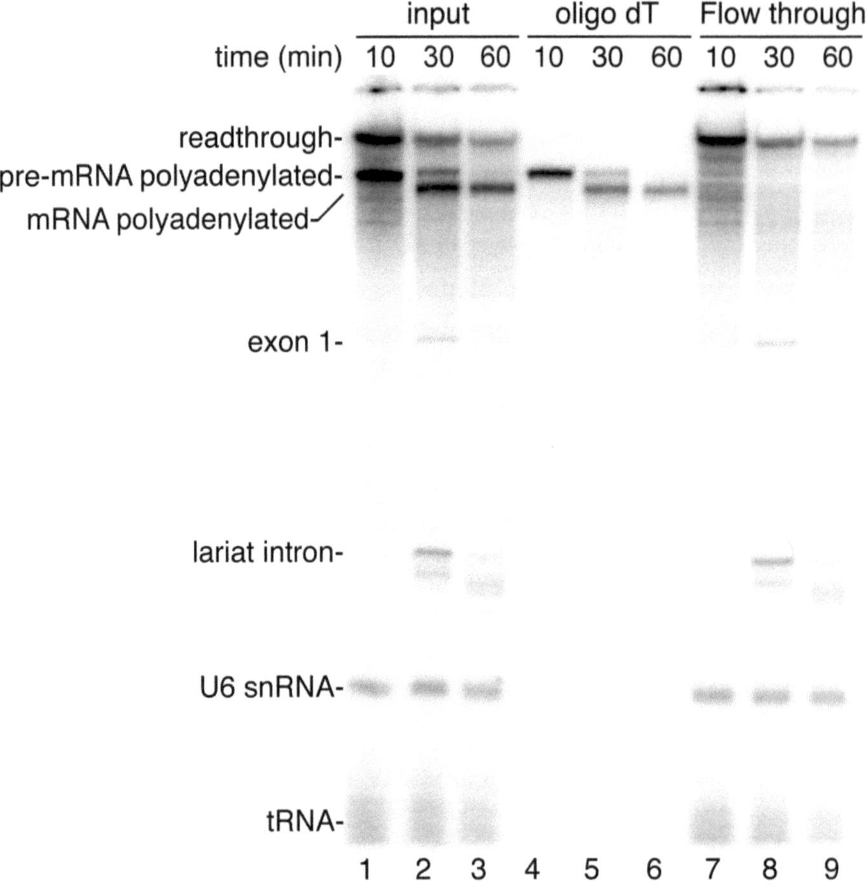

Fig. 2 Coupled RNAP II transcription/splicing/polyadenylation in vitro. ^{32}P-UTP and the CMV-DoF/BGH DNA template were incubated for 20 min to assemble a pre-initiation complex followed by addition of ^{32}P-UTP, ATP, and CrPhos and continued incubation for 10 min. The read-through transcript and polyadenylated pre-mRNA are generated by this time point (*lane 1*). Spliced polyadenylated mRNA is generated at the subsequent time points (*lanes 2* and *3*). Total RNA from aliquots of the reactions shown in *lanes 1–3* were isolated and passed through an oligo-dT column. The polyadenylated pre-mRNA and mRNA bound to the oligo-dT (*lanes 4–6*), whereas the read-through transcript, exon 1, the lariat intron, U6 snRNA, and tRNA are detected in the flow through from the oligo-dT column (*lanes 7–9*)

8. α-Amanitin is used to block transcription, and the time of addition should be optimized for different DNA templates and nuclear extracts.
9. Samples in PK buffer are stable at room temperature and can be stored until all of the time points have been collected. At this step, samples can also be stored at −20 °C overnight and processed further later.
10. Ensure that the pellets are dry before adding the formamide loading dye because any remaining ethanol will add to the

volume of the sample as well as distort the migration of the bands on the gel.

11. To resolve splicing products generated from CMV-DoF DNA template, the gel should be run until the bromophenol blue is at the bottom of the gel.
12. The nuclear extract and CMV-DNA template are incubated with $MgCl_2$ and PVA to assemble a pre-initiation complex (PIC). The PIC is necessary for efficient polyadenylation.
13. An incubation of 2 min is usually sufficient for the first incubation after PIC formation using CMV-DoF. However, this step should be optimized for each preparation of nuclear extract and DNA template.
14. Because the radioactive UTP is usually limiting, addition of cold UTP is used as a chase to generate full-length transcripts.

Acknowledgement

This work was supported by NIH grant GM043375.

References

1. Maniatis T, Reed R (2002) An extensive network of coupling among gene expression machines. Nature 416:499–506
2. Hirose Y, Manley JL (2000) RNA polymerase II and the integration of nuclear events. Genes Dev 14:1415–1429
3. Bentley DL (2005) Rules of engagement: co-transcriptional recruitment of pre-mRNA processing factors. Curr Opin Cell Biol 17:251–256
4. Hicks MJ, Yang CR, Kotlajich MV et al (2006) Linking splicing to Pol II transcription stabilizes pre-mRNAs and influences splicing patterns. PLoS Biol 4:e147
5. Das R, Dufu K, Romney B et al (2006) Functional coupling of RNAP II transcription to spliceosome assembly. Genes Dev 20:1100–1109
6. Das R, Yu J, Zhang Z et al (2007) SR proteins function in coupling RNAP II transcription to pre-mRNA splicing. Mol Cell 26:867–881
7. Yu Y, Das R, Folco EG et al (2010) A model in vitro system for co-transcriptional splicing. Nucleic Acids Res 38:7570–7578
8. Ghosh S, Garcia-Blanco MA (2000) Coupled in vitro synthesis and splicing of RNA polymerase II transcripts. RNA 6:1325–1334
9. Natalizio BJ, Garcia-Blanco MA (2005) In vitro coupled transcription splicing. Methods 37:314–322
10. Rigo F, Kazerouninia A, Nag A et al (2005) The RNA tether from the poly(A) signal to the polymerase mediates coupling of transcription to cleavage and polyadenylation. Mol Cell 20:733–745
11. Rigo F, Martinson HG (2008) Functional coupling of last-intron splicing and 3′-end processing to transcription in vitro: the poly(A) signal couples to splicing before committing to cleavage. Mol Cell Biol 28: 849–862
12. Krainer AR, Maniatis T, Ruskin B et al (1984) Normal and mutant human beta-globin pre-mRNAs are faithfully and efficiently spliced in vitro. Cell 36:993–1005
13. Dignam JD, Lebovitz RM, Roeder RG (1983) Accurate transcription initiation by RNA polymerase II in a soluble extract from isolated mammalian nuclei. Nucleic Acids Res 11: 1475–1489
14. Folco EG, Lei H, Hsu JL et al (2012) Small-scale nuclear extracts for functional assays of gene expression machineries. J Vis Exp 64:e4140. doi:10.3791/4140
15. Reddy R, Henning D, Das G et al (1987) The capped U6 small nuclear RNA is transcribed by RNA polymerase III. J Biol Chem 262: 75–81

Chapter 14

Isolation and Accumulation of Spliceosomal Assembly Intermediates

Janine O. Ilagan and Melissa S. Jurica

Abstract

Isolating spliceosomes at a specific assembly stage requires a means to stall or enrich for one of the intermediate splicing complexes. We describe strategies to arrest spliceosomes at different points of complex formation and provide a detailed protocol developed for isolating intact splicing complexes arrested between the first and second chemical steps of splicing. Briefly, spliceosomes are assembled on a radiolabeled in vitro-transcribed splicing substrate from components present in nuclear extract of HeLa cells. Spliceosome progression is arrested after the first step of splicing chemistry by mutating the pre-mRNA substrate at the 3′ splice site. The substrate also contains binding sites for the MS2 protein, which serve as an affinity tag. Purification of arrested spliceosomes is carried out in two steps: (1) size exclusion chromatography and (2) affinity selection via a fusion of MS2 and maltose-binding protein (MBP). Complex assembly and purification are analyzed by denaturing polyacrylamide gel electrophoresis.

Key words Spliceosome, Affinity purification, Pre-mRNA splicing, MS2:MBP, Nuclear extract, Size exclusion

1 Introduction

The spliceosome is a large macromolecular machine responsible for removing introns in a process known as pre-mRNA splicing. It forms on each intron from over one hundred components including five nuclear ribonucleoproteins (U1, U2, U4, U5, and U6 snRNPs) and many non-snRNP proteins. Spliceosome assembly occurs in a stepwise manner through a series of intermediate splicing complexes that are characterized by their associated components and chemical state of the intron [1]. Briefly, U1 snRNP base pairs with the 5′ splice site in E complex and recruits U2 snRNP. In an ATP-dependent step, U2 snRNP stably base pairs with the branchpoint sequence to form A complex. The addition of tri-snRNP (U5:U4/U6) and Prp19 complex leads to B complex. Several ATP-dependent rearrangements between RNA/RNA and RNA/protein interactions result in loss of U1 and U4 snRNPs and ready the spliceosome for catalysis as B^{act} complex forms. Further

Klemens J. Hertel (ed.), *Spliceosomal Pre-mRNA Splicing: Methods and Protocols*, Methods in Molecular Biology, vol. 1126, DOI 10.1007/978-1-62703-980-2_14, © Springer Science+Business Media, LLC 2014

rearrangements lead to B* complex and first-step splicing chemistry in which the 2′ OH of the branchpoint adenosine in the intron attacks the phosphate bond at the 5′ splice site. This reaction leads to cleavage at the 5′ end of the intron and formation of lariat structure. Additional rearrangements and addition of proteins form C complex lead to second-step chemistry where the 5′ OH of the upstream exon attacks at the 3′ splice site. This reaction cleaves the 3′ end of the intron and ligates the flanking exons. The resulting mRNA and lariat intron are then released from P complex.

Splicing can be recapitulated in vitro using a model pre-mRNA and cellular extract [2, 3]. However, the dynamic nature of spliceosome assembly creates a challenge for capturing intermediate splicing complexes for further biochemical and structural studies. In *S. cerevisiae*, spliceosomes can be arrested and purified at specific stages by genetically manipulating proteins that are required for the next step in assembly [4, 5]. In the human system, spliceosome assembly has been stalled by a variety of means including withholding ATP from the reaction, depleting or inactivating snRNPs with antisense oligonucleotides, and manipulating the pre-mRNA substrate [2, 3, 6–18]. The latter provides the most efficient method to capture spliceosomes at distinct points of splicing catalysis. A pre-mRNA with a polypyrimidine tract less than 10 nt truncated before the 3′ splice site will accumulate B^{act} complex at a point before first-step chemistry [18]. C complex can be stalled after first-step chemistry on a pre-mRNA with a polypyrimidine tract more than 20 nt that is either truncated before the 3′ splice site [7] or that contains a 3′ splice site mutation [11, 19]. Shortening the 3′ exon to less than 25 nt allows accumulation of P complex containing the unreleased splicing products [20].

To purify stalled splicing complexes, the pre-mRNA can be further modified to incorporate an affinity tag. The most commonly used tag consists of three RNA hairpins containing the recognition sequence for the bacteriophage MS2 coat protein. These hairpins serve as a handle for amylose affinity selection by a fusion of MS2 to maltose-binding protein (MBP), and this strategy has been extensively used to isolate and characterize splicing complexes [7, 18–28]. In the protocol outlined below, we detail conditions to assemble C complex spliceosomes in human nuclear extract and isolate the complexes by MS2:MBP affinity purification (Fig. 1). The procedure may also be applied to spliceosome complexes stalled at other assembly intermediates.

2 Materials

To prevent contamination by RNases, all materials and equipment should be handled with gloves. All reagents should be prepared with RNase free water (*see* **Note 1**). Important: Radioactive

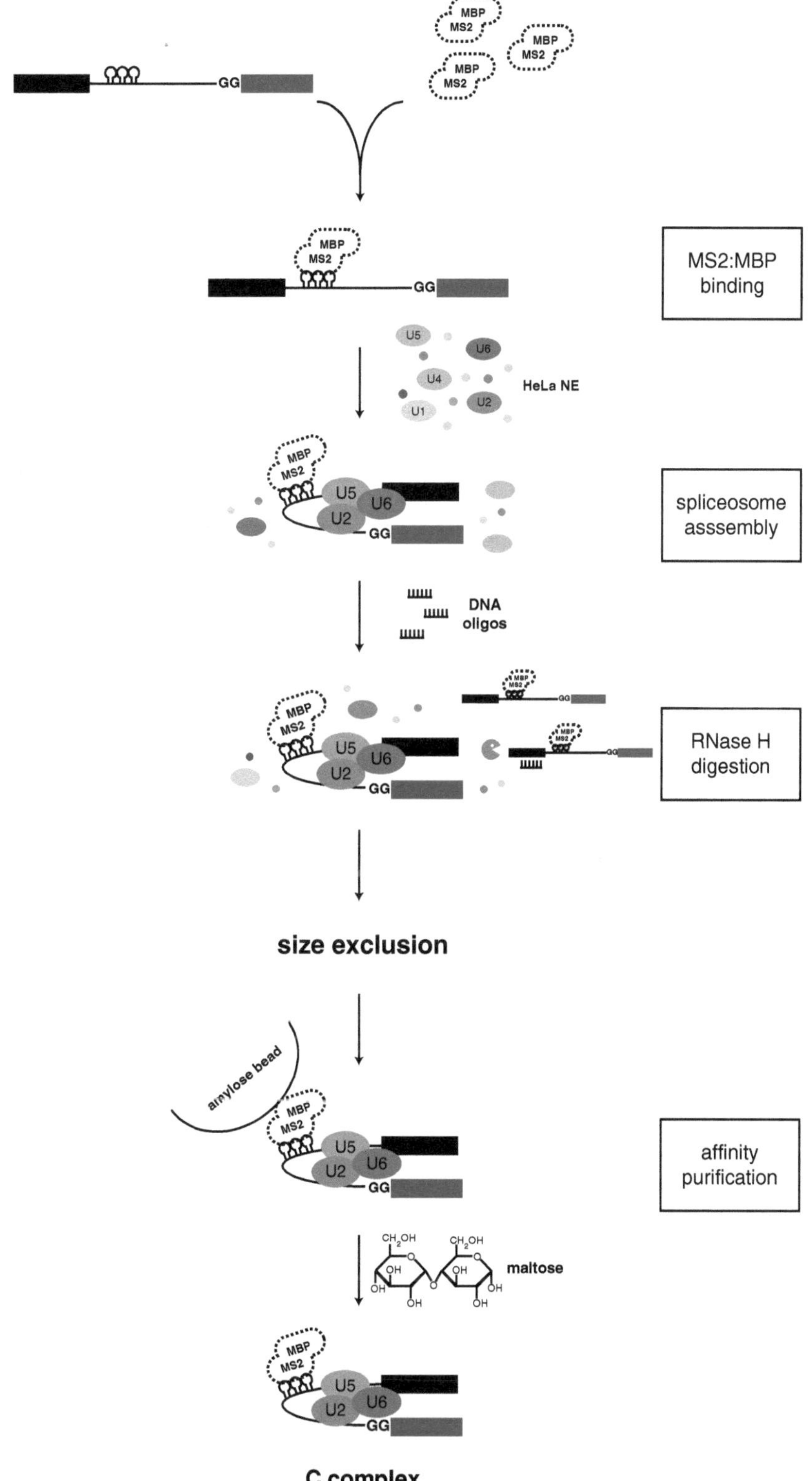

Fig. 1 Schematic of MS2:MBP affinity purification of C complex spliceosomes with tagged pre-mRNA substrate

materials should only be handled by authorized users with protective shielding, proper monitoring, and appropriate attire in compliance with all state and federal regulations. Follow proper waste disposal procedures for all chemicals and radioisotopes.

2.1 In Vitro Transcription

1. DNA template: Linearized plasmid or PCR product containing a T7 promoter sequence followed by pre-mRNA sequence (*see* **Note 2**).
2. Radioactive nucleotide: [α-^{32}P] UTP at 3,000 Ci/mMol (*see* **Note 3**).
3. 5× transcription buffer (usually supplied with T7 polymerase): 200 mM Tris (pH 7.6–8.0), 30–40 mM $MgCl_2$, 10 mM spermidine, 0–250 mM NaCl.
4. Nucleotide stocks: 10 mM ATP, 10 mM CTP, 10 mM GTP, 10 mM UTP.
5. Cap analog: 10 mM G(5′)ppp(5)′G (NEB).
6. 1 M dithiothreitol (DTT).
7. T7 RNA polymerase.
8. RNase Inhibitor (optional).

2.2 Denaturing Polyacrylamide Gel for Pre-mRNA Purification and Analysis of Splicing

1. 1× TBE: 0.09 M Tris-borate, 0.09 M boric acid, 0.0025 M EDTA. Store at room temperature.
2. 15 % denaturing polyacrylamide solution: 7 M Urea, 15 % acrylamide (from AccuGel acrylamide solution 40 % (w/v) 29:1 acrylamide:bis-acrylamide), 1× TBE. Store at 4 °C.
3. 0.8 mm spacers and comb for thick gel and 0.4 mm spacers and comb for thin gel.
4. 20 cm × 27 cm glass plates, one should be notched to fit gel rig (Moliterno).
5. Large binder clips (2 in.).
6. Electrophoresis gel rig (Dan-Kar).
7. High-voltage (>3,000 V) power supply.
8. RNA gel loading buffer: 95 % formamide, 20 mM EDTA, 0.01 % (w/v) bromophenol blue, 0.01 % (w/v) xylene cyanol. Store at −20 °C.
9. Gel extraction buffer: 0.3 M NaAc (pH 4.8), 1 mM EDTA, 10 % phenol (pH 4.5). Store at 4 °C.

2.3 Assembly of Spliceosome Complex

1. HeLa nuclear extract (*see* **Note 4**).
2. 1 M glutamic acid monopotassium salt (KGlu, pH 7.5) (*see* **Note 5**).
3. 100 mM magnesium acetate (MgAc).
4. 100 mM ATP.

5. 250 mM creatine phosphate (CP).
6. 5 mg/mL yeast tRNA.
7. RNase Inhibitor.
8. Pre-mRNA transcript from in vitro transcription.
9. DNA oligonucleotides for RNase H digestion (*see* **Note 6**).
10. 10 mg/mL heparin.
11. 10–50 μM purified MS2-MBP protein (*see* **Note** 7).

2.4 Purification of Spliceosomes

1. Sephacryl S-400 (GE Healthcare).
2. 1.0 cm × 10 cm glass column with stopcock valve.
3. Amylose resin (NEB).
4. Mobicol spin column with small 35 μm filter.
5. Sizing column buffer (SCB-N): 150 mM KCl, 20 mM Tris–HCl (pH 7.9 at 4 °C), 5 mM EDTA, 1 mM DTT, 0.5 % NP-40. Make fresh for each spliceosome purification. Store at room temperature (*see* **Note 8**).
6. Amylose column buffer (ACB): 150 mM KCl, 20 mM Tris–HCl (pH 7.9 at 4 °C), 5 mM EDTA, 1 mM DTT. Make fresh for each spliceosome purification. Keep on ice.
7. Elution buffer: 150 mM KCl, Tris–HCl (pH 7.9 at 4 °C), 5 mM EDTA, 1 mM DTT, 10 mM maltose.
8. Splicing dilution buffer: 100 mM Tris–HCl (pH 7.5), 10 mM EDTA, 1 % SDS, 150 mM NaCl, 0.3 M NaAc (pH 4.8). Store at room temperature.
9. Phenol:chloroform:isoamyl alcohol (25:24:1, pH 4.5).

3 Methods

3.1 In Vitro Transcription

A typical transcription reaction contains 1× transcription buffer, 400 μM ATP, 400 μM CTP, 400 μM UTP, 200 μM GTP, 800 μM Cap analog, 40 ng linearized plasmid DNA template/μL of reaction (or 10–100 ng PCR product template), one-tenth volume [α-^{32}P] UTP, and one-tenth volume T7 RNA polymerase (*see* **Notes 2–3**). A 50 μL reaction usually generates enough pre-mRNA transcript for 2–3 spliceosome preparations. Keep all reagents on ice unless specified.

1. To prepare a 50 μL transcription reaction, mix in order the following ingredients in a 1.5 mL microcentrifuge tube at room temperature: 17.5 μL water, 10 μL 5× transcription buffer, 2 μL 10 mM ATP, 2 μL 10 mM CTP, 2 μL 10 mM UTP, 1 μL 10 mM GTP, 1 μL 1 M DTT, 4.5 μL Cap analog, 3 μL [α-^{32}P] UTP, 2 μL 1 mg/mL linearized plasmid DNA template, and 5 μL T7 RNA polymerase. Mix gently.

2. Dilute 1 μL of the reactions mixture into 100 μL water and set aside. This will be used to calculate transcript concentration the next day in **step 11**.
3. Incubate transcription reaction at 37 °C for 2–4 h.
4. During the incubation time, pour a thick denaturing 5 % polyacrylamide gel. Place 0.8 mm spacers between 20 cm×27 cm glass plates and secure with large binder clips. Seal the bottom of the gel with tape or an additional spacer. In a 50 mL conical tube, mix 15 mL 15 % denaturing polyacrylamide solution and 30 mL 7 M urea in 1× TBE. Add 135 μL 20 % ammonium persulfate and 45 μL TEMED just before pouring the gel. Insert a 0.8 mm comb with wells that can hold up to 60 μL sample. Let gel polymerize for at least 20–30 min on bench top.
5. If PCR product is used as a DNA template, following the reaction incubation, add 1 μL RQ1 DNase to the reaction and incubate for an additional 20 min at 37 °C. Otherwise, skip to **step 6**.
6. Add 50 μL of RNA gel loading buffer to transcription reaction and set aside at room temperature.
7. Remove the bottom seal of the polymerized gel and clamp into an electrophoresis rig with an aluminum heat sink plate. Fill the top and bottom chambers with 1× TBE and be sure to remove any air bubbles at the bottom of the gel. Remove the comb and extensively rinse the wells with buffer using a syringe. Hook up the leads to a high-voltage power supply and run the gel at constant wattage of 45 W for 20 min to prewarm the gel. Meanwhile, heat samples at 95 °C for 2 min and place on ice. Before loading the gel, rinse the wells again. Load 50 μL sample each into to neighboring lanes and run gel for 1 h at constant wattage of 45 W.
8. Carefully take down the gel. Note that most of the unincorporated radioactive nucleotides will be in the bottom chamber buffer. Remove one of the glass plates and cover the gel supported by the other glass plate with plastic wrap. Place glow-in-the-dark stickers on top of plastic wrap to orient the gel after exposure to film. Expose the gel for 1–2 min to X-ray film.
9. Using the X-ray film as a guide, cut out the transcript bands with a clean razor blade and transfer them to a 1.5 mL microcentrifuge tube. Add 400 μL gel extraction buffer and freeze tubes at −80 °C for 10–20 min. Rotate tubes overnight at room temperature.
10. Next day, transfer gel extraction buffer with extracted transcript to 1.5 mL microcentrifuge tube and add 1 mL 100 % ethanol. Discard the gel. Invert the tube a few times to mix and incubate at −80 °C for 30 min. Centrifuge the tube at 14,000 (RCF = 15,700×*g*) for 30 min at 4 °C to pellet the

transcript. Remove ethanol and wash the pellet with 100 μL 70 % ethanol. Remove ethanol and let the pellet air-dry. Resuspend pellet in 50 μL water and store at −20 °C.

11. To quantify the transcript, mix 1 μL with 3 mL of scintillation fluid in a scintillation tube. Repeat with 1 μL of the 1:100 reaction dilution from the previous day. Measure counts with a scintillation counter. Determine the transcript concentration with the following calculation: (cpm of transcript × nmol of cold UTP in reaction × 10^3)/(# of U's in transcript × reaction volume × 100 × cpm of reaction) = concentration of transcript in μM. Dilute the pre-mRNA transcript to 200 nM with water.

3.2 Spliceosome Assembly

A typical splicing reaction contains 1–10 nM pre-mRNA splicing substrate, 0–100 mM KGlu, 0–6 mM MgAc, 2 mM ATP, 5 mM CP, 0.1–0.5 mg/mL tRNA, and 40 % HeLa nuclear extract (*see* **Notes 4–5**). A 1 mL splicing reaction will generate 0.1–0.5 pmol spliceosomes depending on reaction efficiency and RNA degradation in the nuclear extract. Keep all reagents on ice unless specified.

1. For a 1 mL splicing reaction, transfer 50 μL 200 nM pre-mRNA transcript into a microcentrifuge tube and heat at 95 °C for 1 min and then place on ice. Add 50-fold molar excess MS2:MBP fusion protein and incubate on ice for 5 min. For 50 μM MS2:MBP, this is 10 μL.
2. In a separate 1.5 mL microcentrifuge tube, mix in order 410 μL water, 60 μL 1 M KGlu, 20 μL 100 mM MgAc, 20 μL 100 mM ATP, 20 μL 250 mM CP, and 10 μL 5 mg/mL tRNA (*see* **Note 4**). Mix this with the pre-mRNA and MS2:MBP and then add 400 μL HeLa nuclear extract. Splicing efficiency may be increased by splitting 100–200 μL of the reaction into separate 1.5 mL microcentrifuge tubes. Take a 10 μL aliquot into a new 1.5 mL microcentrifuge tube for a zero time point and save on ice.
3. Incubate splicing reaction at 30 °C for 60 min. Take a 10 μL aliquot into a new 1.5 mL microcentrifuge tube for a 60′ time point and save on ice. If the splicing reaction was split into multiple tubes, combine them back into one tube at this point.
4. To digest excess unspliced pre-mRNA, add 10 μL 100 μM DNA oligonucleotides for RNase H digestion to splicing reaction (*see* **Note 6**). Incubate at 30 °C for an additional 20 min. Take a 10 μL aliquot into a new 1.5 mL microcentrifuge tube for an 80′ time point and save on ice.
5. Add 25 μL 10 mg/mL heparin to splicing reaction. Incubate at 30 °C for 5 min and then transfer the splicing reaction to ice (*see* **Note 9**).

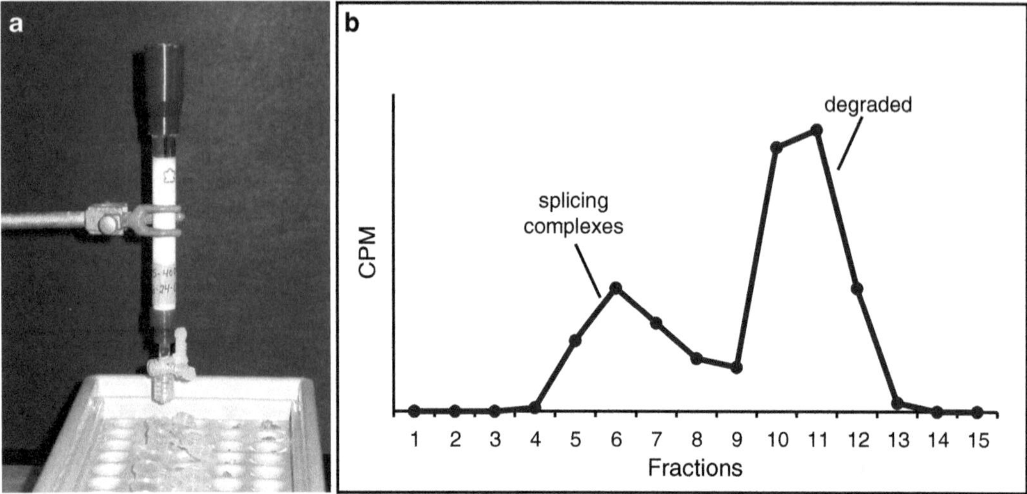

Fig. 2 (**a**) Image of size exclusion column. (**b**) Representative analysis of sizing column fractions. Average cpm is plotted versus fraction number

3.3 Spliceosome Purification

1. Prior to carrying out the purification, pour a 5 mL sizing column of S-400 resin equilibrated in SCB-N into a 1.0 cm × 10 cm glass column (Fig. 2a). Allow the resin to settle by gravity flow. This sizing column can be used multiple times by washing with 10 mL of SCB-N before each use.
2. During the splicing reaction incubation, prepare an amylose column. Fit a small 35 μm filter into a Mobicol column. Add 100 μL of amylose resin equilibrated in ACB into the column (Fig.3a, *see* **Note 10**). Let resin settle by gravity. To get the column flowing, a brief spin in a centrifuge at low speed may be necessary. Keep the column in a 1.5 mL microcentrifuge tube on ice.
3. To start the purification, let buffer flow by gravity from the sizing column until the top of the resin bed is exposed. Carefully load the splicing reaction onto the sizing column being sure to not disturb the resin bed. Let the sample run into the column and then load 500 μL SCB-N onto the resin bed and let it run into the column. Run an additional 10 mL of SCB-N through the column and collect 500 μL fractions on ice.
4. Use a Geiger counter to measure average cpm for each fraction. There should be two peaks of radioactivity (Fig. 2b). The first peak is smaller and usually occurs within the first eight fractions and contains splicing complexes. The second peak is larger and contains degraded pre-mRNA transcript. Take a 10 μL aliquot from the first peak into a new 1.5 mL microcentrifuge tube and save on ice. Pool fractions from the first peak (*see* **Note 11**).

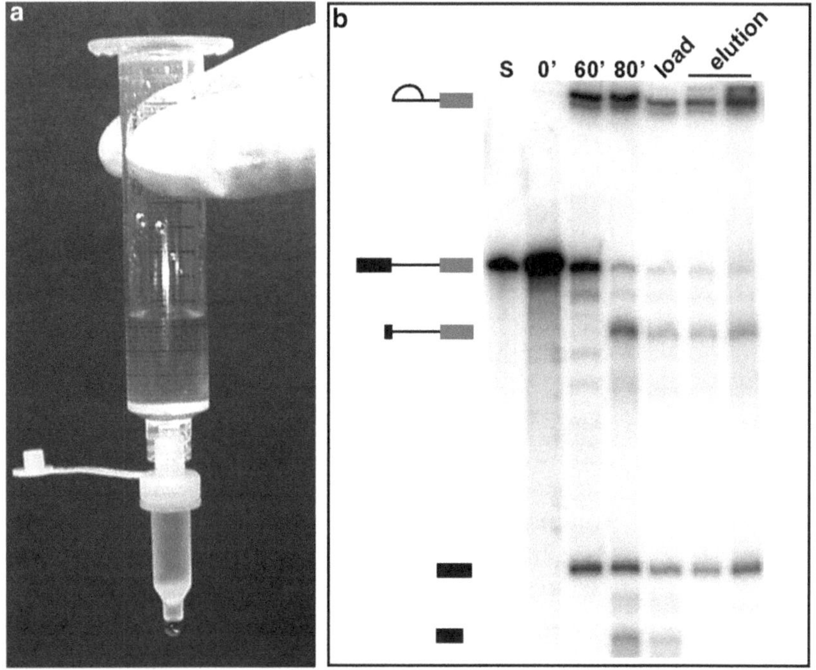

Fig. 3 (**a**) Image of amylose column attached to a syringe for washing. (**b**) Denaturing PAGE analysis of RNA from in vitro splicing and affinity purification of C complex spliceosome. Lanes from left to right are the pre-mRNA standard used for quantification (S), time points taken during the splicing reaction (0 and 60 min) and after RNase H digestion (80 min), size exclusion peak fraction loaded onto amylose column (load), and elution fractions from amylose column (elution). RNA species schematized on the left are, from top to bottom, lariat intermediate, pre-mRNA, 3′ RNase H digestion product, 5′ exon, and 5′ RNase H digestion product

5. Load pooled fractions onto the amylose column by gravity flow and collect the flow-through in tubes on ice. We often reapply the column flow-through two more times to maximize binding (*see* **Note 12**). To wash the column, attach a 10 mL syringe barrel to the top of the column using a luer adaptor cap and place in a 15 ml conical tube. Fill the syringe with 5 mL of cold ACB to wash the column by gravity flow at 4 °C.
6. Elute complexes by applying 30 μL aliquots of elution buffer. Take the drip from the bottom of the column and place into a clean 1.5 mL microcentrifuge tube on ice (*see* **Note 10**). Repeat this 4–5 times. Check the average cpm with a Geiger counter to identify peak fractions containing the purified splicing complexes (*see* **Note 10**).

3.4 Denaturing Gel Analysis of Spliceosome Purification

1. To prepare splicing time point and sizing column peak for denaturing gel analysis, add 90 μL of splicing dilution buffer to each 10 μL sample. Then add 100 μL of phenol:chloroform:isoamyl alcohol. Vortex well and spin for 10 min at 14,000 (RCF = 15,700×*g*) at room temperature.

Take 80 μL from the top aqueous layer, avoiding the interface, and put in a new 1.5 mL microcentrifuge tube. Add 300 μL 100 % ethanol and invert a few times to mix. Incubate at −80 °C for 30 min. Spin for 30 min in a microcentrifuge at 14,000 (RCF = 15,700 × *g*) at 4 °C. Remove the ethanol and let the pellet air-dry. Resuspend with 5 μL of RNA gel loading buffer. For amylose elution fractions, mix 1 μL of each elution fraction with 4 μL RNA gel loading buffer in a 1.5 mL microcentrifuge tube. For pre-mRNA standard, dilute 1 μL of 200 nM pre-mRNA transcript in 39 μL RNA gel loading buffer.

2. Pour a thin denaturing 15 % polyacrylamide gel. Place 0.4 mm spacers between 20 cm × 27 cm glass plates and secure with large binder clips. Seal the bottom of the gel with tape or an additional spacer. In a 50 mL conical tube, take 25 mL 15 % denaturing polyacrylamide solution. Add 75 μL 20 % ammonium persulfate and 25 μL TEMED just before pouring the gel. Insert a 0.4 mm comb with wells that can hold up to 5 μL sample. Let gel polymerize for at least 20–30 min.
3. To run the gel, remove the bottom seal and clamp the gel into an electrophoresis rig with an aluminum heat sink plate. Fill the top and bottom chambers with 1× TBE and be sure to remove any air bubbles at the bottom of the gel. Remove the comb and extensively rinse the wells with buffer using a syringe. Hook up the leads to a high-voltage power supply and run the gel at constant wattage of 30 W for 20 min to prewarm the gel. Meanwhile, heat samples at 95 °C for 1 min and place on ice. Before loading the gel, rinse the wells again. Load 2.5 μL of each splicing time point and sizing column peak and 5 μL elution fraction samples into neighboring lanes. Also load 1 μL of pre-mRNA standard. Run gel for 2 h at a constant wattage of 30 W.
4. Take down the gel. Remove one of the glass plates, lay down a used X-ray film on top of the gel, and press down to adhere the gel to the film. Carefully peel the X-ray film with the gel from the glass plate, and then cover the gel supported by the film with plastic wrap. Place the gel in a phosphorimager cassette and expose overnight.
5. Using the appropriate software to analyze the phosphorimage of the gel, box out bands for pre-mRNA, one of the splicing intermediates and/or splicing products (Fig. 3b). To correct for background, subtract the intensity of an equally sized box of a region in the lane above the band of interest from the band intensity. To normalize for the amount of label in each band, divide the corrected band intensity by the number of uridine residues in the corresponding RNA species. To quantify percentage of splicing efficiency for each lane separately,

divide the intensity of the normalized bands for splicing intermediates or splicing products over the total intensity of bands for pre-mRNA plus splicing intermediates and splicing products. To quantify the concentration of spliceosomes in elution fractions, divide the intensity of a normalized splicing intermediate or splicing product band by the intensity of the normalized pre-mRNA standard band and multiply by 5 nM (or the concentration of pre-mRNA in the standard).

4 Notes

1. Contamination of reagents, tubes, equipment, etc., by RNases is always a concern when handling RNA. Gloves should be worn during the purification and reagents, pipette tips, tubes, etc., should be designated for RNA use only. Bake glassware in a 250 °C oven for at least four hours. We do not use DEPC-treated water, but instead prefer glass distilled water stored in baked glassware.
2. Most commonly, a derivative of the AdML gene product is employed as the pre-mRNA substrate for in vitro splicing in HeLa extract [2]. If the DNA template is contained in a plasmid, the plasmid must be linearized at the desired 3′ end by digestion with the appropriate restriction enzyme. Alternatively, a PCR product may also be used as template.
3. We normally label RNAs with [a-^{32}P] UTP, but other nucleotides can also be used if required. The specific activity of the RNA is controlled by modulating the concentrations of cold and hot UTP in the transcription reaction. We typically use one-tenth the volume of the transcription reaction for hot UTP and 400 μM cold UTP to obtain a "low" label that is sufficient to analyze splicing chemistry and detect complexes during purification.
4. HeLa nuclear extract is prepared as described in refs. [29, 30]. The splicing efficiency of the extract is dependent on the cell source and concentration of potassium and magnesium in the splicing reaction. We purchase HeLa cells that have been cultured for less than 2 weeks and shipped on wet ice from BioVest Intl. After preparing the extract, we freeze it in 200–400 μL aliquots at −80 °C. The extract should be first tested with different concentrations of KGlu and MgAc to determine the best conditions for splicing [31]. We find that the range of optimal conditions lies between 0–100 mM KGlu and 0–6 mM MgAc. The nuclear extract should have at least 20 % splicing efficiency to effectively purify splicing complexes.
5. Filter sterilize KGlu, MgAc, and heparin stocks. We divide these into 1 mL aliquots and store at −20 °C

6. We use two 12 nt DNA oligonucleotides complimentary to the regions between 10 and 30 nt upstream of the 5′ splice site in the AdML pre-mRNA [19]. This region is accessible in unspliced pre-mRNA and the oligos form RNA/DNA hybrids, which allows endogenous RNase H to cleave the RNA. The region is protected from oligo binding in assembled spliceosomes.
7. MS2-MBP protein is expressed in *Escherichia coli* and purified first by amylose affinity followed by heparin chromatography as described in refs. [18, 32].
8. The buffer conditions for the purification were chosen with electron microscopy studies in mind. Often magnesium is thought to be important to stabilize ribonucleoprotein complexes. However, we found that splicing complexes tended to aggregate when 2 mM $MgCl_2$ was present in the buffer, which was alleviated by addition of 5 mM EDTA. Although some proteins disassociate in the presence of EDTA (e.g., SR proteins), most core splicing components remain intact [19]. Different buffer conditions have been successfully used to purify splicing complexes and may be tested as desired.
9. Heparin is added to disrupt nonspecific interactions between protein and nucleic acids and helps prevent splicing complexes from aggregating. However, it may also disrupt weaker specific interactions within the splicing complexes and may be omitted or used at a lower concentration if desired (e.g., 7, 18).
10. By using a small amount of affinity resin in a column geometry and minimizing the elution volume, spliceosomes elute at maximum concentration. We have not found any method to concentrate spliceosomes due to their “stickiness.” To elute splicing complexes in the smallest volume possible, use a pipette to suck out 30 μL elutions from the bottom “nib” of the Mobicol column. Usually the majority of purified spliceosomes peak in the second and third fraction at 5–15 nM concentration.
11. Depending on the downstream application for isolated splicing complexes, we recommend taking only the first half of the splicing complex peak. The second half of the peak appears to contain additional proteins including excess MS2:MBP that we observe as additional background in EM images of the spliceosomes.
12. Reapplying flow-through maximizes binding of splicing complexes to the column. Nevertheless, we find that a significant percentage of the radioactivity does not bind the column.

References

1. Will CL, Luhrmann R (2011) Spliceosome structure and function. Cold Spring Harb Perspect Biol 3(7)
2. Padgett RA, Hardy SF, Sharp PA (1983) Splicing of adenovirus RNA in a cell-free transcription system. Proc Natl Acad Sci USA 80(17):5230–5234
3. Ruskin B, Krainer AR, Maniatis T et al (1984) Excision of an intact intron as a novel lariat structure during pre-mRNA splicing in vitro. Cell 38(1):317–331
4. Lardelli RM, Thompson JX, Yates JR 3rd et al (2010) Release of SF3 from the intron branchpoint activates the first step of pre-mRNA splicing. RNA 16(3):516–528
5. Warkocki Z, Odenwalder P, Schmitzova J et al (2009) Reconstitution of both steps of *Saccharomyces cerevisiae* splicing with purified spliceosomal components. Nat Struct Mol Biol 16(12):1237–1243
6. Barabino SML, Lamond AI (1992) Antisense probes targeted to an internal domain in U2 snRNP specifically inhibit the second step of pre-mRNA splicing. Nucleic Acids Res 20(17):4457–4464
7. Bessonov S, Anokhina M, Will CL et al (2008) Isolation of an active step I spliceosome and composition of its RNP core. Nature 452(7189):846–850
8. Blencowe BJ, Sproat BS, Ryder U et al (1989) Antisense probing of the human U4/U6 snRNP with biotinylated 2′-OMe RNA oligonucleotides. Cell 59(3):531–539
9. Frilander M, Steitz J (1999) Initial recognition of U12-dependent introns requires both U11/5′ splice-site and U12/branchpoint interactions. Genes Dev 13(7):851–863
10. Furman E, Glitz DG (1995) Purification of the spliceosome A-complex and its visualization by electron microscopy. J Biol Chem 270:15515–15522
11. Gozani O, Patton JG, Reed R (1994) A novel set of spliceosome-associated proteins and the essential splicing factor PSF bind stably to pre-mRNA prior to catalytic step II of the splicing reaction. EMBO J 13:3356–3367
12. Lamond AI, Konarska MM, Sharp PA (1987) A mutational analysis of spliceosome assembly: evidence for splice site collaboration during spliceosome formation. Genes Dev 1(6): 532–543
13. Lamond AI, Sproat B, Ryder U et al (1989) Probing the structure and function of U2 snRNP with antisense oligonucleotides made of 2′-OMe RNA. Cell 58(2):383–390
14. Nottrott S, Hartmuth K, Fabrizio P et al (1999) Functional interaction of a novel 15.5 kDa [U4/U6.U5] tri-snRNP protein with the 5′ stem-loop of U4 snRNA. EMBO J 18(21):6119–6133
15. Reed R (1989) The organization of 3′ splice-site sequences in mammalian introns. Genes Dev 3(12B):2113–2123
16. Ryder U, Sproat B, Lamond A (1990) Sequence-specific affinity selection of mammalian splicing complexes. Nucleic Acids Res 18(24):7373–7379
17. Smith CW, Nadal-Ginard B (1989) Mutually exclusive splicing of alpha-tropomyosin exons enforced by an unusual lariat branch point location: implications for constitutive splicing. Cell 56(5):749–758
18. Bessonov S, Anokhina M, Krasauskas A et al (2010) Characterization of purified human Bact spliceosomal complexes reveals compositional and morphological changes during spliceosome activation and first step catalysis. RNA 16(12):2384–2403
19. Jurica M, Licklider L, Gygi S et al (2002) Purification and characterization of native spliceosomes suitable for three-dimensional structural analysis. RNA 8:426–439
20. Ilagan JO, Chalkley RJ, Burlingame AL et al (2013) Rearrangements within human spliceosomes captured after exon ligation. RNA 19:400–412
21. Behzadnia N, Golas MM, Hartmuth K et al (2007) Composition and three-dimensional EM structure of double affinity-purified, human prespliceosomal A complexes. EMBO J 26(6):1737–1748
22. Deckert J, Hartmuth K, Boehringer D et al (2006) Protein composition and electron microscopy structure of affinity-purified human spliceosomal B complexes isolated under physiological conditions. Mol Cell Biol 26(14):5528–5543
23. Herold N, Will CL, Wolf E et al (2009) Conservation of the protein composition and electron microscopy structure of *Drosophila melanogaster* and human spliceosomal complexes. Mol Cell Biol 29(1):281–301
24. Ilagan J, Yuh P, Chalkley RJ et al (2009) The role of exon sequences in C complex spliceosome structure. J Mol Biol 394(2):363–375
25. Merz C, Urlaub H, Will CL et al (2007) Protein composition of human mRNPs spliced in vitro and differential requirements for mRNP protein recruitment. RNA 13(1): 116–128
26. Sharma S, Falick AM, Black DL (2005) Polypyrimidine tract binding protein blocks the 5′ splice site-dependent assembly of U2AF and the prespliceosomal E complex. Mol Cell 19(4):485–496

27. Sharma S, Kohlstaedt LA, Damianov A et al (2008) Polypyrimidine tract binding protein controls the transition from exon definition to an intron defined spliceosome. Nat Struct Mol Biol 15(2):183–191
28. Zhou Z, Licklider LJ, Gygi SP et al (2002) Comprehensive proteomic analysis of the human spliceosome. Nature 419(6903): 182–185
29. Abmayr SM, Reed R, Maniatis T (1988) Identification of a functional mammalian spliceosome containing unspliced pre-mRNA. Proc Natl Acad Sci USA 85:7216–7220
30. Dignam JD, Lebovitz RM, Roeder RD (1983) Accurate transcription initiation by RNA polymerase II in a soluble extract from isolated mammalian nuclei. Nucleic Acids Res 11:1475–1489
31. Reichert V, Moore MJ (2000) Better conditions for mammalian in vitro splicing provided by acetate and glutamate as potassium counterions. Nucleic Acids Res 28(2):416–423
32. Jurica MS, Moore MJ (2002) Capturing splicing complexes to study structure and mechanism. Methods 28(3):336–345

Chapter 15

Complementation of U4 snRNA in *S. cerevisiae* Splicing Extracts for Biochemical Studies of snRNP Assembly and Function

Martha R. Stark and Stephen D. Rader

Abstract

Pre-messenger RNA splicing is a surprisingly complex and dynamic process, the details of which remain largely unknown. One important method for studying splicing involves the replacement of endogenous splicing components with their synthetic counterparts. This enables changes in protein or nucleic acid sequence to be tested for functional effects, as well as the introduction of chemical moieties such as cross-linking groups and fluorescent dyes. To introduce the modified component, the endogenous one must be removed and a method found to reconstitute the active splicing machinery. In extracts prepared from *S. cerevisiae*, reconstitution has been accomplished with the small, nuclear RNAs U6, U2, and U5.

We describe a comparable method to reconstitute active U4 small, nuclear RNA (snRNA) into a splicing extract. In order to remove the endogenous U4 it is necessary to target it for oligo-directed RNase H degradation while active splicing is under way, i.e., in the presence of a splicing transcript and ATP. This allows complete degradation of endogenous U4 and subsequent replacement with an exogenous version. In contrast to the procedures described for depletion of U6, U2, or U5 snRNAs, depletion of U4 requires concurrent active splicing. The ability to reconstitute U4 in yeast extract allows a variety of structural and functional studies to be carried out.

Key words U4 snRNP, Splicing extract, *S. cerevisiae*, Functional complementation, Functional reconstitution, Pre-mRNA splicing, snRNA, snRNP

1 Introduction

Alternative splicing of pre-messenger RNA (pre-mRNA) gives rise to much of the amazing diversity of human proteins, despite the relative paucity of actual genes. Pre-mRNA splicing is also important for human gene expression because even small errors in splicing can have catastrophic consequences, as illustrated by the vast number of diseases whose cause can be found in splicing errors (reviewed in refs. 1, 2). It has recently been proposed that up to 60 % of all hereditary human diseases may be caused by the disruption of normal splicing patterns [3]. The large number of proteins

Klemens J. Hertel (ed.), *Spliceosomal Pre-mRNA Splicing: Methods and Protocols*, Methods in Molecular Biology, vol. 1126, DOI 10.1007/978-1-62703-980-2_15,

involved in pre-mRNA splicing as well as the many macromolecular rearrangements that must occur during the process have hampered our understanding of the detailed molecular mechanism by which the splicing machinery carries out its functions.

The chemical steps of pre-mRNA splicing are relatively straightforward: in the first step, the 2′ hydroxyl of the branch point adenosine reacts with the 5′ splice site, breaking the phosphodiester bond between the last exonic nucleotide and the first intronic one. This results in the formation of the so-called lariat intron intermediate with the branched adenosine connected to the intronic loop on one side and the downstream exon on the other. The 5′ exon is not covalently attached to the remainder of the transcript after the first chemical step. In the second step, the 3′ hydroxyl of the upstream (5′) exon reacts with the 3′ splice site, thereby ligating the two exons together and releasing the intervening intron as a lariat.

In contrast to the simplicity of the chemical steps, the assembly, catalytic mechanism, and regulation of splicing are so complicated that after 30 years of study we still understand only the broadest outlines of these processes. Assembly of the splicing complex, known as the spliceosome, appears to happen as an ordered series of reversible events (reviewed in ref. 4). The spliceosome is composed of five small, nuclear ribonucleoproteins (snRNPs), each consisting of a small, nuclear RNA (snRNA) and a set of proteins, as well as a number of other proteins and protein complexes. Two of these snRNPs, U1 and U2, recognize the 5′ splice site and branch point of the pre-mRNA, respectively, via direct base pairing between snRNAs and transcript. The other three, U4, U5, and U6, join the assembling spliceosome together as a preassembled tri-snRNP. After a number of rearrangements, the active spliceosome, consisting of the U2, U5, and U6 snRNPs, remains to catalyze the chemical steps of splicing.

General functions for U5 (in holding the exons together) and U6 (in catalyzing the chemical steps) have been suggested, but U4's role in the splicing process remains enigmatic, as it dissociates prior to the chemical steps. U4 is known, however, to be closely associated with U6 via extensive base pairing. This association is necessary for the introduction of U6 into the assembling spliceosome, but whether it serves some additional regulatory function and what prevents U6 from assembling by itself are unknown.

A number of genetic and biochemical studies have suggested that the main role of U4, aside from base pairing to U6, is to facilitate assembly with U5 to form the tri-snRNP [5]. In addition, recent clinical work has linked mutations in U4 to a type of congenital dwarfism [6, 7]. Nevertheless, many aspects of U4's functions remain to be worked out, including the mechanism by which it associates with U6, its role in promoting assembly of the tri-snRNP, and the mechanism by which it dissociates from the spliceosome.

The method presented here provides a powerful tool for studying this enigmatic molecule.

The normal method for reconstituting snRNAs involves either DNA oligonucleotide-targeted degradation of the endogenous RNA by RNase H [8–14] or the removal of the RNA using streptavidin-agarose affinity selection with 2′-*O*-methyl RNA oligos complementary to the snRNA [15–18]. This is followed by addition of an exogenous version of the snRNA. The first challenge is to identify a region of the snRNA that is accessible to the targeting oligonucleotide, and frequently a number of oligos must be tested before an effective one is found. The second complication is that snRNAs may exist in more than one form in a static extract (i.e., an extract in which nothing is happening biochemically), some of which may be accessible to oligo binding and others not. Third, and finally, the added exogenous snRNA may not assemble properly with other splicing components.

In the method presented here, the region of U4 targeted for degradation is the 5′ end of the molecule, the part that forms the majority of the base pairing interactions with U6 (Fig. 1). Degradation of this region most effectively eliminates U4 function [19] but is not accessible in the majority of U4 molecules, as they are base paired to U6. Consequently, to make this region accessible it is necessary to ensure that U4 is actively cycling in and out of the spliceosome. In yeast extract this requires the addition of ATP as

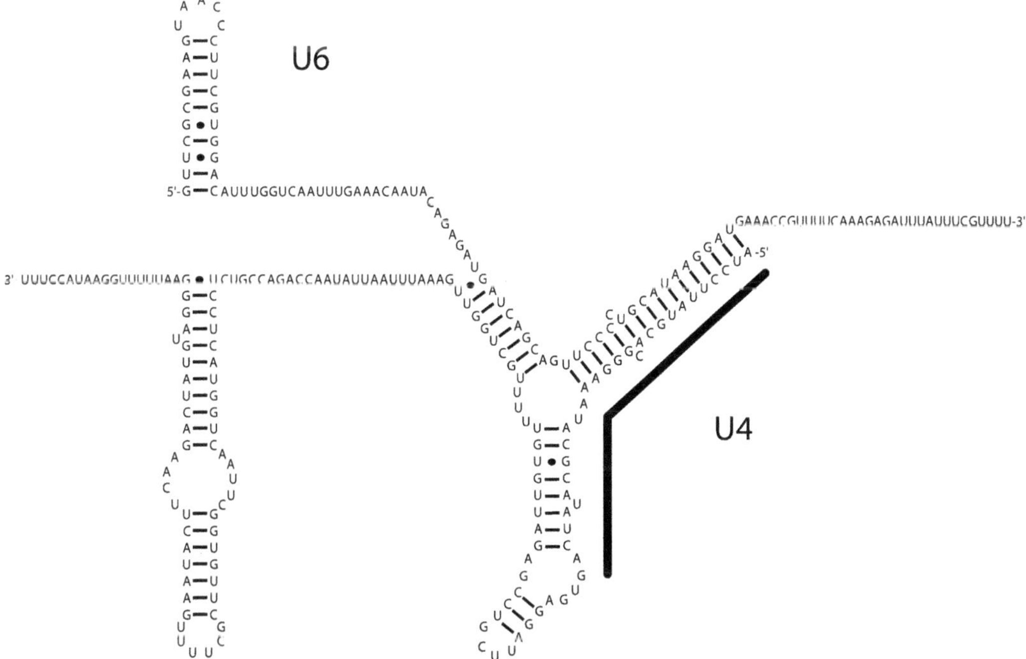

Fig. 1 Base pairing between U6 snRNA (*top*) and U4 snRNA (*bottom*). The location of the DNA oligo used in this study to degrade U4 is indicated by the *thick, black line*

well as an intron-containing RNA transcript. Once endogenous U4 has been sufficiently depleted, exogenous U4 is added to reconstitute splicing activity. A large excess of U4 relative to other splicing components was found to be necessary for maximal splicing, in part due to degradation of the exogenous U4 and perhaps due to its misfolding as well [5].

2 Materials

Take all possible precautions to avoid nuclease contamination: purchase nuclease-free plasticware, use ultrapure water in all buffers and solutions, and filter all solutions that will be added to reactions containing RNA through high-protein-binding nitrocellulose filters.

2.1 Prepare Splicing Extract

Make yeast splicing extract from protease-deficient strain BJ2168 [20] following the protocol in Chapter 9.

2.2 Prepare Unlabeled U4 snRNA and Actin Pre-mRNA

1. Plasmids containing the T7 promoter followed by a partial actin gene (pJPS149; [21] (*see* **Note 1**)) and the U4 gene [22].
2. Restriction enzymes HindIII and StyI, and appropriate buffers.
3. MEGAshortscript Kit for in vitro transcription (Invitrogen).
4. Vertical gel system; glass gel plates approximately 16 cm × 14 cm, 0.75 mm spacers.
5. 20× Tris–borate–EDTA (TBE) gel running buffer (1.8 M Tris base, 1.8 M boric acid, 25 mM EDTA). Autoclave.
6. 6 % (19:1), 7 M urea polyacrylamide gel.
7. Formamide.
8. Fluor-coated thin-layer chromatography plate.
9. Handheld UV light.
10. Disposable 1.5 mL microcentrifuge tube pestle (Kontes Scientific).
11. DTR gel filtration cartridge (Edge BioSystems).
12. 20 mg/mL glycogen in water.
13. 3 M NaOAc, pH 5.2.
14. 70 and 100 % EtOH.

2.3 Prepare Radioactive Actin Pre-mRNA and U4 Northern Probe

1. U4 Northern probe (14B), 200 μM stock in water: 5′AGGTATTCCAAAAATTCCCTAC3′.
2. T4 polynucleotide kinase (PNK) and 10× buffer.
3. T7 RNA polymerase and 10× buffer.
4. Superasin RNase inhibitor (Invitrogen).
5. 100 mM ATP, CTP, GTP, UTP.

6. α^{32}P-GTP and γ^{32}P-ATP, 10 mCi/mL, 3,000 Ci/mmol.
7. TE, pH 7.5 (10 mM Tris–HCl, pH 7.5, 1 mM EDTA). Filter sterilize.
8. G-25 spin columns.
9. Scintillation counter.

2.4 Deplete Endogenous U4

1. U4-targeting oligonucleotide, 200 μM stock in water: 5′CTGATATGCGTATTTCCCGTGCATAAGGAT3′.
2. 5× splicing buffer (12.5 mM $MgCl_2$, 300 mM KPO_4, pH 7.0, 15 % PEG 8000). Make fresh each time from stocks that have been autoclaved, filter sterilized, and stored at −80 °C.
3. 100 mM ATP.

2.5 Test Effectiveness of Depletion

1. Splicing stop buffer (300 mM NaOAc, pH 5.2, 1 mM EDTA, 0.1 % SDS, 34 μg/mL *E. coli* tRNA). Filter sterilize. Store at room temperature. Add the tRNA just before use.
2. Phenol/chloroform, pH 6.7 and chloroform.
3. Urea loading buffer (4.2 g urea, 500 μL 20× TBE, 20 μL 0.5 M EDTA, 2.5 mg xylene cyanol, 2.5 mg bromophenol blue, water to 10 mL). Filter sterilize. Aliquot and store at −20 °C.
4. Hybond + nylon membrane (GE Healthcare).
5. Whatman paper.
6. Semidry electroblotter (e.g., Owl).
7. UV cross-linking apparatus (e.g., Stratalinker).
8. Rapid Hyb hybridization buffer (GE Healthcare).
9. Hybridization oven.
10. 20× SSC (3 M NaCl, 0.3 M sodium citrate; pH to 7.0 with a few drops of concentrated HCl).
11. Northern wash buffer (6× SSC, 0.2 % SDS).
12. Phosphorimager, screen, and cassette.

2.6 Reconstitution with Exogenous U4

Same materials as in Subheadings 2.2–2.5.

2.7 Measure Pre-mRNA Splicing

Same materials as in Subheadings 2.2–2.5.

3 Methods

3.1 Prepare Splicing Extract

See Chapter 9 in this volume.

3.2 Prepare Unlabeled U4 snRNA and Actin Pre-mRNA

1. Linearize the U4 IVT template. Digest 10 μg of the pT7U4 plasmid by mixing the DNA with 10 μL 10× restriction buffer, 10 μL StyI, 1 μL 10 mg/mL BSA, and water to 100 μL. Incubate for 4 h at 37 °C (*see* **Notes 2** and **3**). Add 100 μL water to the digestion reaction, phenol/chloroform extract, and precipitate (*see* **Note 4**). Resuspend pellet in 12 μL water and determine the concentration by measuring the A_{260} on a spectrophotometer.
2. Linearize the actin IVT template. Digest 10 μg of the pJPS149 (actin pre-mRNA template) plasmid by mixing the DNA with 3 μL 10× restriction buffer, 1.5 μL HindIII, 3 μL 1 mg/mL BSA, and water up to 30 μL. Incubate for 1 h at 37 °C. Prepare DNA for transcription by extraction and precipitation, as above.
3. Transcribe U4 RNA and actin pre-mRNA using an in vitro transcription kit (e.g., MEGAshortscript) with the linearized DNA templates from **steps 1** and **2** (*see* **Note 5**). Use approximately 2–4 μg linearized pT7U4 and 1 μg pJPS149 per 20 μL reaction. Incubate for 3–4 h at 37 °C.
4. Make the denaturing polyacrylamide gel once the transcription reactions have been set up (*see* **Note 6**). Near the end of the transcription incubation step, pre-run the gel for 15–30 min at 400 V (*see* **Note 7**).
5. Separate full-length IVT products from any truncated RNAs. Add 20 μL formamide to each transcription reaction and heat for 3 min at 65 °C. Load samples onto the prepared gel (*see* **Note 8**). Run for 1 h at 400 V.
6. Visualize the RNA transcripts by UV shadowing (*see* **Note 9**). Move the wrapped gel onto a clean, scratchproof surface. Using a sterile scalpel, cut the marked fragment of gel away from the remainder and place in a microcentrifuge tube.
7. Elute the RNA from the gel fragment. Crush with a disposable pestle, add 400 μL water, crush some more, and then heat at 70 °C for 10 min.
8. Remove the acrylamide from the eluted RNA. While heating the gel solution, pre-spin the DTR cartridge at 850 × *g* for 3 min in a microcentrifuge. Transfer the cartridge to a new tube. Load the entire gel solution onto the column and spin again for 3 min. Discard the cartridge containing the acrylamide.
9. Precipitate the eluted RNA with 0.01 volume 3 M NaOAc, pH 5.2 plus 2.5 volumes cold 100 % EtOH, adding 15 μg glycogen as a carrier (*see* **Note 4**). Resuspend the RNA in 25–50 μL water and determine the concentration by UV spectrophotometry (*see* **Note 10**).

3.3 Prepare Radioactive Actin Pre-mRNA and U4 Northern Probe

1. In vitro transcribe radiolabeled actin. To 500 ng linearized pJPS149 (in a maximum volume of 4.5 μL) add 1 μL 10× T7 RNA polymerase buffer, 0.5 μL 10 mM each NTP (ATP, CTP, UTP), 0.5 μL 0.5 mM GTP, 0.5 μL Superasin, 2.5 μL α^{32}P-GTP, 0.5 μL T7 RNA polymerase, and water to 10 μL (*see* **Note 11**). Incubate for 1.5 h at 37 °C.
2. 5′-end radiolabel the U4 DNA oligo. To 25 pmol U4 14B oligo add 2.5 μL 10× PNK buffer, 2.5 μL γ^{32}P-ATP, 1.5 μL PNK, and water to 25 μL. Incubate for 1 h at 37 °C.
3. Remove unincorporated nucleotides from the IVT and kinasing reactions, and calculate the efficiency of ^{32}P incorporation. Pre-spin 2 G-25 spin columns according to the manufacturer's instructions. Dilute the labeling reactions to 50 μL with TE, pH 7.5. Count 1 μL in a scintillation counter. Load the remainder onto the G-25 column and spin according to instructions. Count 1 μL of the flow through in a scintillation counter. Determine the percent ^{32}P-GTP incorporation into the actin transcript and the cpm/fmol actin (*see* **Note 12**). Dilute the actin to 4 fmol/μL in TE.

3.4 Deplete Endogenous U4

1. To 4 μL of yeast splicing extract add 1.6 μL 5× splicing buffer, 0.64 μL 10 μM U4-targeting oligo, 15–80 fmol unlabeled IVT actin pre-mRNA, 0.8 μL 100 mM ATP, and water to 8 μL (*see* **Note 13**). At the same time set up a mock-depleted reaction, leaving out the U4 oligo. Incubate for 30 min at 30 °C (*see* **Note 14**).

3.5 Test Effectiveness of Depletion

1. Terminate reactions by adding 200 μL splicing stop buffer. Phenol/chloroform extract to remove the proteins from the splicing extract, and precipitate the RNA with 3 volumes of cold 100 % EtOH (*see* **Note 4**).
2. Resuspend pellets in 8 μL urea loading buffer. Heat for 3 min at 65 °C and load on a pre-run 6 %, 7 M urea denaturing polyacrylamide gel. Run in 1× TBE 400 V for 30 min.
3. Transfer RNA to a nylon membrane using a semidry electroblotter (*see* **Note 15**). Transfer in 1× TBE for 20 min at 2.5 mA/cm^2 of membrane.
4. Cross-link RNA to membrane using a UV cross-linker (approximately 120,000 μJ using a 254 nm light source for 25–50 s).
5. Pre-hybridize the cross-linked membrane in 5–10 mL Rapid hyb buffer in a roller tube in a hybridization oven at room temperature for 30 min.
6. Add 20 μL ^{32}P-labeled U4 14B probe, and hybridize the probe to the RNA at room temperature for 1–2 h.

7. Pour off the probe and wash 3 × 5 min in Northern wash buffer to remove any probe that has bound nonspecifically to the membrane (*see* **Note 16**).
8. Remove membrane from tube with forceps, wrap in plastic wrap, and expose to phosphor screen overnight.
9. Scan the phosphor screen in a phosphorimager, and determine the efficiency of U4 degradation by quantitatively comparing the intensity of full-length U4 in the depleted reaction to that in the mock-depleted reaction (Fig. 2a).

3.6 Reconstitution with Exogenous U4

1. To 8 μL of depleted splicing extract (**step 1**, Subheading 3.4) add either 0.48 μL of water (control) or 5 μM U4 IVT (283 nM final concentration). Incubate for 12 min at 23 °C to allow time for snRNP assembly.

3.7 Measure Pre-mRNA Splicing

1. Add 4 fmol internally ^{32}P-GTP-labeled actin pre-mRNA transcript and incubate for an additional 30 min at 23 °C.

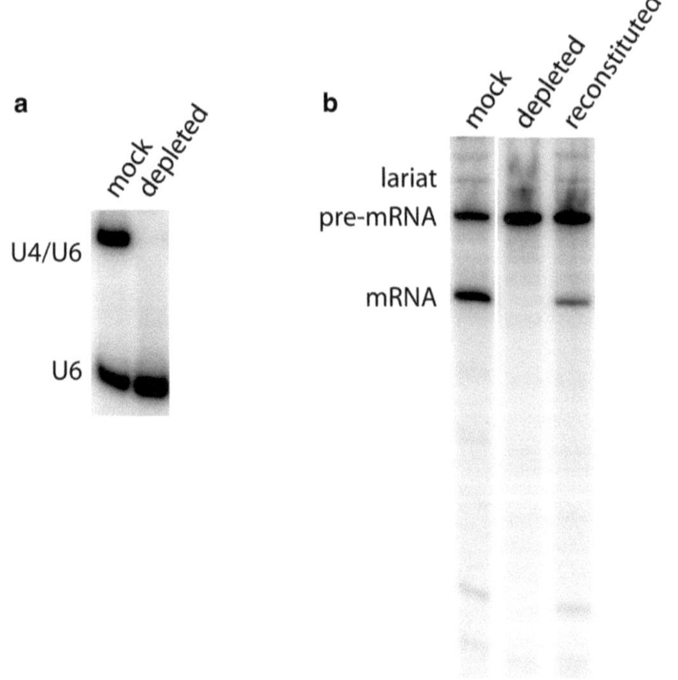

Fig. 2 Example of U4 depletion and reconstitution of splicing. (**a**) Non-denaturing northern blot of mock-depleted (*lane 1*) and U4-depleted (*lane 2*) splicing extract probed for U6. In the absence of U4, all of the U6 snRNA runs as free U6. (**b**) Denaturing autoradiogram of pre-mRNA splicing in mock-depleted (*lane 1*), U4-depleted (*lane 2*), and reconstituted (*lane 3*) extract. Positions of pre-mRNA, mRNA, and lariat intron are marked on the *left*

2. Stop the reaction with 200 μL splicing stop buffer, extract with phenol/chloroform, and precipitate with 3 volumes of cold 100 % EtOH (*see* **Note 4**).
3. Resuspend the pellets in 8 μL urea loading buffer, heat for 3 min at 65 °C, and electrophorese through a 6 % (19:1) 7 M urea denaturing polyacrylamide gel for 1 h at 400 V.
4. Remove the top plate from the gel. Press a piece of Whatman paper onto the gel, and carefully peel the gel off of the gel plate. Wrap the gel in plastic wrap and expose to a phosphor screen at −80 C for a few hours.
5. Scan the phosphor screen (Fig. 2b), and calculate the splicing efficiency (*see* **Note 17**).

4 Notes

1. We have experienced problems with the pJPS149 plasmid when purified from a glycerol stock of DH5α. Consequently, always freshly transform DH5α cells from the original plasmid stock when more plasmid is needed.
2. Use a large excess of StyI and digest for 4 h–overnight as the enzyme cuts very inefficiently.
3. Analyze 200 ng of the cut plasmid alongside 200 ng of uncut plasmid on a 0.8 % agarose gel containing 0.5 μg/mL EtBr in 1× TBE to make sure that the plasmid is completely linearized.
4. Extract one time with an equal volume of phenol/chloroform. Back extract with 1 volume of chloroform and then precipitate with 0.1 volume of 3 M NaOAc, pH 5.2 and 2 volumes of cold 100 % EtOH. Spin for 30 min at 4 °C, aspirate the supernatant, wash pellet with 70 % EtOH, spin for 5 min, and aspirate supernatant. Allow pellet to air-dry for 5 min.
5. The IVT U4 and actin RNAs are 163- and 590-nt long, respectively. The MEGAshortscript kit has been optimized for high yields of RNA in the 20–500 nt range, but we have found it to also give excellent yields of the slightly larger actin RNA. The transcription reaction can also be carried out without a kit, although the RNA product yields may be lower, especially for the small U4 RNA.
6. Mix 2.25 mL 40 % (19:1) acrylamide, 750 μL 20× TBE, 6.3 g urea, and 7.3 mL water in a 50 mL glass beaker with a stir bar. Stir until the urea has completely dissolved. Add 150 μL 10 % ammonium persulfate and 15 μL TEMED. Mix and pour into gel cassette. Push comb most of the way into the gel in order to make deep wells. Allow to solidify for at least 1 h.
7. Remove the comb from the polyacrylamide gel, and place the gel in the gel box. Fill with 1× TBE buffer. Use a syringe to rinse

out each well with running buffer, making sure that there are no air bubbles trapped in the wells or in the space under the gel.

8. Rinse the wells again before loading the samples. Load the entire sample into one well. Leave an empty well or two between the samples to prevent any spillover contamination of the transcripts.
9. Remove the glass plates from the gel, and wrap the gel in plastic wrap. Place gel on a fluorescent thin-layer chromatography plate. Using a handheld shortwave UV lamp, identify the location in the gel of the desired RNA product by looking for the shadow cast by the RNA. Mark the location of the band by drawing a box on the plastic wrap with a permanent marker.
10. Store RNA transcripts at −80 °C. Dilute some of the stock to a working concentration and aliquot into several tubes so that the stock tube is not repeatedly frozen/thawed.
11. Assemble the IVT reaction at room temperature. The spermidine in the T7 polymerase buffer can cause the DNA template to precipitate at 4 °C.
12. Calculate the molar activity of the transcript.

    Proportion of GTP incorporated into actin:

    (cpm/μL after G25 column)/(cpm/μL before column) = % ^{32}P GTP incorporated.

    Total moles of ^{32}P GTP in the IVT reaction = 8.3×10^{-9} mmol (8.3 pmol).

    Total moles of unlabeled GTP in the IVT reaction = 2.5×10^{-7} mmol (250 pmol).

    Total moles of GTP in the IVT reaction:

    8.3 pmol + 250 pmol = 2.58×10^{-7} mmol (258 pmol).

    Total moles of GTP incorporated into actin:

    Total GTP in reaction x% incorporated = 2.58×10^{-7} mmol GTP x% ^{32}P GTP incorporated.

    Moles of actin synthesized:

    Total moles of GTP incorporated/(moles of G/mole of actin) = total moles of GTP incorporated into actin/(118 moles G nucleotides per mole actin).

    Molar activity of actin transcript:

    Total cpm for IVT reaction/moles of actin (expressed in cpm/fmol).

13. Varying amounts of pre-mRNA transcript are required for complete degradation of U4, dependent on the specific splicing extract being depleted. Titrate the transcript in the depletion reaction to determine the lowest amount necessary for efficient degradation of U4 and inhibition of splicing.

14. The splicing extract contains sufficient endogenous RNase H activity to degrade the DNA oligo:RNA duplex. Therefore, it is not necessary to add any exogenous RNase H.
15. Cut a piece of nylon membrane to cover the portion of the gel containing the RNA. The U4 will be in the top half of the gel, so it is not necessary to transfer the bottom half. Label the back of the membrane with a pencil. Cut six pieces of Whatman paper slightly larger than the membrane. Remove the top plate from the gel, and press one piece of Whatman paper onto the portion of the gel for transfer. Carefully peel the Whatman paper with the gel stuck to it off of the bottom gel plate. Make a sandwich on the electroblotter platform consisting of two pieces of Whatman paper wetted in 1× TBE buffer, followed by the Whatman/gel (gel facing up; if the Whatman paper on the gel does not get completely wet, add more buffer underneath it; gently roll a pipet over the surface of the gel to remove any trapped air bubbles), and then three more wet pieces of Whatman paper on top. Carefully place electroblotter lid on top of stack, and begin transfer.
16. The probe in Rapid hyb buffer can be saved at 4 °C and used up to three times.
17. To calculate percent splicing, divide the intensity of bands corresponding to product (mRNA and lariat) by the total intensity of the starting material plus the products (pre-mRNA, lariat, and mRNA):

Splicing efficiency = products/total = (mRNA + lariat)/(pre-mRNA + lariat + mRNA).

Acknowledgments

This work was supported by NSERC Discovery Grant 298521 and UNBC Office of Research awards to SDR.

References

1. Faustino NA, Cooper TA (2003) Pre-mRNA splicing and human disease. Genes Dev 17:419–437
2. Wang G-S, Cooper TA (2007) Splicing in disease: disruption of the splicing code and the decoding machinery. Nat Rev Genet 8:749–761
3. López-Bigas N, Audit B, Ouzounis C et al (2005) Are splicing mutations the most frequent cause of hereditary disease? FEBS Lett 579:1900–1903
4. Will CL, Lührmann R (2011) Spliceosome structure and function. Cold Spring Harbor Perspect Biol 3:1–23
5. Hayduk AJ, Stark MR, Rader SD (2012) In vitro reconstitution of yeast splicing with U4 snRNA reveals multiple roles for the 3′ stem-loop. RNA 18:1075–1090
6. Edery P, Marcaillou C, Sahbatou M et al (2011) Association of TALS developmental disorder with defect in minor splicing component U4atac snRNA. Science 332: 240–243
7. He H, Liyanarachchi S, Akagi K et al (2011) Mutations in U4atac snRNA, a component of the minor spliceosome, in the developmental disorder MOPD I. Science 332:238–240

8. McPheeters DS, Fabrizio P, Abelson J (1989) In vitro reconstitution of functional yeast U2 snRNPs. Genes Dev 3:2124–2136
9. Fabrizio P, McPheeters DS, Abelson J (1989) In vitro assembly of yeast U6 snRNP: a functional assay. Genes Dev 3:2137–2150
10. O'Keefe RT, Norman C, Newman AJ (1996) The invariant U5 snRNA loop 1 sequence is dispensable for the first catalytic step of pre-mRNA splicing in yeast. Cell 86:679–689
11. Pan ZQ, Ge H, Fu XY et al (1989) Oligonucleotide-targeted degradation of U1 and U2 snRNAs reveals differential interactions of simian virus 40 pre-mRNAs with snRNPs. Nucleic Acids Res 17:6553–6568
12. Pan ZQ, Prives C (1988) Assembly of functional U1 and U2 human-amphibian hybrid snRNPs in Xenopus laevis oocytes. Science 241:1328–1331
13. Hamm J, Dathan NA, Mattaj IW (1989) Functional analysis of mutant Xenopus U2 snRNAs. Cell 59:159–169
14. Hamm J, Dathan NA, Scherly D et al (1990) Multiple domains of U1 snRNA, including U1 specific protein binding sites, are required for splicing. EMBO J 9:1237–1244
15. Wersig C, Bindereif A (1992) Reconstitution of functional mammalian U4 small nuclear ribonucleoprotein: Sm protein binding is not essential for splicing in vitro. Mol Cell Biol 12:1460–1468
16. Wolff T, Bindereif A (1992) Reconstituted mammalian U4/U6 snRNP complements splicing: a mutational analysis. EMBO J 11:345–359
17. Ségault V, Will CL, Sproat BS, Lührmann R (1995) In vitro reconstitution of mammalian U2 and U5 snRNPs active in splicing: Sm proteins are functionally interchangeable and are essential for the formation of functional U2 and U5 snRNPs. EMBO J 14:4010–4021
18. Will CL, Rümpler S, Klein Gunnewiek J et al (1996) In vitro reconstitution of mammalian U1 snRNPs active in splicing: the U1-C protein enhances the formation of early (E) spliceosomal complexes. Nucleic Acids Res 24: 4614–4623
19. Black DL, Steitz JA (1986) Pre-mRNA splicing in vitro requires intact U4/U6 small nuclear ribonucleoprotein. Cell 46:697–704
20. Ansari A, Schwer B (1995) SLU7 and a novel activity, SSF1, act during the PRP16-dependent step of yeast pre-mRNA splicing. EMBO J 14:4001–4009
21. Vijayraghavan U, Parker R, Tamm J et al (1986) Mutations in conserved intron sequences affect multiple steps in the yeast splicing pathway, particularly assembly of the spliceosome. EMBO J 5:1683–1695
22. Ghetti A, Company M, Abelson J (1995) Specificity of Prp24 binding to RNA: a role for Prp24 in the dynamic interaction of U4 and U6 snRNAs. RNA 1:132–145

Chapter 16

Expression and Purification of Splicing Proteins from Mammalian Cells

Eric Allemand and Michelle L. Hastings

Abstract

Pre-mRNA splicing is a complex process that is carried out by a large ribonucleoprotein enzyme, termed the spliceosome, which comprises up to 200 proteins. Despite this complexity, the role of individual spliceosomal proteins in the splicing reaction has been successfully investigated using cell-free assays. In many cases, the splicing factor of interest must be expressed and purified in order to study its function in vitro. Posttranslational modifications such as phosphorylation, methylation, acetylation, and ubiquitination of splicing factors are important for activity. Thus, their purification from mammalian cells presents numerous advantages. Here, we describe a method for expression and purification of splicing proteins from mammalian cells.

Key words Pre-mRNA, Splicing factors, Recombinant protein, YB-1, SRSF1

1 Introduction

Splicing factors are proteins that have been identified in purified spliceosomes and play a role in pre-mRNA splicing. Assigning function to spliceosomal proteins has been aided enormously by cell-free systems including in vitro splicing (coupled or uncoupled from transcription) and binding assays (RNA and proteins). In many cases, these techniques require the expression and purification of the splicing factor. Currently, recombinant proteins can be produced from numerous sources, although bacteria have been the most traditional source historically. Proteins generated in bacteria are limited in their ability to perform the posttranslational protein modifications that are critical for activity of many splicing factors [1–6]. Eukaryotic systems such as yeast and insect cell infection with baculovirus are advantageous in that they can be scaled up and they are capable of posttranslational modifications. However, both of these cell systems are subject to unique posttranslational modifications and protein-processing events that may compromise their activity in heterogeneous cell-free assays.

Klemens J. Hertel (ed.), *Spliceosomal Pre-mRNA Splicing: Methods and Protocols*, Methods in Molecular Biology, vol. 1126, DOI 10.1007/978-1-62703-980-2_16, © Springer Science+Business Media, LLC 2014

Production of proteins in mammalian cells is beneficial because the cells can generate the recombinant factor with the proper state of maturation. In the case of splicing proteins, in particular those with arginine- and serine-rich (RS) domains, the unphosphorylated proteins are often insoluble and difficult to purify. Thus, using the mammalian expression system in which posttranslational modifications such as phosphorylation occur naturally greatly improves the solubility and ease of purification. Moreover, it is relatively easy, now, to scale up production from mammalian cell cultures and obtain large enough quantities for many experiments. To aid in the purification of proteins from mammalian cells, an amino (N)- or a carboxy (C)-terminal epitope or a histidine tag is incorporated and used for specific enrichment by affinity chromatography. The primary drawback with this approach is the potential co-purification of endogenous partners that could affect the activity of the splicing factor. Thus, if necessary, additional purification steps such as ammonium sulfate precipitation, cesium chloride and glycerol gradient centrifugation, gel filtration, or ion exchange chromatography can be employed to yield recombinant splicing proteins of high purity and potentially greater activity.

Numerous approaches have been developed for the expression and purification of recombinant proteins from transiently transfected mammalian cells [7–9]. Various protein tags have been used for their purification including, but not limited to, His, T7, GST, and FLAG [8–11]. In addition, cells grown in monolayers or in suspension can be used to express proteins [8, 9]. Here we describe the purification of a histidine-tagged splicing factor from HEK-293-EBNA1 (HEK293E) cells grown in suspension using nickel-NTA beads. In this procedure, we used a cesium gradient to remove nucleic acids; however, in the purification of other splicing proteins (like SRSF1) this step is replaced by an additional salt precipitation to achieve greater purity by making the crude extract less complex before the start of the purification protocol. HEK293E cells are easy to grow in serum-free suspension cultures and have high transfection efficiency using polyethylenimine (PEI), which are two cost-effective advantages for very-large-scale production of recombinant proteins.

Many splicing proteins have been purified using this method [10–14]. Here, we use as an example the expression and purification of splicing factors YB-1 [15] and SRSF1 [1] to demonstrate a typical approach to purifying splicing factors from mammalian cells (Fig. 3). A general scheme for expression and purification is outlined in Fig. 1. First, an expression plasmid must be constructed in a suitable mammalian expression vector. Second, cells are transfected with the expression plasmid and allowed to grow for 3–4 days. Third, cells are harvested and lysed, and then total protein extract is subjected to cesium chloride gradient or a low-percentage salt precipitation (optional procedure). Fourth, tagged, recombinant protein is purified by passing lysate through an affinity column or incubating lysate with affinity resin in batch. Fifth, the resin is washed to remove cellular proteins and recombinant protein is eluted.

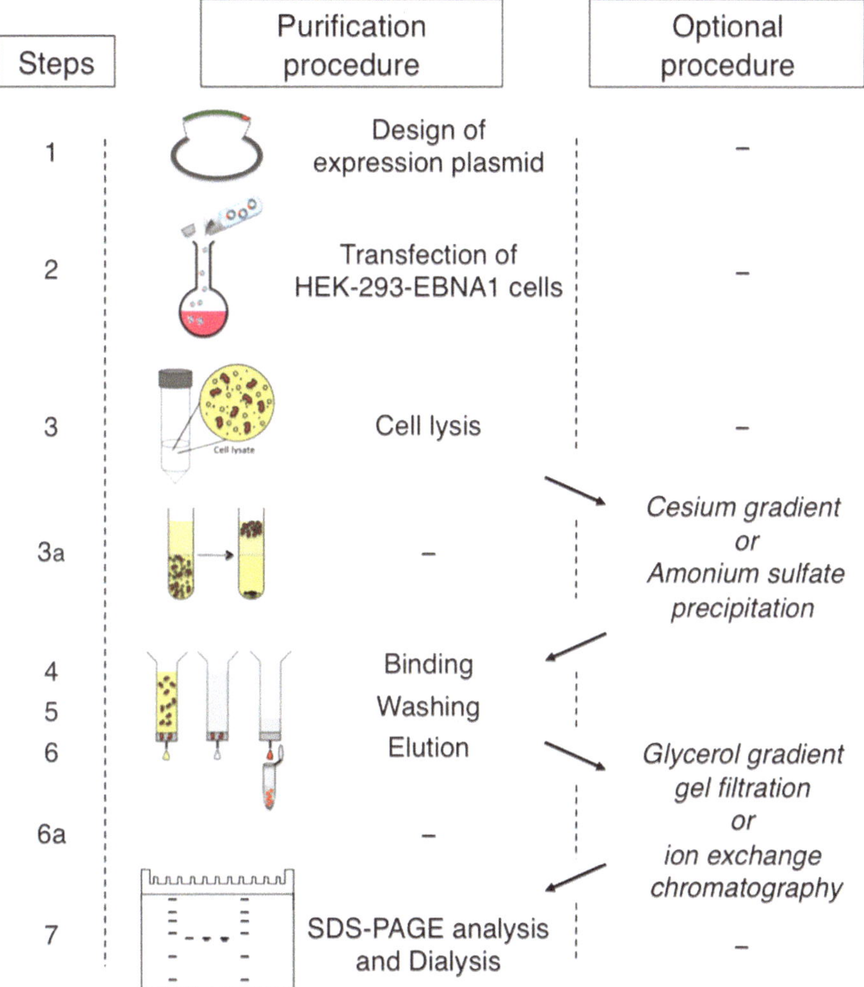

Fig. 1 Schematic procedure to purify splicing proteins from HEK-293-EBNA1 cells

Sixth, eluted proteins are analyzed for purity by SDS-PAGE and, depending on their quality, can be processed for an additional step of enrichment (glycerol gradient, gel filtration, or ion exchange chromatography). Finally, if necessary, the protein solution is exchanged into the appropriate storage or reaction buffer by dialysis.

2 Materials

2.1 Transient Transfection of Suspension Cells

1. 293-EBNA cells which stably express the Epstein–Barr virus Nuclear Antigen 1 (293 c18; CRL-10852 American Type Culture Collection).
2. MEM Joklik's suspension modification medium with L-GLUTAMINE supplemented with 5 % calf serum and penicillin/streptomycin (*see* **Note 1**).

3. Incubator preset to 37 °C, humidified, 5 % CO_2.
4. 125–12,000 ml Pyrex Florence Flask (Corning) with matching rubber stopper wrapped in foil and autoclaved (*see* **Note 2**).
5. Acid-washed stir bar.
6. PEI, linear MW = 25 kDa.
7. High-quality plasmid DNA prepared using a MAXIprep column or a similar plasmid purification protocol.

2.2 Expression Analysis and Extract Preparation

1. Phosphate-buffered saline (1× PBS).
2. Lysis buffer: 50 mM Tris–HCl pH8, 0.1 % NP-40 (added after sonication), 5 mM imidazole, 500 mM NaCl (*see* **Note 3**), 5 mM DTT or β-mercaptoethanol, protease inhibitor cocktail.
3. Sonicator (Cell disruptor W-225R, Ultrasonics Inc.).
4. Oak Ridge centrifuge tubes.
5. Protein loading solution 6×: Add 5.91 g Tris–HCl, 6 g SDS, 48 ml 100 % glycerol, 9 ml 14.7 M β-mercaptoethanol, and 30 mg bromophenol blue. Bring to 100 ml with deionized water.
6. Protein analysis apparatus (SDS-PAGE gel and protein transfer).
7. Prestained protein ladder.
8. Protein transfer solution (1 l): Add 2.21 g of CAPS, NaOH pH11, 10 % ethanol.
9. Nitrocellulose membrane.
10. Nonfat dry milk.
11. 1× TBST: 50 mM Tris–HCl pH7.6, 150 mM NaCl, 0.05 % Tween 20.
12. Monoclonal anti-poly-Histidine antibody (Sigma).
13. Coomassie blue staining solution: 0.05 % (w/v) Brilliant Blue R (Sigma), 45 % (v/v) methanol, 10 % (v/v) acetic acid, 45 % deionized water.
14. Destain solution (1 l): 45 % (v/v) ethanol, 10 % (v/v) acetic acid, 45 % deionized water.
15. SS-34 rotor (Beckman).

2.3 Cesium Chloride Density Gradient (Optional Procedure)

1. Dense CsCl solution: 10.04 g of CsCl for 10 ml of the protein extract.
2. Low-density solution: 3.9 g of CsCl in 10 ml lysis buffer solution.
3. Ultracentrifuge tubes: 25 mm × 89 mm polycarbonate tube (Beckman).
4. Ultracentrifuge Rotor Beckman 60 Ti.
5. Slide-A-Lyzer dialysis cassette 3–12 ml and cutoff 7kDa (Thermo Scientific).

2.4 Ammonium Sulfate Precipitation (Optional Procedure)

1. Ammonium sulfate.
2. Oak Ridge centrifuge tubes.
3. SS-34 rotor (Beckman).

2.5 Protein Purification

1. Poly-prep column and two-way stopcock (Bio-Rad).
2. Ni-NTA Superflow resin (Qiagen).
3. Wash buffer: 50 mM Tris–HCL pH8, 500 mM NaCl, 5 mM imidazole.
4. Elution buffer: 50 mM Tris–HCl, 500 mM NaCl, 0.5 M imidazole.
5. Protein Assay Dye Reagent Concentrate (Bio-Rad).
6. Modified buffer D (dialysis buffer): 20 mM Hepes–KOH pH8 (Sigma), 400 mM KCl, 0.2 mM EDTA, 20 % (v/v) glycerol.
7. Slide-A-Lyzer dialysis cassette 0.5–3 ml and cutoff 7 kDa (Thermo Scientific).

3 Methods

3.1 Expression Plasmid Construction

There are several affinity tags that can be added to the cDNA of splicing factors (*see* **Note 4**); here, we report the expression and purification of His-tagged proteins. To generate such constructs, we classically used the pTT3 expression vector, which was designed by Durocher et al. and is derived from pcDNA3.1 [8]. The splicing factor sequence is cloned in pTT3 using designed PCR primer that adds an optimal Kozak sequence around the translation initiation codon and the sequence coding for six histidines in N- or C-terminus of the cDNA (*see* schema in Fig. 2). We generally prefer to add the tag at the C-terminal end of the protein because it will ensure purification of only the full-length protein. However, in some cases C-terminal tag impairs the protein function and necessitates tagging at the N-terminus of the splicing factor.

3.2 Protein Expression Mammalian Cells

1. Thaw 293-EBNA cells in 100 ml spinner flask with 20 ml of medium. Rotate stir bar slowly. The cells are grown and diluted several times to reach an exponential expansion and optimize the transfection efficiency (*see* **Note 2**).
2. 24 h before transfection, transfer cells to large 1 l flask with 250 ml of media at a final cell concentration of ~2.5×10^5 cells/ml (*see* **Note 5**).
3. Transfect cells by mixing plasmid DNA (1 μg/ml of cells) and PEI (2 mg PEI/1 mg plasmid) slowly to 25 ml (10 % of final volume) culture medium in a disposable 50 ml centrifuge tube (*see* **Note 6**). Vortex solution and incubate at room temperature for at least 10 min, and then add PEI/DNA mixture to cells in culture.

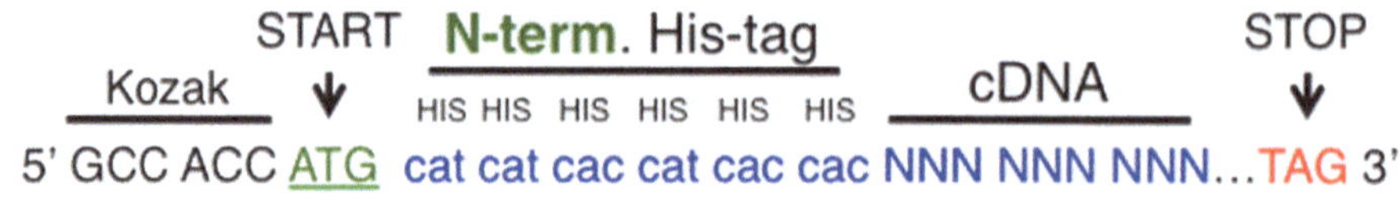

Fig. 2 Typical design of N- or C-terminal His-tagged protein expression construct that can be cloned into a mammalian expression vector

4. Analyze the production of recombinant protein at 24 and 48 h after transfection. Remove 1 ml of cells from the culture in a 1.5 ml tube, and pellet the cells by centrifugation (1,000×g for 10 min). Wash the pellet with 1 ml of PBS, and repeat the centrifugation step. The cells are resuspended in 100 μl of 1× PBS, lysed by adding 6× protein loading solution, and then boiled for 15 min. Finally, 5 and 20 μl of sample is loaded on to a 10–12 % SDS-PAGE gel in duplicate such that the gel can be cut in half and analyzed by Coomassie blue staining to assess abundance of the over-expressed protein (*see* **Note 7**) and western blot to assess the presence of expressed protein (*see* Fig. 3a, b).

3.3 Cell Lysis and Extract Preparation

1. Collect the cells 48–72 h after transfection, centrifuge (2,400×g), and wash them once with PBS. A sample of cells should be prepared as in Subheading 3.2, **step 4**, and store in −20 °C for a final analysis of protein purification.
2. Resuspend cell pellet in lysis buffer (one-tenth of starting culture volume) by gentle swirling (*see* **Note 3**). Then, sonicate the lysate three times for 30 s using a large, flat tip with 1min wait on ice to cool sample between each sonication time (*see* **Note 8**). Add NP-40 to 0.1 % (v/v) and rock or rotate in cold room for 30 min.
3. Transfer the lysate to Oak Ridge centrifuge tubes, and centrifuge the sonicated extract at 16,500 rpm (32,500×g) in the SS-34 rotor for 20 min. The supernatant is collected in a fresh tube on ice and 50 μl are removed for protein analysis in order to estimate the solubility of recombinant splicing factor by comparison with total protein extract (*see* **Note 7**).

3.3.1 Optional Steps to Enhance Purity

1. Cesium gradient separation (used to purify the splicing factor YB-1-His): Add CsCl to the soluble supernatant in order to make the dense CsCl solution (A) and rock at 4 °C for 5 min to dissolve CsCl (*see* **Note 9**). Place 10 ml of low-density solution

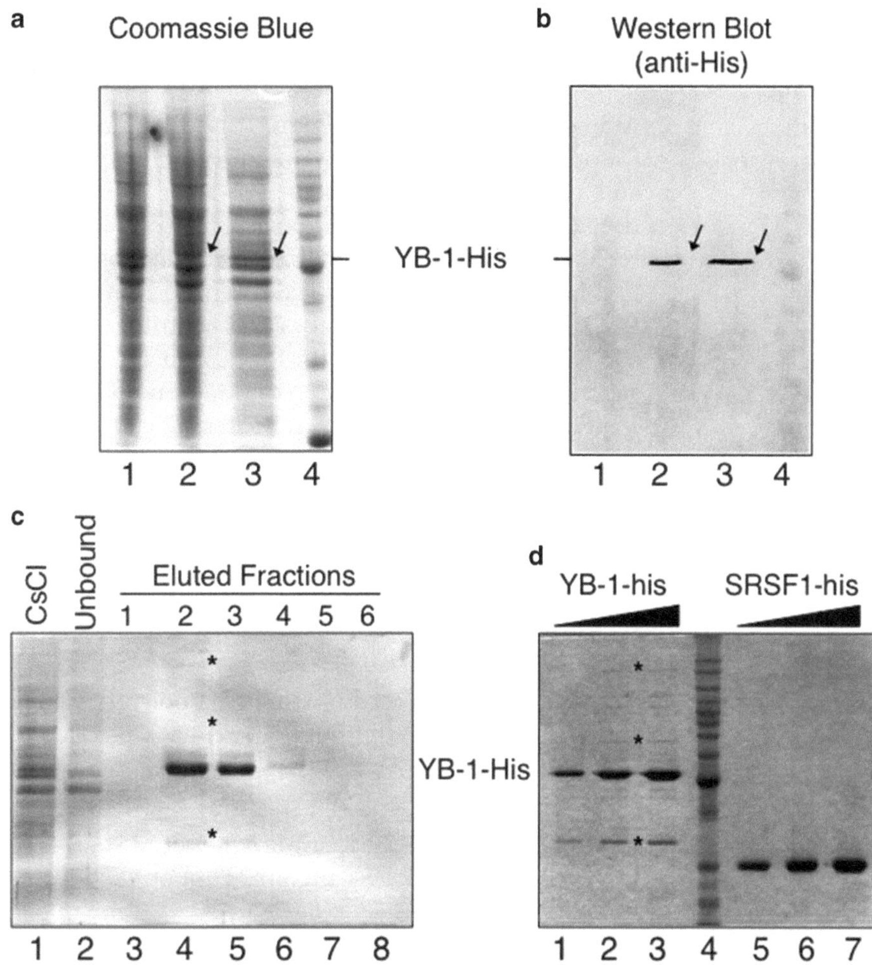

Fig. 3 Analysis of protein purification. (**a**) Coomassie blue-stained SDS-PAGE of untransfected cells (*lane 1*), YB-1-transfected cells (*lane 2*), and YB-1-transfected cell lysate before cesium chloride (CsCl) gradient centrifugation (*lane 3*). Marker (*lane 4*). (**b**) Western blot analysis of the gel from panel (**a**), probed with anti-His antibody. His-tagged YB-1 is indicated with an *arrow*. (**c**) Coomassie blue-stained SDS-PAGE of the input (CsCl isolated fraction, *lane 1*), the column flow-through containing proteins that did not bind to the column (*lane 2*), and the six fractions collected from the column after elution, containing the recombinant protein released from the resin (*lanes 3–8*). Co-purified additional proteins are marked with *asterisk*. (**d**) Coomassie blue-stained SDS-PAGE of purified YB-1-His and SRSF1-His proteins after dialysis. YB-1-His was purified as described here, and SRSF1 was purified using the optional procedure of ammonium sulfate precipitation. Additional bands co-purified with YB-1-His which were ultimately removed by ion exchange chromatography. These two purified splicing factors highlight the types of results that can be expected after this procedure. Each splicing factor may require protein-specific modifications to the method as noted in the text

(B) into an ultracentrifuge polycarbonate tube. Then, carefully add 13 ml of solution A under solution B using a pipette that can penetrate through solution B. In this way, solution A slowly displaces solution B to the top of the tube to form a gradient with two phases. Next, centrifuge the gradient at 48,000 rpm (170,000 × *g*) using a 60 Ti rotor for 24 h at 4 °C (*see* **Note 10**).

Very carefully collect 5 ml from the top of every centrifuged gradient, and dialyze this fraction twice for 4 h each in the lysis buffer before going on to the purification, Subheading 3.4.

2. Ammonium sulfate precipitation (used to purify the splicing factor SRSF1-His): Note the volume of soluble protein extract, and calculate the amount of crystalline ammonium sulfate needed to reach 40 % saturation at 4 °C (*see* **Note 11**). Add the required amount of ammonium sulfate, and dissolve it by rotation in the cold room for at least 30 min. Transfer the solution to Oak Ridge centrifuge tubes and centrifuge at 16,500 rpm (32,500 × *g*) in an SS-34 rotor for 30 min at 4 °C. Transfer the supernatant to a fresh tube, continue on to the purification step, and discard the pellet (*see* **Notes 12** and **13**).

3.4 Affinity Purification

1. Before incubation of soluble protein extract with the Ni-NTA resin, the beads are equilibrated in batch by performing two washes with the lysis buffer. We typically use 0.5 ml of bead slurry (0.25 ml of packed beads) for 250 ml of initial cell culture (*see* **Note 14**). Next, add the soluble protein extract to the beads and rotate for 1 h at 4 °C.
2. Transfer the lysate/bead mixture to a polypropylene column, and collect the flow-through material, which contains the unbound proteins. At this step, 20 μl of the flow-through should be saved for later analysis (Fig. 3c, lane 2).
3. Wash the column with at least 50× the bead volume (12.5 ml) with wash buffer. Collect the first 1 ml of wash from the column for later analysis (Fig. 3c, lane 3).
4. Elute the recombinant protein by adding 1 ml of elution buffer. The eluate is collected in 0.1–0.2 ml fractions into separate microcentrifuge tubes. The stopcock attached to the column can be used to control the flow of the eluate. In this way, the protein is eluted in a minimal volume and maximum protein concentration. The protein-containing fractions (*see* **Note 15**) are next analyzed in SDS-PAGE gel (Fig. 3c, lanes 4–6) in order to assess the purity of the recombinant splicing factor.
5. In most cases, the recombinant splicing factor is sufficiently purified using this protocol. However, if contaminating proteins are co-purified in abundance, the recombinant splicing factor can be further purified using additional methods such as glycerol gradient centrifugation, gel filtration, or ion exchange chromatography (*see* **Note 16**).

3.5 Buffer Exchange

1. Prepare modified buffer D (2 × 2 l) in 2–4 l beaker and chill to 4 °C.
2. Pool all selected fractions together into a Slide-a-Lyzer cassette, and dialyze the purified protein at 4 °C with stirring;

after 3 h change to fresh 2 l of modified buffer D for a final dialysis of 3–12 h at 4 °C.

3. Recover sample from Slide-a-Lyzer and centrifuge for 30 min at 4 °C with a benchtop centrifuge at 15,000 rpm (16,800 × *g*) or SS-34 rotor at 16,500 rpm (32,500 × *g*) depending on the volume. Next, the supernatant is stored at −80 °C in small fractions. It is recommended to analyze recombinant protein by SDS-PAGE after thawing to ensure that protein has not precipitated during dialysis or freeze–thaw cycle (Fig. 3d, lanes 5–7).

4 Notes

1. Other culture medias have been recommended that may improve transfection efficiency including FreeStyle 293 Expression Medium (Life Technologies) and SFM4 HEK293 medium (HyClone) [16] and serum-free F17 medium (Life Technologies) supplemented with 0.1 % Pluronic-F68 and 4 mM glutamine [17].
2. This cell line has a tendency to form clumps, which can be dislodged and broken up by repetitive pipetting when the volume of culture is still low (20–50 ml). The optimal rotation speed to reduce clump formation is difficult to predict, as it can be variable depending on the equipment. Florence flasks work well for culturing and are available in large sizes for cell cultures of 8 l or more. A sterile, foil-wrapped rubber stopper set loosely on the opening of flask allows for aeration. Other suspension culture vessels can be used including roller bottles, spinner flasks, and Erlenmeyer flasks with vent.
3. The concentration of sodium chloride in the lysis buffer can be modified from 0 to 1 M. For most splicing proteins tested, 0.5–1.0 M NaCl is optimal, though care should be taken to adjust salt concentration prior to incubation with Ni-NTA beads in order to comply with the maximal salt tolerance of the column (*see* **Note 11** below).
4. The most common tags that are used for protein purification are FLAG, HA, T7, and 6× histidine. Two different tags can also be added to the splicing protein. Though multiple tags generate more steric hindrance and may increase the probability of an effect on protein activity, this strategy can improve the purity of the recombinant factors and decrease contaminants co-purified with the first affinity tag.
5. For first-time analysis of a protein, we recommend also setting up a control culture that will allow the level of expression of your protein to be compared to untransfected cells (Fig. 3a).

6. The optimal DNA:PEI ratio should be tested, as transfection efficiency can vary between different PEI sources [18]. A ratio of 1:2 or 1:3 is often ideal in our hands. The PEI transfection procedure requires a large amount of plasmid. To purify large quantities of high-quality plasmid DNA (over 10 mg) we find that purification using a classical cesium chloride gradient provides a superior production yield.
7. A darker band at the size of the transfected protein predicted molecular weight should be observed if expression is high (Fig. 3a). Alternatively, a western blot of the samples can be performed to confirm expression (Fig. 3b). Performing this step prior to collection and purification steps can save time if expression was not successful. The presence of the overexpressed protein may not be detectable by Coomassie staining, in which case western blot analysis should be employed.
8. The precise amplitude and power of sonication are variable depending on the sonicator. In general, the lysate is sonicated between three to six times. However, the experimenter can ensure sufficient sonication by assessing the lysate in the tube. A visible decrease in viscosity occurs with nucleic acid breaks an indication that lysis is complete.
9. Adding CsCl to the soluble protein extract generates an endothermic reaction that makes the solution colder. The tube should be rocked to solubilize the CsCl.
10. During the centrifugation of the CsCl gradient, the proteins will migrate to the top of the gradient while nucleic acid will pellet at the bottom of the tube. Thus, the tubes should be inserted and removed from the rotor carefully in order not to disturb the gradient and protein separation.
11. The solubility of ammonium sulfate depends upon temperature, which can cause large variations in volume. Thus, the quantity of crystalline ammonium sulfate required to reach 40 % will be different at 4 and 20 °C. To calculate the quantity needed to achieve the correct percentage, a pre-calculated table can be referenced (available on the Internet; e.g., http://www.encorbio.com/protocols/AM-SO4.htm).
12. Adding ammonium sulfate to 40 % saturation increases the salt concentration to 1.85 M (based on the use of lysis buffer with 0.5 M sodium chloride), which is just below the limit that allows efficient binding of His-tagged proteins to Ni-NTA resin (2 M). In case the soluble protein extract is prepared with a higher salt concentration in the lysis buffer, the soluble supernatant must be diluted before adding ammonium sulfate.
13. The first time a protein is analyzed in this way, western blot analysis can be performed to verify that the recombinant protein does not precipitate during ammonium sulfate precipitation.

14. Typical yields of His-tagged recombinant protein are 0.5–2.0 mg/100 ml of starting culture. The binding capacity of the Ni-NTA beads is 50 mg/ml; thus, 0.5 ml of bead slurry (0.25 ml of beads) is sufficient for most purifications from a 250 ml starting culture.
15. To restrict the analysis of eluted fractions to those that contain protein, each fraction is tested in a Bradford assay by mixing 1 μl of the fraction into 50 μl of diluted protein dye reagent solution. In order to determine sensitivity of the protein dye reagent, test various dilutions of the reagent with the elution buffer and 0.5 μg of bovine serum albumin (BSA).
16. The recombinant splicing protein can be further purified on a linear glycerol gradient (15–40 %) or by gel filtration chromatography (Superose 12 or Superdex 75, GE Healthcare). Such a procedure will help to remove imidazol, elution peptides, and co-purified proteins that do not interact directly with the recombinant factor. We have also experimented with ion exchange chromatography to separate partners that are in association with the recombinant splicing factor. For this purpose, we recommend the use of a cation exchanger, in particular the SP sepharose fast flow (GE Healthcare), although each specific splicing factor should be tested to determine interactions with ion exchangers. Finally, the experimenter should keep in mind that the primary drawback with this optional step is the potential for dilution of the recombinant protein.

Acknowledgements

We thank Judy Potashkin for comments on the manuscript. This work was supported by NIH grant NS069759 to MLH.

References

1. Sinha R, Allemand E, Zhang Z et al (2010) Arginine methylation controls the subcellular localization and functions of the oncoprotein splicing factor SF2/ASF. Mol Cell Biol 30: 2762–2774
2. Bedford MT, Clarke SG (2009) Protein arginine methylation in mammals: who, what, and why. Mol Cell 33:1–13
3. Stojdl DF, Bell JC (1999) SR protein kinases: the splice of life. Biochem Cell Biol 77:293–298
4. Stamm S (2008) Regulation of alternative splicing by reversible protein phosphorylation. J Biol Chem 283:1223–1227
5. Edmond V, Moysan E, Khochbin S et al (2011) Acetylation and phosphorylation of SRSF2 control cell fate decision in response to cisplatin. EMBO J 30:510–523
6. Bellare P, Small EC, Huang X et al (2008) A role for ubiquitin in the spliceosome assembly pathway. Nat Struct Mol Biol 15:444–451
7. Rhodes N, Gilmer TM, Lansing TJ (2001) Expression and purification of active recombinant ATM protein from transiently transfected mammalian cells. Protein Expr Purif 22:462–466
8. Durocher Y, Perret S, Kamen A (2002) High-level and high-throughput recombinant protein production by transient transfection of suspension-growing human 293-EBNA1 cells. Nucleic Acids Res 30:E9

9. Cazalla D, Sanford JR, Caceres JF (2005) A rapid and efficient protocol to purify biologically active recombinant proteins from mammalian cells. Protein Expr Purif 42:54–58
10. Hastings ML, Allemand E, Duelli DM et al (2007) Control of pre-mRNA splicing by the general splicing factors PUF60 and U2AF(65). PLoS One 2:e538
11. Zhang Z, Krainer AR (2007) Splicing remodels messenger ribonucleoprotein architecture via eIF4A3-dependent and -independent recruitment of exon junction complex components. Proc Natl Acad Sci USA 104: 11574–11579
12. Shaw SD, Chakrabarti S, Ghosh G et al (2007) Deletion of the N-terminus of SF2/ASF permits RS-domain-independent pre-mRNA splicing. PLoS One 2:e854
13. Lopez-Mejia IC, Vautrot V, De Toledo M et al (2011) A conserved splicing mechanism of the LMNA gene controls premature aging. Hum Mol Genet 20:4540–4555
14. Eshar S, Allemand E, Sebag A et al (2012) A novel *Plasmodium falciparum* SR protein is an alternative splicing factor required for the parasites' proliferation in human erythrocytes. Nucleic Acids Res 40(19):9903–9916
15. Allemand E, Hastings ML, Murray MV et al (2007) Alternative splicing regulation by interaction of phosphatase PP2Cgamma with nucleic acid-binding protein YB-1. Nat Struct Mol Biol 14:630–638
16. Hopkins RF, Wall VE, Esposito D (2012) Optimizing transient recombinant protein expression in mammalian cells. Methods Mol Biol 801:251–268
17. Raymond C, Tom R, Perret S et al (2011) A simplified polyethylenimine-mediated transfection process for large-scale and high-throughput applications. Methods 55:44–51
18. Tom R, Bisson L, Durocher Y (2008) Transfection of HEK293-EBNA1 cells in suspension with linear PEI for production of recombinant proteins. CSH Protoc pdb prot4977

Chapter 17

Single Molecule Approaches for Studying Spliceosome Assembly and Catalysis

Eric G. Anderson and Aaron A. Hoskins

Abstract

Single molecule assays of splicing and spliceosome assembly can provide unique insights into pre-mRNA processing that complement other technologies. Key to these experiments is the fabrication of fluorescent molecules (pre-mRNAs and spliceosome components) and passivated glass slides for each experiment. Here we describe how to produce fluorescent RNAs by splinted RNA ligation and fluorescent spliceosome subunits by SNAP-tagging proteins in cell lysate. We then depict how to passivate glass slides with polyethylene glycol for use on an inverted microscope with objective-based total internal reflection fluorescence (TIRF) optics. Finally, we describe how to tether the pre-mRNA onto the passivated slide surface and introduce the SNAP-tagged cell lysate for analysis of spliceosome assembly by single molecule fluorescence.

Key words Single molecule, CoSMoS, TIRF, Colocalization, Fluorescence, Microscope, Spliceosome, Assembly, Splicing, RNA, Ligation, SNAP tag

1 Introduction

The application of single molecule techniques to analyze biochemical processes has become increasingly prevalent for elucidating complex reaction pathways and has been applied to a wide range of systems including replication, transcription, and translation [1]. These methods often measure fluorescence light emitted from a dye-labeled biomolecule (e.g., single molecule colocalization or fluorescence resonance energy transfer (FRET) experiments) or the response of a biomolecule to force (e.g., optical traps, magnetic tweezers, or atomic force microscopy) [1]. The ability to follow reaction trajectories of individual biomolecules is an extremely powerful approach for studying biochemical reactions, particularly when transient or low-abundance intermediates cannot be observed in bulk assays due to ensemble averaging [2]. Additionally, the elaborate assembly pathways for macromolecular machines can be easily deconvoluted using single molecule fluorescence colocalization assays to follow construction of a single complex from start to finish [3–5].

Klemens J. Hertel (ed.), *Spliceosomal Pre-mRNA Splicing: Methods and Protocols*, Methods in Molecular Biology, vol. 1126, DOI 10.1007/978-1-62703-980-2_17,

Recently, single molecule approaches have begun to shed new light on the mechanisms of spliceosome assembly and the splicing of precursor mRNAs (pre-mRNAs) [6]. Unlike most single molecule experiments that are carried out using highly purified components, single molecule splicing reactions to date have been carried out in yeast whole cell extract (WCE). This has presented a set of unique challenges at nearly each stage of the single molecule experiment—from fluorophore labeling in WCE to image acquisition to data analysis. Despite the experimental complexity, single molecule methods have been used to study both pre-mRNA conformational changes during splicing [7] and the dynamic interactions of spliceosome subcomplexes [the U1 and U2 small nuclear ribonucleoproteins (snRNPs), the U4/U6.U5 tri-snRNP, and the Prp19-associated complex (NTC)] with pre-mRNA [8]. Both sets of experiments showed that a number of transitions along the splicing pathway appear readily reversible. Analysis of spliceosome assembly reaction kinetics further revealed that steps in this process are highly ordered on the RP51A substrate and no single step appeared to limit the rate of the overall assembly reaction [8].

The results described above concerning spliceosome assembly were obtained by CoSMoS—Colocalization Single Molecule Spectroscopy [8, 9]. In CoSMoS experiments of splicing, the pre-mRNA is often attached to the surface, and fluorescent spliceosome subcomplexes or splicing factors are monitored as they bind to and release from the pre-mRNA (Fig. 1). These proteins or subcomplexes are labeled with different colors of fluorophores such that each species can be individually tracked and distinguished from the tethered pre-mRNAs.

CoSMoS experiments are enabled by both surface tethering of the pre-mRNA and total internal reflection fluorescence (TIRF) illumination. The evanescent wave of energy that is used to excite fluorophores by TIRF dissipates rapidly from the glass/water interface. This means that only fluorophores within ~100 nm of the surface become excited and emit light. Additionally, molecules must remain fixed in position for a time period comparable with the camera frame rate to be observed in a CoSMoS experiment, thus, necessitating surface tethering of one of the fluorescent components for viewing discrete "spots" of single molecule fluorescence. Molecules that transiently pass through the evanescent field cannot be discerned as discrete "spots" but blur into the background since the camera frame rate in most microscopy experiments (maximum speed of ~500 fr/s) is much slower than diffusion. Consequently, experiments can be performed with free, fluorescent molecules in solution at concentrations <100 nM. The surface tethering is accomplished either by direct attachment of a biomolecule to a surface or by indirectly binding interactions between a fluorescent biomolecule in solution and its immobilized partner. It is critical that the surface be sparsely populated with biomolecules to ensure that fluorescent spots of single molecules are being separately observed.

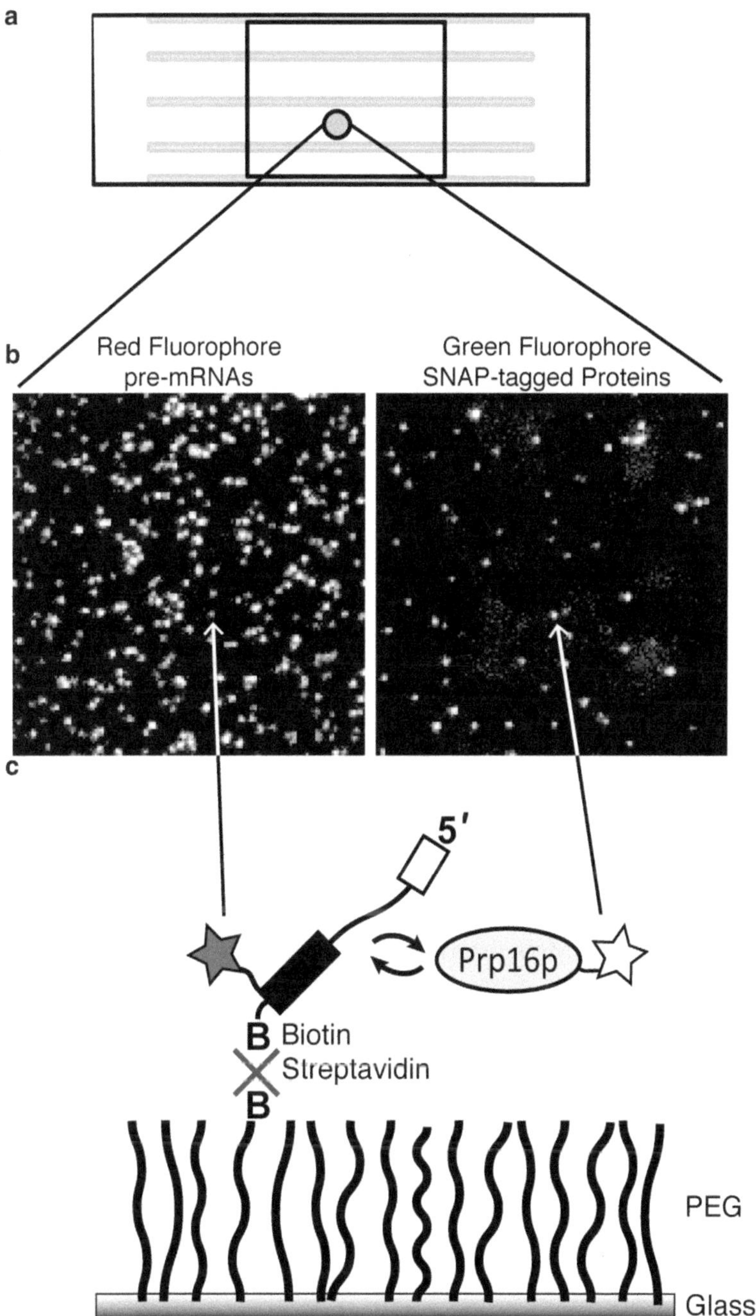

Fig. 1 Schematic overview of a CoSMoS experiment. (**a**) Drawing of a flow chamber with four lanes; the *gray circle* depicts an area imaged during an experiment (~400 μm^2). (**b**) Magnified portion of a field of view from a 2-color CoSMoS experiment. The *left square* shows single molecules (*spots*) of pre-mRNA labeled with a red fluorophore and fluorescence imaged at >635 nm. The *right square* shows the same field of view with single molecules of a SNAP-labeled spliceosome protein bound to the surface-tethered pre-mRNA and imaged with a green fluorophore at <635 nm. (**c**) Drawing of a SNAP-labeled Prp16p molecule interacting with a surface-tethered pre-mRNA. The pre-mRNA is attached to the slide through a biotin:streptavidin:biotin linkage

Key to the implementation of a multicolor CoSMoS experiment is the efficient detection of photons emitted from different fluorophores excited by lasers of different wavelengths. This can be accomplished using a TIRF microscope design pioneered by the Gelles laboratory called micromirror TIRF (mmTIRF) [9]. In mmTIRF, small broadband mirrors (mm in size) are used to direct the excitation laser light into and out of the microscope objective. This leaves the center of the objective free for fluorescence emission and unobstructed by dichroic mirrors found in other designs. Details on the construction of an mmTIRF microscope for CoSMoS as well as other TIRF microscope configurations have been published elsewhere [9, 10].

In this chapter, we focus on the preparation of three components of a typical CoSMoS splicing assay: synthesis of a fluorescent and biotinylated pre-mRNA, labeling of spliceosome proteins in a yeast WCE with the SNAP tag, and assembly of flow chambers for an objective-based TIRF microscope. We then provide a protocol for putting these components together to perform a CoSMoS experiment between a surface-tethered pre-mRNA and a single, SNAP-labeled spliceosome component. Due to the diversity in microscope designs and software implementation, we do not focus in this chapter on the specifics of image acquisition and processing, as this will vary lab-to-lab.

2 Materials

2.1 Design of Fluorescently Labeled Pre-mRNA Substrates

In order to monitor spliceosome assembly and/or RNA splicing by CoSMoS, pre-mRNA substrates are immobilized on a streptavidin-coated glass surface and their locations determined by fluorescence. We typically construct two types of fluorescent pre-mRNAs for CoSMoS: location reporters and splicing reporters. Location reporters contain a single fluorophore and biotin located near the 3′ end of the pre-mRNA. Splicing reporters contain fluorophores located in either the 5′ or 3′ exon or intron in addition to a biotin modification at the 3′ end. The construction of splicing reporters has previously been described in detail [11–13] and is beyond the scope of this article. We will instead focus on the more straightforward construction of location reporters. We have found that the most versatile approach is to incorporate a fluorophore and biotin into a short oligonucleotide (oligo) or "handle" that is ligated to the pre-mRNA 3′ end in a single step rather than direct modification of the RNA transcript. This approach allows a great deal of flexibility in the choice of fluorophore and pre-mRNA substrate.

Similar to ensemble experiments, relatively few pre-mRNA substrates have been used in single molecule studies. For studying the yeast spliceosome, we often use the RP51A, UBC4, or ACT1 pre-mRNAs, all of which splice well in vitro. The pre-mRNA is prepared by transcription with T7 RNA polymerase using a PCR-generated template (*see* **Note 1**). The pre-mRNA can be capped during transcription by the addition of cap analog dinucleotide or after transcription using an enzymatic capping system. The transcript may also be trace-labeled with radioactive α-[^{32}P]-UTP. This facilitates accurate quantification of the pre-mRNA and eliminates the need for exposing the RNA to potentially damaging UV radiation [14]. The levels of radiation used are miniscule and often undetectable with a Geiger counter in a single molecule assay. Methods for preparation of pre-mRNA substrates by transcription have been published elsewhere [11, 12, 15].

For the "handle" that will be ligated to the 3′ end of transcript, we use a commercially prepared 27-nucleotide oligonucleotide containing ribose 2′-*O*-methyl modifications to prevent degradation and triggering of RNaseH cleavage during the splicing assay. The biotin is incorporated during synthesis at the 3′ end, and the oligo includes a 5-(2-aminoallyl)uridine to facilitate labeling with *N*-hydroxysuccinimide (NHS) activated fluorophores. These oligos can be purchased from a number of commercial suppliers such as IDT or Dharmacon/Thermo Scientific. Conditions for labeling the oligo with fluorophores and purification of the labeled oligo have been previously described [12].

The fluorescent handle oligo is 5′ phosphorylated using polynucleotide kinase (PNK) and joined to the pre-mRNA via splinted ligation with RNA ligase. Either T4 RNA Ligase 1 or 2 (RNL1 or 2) can be used, though we often use RNL1 and a protocol developed by Stark et al. for this particular junction [16]. Since RNL1 is a single-stranded ligase, the splint is designed such that the 3′ and 5′ ends of the RNA and biotin handle, respectively, are free but in close proximity (Fig. 2). If RNL2 is used, then the splint is designed to directly about the two ends being joined (Fig. 2). Protocols for using RNL1 and RNL2 are similar and may need to be optimized for each junction by adjusting the ligation time, temperature, amount of enzyme, or ratios of the RNA fragments and splint oligo.

2.2 Ligation of a Fluorescent Biotin Handle to a Pre-mRNA

1. [^{32}P]-labeled pre-mRNA transcript, gel purified (1 equivalent, 28 pmol in ≤8 μL of H_2O).
2. Fluorescent biotin "handle" oligonucleotide (2 equivalents, 56 pmol in ≤3 μL of H_2O).
3. DNA splint oligonucleotide (1.5 equivalents, 39 pmol in ≤2 μL of H_2O).
4. T4 Polynucleotide Kinase and 10× PNK Buffer.
5. ATP (700 μM and 20 mM stocks in H_2O, each prepared fresh from an aliquot of a 100 mM stock solution at pH ~7).
6. RNasin Plus RNase Inhibitor (optional).
7. T4 RNA Ligase 1 (RNL1) and 10× RNL1 Buffer.
8. RNase-free deionized H_2O, not DEPC treated.

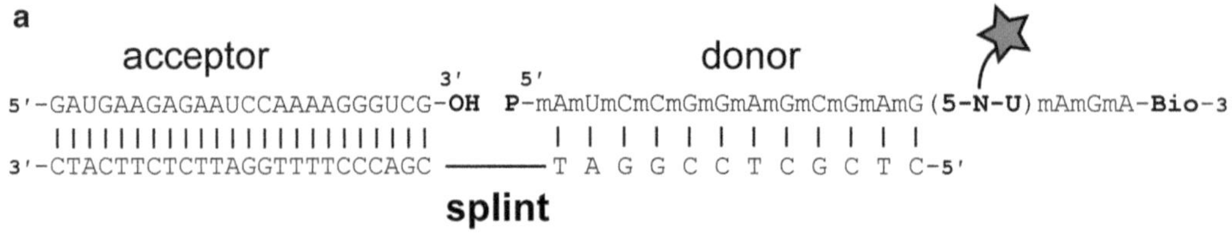

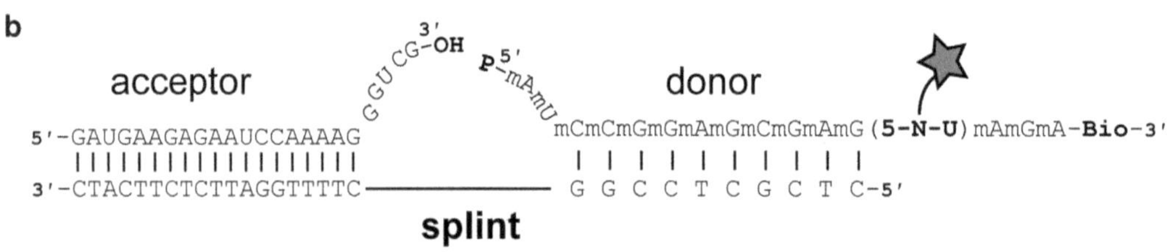

Fig. 2 Schemes for splinted ligation to prepare fluorescent, biotinylated pre-mRNAs. The *bold* P and OH indicate the 5′ phosphate 3′ hydroxyl groups of the donor biotin handle and acceptor pre-mRNA, respectively. Sequences shown are for the RP51A pre-mRNA substrate. The biotin handle contains 2′ *O*-methyl residues (*lowercase m*), a fluorophore (*star*) attached to a 5-aminoallyl-uridine (5-N-U), and a 3′ biotin (Bio). The DNA splints are continuous, but here represented as two distinct portions separated by lines. (**a**) Splint and junction design for ligation with RNA ligase 2 (RNL2), a double-stranded RNA ligase. (**b**) Splicing and junction design for ligation with RNA ligase 1 (RNL1), a single-stranded RNA ligase. Figure adapted from ref. 16

9. 1× TBE: 8.9 mM Tris-base, 8.9 mM boric acid, 2 mM EDTA.
10. Denaturing polyacrylamide gel: 6 %, acrylamide:bis 19:1, 8 M urea, 1× TBE, dimensions ~20×26 cm with ~0.8 mm thick spacers.
11. Gel loading buffer: 90 % deionized formamide, 1 mM EDTA, 0.03 % w/v bromophenol blue, 0.03 % w/v xylene cyanol.
12. Dye-free gel loading buffer: 90 % deionized formamide, 1 mM EDTA.
13. Gel elution buffer: 300 mM sodium acetate pH 5.2, 1 mM EDTA, 10 % v/v phenol pH 4.3.
14. 100 % Ethanol, ice cold.
15. 70 % v/v Ethanol with RNase-free H_2O, ice cold.
16. PCR machine and RNase-free PCR tubes.
17. Vertical gel apparatus and high voltage power supply.
18. Scintillation counter, vials, and scintillation fluid.
19. X-ray film and developer or phosphorimager screen and scanner.
20. Whatman filter paper.

2.3 Generation of Yeast Extracts Containing Tagged Proteins

CoSMoS splicing experiments often require that spliceosome proteins be fluorescently labeled. This represents a considerable challenge, particularly if the experiments are to be conducted in a WCE. While it is possible to add recombinant proteins back to a WCE, many spliceosome proteins are difficult to work with and these experiments could necessitate prior depletion of the endogenous protein to avoid competition between the native and fluorescent molecules. As an alternative approach, chemical tools can be combined with yeast genetics to modify spliceosome proteins with N- or C-terminal protein tags in vivo by homologous recombination [17]. Small molecule fluorophores can then be added to a WCE containing the tagged proteins to obtain fluorescent spliceosome components.

Tagging endogenous proteins by homologous recombination offers several advantages. First, if the yeast strains employed are haploid and proteins essential for viability are tagged, then survival of the yeast strains is a good indicator that the tagged proteins are functional in vivo. The effect of the tag on yeast growth can also be monitored and compared to the parental strain, as can the in vitro splicing activity of the WCE. Often, but not always, we have found that strains which grow poorly also produce WCE with poor splicing activity. Another advantage of tagging endogenous proteins is that it eliminates complicated procedures involving expression, purification, and labeling of recombinant proteins. The specific activity of the recombinant protein may be difficult to determine, and high concentrations are often needed to restore splicing activity in depleted extracts. These concentrations may exceed the limit imposed by the TIRF measurement (<100 nM fluorophore in solution).

The limit on solution fluorophore concentration also impacts which proteins can be tagged in the WCE by homologous recombination and subsequently visualized by CoSMoS. We have found that many spliceosome proteins are likely present in concentrations of <10 nM in a WCE and many can easily be tagged and visualized by CoSMoS. Other proteins (such as those involved in translation) are present at much higher concentrations and these fluorophore-labeled proteins could increase the background in the experiment to the point that the surface can no longer be discerned. The yeast GFP database [18] is an excellent resource for determining the suitability of a protein target for tagging and subsequent CoSMoS experiments. If a highly abundant WCE protein is being studied, there are several single molecule technologies that can make these experiments possible [19–21].

While there is a plethora of protein labeling technologies available, aspects of both in vitro splicing and the CoSMoS experiment itself place constraints on the applicable methods. Since proteins will be labeled in WCE, labeling must be highly specific for the protein of interest. If proteins are present at low levels in the WCE,

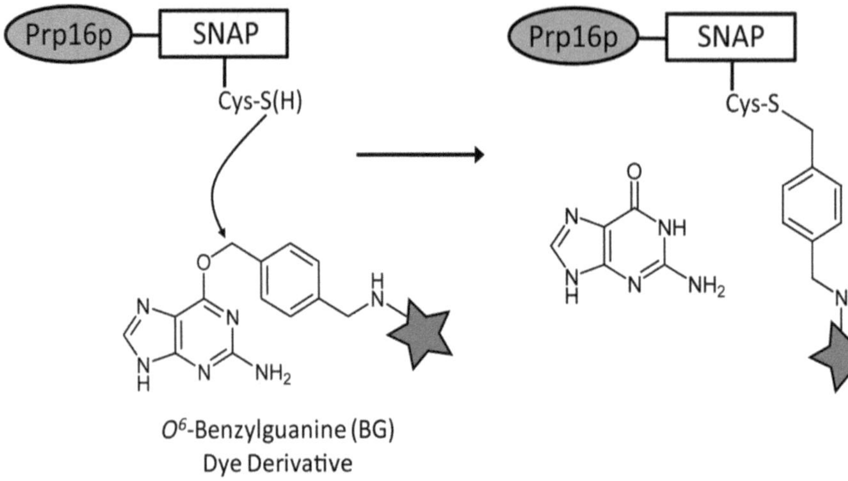

Prp16p SNAP Cys-S(H) → Prp16p SNAP Cys-S

O^6-Benzylguanine (BG) Dye Derivative

Fig. 3 Reaction scheme for SNAP-tag protein labeling. The SNAP-tag protein reacts with an O^6-benzylguanine dye substrate, transferring the fluorophore to an active site cysteine

the labeling method must be sufficiently rapid to ensure a high degree of protein derivatization (ideally quantitative) under conditions that will retain the splicing activity of the extract (~30 min at room temperature or ~3 h at 4 °C). Finally, CoSMoS experiments are greatly facilitated by the bright and stable fluorescence signals observed from organic fluorophores. Many fluorescent proteins either photobleach too rapidly for studies of splicing lasting tens of minutes, are too dim to observe easily as single molecules, or display unfavorable photophysical properties (i.e., blinking) that can confound analysis.

With these constraints in mind, we empirically determined that two component systems relying on a protein tag and small molecule ligand were the best suited for labeling spliceosome proteins in WCE. In our laboratories, we have often used either the *E. coli* dihydrofolate reductase (EcDHFR) tag developed by the Cornish laboratory (marketed by Active Motif) [22], the SNAP or CLIP tags developed by Johnsson and coworkers (marketed by New England Biolabs) [23], or the Halo tag developed and marketed by Promega [24]. For the purposes of this chapter, we focus on the SNAP tag.

SNAP tag labeling utilizes a modified human DNA repair enzyme (O^6-alkylguanyl-*S*-transferase) that becomes alkylated at an active site thiolate in the presence of O^6-benzylguanine (bG) derivatives (Fig. 3). We have found that for the SNAP tag protein and most bG fluorophores efficient labeling occurs with 2 μM bG fluorophore in 30 min at 20 °C in WCE. However, this rate can vary for different dye substrates and needs to be determined experimentally for each dye/tag pair. Recently a variant SNAP tag with improved reaction kinetics has been reported ("fast SNAP" or $SNAP_f$) [25].

With the $SNAP_f$ tag, efficient labeling can often be achieved with 500 nM dye in ~15 min at 20 °C. In addition to the SNAP tag, the CLIP tag has also been developed along with its "fast CLIP" counterpart, $CLIP_f$ [26]. Rather than being reactive towards bG derivatives, the CLIP tags react with benzylcytosine (bC). We have used the CLIP tags successfully for labeling spliceosome proteins albeit with slower kinetics and higher substrate requirements than for the SNAP tags.

A variety of fluorophore bG derivatives are available from New England Biolabs or can be synthesized easily from bG-amine building blocks. In practice, we have found that unpredictable interactions of the fluorophores with the yeast extract, stickiness of the fluorophore to glass surfaces used during microscopy, or unwanted photophysical properties (e.g., blinking) have limited the choice of fluorophores that can be practically used in a CoSMoS experiment. In general, the bG derivatives of Atto-488, DY549, and DY647 (sold as SNAP-Surface® 488, 549, and 647, respectively, by New England Biolabs) work well in WCE and for CoSMoS experiments.

During SNAP labeling in WCE, the bG fluorophore is in excess over the spliceosome protein being labeled. The free dye must then be removed prior to the CoSMoS experiment. We have concluded that dialysis is completely ineffective for dye removal from labeled extracts, therefore size-exclusion chromatography (SEC) is used. We developed an SEC method in which efficient dye removal was balanced against maintaining splicing activity by minimizing extract dilution. This SEC method is technically a group separation or desalting step, with the extract collected in the column void volume while the free dye remains in the column. With practice, this method can reduce free bG fluorophore background to <10 nM while maintaining the splicing activity of the WCE.

2.4 Labeling of Spliceosome Proteins with Fluorophores in Whole Cell Extract

1. Yeast WCE from a SNAP-tagged strain (1.2 mL aliquot, *see* **Note 2**).
2. SNAP-tag bG dye substrate (~1 mM in DMSO, New England Biolabs, *see* **Note 3**).
3. SEC Buffer: 25 mM HEPES–KOH pH 7.9, 50 mM KCl, 1 mM DTT, 10 % v/v glycerol.
4. Sephadex G-25, 50 % slurry in water.
5. Low pressure liquid chromatography (LPLC) column (e.g., Kontes Flex Column, 0.7 × 15 cm).
6. LPLC luer-lock fittings (2 × 3-way stopcocks, 1× barbed adapters for pump tubing).
7. Peristaltic pump (e.g., Pump P-1, GE Life sciences).
8. Liquid N_2 and dewar.

2.5 Design of Flow Chambers for TIRF Microscopy and Single Molecule Experiments

Like many other microscopy techniques, the single molecule CoSMoS splicing assay is carried out on glass slides. These assays can be conducted on TIRF microscopes employing either a prism-based or objective-based illumination scheme. With prism-based illumination, the excitation source is directed onto the sample via a prism positioned on top of the glass slide. When an objective-based illumination scheme is employed, the excitation laser is directed onto the sample from the bottom of the glass slide. Due to these differences in geometry, different styles of flow chambers must be used for each illumination scheme. The preparation of flow chambers for prism-based TIRF microscopy has been described elsewhere [10, 27]. Here we illustrate how to manufacture simple flow chambers for an objective-TIRF microscope.

For colocalization experiments, we manufacture flow chambers using disposable glass microscope slides. It is critical that the glass slide closest to the objective be of the proper thickness (typically No. 1.5 cover glass which is ~0.17 mm thick) to obtain images of high quality. For experiments in which FRET is also monitored, then the slide should be made from quartz or fused silica to eliminate the high background signal due to impurities found in lower grades of glass. Fused silica slides are extremely fragile, expensive, and non-disposable. Instructions for the preparation of No. 0 fused silica slides for CoSMoS can be found in other resources [11]. Note that these fused silica slides are both thinner and possess a different refractive index than No. 1.5 cover glass; consequently, adjustments in the TIR angle and microscope optics are necessary to obtain high-quality images. Fused silica slides and coverslips can be purchased from suppliers such as SPI Supplies.

Whether glass or fused silica slides are used in the experiment, the slide and coverslip must be scrupulously cleaned before use. We often employ a sonic water bath for this purpose. Alternatively, a plasma cleaner can be used if one is available [10]. It is important that the cleaned glass is protected from airborne dust particles, and in some environments the slides may need to be cleaned and assembled under HEPA air filtration such as a PCR workstation or in a hood with horizontal laminar flow (AirClean Systems). As a part of the cleaning procedure, the glass is activated for silanization and derivatization with amine reactive polyethylene glycol (PEG) reagents. Adequate passivation of the slide with PEG or other molecules is essential for single molecule experiments to prevent nonspecific binding of biomolecules to the glass surface. In addition, biotin can be incorporated onto the slide surface at this stage for biomolecule attachment via PEG–biotin conjugates and streptavidin. While the methods described below have been found to facilitate a great number of experiments, occasionally passivation protocols must be optimized by varying the surface chemistry, addition of BSA or nucleic acids, or altering the PEG molecule length in order to obtain sufficient passivation [28, 29].

Once assembled and passivated, the slides can be stored at 4 °C for up to 1 week. We find that the lowest amount of nonspecific binding typically occurs on slides <24-h old, and the degree of nonspecific binding increases with slide age. In some cases the derivatized slides can be washed, dried, and stored at −80 °C with desiccant for longer periods of time. These slides are often adequate for day-to-day experiments but may possess a higher degree of nonspecific surface binding compared to freshly prepared slides.

For many experiments, flow chambers can be constructed from simply two pieces of glass separated by a thin layer of vacuum grease as described below. In this configuration, capillary action is used to introduce liquid into each chamber and liquids are wicked out using filter paper. These slides typically have four chambers, each with a volume of ~20 μL. More complicated flow chambers with altered geometries, attachment points for inlet and outlet tubing, or that use alternative materials to vacuum grease [e.g., polydimethylsiloxane (PDMS)] can also be constructed. However, we find that the chamber configuration described below is often suitable for a wide range of CoSMoS experiments.

After the flow chambers and fluorescent RNAs and extracts have been prepared, the CoSMoS experiment is ready to be conducted. The exact protocol will depend on the configuration of the microscope and flow chamber as well as the nature of the experiment. In this chapter, we describe only the fundamentals of surface attachment of a fluorescent RNA and introduction of the WCE. It should be noted that these experiments often require the addition of an enzymatic oxygen scavenging system and triplet state quenchers to extend fluorophore lifetime and limit blinking [12, 30–33] (*see* **Note 4**).

The image acquisition protocol, microscope controls, and data processing routines will vary depending on the software and hardware preferences of each laboratory. While many laboratories utilize custom software to analyze single molecule data, several computer programs have recently become available to facilitate this stage of the experiment [34–37]. The nuances of interpreting single molecule data have been described in detail in many publications and great care must be taken to account for the observation of a single molecule event as well as the *probability* of having seen that particular event during the experiment [38, 39].

2.6 Preparation of Flow Chambers for Objective-TIRF Microscopy

1. Micro-90 Cleaning Solution (M-9050-12, International Products Corp.).
2. KOH (100 mM solution in MilliQ H_2O).
3. Ethanol (200 proof).
4. MilliQ H_2O.
5. Vectabond (SP-1800, Vector Laboratories) (*see* **Note 5**).

6. Acetone (spectrophotometric grade, 99+%).
7. Sodium bicarbonate.
8. Vacuum grease (e.g., Dow Corning high vacuum grease).
9. Compressed nitrogen gas (ultra high purity).
10. Biotin–PEG (biotin-PEG-SVA MW 5000, Laysan Bio) (*see* **Note 5**).
11. PEG succinimidyl valerate MW 5000 (MPEG-SVA-5000, Laysan Bio) (*see* **Note 5**).
12. Clear nail polish (optional).
13. 25×25 mm cover glass (Corning No. 1.0 or 1.5).
14. 24×60 mm cover glass (Gold Seal No 1.5, No. 3423).
15. Plastic disposable luer lock syringes (3–10 mL).
16. Empty pipette tip boxes with inserts (e.g., TipOne boxes, USA Scientific).
17. Razor blades or metal slide holders.
18. Sonic water bath (e.g., VWR Symphony).
19. Slide Mailers ×5 (Fisher Scientific).
20. Cover glass forceps.
21. Wafergard GN Gas Filter Gun (Entegris).
22. Syringe filters (0.20 μm, regenerated cellulose).

2.7 Assembling a Single Molecule Splicing Experiment

1. Biotinylated, fluorescent pre-mRNA.
2. SNAP-labeled yeast WCE.
3. Cleaned and passivated glass slide.
4. Dithiothreitol (DTT, 1 M stock solution, 0.20 μm filtered, aliquoted and stored at −20 °C).
5. PEG 8000 (15 % w/v solution in MilliQ H_2O, 0.20 μm filtered).
6. Potassium phosphate (500 mM in MilliQ H_2O, pH 7.3, 0.20 μm filtered).
7. 3,4-Dihydroxybenzoic acid (PCA, 50 mM in MilliQ H_2O, aliquoted and stored at −80 °C, the PCA may be recrystallized from hot MilliQ H_2O to increase solubility).
8. Protocatechuate 3,4-dioxygenase from *Pseudomonas* (PCD, resuspended to a concentration of 24–48 U/mL in 50 mM Tris-base pH 8.0, aliquoted and stored at −80 °C).
9. (±)-6-Hydroxy-2,5,7,8-tetramethylchromane-2-carboxylic acid (trolox).
10. 200× Triplet quencher master mix in DMSO (optional): 100 mM propyl gallate, 200 mM 4-nitrobenzyl alcohol, 200 mM cyclooctatetraene, aliquoted and stored at −80 °C.

11. Streptavidin (10 mg/mL aliquots in PBS, stored at −80 °C, Prozyme, SA10).
12. Nuclease-Free Bovine Serum Albumin (100 mg/mL).
13. ATP (100 mM, pH 7, aliquoted and stored at −80 °C, optional).
14. $MgCl_2$ (1 M stock solution, 0.20 μm filtered).
15. RNasin Plus RNase Inhibitor (Promega, optional).
16. Filter paper (Whatman 1, 90 mm).
17. Plastic disposable luer lock syringes.
18. Syringe Filters (0.20 μm, regenerated cellulose).
19. Low adhesion pipette tips and microcentrifuge tubes.

3 Methods

3.1 Ligation of a Fluorescent Biotin Handle to a Pre-mRNA Transcript

1. In a small PCR tube, combine the fluorescently labeled biotin "handle" oligo (56 pmol) with 0.5 μL of 10× PNK buffer, 0.5 μL of 700 μM ATP, 0.5 μL of T4 PNK (5 U), and RNase-free H_2O for a final volume of 5 μL. Avoid prolonged exposure of fluorescent materials to light.
2. Incubate in a PCR machine at 37 °C for 60 min to phosphorylate the biotin "handle". Heat inactivate PNK by incubation for 20 min at 65 °C. Spin down the PCR tube to collect liquids in the bottom.
3. In a separate PCR tube, combine the pre-mRNA transcript (28 pmol), the DNA splint oligo (39 pmol), the phosphorylated biotin "handle" (56 pmol), and RNase-free H_2O to a final volume of 14–15 μL.
4. Anneal the pre-mRNA and biotin handle to the splint by incubating at 65 °C for 3 min in a PCR machine followed by a room temperature incubation for 5 min.
5. Add 1 μL RNasin, 2 μL 10× RNL1 Buffer, 1 μL of 20 mM ATP, and 1–2 μL RNL1 (10–20 U) to a final volume of 20 μL.
6. Incubate at 37 °C for 60 min in a PCR machine.
7. While the ligation reaction is incubating, pre-run a 6 % acrylamide denaturing gel for at least 30 min to an operating temperature of ~50 °C. It is best to run the gel either in a darkened room or inside a large cardboard box to protect the ligated fluorescent RNA from light.
8. Once the ligation reaction has been completed, add an equivalent volume of dye-free loading buffer (20 μL) to the reaction and load onto the pre-run gel. Include loading buffer containing bromophenol blue and xylene cyanol in an adjacent lane in order to track the progress of the electrophoresis.

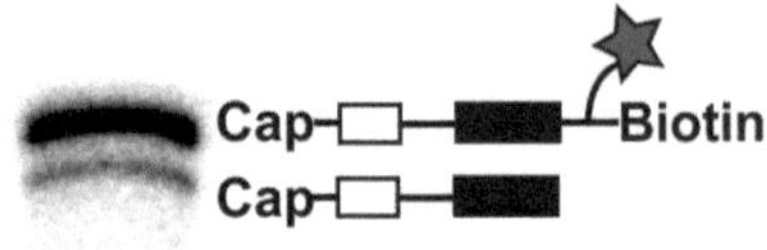

Fig. 4 Denaturing polyacrylamide gel purification of the pre-mRNA-biotin handle following ligation. The ligation product (*upper band*) is well separated from the unmodified transcript (*lower band*). The *dark polygon* on the *right* originated from a radioactive marker used to determine the location of the ligation product before it was excised from the gel (Gel courtesy of Joshua Larson, U. Wisconsin-Madison)

9. Carry out gel electrophoresis at constant power (35 W) for 1.5–2.5 h, depending on the size of the RNA transcript and the acrylamide gel being used.
10. Dismantle the gel from the electrophoresis apparatus and remove one of the glass plates. Cover the gel and remaining glass plate with plastic wrap. To locate the ligation product, cut small shapes (3–4) out of Whatman filter paper and soak each in a radioactive solution of similar activity to the trace-labeled RNA. Using clear tape, adhere the filter papers to the plastic wrap in an asymmetric pattern around the lane that contains the ligation product. Expose the gel to either X-ray film (~1–5 min) or a phosphorimager screen (~1–2 min). From the developed X-ray film or phosphorimage, create a 1:1 replica copy of the gel on paper and cut out the locations of the filter paper shapes and the RNA ligation product (usually the upper of two RNA bands) (Fig. 4).
11. Using the paper template, excise the ligated pre-mRNA product and place into a 1.5 mL microcentrifuge tube. It is often beneficial to cut the RNA band into small pieces (~1 mm^2) and to avoid including any excess acrylamide. Add 400 μL of gel elution buffer to the tube and centrifuge or vortex briefly to immerse the gel slice in the buffer. Freeze the gel slice with dry ice (~5–10 min), wrap the tube with aluminum foil to protect the RNA from light, and rotate the tube end-over-end for ~16 h at room temperature.
12. After incubation, briefly centrifuge the tube to pellet the acrylamide gel slices (13,000 × *g*, ~1 min). Transfer the extracted RNA (supernatant) to a new tube and add 1.2 mL ice-cold EtOH.

13. Incubate at −80 °C for ≥1 h, then centrifuge at 13,000–18,000 × g for 30 min at 4 °C. Remove the supernatant and wash the pellet with 70 % EtOH and centrifuge as before for 5 min.
14. Remove the supernatant and air-dry the pellet for 5 min at room temperature.
15. Resuspend the pellet in 50 μL nuclease-free water. Store the RNA in ~10 μL aliquots at −80 °C. Protect the RNA from light using amber tubes for storage and/or a light proof container.
16. The pre-mRNA concentration can be determined by scintillation counting using a sample of known concentration as the reference (often a sample of the trace-labeled transcript before ligation or an aliquot of the transcription reaction used to produce the RNA).

3.2 Labeling of SNAP Proteins in WCE

1. Assemble a low pressure chromatography column with a stopcock and attach to a ring stand in a 4 °C cold room (Fig. 5a). With the stopcock closed add 3.8 mL of water and mark the height of the water on the side of the column as a reference line. Add 0.2 mL of additional water and mark a second reference line (4.0 mL). Drain the water from the column.
2. Add 7–8 mL of a ~50 % Sephadex G25 suspension to the column. Open the stopcock and allow the column to begin to drain by gravity. Do not let the column run dry and avoid cracks or channels in the resin.
3. Set up and prime a peristaltic pump with SEC buffer, turn off the pump, connect the tubing to the inlet at the top of the column, and turn the pump back on to maintain a flow rate of ~0.4 mL/min once the resin has packed.
4. Allow the column bed to compact. If necessary, adjust the bed height to a position between the 3.8 and 4 mL reference marks. This can be done by using a glass pipette or tuberculin syringe to resuspend the upper portion of the resin bed and adding or removing resin as needed.
5. Equilibrate the column in SEC buffer for at least 2 h.
6. Thaw a 1.2 mL aliquot of yeast extract containing SNAP-tagged proteins on ice.
7. Add SNAP-tag bG fluorophore substrate to a final concentration of 1 μM for the $SNAP_f$ tag or 2 μM for the SNAP tag. Mix the solution well by inversion.
8. Incubate in the dark for 30 min at room temperature, mixing every 10 min. After 30 min, place the extract on ice and immediately proceed to the column purification.
9. Stop the peristaltic pump. Using a glass pipette or tuberculin syringe, gently remove the buffer above the column bed, taking care not to disturb the resin. Allow any remaining buffer on top of the resin to drain from the column. Once all of

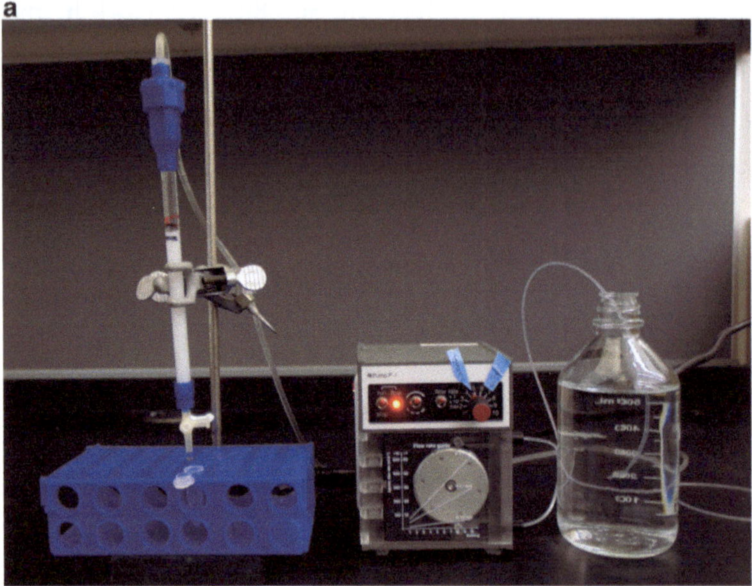

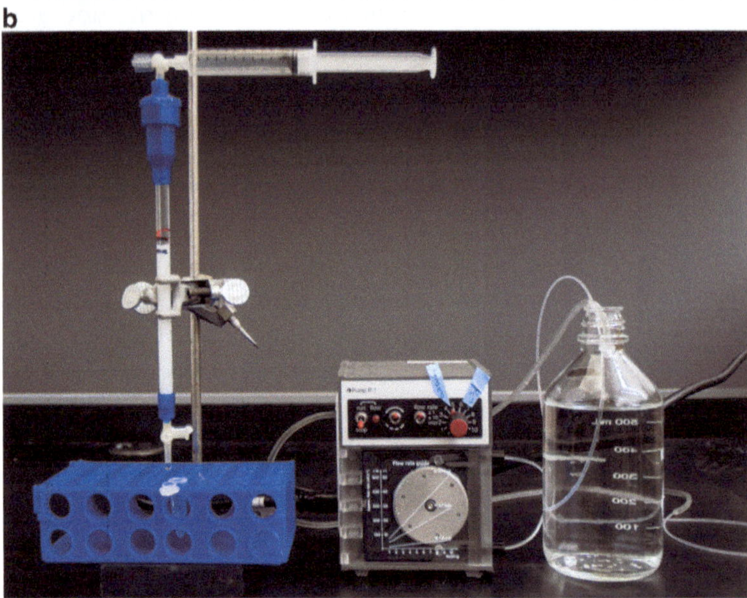

Fig. 5 SEC apparatus used to remove excess bG dye from SNAP-labeled WCE. (**a**) During equilibration and elution, the column is connected to a peristaltic pump and fractions collected into 1.5 mL microfuge tubes. (**b**) During WCE loading, a syringe is used to apply pressure to the column to increase the flow rate. Photography credit Robin Davies

the buffer has entered the resin, close the stopcock to prevent the column from drying.

10. Slowly and gently add the labeled yeast extract to the column directly on top of the resin bed being careful to disturb it.

11. Open the stopcock and allow the extract to drain into the resin. Use a 10 mL syringe fitted with a 3-way stopcock to apply manual pressure to the top of the column to increase the flow rate to ~0.5 mL/min (Fig. 5b). The resin will pack (shrink) during application of the extract. After ~1 mL of extract has entered the resin, begin collecting the column eluate in 1.5 mL microcentrifuge tubes (0.4–0.5 mL per fraction).
12. Once the extract has entirely entered the resin, release the pressure, close the stopcock at the base of the column, and remove the syringe and column top. Carefully add SEC buffer on top of the resin and continue adding SEC buffer till the column has nearly filled (~4–5 mL). Reattach the column top and reconnect to the peristaltic pump.
13. Turn the peristatlc pump on to maintain a flow rate of ~0.25 mL/min. Continue to fill the microcentrifuge tube from **step 3** until it has reached a volume of 0.4–0.5 mL. This will be fraction #1. Continue to collect four additional fractions as the extract elutes from the column. Active extract will typically elute in fractions #2 and #3. Avoid prolonged exposure of the labeled extract to light. If possible elute the extract in a darkened cold room while using a flashlight to monitor each fraction. Keep the fractions on ice.
14. Fractions containing the splicing extract should be noticeably yellow in color. Aliquot these fractions in 20–50 μL portions, freeze in liquid N_2, and store at −80 °C.
15. Assay each fraction for in vitro splicing activity and compare with an unlabeled control extract. Confirm labeling of the SNAP protein by SDS-PAGE of each fraction followed by fluorescence imaging of the unstained gel (Fig. 6).

3.3 Flow Chamber Cleaning and Assembly

1. Place 1–5 large glass coverslips (24 × 60 mm) into a clean slide mailer and place 1–5 small glass coverslips (25 × 25 mm) into a second mailer. If slides appear dusty, they can be rinsed with MilliQ H_2O using a squirt bottle beforehand. Due to the like lihood of breaking or dropping a slide during the following steps, it is best to clean more slides at this stage than are needed for subsequent experiments.
2. Fill each slide mailer with a 0.2 % v/v Micro-90 solution in MilliQ H_2O. Close the mailer and secure the lid with Parafilm. Float in a sonic water bath for 60 min.
3. Remove the slide mailers from the sonic water bath and pour out the Micro-90 solution. Fill the mailer several times with MilliQ H_2O and pour out the water. Fill the mailer with 100 % ethanol, close the lid, and secure with Parafilm. Float in a sonic water bath for 60 min.
4. Remove the mailers from the sonic water bath and pour out ethanol. Fill the mailer several times with MilliQ H_2O and

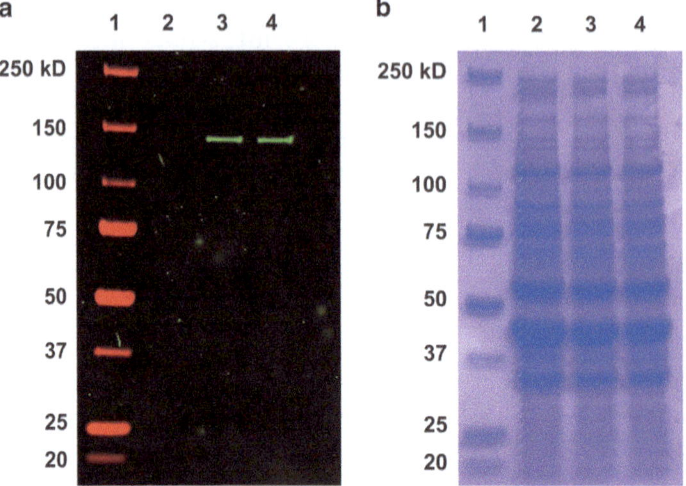

Fig. 6 Results from SNAP labeling of the spliceosomal Prp16 protein in yeast WCE with bG-DY549. (**a**) Representative SDS-PAGE gel showing SNAP labeling of Prp16 in WCE and visualized by in-gel fluorescence. *Lane 1*, protein ladder; *lane 2*, unlabeled wild-type yeast extract; *lanes 3* and *4*, fractions obtained by SEC after labeling the Prp16-SNAP extract. (**b**) The same gel as in (**a**) after Commassie blue staining

pour out the water. Fill the mailer with 100 mM KOH in MilliQ H_2O, close the lid, and secure with Parafilm. Float in a sonic water bath for 30 min.

5. Remove the mailers from the sonic water bath and pour out the KOH solution. Fill the mailer several times with MilliQ H_2O and pour out the water. Fill the mailer with MilliQ H_2O, close the lid, and secure with Parafilm. Float in a sonic water bath for 60 min. At the end of sonication, leave the slides submerged in MilliQ H_2O. It is best to proceed directly to derivatization once the slides have been cleaned.
6. Pour out the MilliQ H_2O from the mailer. While wearing clean gloves, carefully remove a slide from the mailer with forceps. Grasp the slide securely by the edges while avoiding contact with the center of the slide and dry with a stream of high-purity N_2 from a "filter gun" attached to a gas cylinder. Once the slide is completely dry, transfer it to a new, dry slide mailer. Repeat for each large slide and smaller coverslip. Up to five dried slides can be derivatized in each mailer, but the slides and coverslips should kept be in separate containers.
7. Mix 300 μL of Vectabond with 30 mL of acetone in a plastic 50 mL conical vial. Add the solution to the slide mailer containing the larger slides. Incubate for 5 min.
8. After 5 min, pour the acetone solution back into the 50 mL conical vial and immediately fill the mailer containing the larger slides with fresh MilliQ H_2O. Empty the mailer and fill again

with MilliQ H_2O. Repeat the water rinse two more times. Leave the mailer empty after the final rinse.

9. Repeat **steps 7** and **8** with the mailer containing the smaller coverslips. The same acetone/Vectabond solution can be reused for the coverslips.
10. While wearing clean gloves, carefully remove a slide from the mailer with forceps and dry with a N_2 stream as in **step 6**. Once the slide is completely dry, place on a slide holder. We often use either two razor blades or a metal slide holder to secure the slide in an empty pipette tip box while avoiding contact with either the top or bottom of the slide (Fig. 7a, b). The slide and/or holder should be gently taped to the empty pipette tip insert rack to prevent movement during subsequent steps or during transportation. The slide can be secured to the holder with tape, a small amount of vacuum grease, or with clear nail polish. To prevent the PEGylated slides from drying out, a small piece of sponge or paper towel soaked in MilliQ H_2O can be placed in the bottom of the pipette tip box to create a humid environment.
11. Using a syringe filled with vacuum grease and fitted with a 200 μL plastic pipette tip, draw five grease lanes horizontally across the slide to divide the slide into four chambers (Fig. 7a).
12. While wearing clean gloves, carefully remove a coverslip from the mailer with forceps and dry with an N_2 stream as in **step 6**. Once the coverslip is completely dry, carefully position it above the center of the slide and grease lanes made in **steps 10** and **11**. Gently push the coverslip into the grease to secure it to the slide and to seal each chamber (Fig. 7b).
13. Prepare a fresh solution (10 mL) of 100 mM sodium bicarbonate in MilliQ H_2O. Filter the solution through a 0.2 μm syringe filter.
14. Briefly centrifuge a 1 mg aliquot of biotin–PEG to spin any solids down to the bottom of the tube. Dissolve the biotin–PEG in 400 μL of the sodium bicarbonate solution.
15. Briefly centrifuge a 40 mg aliquot of PEG to spin any solids down to the bottom of the tube. Dissolve the PEG in 160 μL of the biotin–PEG solution. Briefly (2 s) place the solution in a sonic water bath to help the PEG dissolve. Vortex the solution and then briefly centrifuge. Mix the resulting PEG solution thoroughly by aspirating up and down several times with a pipette.
16. Add the PEG to the side of each chamber. It should enter the chamber by capillary action. Avoid getting any of the PEG solution outside of the chamber. One 40 mg aliquot of PEG is typically enough to derivatize two slides each containing four 20 μL chambers.

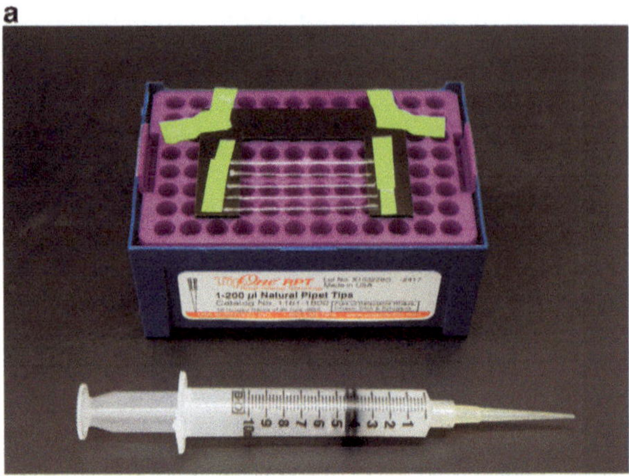

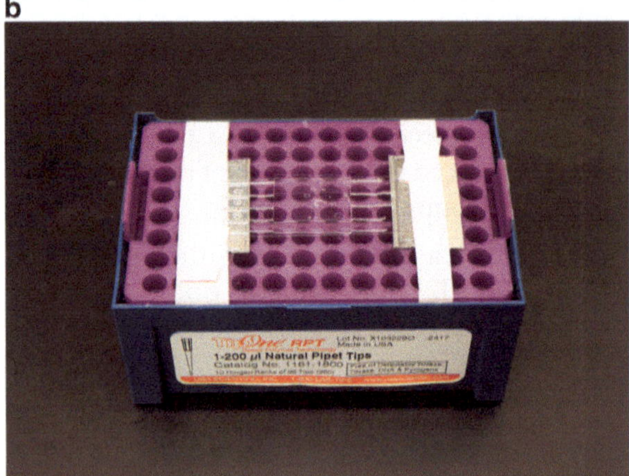

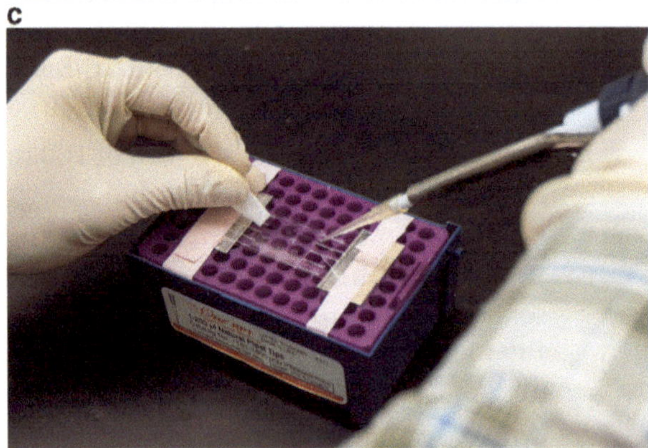

Fig. 7 Construction of flow chambers for CoSMoS experiments. (**a**) A glass slide is placed onto a slide holder (or a pair of razor blades) and held in place in a pipette tip box with lab tape. A syringe is then used to draw five thin lines of vacuum grease horizontally across the slide to create four chambers. (**b**) A clean coverslip is gently pressed on top of the lanes. The chamber is completed with the addition of a PEG:PEG–biotin mixture. The completed slide is also shown schematically in Fig. 1a. (**c**) To wash or add sample to a flow chamber, a pipette is used to dispense fluid to one side of the chamber, while the liquid is drawn through to the other side with a small piece of folded filter paper. Photography credit Robin Davies

17. Incubate the slide at room temperature for at least 3 h and up to 16 h.
18. After the room temperature incubation, the slide can be stored at 4 °C for up to 1 week.

3.4 Pre-mRNA Immobilization and Addition of WCE to the Flow Cell

1. If the derivatized slide has been kept at 4 °C, bring the slide to room temperature and allow to equilibrate for ~30–60 min. Warming the slide to room temperature greatly facilitates removal of the viscous PEG solution in each lane as well as limits the risk of condensation appearing on the slide and interfering with imaging.
2. Prepare 5–10 mL of 2× Wash Buffer (200 mM KPi pH 7.3, 6 % PEG 8000, 2 mM DTT) and 2.5 mL of a 2× Splicing Assay Buffer (200 mM KPi pH 7.3, 6 % PEG 8000, 2 mM DTT, 10 mM PCA, 5 mM $MgCl_2$, 2 mM Trolox). It may be necessary to briefly immerse the 2× Splicing Assay Buffer in a sonic water bath in order to fully dissolve the trolox. Alternatively, trolox may be added from a 200 mM stock solution in DMSO or methanol. These stock solutions should be stored at −80 °C in single-use aliquots. Filter the 2× Wash and Splicing Assay Buffers through 0.2 μm syringe filters.
3. Prepare a stockpile of "filter paper triangles." Cut each filter paper into 6–8 triangles, producing pizza slice-shaped pieces. Fold each triangle two times by first bringing the pointy end (the bottom of the pizza slice) up to the top and then folding in half again along an axis perpendicular to the first fold. We often prepare many dozens of these folded triangles at once and store them in empty pipette tip boxes.
4. Prepare 2 mL of 1× Wash Buffer using the 2× stock and RNase-free MilliQ H_2O. Flush each channel on the slide three times with 100 μL of the 1× Wash Buffer. The washing can be carried out on the lab bench using a pipette on one end of the channel and using a filter paper triangle on the other end to draw the liquid through the channel (Fig. 7c). The PEG solution is quite viscous and will move slowly through the channel until it has been cleared.
5. The next steps can either be done on the laboratory bench, or if permitted by the microscope configuration, on the microscope stage itself. If work is being done on the microscope, mount the slide to the stage and position the objective appropriately. Extreme care should be taken not to allow liquids to come into contact with the objective or other sensitive components.
6. Immediately prior to each experiment, streptavidin is added to a slide channel. For optimal binding of the RNA added subsequently, we recommend adding streptavidin to only one

channel at a time just before conducting an experiment in that channel. Using low adhesion pipette tips and tubes, prepare 50 μL of a 0.2 mg/mL streptavidin solution in 1× Wash Buffer. Add the entire streptavidin solution to a channel on the slide and draw the solution through using a filter paper triangle. Let the channel incubate with streptavidin for 2 min and then flush with 200 μL of 1× Wash Buffer. Proceed immediately to **step 6**.

7. Working with low adhesion pipette tips and tubes, prepare a 1 nM stock of the biotinylated, fluorescent pre-mRNA in 1× Wash Buffer that also includes 40 U of RNasin and 0.1 mg/mL nuclease-free BSA. Store the RNA stock solution on ice. From this stock solution, prepare 50 μL of a 200 pM RNA stock in 1× Wash Buffer.
8. Introduce the entire 200 pM RNA stock solution into the slide channel. If this is being done while the slide is mounted on the microscope, the accumulation of the RNA on the slide surface can be monitored in real time. It is imperative that the microscopic field of view does not become too saturated with RNA molecules and individual molecules are easily resolved and separated from one another. Typically an appropriate surface density is reached in 2–3 min, although this can vary. Stop the RNA accumulation on the surface by flushing the channel with 100 μL of 1× Wash Buffer.
9. Prepare 100 μL of a 1× Splicing Assay Buffer solution by combining the 2× Buffer with ATP (if needed), 0.1 U of PCD, and RNase-free water. Flush the slide chamber with this solution.
10. At this stage, the slide is ready for WCE to be added. We typically position the slide appropriately for the experiment and optimize laser powers and TIR at this time. Once the microscope has been appropriately configured for the experiment, proceed to **step 11**.
11. Prepare 100 μL of a splicing assay mixture by combining the 2× Buffer with 2 mM ATP (if needed), 35–40 μL of the labeled WCE, 0.1 U of PCD, and RNase-free water. Introduce the assay mixture to the slide channel and begin image acquisition. With practice, this can be done by hand with a deadtime of ~30 s. For studying events that may occur within that deadtime, a syringe and modified flowcell can be used to pull samples into the slide channel during data acquisition [8, 11].
12. We typically acquire data for 30–90 min at intervals of 2–10 s between frames (1 s/fr). The use of timelapse recording is beneficial for reducing photobleaching of the fluorophores during long experiments and to prevent laser-induced accumulation of fluorescent molecules on the slide surface.

4 Notes

1. For efficient ligation, the transcripts should possess homogenous 3′ ends and +1 (or greater) non-templated addition productions should be avoided. This can easily be accomplished by using DNA primers containing 2′-methoxy substituents during preparation of the transcription template by PCR [40]. Alternatively, homogenous ends can be generated by targeted RNaseH cleavage after transcription [41].
2. We typically prepare yeast WCE using the method of Ansari and Schwer [42] with the exception of using a ball mill (Retsch) to lyse the yeast cells. The yeast WCE can be aliquoted (1.2 mL) and frozen at −80 °C immediately after high-speed centrifugation at 166,000 ×g and before dialysis with no effect on splicing activity. For single molecule assays, it is extremely beneficial if the splicing activity of the WCE is as high as possible with at least 20 % of the pre-mRNA being converted to mRNA in 30–45 min at room temperature for yeast WCE.
3. SNAP tag substrates can be resuspended in DMSO, aliquoted, and stored at −20 °C. These aliquots retain labeling activity for many months. It is critical that the concentration of the bG substrate be quantified accurately by UV–Visible spectroscopy so that neither too much nor too little is added to the yeast WCE during labeling.
4. Great care must be taken in choosing the appropriate combination of oxygen scavengers, reducing agents, and triplet quenchers for the single molecule experiment. These components should be tested for possible inhibition or interactions with the biomolecules under study including detrimental RNase or DNase activity. Additionally, some oxygen scavenging systems may influence the pH of poorly buffered assay mixtures and this should be studied prior to setting up the single molecule assay.
5. To maintain surface attachment chemistry, both the Vectabond and PEG solutions should be carefully aliquoted. Vectabond can be stored in 300 μL aliquots in 1.5 mL microfuge tubes that have been backfilled with dry nitrogen or argon. We store the Vectabond at room temperature and protected from light. Aliquots retain activity for several weeks; however, we only aliquot one stock solution of Vectabond at a time. It is critically important that PEG and biotin–PEG aliquots be made when the bottles are first opened. We typically allow the stock bottles to come to room temperature (to avoid condensation) and make ~40 mg aliquots of PEG and ~1 mg aliquots of biotin–PEG in separate 0.5 mL microfuge tubes. We backfill each tube with dry nitrogen and store at −20 °C in a container with a tight fitting lid and with desiccant. These aliquots will retain reactivity for many months.

Acknowledgements

Eric Anderson is supported by funding to the laboratories of Melissa J. Moore (University of Massachusetts Medical School, HHMI Investigator and NIH GM053007) and Jeff Gelles (Brandeis University, NIH GM43369 and GM81648). Aaron Hoskins is funded by startup funds from the University of Wisconsin-Madison and the Wisconsin Alumni Research Foundation and NIH R00 GM079971. We thank Tucker Carrocci, Joshua Larson, and Inna Shcherbakova for critical reading of the manuscript.

References

1. Dulin D, Lipfert J, Moolman MC et al (2012) Studying genomic processes at the single-molecule level: introducing the tools and applications. Nat Methods 14:9–22
2. Weiss S (2000) Measuring conformational dynamics of biomolecules by single molecule fluorescence spectroscopy. Nat Struct Biol 7: 724–729
3. Friedman LJ, Gelles J (2012) Mechanism of transcription initiation at an activator-dependent promoter defined by single-molecule observation. Cell 148:679–689
4. Tsai A, Petrov A, Marshall RA et al (2012) Heterogeneous pathways and timing of factor departure during translation initiation. Nature 487:390–393
5. Joo C, Fareh M, Narry Kim V (2013) Bringing single-molecule spectroscopy to macromolecular protein complexes. Trends Biochem Sci 38:30–37
6. Hoskins AA, Gelles J, Moore MJ (2011) New insights into the spliceosome by single molecule fluorescence microscopy. Curr Opin Chem Biol 15:864–870
7. Abelson J, Blanco M, Ditzler MA et al (2010) Conformational dynamics of single pre-mRNA molecules during in vitro splicing. Nat Struct Mol Biol 17:504–512
8. Hoskins AA, Friedman LJ, Gallagher SS et al (2011) Ordered and dynamic assembly of single spliceosomes. Science 331:1289–1295
9. Friedman LJ, Chung J, Gelles J (2006) Viewing dynamic assembly of molecular complexes by multi-wavelength single-molecule fluorescence. Biophys J 91:1023–1031
10. Selvin PR, Ha T (2008) Single Molecule Techniques. Cold Spring Harbor Laboratory Press, Cold Spring Harbor
11. Crawford DJ (2010) Single molecule fluorescence studies of Saccharomyces cerevisiae pre-mRNA splicing. Ph.D. dissertation, Brandeis University, Waltham, MA
12. Crawford DJ, Hoskins AA, Friedman LJ et al (2008) Visualizing the splicing of single pre-mRNA molecules in whole cell extract. RNA 14:170–179
13. Abelson J, Hadjivassiliou H, Guthrie C (2010) Preparation of fluorescent pre-mRNA substrates for an smFRET study of pre-mRNA splicing in yeast. Methods Enzymol 472:31–40
14. Greenfeld M, Solomatin SV, Herschlag D (2011) Removal of covalent heterogeneity reveals simple folding behavior for P4–P6 RNA. J Biol Chem 286:19872–19879
15. Moore MJ, Query C (1998) Uses of site-specifically modified RNAs constructed by RNA ligation. In: Smith CWJ (ed) RNA:protein interactions. Oxford University Press, New York, pp 75–108
16. Stark MR, Pleiss JA, Deras M et al (2006) An RNA ligase-mediated method for the efficient creation of large, synthetic RNAs. RNA 12:2014–2019
17. Amberg DC, Burke D, Strathern JN (2005) Methods in yeast genetics: a Cold Spring Harbor laboratory course manual. Cold Spring Harbor Laboratory Press, Cold Spring Harbor
18. Huh W-K, Falvo JV, Gerke LC et al (2003) Global analysis of protein localization in budding yeast. Nature 425:686–691

19. Uemura S, Aitken CE, Korlach J et al (2010) Real-time tRNA transit on single translating ribosomes at codon resolution. Nature 464: 1012–1017
20. Leslie SR, Fields AP, Cohen AE (2010) Convex lens-induced confinement for imaging single molecules. Anal Chem 82:6224–6229
21. Loveland AB, Habuchi S, Walter JC et al (2012) A general approach to break the concentration barrier in single-molecule imaging. Nat Methods 9:987–992
22. Miller LW, Cai Y, Sheetz MP et al (2005) In vivo protein labeling with trimethoprim conjugates: a flexible chemical tag. Nat Methods 2:255–257
23. Keppler A, Gendreizig S, Gronemeyer T et al (2003) A general method for the covalent labeling of fusion proteins with small molecules in vivo. Nat Biotechnol 21:86–89
24. Ohana RF, Encell LP, Zhao K et al (2009) HaloTag7: a genetically engineered tag that enhances bacterial expression of soluble proteins and improves protein purification. Protein Expr Purif 68:110–120
25. Sun X, Zhang A, Baker B et al (2011) Development of SNAP-tag fluorogenic probes for wash-free fluorescence imaging. ChemBioChem 12:2217–2226
26. Gautier A, Juillerat A, Heinis C et al (2008) An engineered protein tag for multiprotein labeling in living cells. Chem Biol 15:128–136
27. Zhao R, Rueda D (2009) RNA folding dynamics by single-molecule fluorescence resonance energy transfer. Methods 49:112–117
28. Alemán EA, Pedini HS, Rueda D (2009) Covalent-bond-based immobilization approaches for single-molecule fluorescence. ChemBioChem 10:2862–2866
29. Revyakin A, Zhang Z, Coleman RA et al (2012) Transcription initiation by human RNA polymerase II visualized at single-molecule resolution. Genes Dev 26:1691–1702
30. Swoboda M, Henig J, Cheng H-M et al (2012) Enzymatic oxygen scavenging for photostability without pH drop in single-molecule experiments. ACS Nano 6:6364–6369
31. Aitken CE, Marshall RA, Puglisi JD (2008) An oxygen scavenging system for improvement of dye stability in single-molecule fluorescence experiments. Biophys J 94:1826–1835
32. Rasnik I, Mckinney SA, Ha T (2006) Nonblinking and long-lasting single-molecule fluorescence imaging. Nat Methods 3:891–893
33. Dave R, Terry DS, Munro JB et al (2009) Mitigating unwanted photophysical processes for improved single-molecule fluorescence imaging. Biophys J 96:2371–2381
34. Mckinney SA, Joo C, Ha T (2006) Analysis of single-molecule FRET trajectories using hidden Markov modeling. Biophys J 91: 1941–1951
35. Bronson JE, Fei J, Hofman JM et al (2009) Learning rates and states from biophysical time series: a Bayesian approach to model selection and single-molecule FRET data. Biophys J 97:3196–3205
36. Greenfeld M, Pavlichin DS, Mabuchi H et al (2012) Single molecule analysis research tool (SMART): an integrated approach for analyzing single molecule data. PLoS ONE 7:e30024
37. Milescu LS, Nicolai C, Bannen J, 2000–2013 QuB software
38. Sakmann B, Neher E (2009) Single-channel recording. Springer, New York
39. Schnitzer MJ, Block SM (1995) Statistical kinetics of processive enzymes. Cold Spring Harb Symp Quant Biol 60:793–802
40. Kao C, Zheng M, Rüdisser S (1999) A simple and efficient method to reduce nontemplated nucleotide addition at the 3′ terminus of RNAs transcribed by T7 RNA polymerase. RNA 5(9):1268–1272
41. Stone MD, Mihalusova M, O'Connor CM et al (2007) Stepwise protein-mediated RNA folding directs assembly of telomerase ribonucleoprotein. Nature 446:458–461
42. Ansari A, Schwer B (1995) SLU7 and a novel activity, SSF1, act during the PRP16-dependent step of yeast pre-mRNA splicing. EMBO J 14:4001–4009

Chapter 18

Cell-Based Splicing of Minigenes

Sarah A. Smith and Kristen W. Lynch

Abstract

Cell-based splicing of minigenes is used extensively in the analysis of alternative splicing events. In particular, such assays are critical for identifying or confirming the in vivo relevance of *cis*- and *trans*-acting factors in the regulation of particular splicing patterns. Here we provide detailed information on the methods specific to the cell-based analysis of minigene splicing. In addition, we discuss some of the theoretical considerations that must be given to the design of the minigene and subsequent experimental conditions.

Key words Minigene, RNA isolation, RT-PCR, Transfection, Stable cell lines, Alternative splicing, Exon, Intron

1 Introduction

The use of minigenes has long been a central tool in the characterization of splicing regulation and mechanisms. As the name implies, a "minigene" is a simplified version of an endogenous pre-mRNA. Most pre-mRNAs are vastly too long for ready manipulation, and mutation of endogenous genes in living cells is both inefficient and potential toxic. Therefore, use of a simplified model of a pre-mRNA, or "minigene," opens the door to lines of investigation not otherwise feasible. Indeed, the vast majority of known sequence elements that control pre-mRNA splicing were identified and/or characterized through minigene studies (e.g., [1–3]). Minigenes are also widely used to report on splicing patterns in cell-based screens for trans-acting proteins and regulatory pathways (e.g., [4–6]).

At the point at which one has identified a splicing event of interest—perhaps a change in the alternative splicing pattern of a particular gene in normal and diseased cells—the next step in understanding how the splicing pattern is regulated is most typically the development and characterization of a minigene. In vitro analysis of minigenes, as described in Chapter 11, can answer many questions. However, splicing competent extracts have only been generated from a handful of cell types and thus cannot recapitulate

Klemens J. Hertel (ed.), *Spliceosomal Pre-mRNA Splicing: Methods and Protocols*, Methods in Molecular Biology, vol. 1126, DOI 10.1007/978-1-62703-980-2_18, © Springer Science+Business Media, LLC 2014

many biologically important splicing events. Moreover, even in cases in which in vitro splicing is possible, analysis of splicing in living cells represents a powerful complementary approach.

Importantly, cell-based analysis of minigene splicing is relatively straightforward and widely applicable to almost any cell type and splicing event of interest. In brief, such assays involve (1) appropriate design of the minigene, (2) transfection and expression of the minigene in a suitable cell line, (3) harvest of RNA, and (4) analysis of splicing pattern by RT-PCR. This chapter covers each of these aspects in turn, discussing both the theoretic considerations and providing protocols for those aspects of the assay that are most unique. Common techniques such as PCR and subcloning, as well as cell-type specific methods for cell transfection and maintenance, are not covered here but references to other resources are provided.

2 Materials

2.1 Design and Construction of the Minigene

1. Genomic DNA.
2. Primers with appropriate restriction sites and complementarity to the gene region of interest.
3. Standard reagents for high-fidelity PCR, electrophoresis, subcloning, and plasmid preparation.
4. Expression vector for cloning sequence of interest.

2.2 Transfection and Expression of Minigene

1. Tissue culture cell line.
2. Tissue culture media, serum, and antibiotics.
3. Transfection reagent, such as Lipofectamine 2000 (Invitrogen) or 0.4 cm electroporation cuvettes (USA Scientific).
4. Purified minigene vector DNA (10 μg in 10 μl DI (distilled and deionized) H_2O) (*see* **Note 1**).
5. Incubator with CO_2.
6. Tissue culture plastic ware including sterile flasks, plates, and pipettes.
7. Low-speed centrifuge with capacity for 15 ml conical tubes (*see* **Note 2**).

2.3 Harvest of Total RNA

1. Low-speed centrifuge with capacity for 15 ml conical tubes (*see* **Note 2**).
2. RNase-free microcentrifuge tubes and pipette tips.
3. Refrigerated microcentrifuge or microcentrifuge in 4 °C room.
4. RNA-Bee (TelTest, Inc., *see* **Note 3**).
5. Chloroform.

6. Ice-cold 70 % ethanol.
7. RNase-free DI (distilled and deionized) H_2O.

2.4 RT-PCR Reagents for Analysis of Splicing Pattern

1. Vector-specific forward and reverse primers (*see* **Note 4**).
2. ^{32}P-gamma-ATP.
3. Polynucleotide kinase (PNK) and accompanying buffer.
4. Phenol:chloroform:isoamylalcohol.
5. 100 % Ethanol and ice-cold 70 % ethanol.
6. 5× Hyb buffer (1.5 M NaCl, 50 mM Tris–HCl pH 7.5, 10 mM EDTA).
7. 1.25× RT-mix (1.25 mM each dNTPs, 12.5 mM DTT, 12.5 mM Tris–HCl pH 8.0, 7.5 mM $MgCl_2$) (*see* **Note 5**).
8. MMLV reverse transcriptase (*see* **Note 6**).
9. RT-PCR buffer (0.5 M KCl, 0.1 M Tris–HCl pH 8.3, 15 mM $MgCl_2$, 0.01 % gelatin).
10. Taq polymerase.
11. Formamide buffer (45 ml formamide, 2.5 ml 0.5 M EDTA, 0.01 g bromophenol blue, 0.01 g xylene cyanol).
12. Thermocycler.
13. Mineral oil (*see* **Note** 7).

3 Methods

3.1 Design and Construction of the Minigene

Minigene design can vary depending on the question to be addressed. Typically, one aims to create a minigene that mimics the endogenous splicing pattern of a given exon or intron. The following protocol gives several examples of how this is done.

1. Using standard PCR methods [7–9] isolate a fragment of genomic DNA encompassing the sequencing event of interest (*see* Fig. 1). Include in the PCR primers restriction sites as needed for **step 2**.
2. Digest the PCR product with restriction enzymes and clone into a suitable expression vector such as shown in Fig. 2. *See* [8, 9] if unfamiliar with standard subcloning methods.
3. Confirm minigene sequence and prepare DNA by a method that generates sufficiently pure and concentrated DNA for cell transfection.

3.2 Transfection and Expression of Minigene

The methods for transfection and cell grown/minigene expression are highly dependent on the choice of cell line, which in turn, is highly dependent on the splicing event one wishes to study. Here, we provide a protocol for transient transfection of the commonly

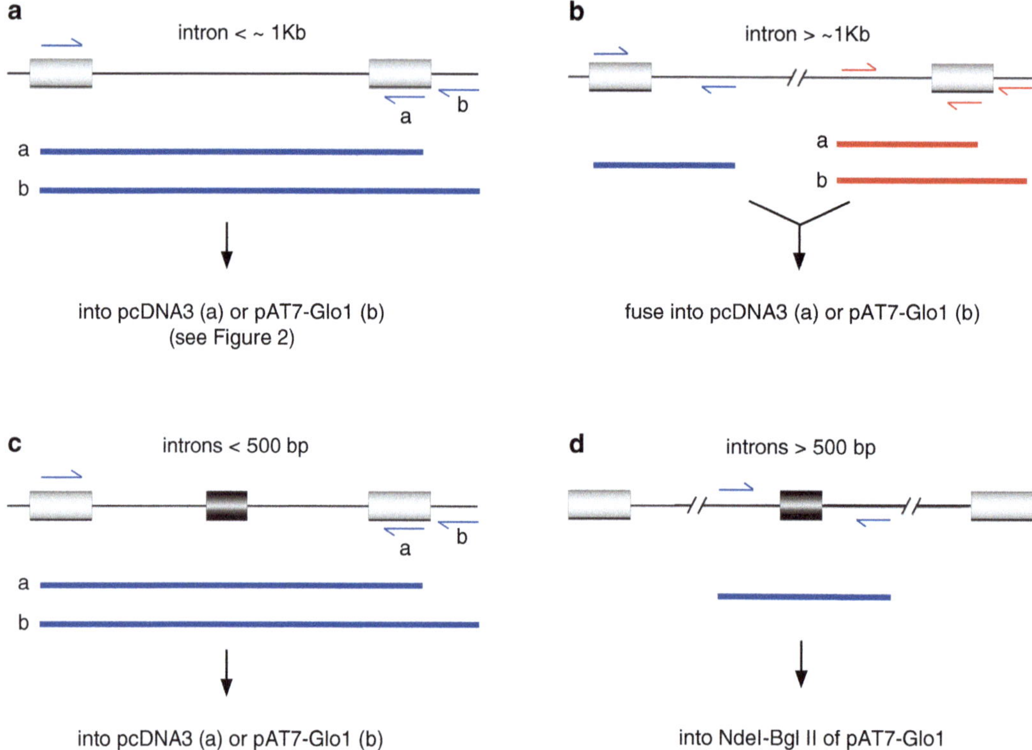

Fig. 1 Genomic segments used to generate minigenes. Typical range of genomic sequence isolated to generate minigenes to test splicing of a single intron (**a**, **b**) or inclusion of a cassette exon (**c**, **d**). Exons are indicated by *boxes*, introns by *black lines*, PCR primers by *blue* and *red single arrowhead*, PCR products by *blue* and *red lines*. Examples are given for strategies typical for large (**b**, **d**) or small (**a**, **c**) introns. In scenarios (**a**–**c**) the minigene could include only sequences from the endogenous gene (*a*) cloned into an empty mammalian expression construct such as pcDNA3, or the downstream exon could be followed by its intron in the PCR fragment (*b*) and then fused to a test intron/exon such as that from β-globin as found in pAT7-Glo1 (*see* Fig. 2). The latter construction has the advantage that splicing of the final exon to the β-globin exon functions as an internal positive control in the ultimate cell-based assay and abrogates concerns of false results from contaminating DNA (*see* **Note 30**)

used HEK293 cell line (Subheading 3.2.1), and a protocol for transfection and establishment of stable minigene expressing lines (Subheading 3.2.2), which is suitable for hard-to-transfect cell lines. Discussion of considerations for choosing an appropriate cell line and method is given in **Notes 8** and **9**.

3.2.1 Transient Transfection of HEK293 Cells

1. For each transfection, plate 3×10^5 HEK293 cells in one well of a 6-well plate in a volume of 2 ml DMEM plus serum.
2. Let attach for 24 h.
3. Carefully remove DMEM and overlay cells with 1.5 ml pre-warmed Opti-Mem (Invitrogen).
4. Mix 1 μg DNA (1 μg/μl) with 250 μl Opti-Mem and let sit 5 min at room temperature.

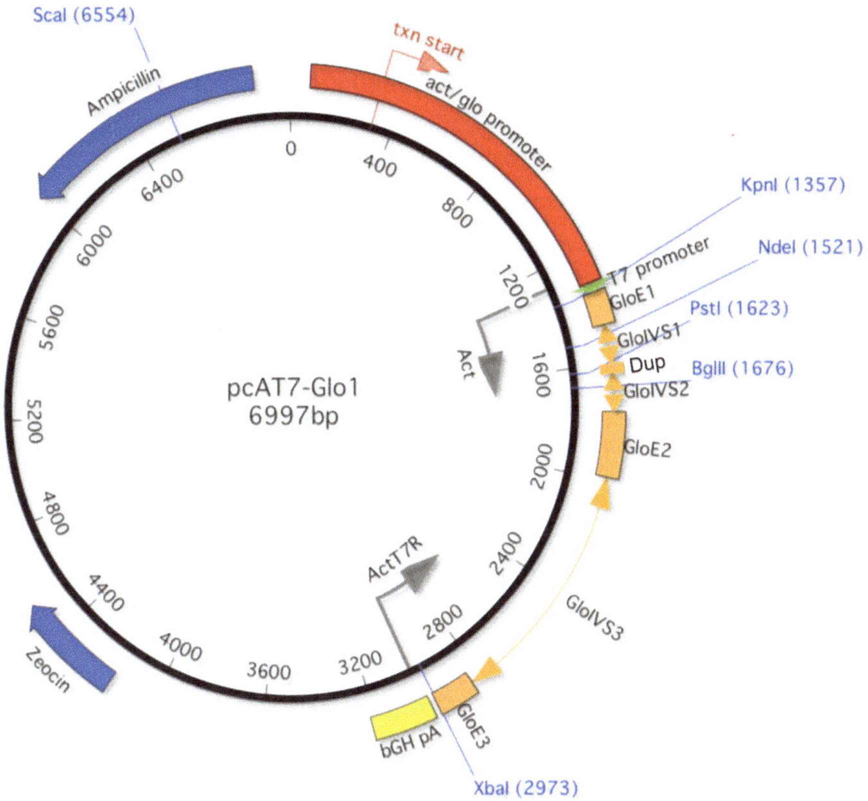

Fig. 2 pAT7-Glo1: A convenient backbone for minigene analysis. Plasmid map of pAT7-Glo1, which contains a modified version of the β-globin gene. The human β-globin gene is one of two splicing modules that has been extensively characterized and is often used as a background for minigenes (the other is the Adenovirus Major Late gene or AdML; [10]). pAT7-Glo1 contains the three exons endogenous to β-globin (*gold boxes*) plus a small test exon (*Dup, gold box*) that contains duplicated splice sites from the first and second endogenous exons [1]. Additional restriction sites have been engineered to facilitate subcloning of appropriate fragments of genomic DNA as described in Fig. 1. Other salient features are selectable markers (*blue*) for bacterial (Amp) and mammalian (Zeocin) selection, a eukaryotic promoter (act/glo, *red*) that functions in all mammalian cell lines tested, and a T7 promoter (*green*). Location of primers used for RT-PCR (Act, ActT7R) are also indicated

5. Mix 5 μl Lipofectamine 2000 with 250 μl Opti-Mem and let sit 5 min at room temperature (*see* **Note 10**).
6. Combine mixtures from **steps 4** and **5**, flick with finger to mix and let sit 20 min at room temperature for precipitate to form (*see* **Note 11**).
7. Add mixture from **step 6** in a drop-wise manner overtop of cells.
8. Harvest cells after 48 h to analyze splicing as described below (Subheadings 3.3 and 3.4).

3.2.2 Generating Stable Clonal Transfectants of Jurkat Cells

1. Split 10 million Jurkat cells per transfection to 0.5 million per ml. Let cells grow for about 24 h so that they are in mid-log phase for transfection (0.8–1.2 million per ml).
2. The next day, spin down 10–20 million Jurkat cells for each construct to be transfected.
3. Wash cells twice in serum-free antibiotic-free medium.
4. Resuspend cells in 400 μl serum-free antibiotic-free medium per transfection. Transfer 400 μl cells to an electroporation cuvette (*see* **Note 12**).
5. Add 10 μl plasmid DNA (1 μg/μl). Flick cuvette to mix well. Let stand up to 5 min.
6. Set electroporator for 250 mV, capacitance = 960 and time to constant. Place each cuvette into holder and electroporate sample (*see* **Note 13**).
7. Flick cuvette vigorously to mix pH gradient that has been formed and let stand for 5 min.
8. Remove cells from cuvette, being careful to avoid transferring the clump of dead cells and debris. Add cells to a well of a 6-well plate to which 6 ml of medium plus 10 % serum has been added.
9. After 48–72 h, serially dilute cells to achieve 20 ml each of cells diluted to 1×10^5, 3×10^4, and 1×10^4 per ml into medium containing serum and antibiotics for selection (*see* **Note 14**).
10. Aliquot each dilution into a full 96-well plate, using 200 μl diluted cells per well.
11. Allow 14–21 days for colonies to appear. Slowly expand wells that contain a single colony eventually to a 6-well plate (*see* **Note 15**).
12. Harvest a 5 ml sample of cells for each clone using Subheading 3.3 below (leaving sufficient cells continuing to grow to maintain clone) and perform RT-PCR (Subheading 3.4) to screen for minigene expression and splicing (*see* **Note 16**).

3.3 Harvest of RNA

1. Harvest up to 20 million cells into an appropriate conical or microcentrifuge tube and collect by centrifugation (*see* **Note 2**).
2. Remove and discard supernatant.
3. Resuspend cell pellet with 1 ml PBS and transfer to microcentrifuge tube.
4. Spin in a microcentrifuge for 1 min at 900 ×g (revolutions per minute) (*see* **Note 2**).
5. Remove and discard supernatant (aspiration works well here).
6. Resuspend cell pellet in 800 μl RNA-Bee (*see* **Note 17**) and place tubes on ice (*see* **Note 18**).

7. Add 200 μl chloroform. Invert 2–3 times and vortex for 5–10 s (*see* **Note 19**).
8. Hold on ice for 10 min inverting occasionally.
9. Spin in a refrigerated (4 °C) microcentrifuge for 12 min at 17,000 ×*g* (*see* above).
10. Meanwhile, label fresh RNase-free tubes and add 600 μl isopropanol. Place tubes on ice to chill alcohol.
11. Remove tubes from microcentrifuge and place at room temperature (*see* **Note 20**). Transfer clear supernatant to tubes with isopropanol. Do not carry over white interface that contains DNA.
12. Vortex tubes thoroughly and spin in a refrigerated (4 °C) microcentrifuge for 12 min at 17,000 ×*g* (*see* above).
13. Decant supernatant and add 800 μl ice-cold 70 % ethanol (*see* **Note 21**).
14. Spin in a refrigerated (4 °C) microcentrifuge for 2 min at 17,000×*g* (see above).
15. Decant supernatant being careful not to lose pellet and repeat **steps 13** and **14** twice more (*see* **Note 22**).
16. Remove all liquid with a pipet tip (*see* **Note 23**)
17. Resuspend in 12 μl RNase-free DI H_2O.
18. Check concentration by OD_{260} and adjust to 0.5 mg/ml in RNase-free DI H_2O.

3.4 Analysis of Splicing Pattern

The most widely used and robust assay to analyze splicing patterns in cells is low-cycle RT-PCR. The following protocol provides a highly reproducible assay to quantify changes in isoform expression between two conditions (i.e., cell growth conditions, presence or absence of a trans-acting factor, or between wild-type and mutant minigenes) (*see* **Note 24**).

3.4.1 Making 5′ ^{32}P-End Labeled PCR Primer

1. Mix together 77 μl dH_2O, 10 μl ^{32}P-gamma-ATP, 10 μl 10× PNK buffer, 2 μl (100 ng/μl) downstream primer, and 1 μl PNK enzyme.
2. Incubate at 37 °C for 30 min.
3. PCA extract, precipitate with ethanol and wash in 70 % ice-cold ethanol (*see* **Note 25**).
4. Resuspend in 80 μl RNase-free DI H_2O for a final concentration of ~2.5 ng/μl.

3.4.2 RT-PCR

1. To a PCR tube on ice, add 2 μl RNA at 0.5 μg/μl, 1 μl 5× Hyb buffer, and 1 μl downstream primer (1 ng/μl) (*see* **Note 4**).
2. Using a thermocycler, heat RNA/primer/Hyb buffer mix to 90 °C for 20 s then cool slowly to 43 °C by decreasing

temperature in 1 °C increments every 20 s then holding at 43 °C. This denatures the RNA and anneals the primer.

3. Add 19.5 μl RT-mix plus 0.5 μl MMLV, which have been premixed and warmed to 43 °C (*see* **Note 26**).
4. Continue incubation at 43 °C for 30 min followed by 94 °C for 5 min, and rapid cool to 4 °C. Hold at 4 °C until next step.
5. Set up hot PCR reaction by adding 1 μl downstream primer (5 ng/μl), 1 μl upstream primer (2.5 ng/μl), 1 μl ^{32}P-labeled upstream primer (from Subheading 3.4.1. **step 4**), 1.5 μl RT-PCR buffer, 10.3 μl dH_2O, and 0.2 μl Taq DNA polymerase. Mix well. Add a drop of mineral oil onto sample (*see* **Note 7**).
6. Set Thermocycler to run a program such as the following: 94 °C 2 min; *X* cycles of (94 °C, 1 min; *Y* °C, 1 min; 72 °C, *Z* min); 72 °C, 7 min, 4 °C and hold. Annealing temperature (*Y*) and extension time (*Z*) should be determined by your primers and length of predicted product. Cycle number (*X*) must be determined empirically for each transcript to provide signal that is linear with respect to input RNA and is dependent on expression level of the RNA, but is typically 20–25 cycles.
7. Add 15 μl formamide buffer. Store samples at −20 °C.

3.4.3 Denaturing Gel Electrophoresis

1. Prepare and pre-run a 5 % denaturing TBE–urea–polyacrylamide gel.
2. Boil reactions from Subheading 3.4.2, **step 8** for 5 min and place immediately on ice.
3. Turn off power to gel. Load 5 μl of each reaction per lane plus an appropriate marker in an additional lane (*see* **Note 27**). Restore power to gel (*see* **Note 28**).
4. Run gel an appropriate length of time to resolve bands.
5. Disassemble gel plates, submerge gel on one plate in 10 % acetic acid + 10 % methanol for 15 min to fix.
6. Transfer gel to Whatmann paper. Use an additional sheet of Whatmann underneath and overlay with saran-wrap (don't wrap saran-wrap underneath Whatmann paper).
7. Dry on gel dryer, then expose to phosphorimage screen (*see* **Notes 29** and **30**).

4 Notes

1. For stable transfections, it is optimal to linearize the minigene expression plasmid prior to transfection. This provides free ends that increase the efficiency of chromosomal integration. Moreover, if the plasmid is linearized by restriction digest prior to transfection, one can ensure that the cut is made *outside* of

the minigene sequence and its promoter, thereby decreasing the number of drug-resistant clones that have lost the ability to express the minigene due to random nicking. Typically, one uses a restriction site that is unique to the bacterial resistance gene (i.e. Amp^r) as the integrity of this gene is not relevant for growth in mammalian cells. After linearization in a standard restriction digest, the DNA should be repurified by PCA extraction/ethanol precipitation (*see* **Note 25**) and resuspended in DI H_2O.

2. Mammalian cultured cells are collected by low-speed centrifugation (<1,000 × *g*). This typically is done by spinning at 3K rpm for 1 min in a microcentrifuge or 5 min in a typical tabletop centrifuge at 1K rpm (~200 × *g*). The choice of centrifuge will depend on the volume of media from which cells are being collected. Centrifugation at a force higher than 1,000 × *g* will cause cells to lyse.
3. The use of RNA-Bee and the above protocol for RNA harvest is easy, fast and yields high-quality RNA that supports reproducible results in subsequent applications. RNA-Bee is a monophase solution containing phenol and guanidine thiocyanate. Many similar reagents are also available commercially, such as Trizol (Invitrogen), while some labs chose to make their own solutions. Other methods for RNA isolation include column-based kits from companies such as Qiagen and Ambion.
4. Appropriate design of primers is essential for the RT-PCR to be robust and quantitative. First, the primers need to be specific to the minigene expression construct and not cross-reactive to the endogenous gene. Typically, we use a forward primer that is complementary to the first 20–30 transcribed nucleotides of the minigene, which includes significant sequence from the vector cloning sites. Similarly, the reverse primer is optimally complementary to some constitutive portion of the minigene, such as the final exon in the sample vector shown in Fig. 2 (Act and ActT7R). Use of a poly-dT reverse primer should be avoided, as this does not have a sufficiently high Tm to allow for stringent RT-PCR conditions such that the results obtained with such a primer can be highly variable. The minimal Tm of the primer for stringent RT-PCR is 60 °C. Whenever possible, we design primers to anneal at 70 °C with the 3′ terminal 2–3 nucleotides consisting of a G or C.
5. The 1.25× RT-mix should be stored in single or double-use aliquots at −80 °C to avoid repeated freeze–thaw cycles that decrease the stability of the dNTPs.
6. Several RT (reverse transcriptase) enzymes are commercially available. In our hands MMLV is the most robust and is most heat-stable such that the RT reaction can be done at a temperature that limits the existence of RNA secondary structure.

However, for one sequence we have found AMV to be more robust. By contrast, the "Superscript" family of RT enzymes yield highly variable results and their use in quantitative assays such as those outlined here should be avoided.

7. Oil should be used even with "hot-lid" machines, as ^{32}P can be volatile. Also in our experience "hot-lids" can alter the accuracy of the temperature of the reaction, occasionally causing problems with reproducibility in the RT-PCR reaction.
8. Any standard method of transfection and gene expression can be used for the cell-based analysis of splicing. HeLa, HEK293, and COS cells have been widely used for the analysis of general splicing events due to the ease of transfection. However, understanding tissue-specific, disease-specific or pathway-specific splicing events often requires use of more specialized cell lines. Thus, researchers should use the cell-line that best fits the biology of the system, and chose a method for transfection and gene expression that is most optimal for the specified cell type.
9. Transient transfection/expression of a minigene allows analysis of splicing within 2–4 days, versus the 3–4 weeks required for the creation of a stable cell line. However, establishing stable, clonal cell lines is advantageous in the instances where transfection efficiency is lower, or when the minigene is to be used for cell-based screening. In our experience transient transfections also at times yield highly variable results due to the unstable nature of the expression of the minigene. Therefore, while transient transfection is a good "quick and dirty" method, the total time required to obtain statistically significant results is often similar for stable versus transient transfection.
10. The exact ratio of DNA to Lipofectamine can have significant impact on the efficiency of transfection. The optimal ratio must be determined empirically for each cell type. See the Invitrogen protocol for more detail (http://www.invitrogen.com).
11. Note that the precipitate is often not visible to the naked eye. This does not alter the efficiency of transfection.
12. The efficiency of electroporation is highly sensitive to volume but not to total cell number. It is important to use exactly 400 μl of cells in each cuvette even if this means using less than the optimal 10–20 million cells.
13. Settings may vary for different machines and different strains of Jurkats or other suspension cells.
14. The given plating densities are appropriate for Jurkat transfections selected with Zeocin (250 μg/ml) or G418 (2 mg active compound/ml). Optimal plating densities for other cell lines or drugs should be determined empirically. The goal is to obtain at least one 96 well plate in which ~30 % of wells have colonies,

as statistically most of these will be single clones and this will provide enough clones for further expansion and analysis.

15. Expand by first diluting the 200 μl from the 96 well plate into 2 ml, then eventually 6–10 ml. Overly diluting cells at this stage often causes cell death and loss of clone.
16. If the minigene DNA has been linearized prior to transfection (*see* **Note 1**) generally 80 % or greater of drug-resistant clones will express the minigene. This success rate typically drops to 25 % if a circularized vector is used. The transfection may yield many more single clones than one wishes to pursue for further study. It is prudent to screen 2–4 times more clones than one wants to ensure that a sufficient number is obtained to ensure statistical significance of the final data (3–4 clones if the minigene is spliced consistently in all clones).
17. The final quantity and quality of the RNA is highly dependent on efficient cell lysis at this step. Solubilize the cell pellet in RNA-Bee by repeated pipeting (~10×). A sign that the cells have lysed will be that the solution will become increasingly viscous, air bubbles will not dissipate as quickly and one may notice a "schlurpping" sound.
18. If it is not convenient to do the entire RNA harvest in 1 day, the process can be stopped after the cells are solublized in RNA-Bee and the tubes stored at −80 °C. Similarly, the RNA can be stored at −80 °C after isopropanol is added in **step 10**.
19. Do not vortex too aggressively at this stage. The component should be well mixed, but if vortexed too long an emulsion will form between the chloroform and RNA-Bee that cannot be separated by subsequent centrifugation. If this happens no RNA will be able to be obtained from the sample.
20. At this point there should be a clear liquid phase (containing RNA) over a clear blue phase (containing protein) with a white solid interface (containing DNA). Putting tubes on ice at this point will cause the clear phases to become turbid, making discrimination of RNA phase from DNA phase more difficult.
21. Following centrifugation a small, white pellet of RNA should be visible at the bottom on the tube. Be careful not to dislodge while decanting supernatant. Use a pipet to remove the liquid if this is a problem.
22. Thorough washing of the RNA pellet is required to ensure no carry over of contaminants that can hinder the subsequent RT-PCR reaction.
23. If over-dried the RNA pellet will become resistant to being solublized in H_2O. Do not use a speed-vac to dry. Simply remove any visible liquid with a pipet, allow to air-dry for no more than 5 min and immediately add 12 μl RNase-free DI H_2O.

24. Radiolabeled RT-PCR (as in the protocol here) is superior to Northern blots, real-time RT-PCR or non-labeled RT-PCR for the analysis of splicing for several reasons. Ethidium bromide staining of RT-PCR (in the absence of radiolabel) is much less sensitive and quantitative and can only be used to roughly judge qualitative differences in splicing. Real-time RT-PCR requires use of distinct primers to measure alternate isoforms, adding variability to the results. In addition this method can only measure predicted isoforms and does not report on the presence of unanticipated or cryptic products. Northern blots are highly quantitative, but cumbersome and often don't have the size resolution needed.

25. PCA extraction and ethanol precipitation are standard techniques in any molecular biology lab. PCA is a 25:24:1 mixture of phenol:chloroform:isoamyl alcohol that is used at equal volume to the aqueous reaction to extract protein. RNA/DNA is then precipitated using 0.1 volumes of 3 M NaOAC and 2.5 volumes of 100 % ethanol. *See* [9] for more details.

26. Pre-warm the RT + MMLV mix to 43 °C for 2 min before adding to the RNA/primer mix. This prevents the RNA from cooling below 43 °C at any point and thus limits the formation of RNA secondary structure, which can inhibit the reverse transcription reaction. We have seen significantly greater reproducibility in the RT-PCR results when the RNA is kept at 43 °C or greater versus conditions that cool the reaction to 37 °C.

27. Any radiolabeled markers can be used here. We typically use the pBR322-MspI digest markers from NEB. These cover a size range that is typically appropriate (622 bp and smaller) and can be easily radiolabeled by filling in the fragment ends with Klenow polymerase fragment and ^{32}P-dCTP (*see* www.neb.com).

28. Never handle or load a gel with the power supply still attached and/or current still running as this can lead to electrocution.

29. Quantification of splicing is done by using a phosphorimager to quantify the intensity of the spliced products compared to the background lane. Splicing is then quantified as either % variable exon inclusion (intensity of product including the variable exon/total intensity of all products), or alternatively, when comparing the splicing of minigenes grown under different cellular conditions it is often advantageous to calculate a fold difference in isoform ratio by ([included/excluded]$_{\text{condition 1}}$/[included/excluded]$_{\text{condition 2}}$ as described in [1].

30. Designing the experiment to quantify two alternative spliced products (i.e. variable exon inclusion versus exclusion between two constitutive exons) and/or including at least

one constitutively spliced intron in the minigene (i.e. β-globin exon 2-exon 3; *see* Fig. 2) has the significant advantage that one does not have to differentiate between unspliced message and vector DNA. Therefore DNase treatment and extensive "no-RT" controls are not necessary.

References

1. Rothrock C, Cannon B, Hahm B et al (2003) A conserved signal-responsive sequence mediates activation-induced alternative splicing of CD45. Mol Cell 12(5):1317–1324
2. Xie J, Black DL (2001) A CaMK IV responsive RNA element mediates depolarization-induced alternative splicing of ion channels. Nature 410(6831):936–939
3. Zhu J, Mayeda A, Krainer AR (2001) Exon identity established through differential antagonism between exonic splicing silencer-bound hnRNP A1 and enhancer-bound SR proteins. Mol Cell 8(6):1351–1361
4. Topp JD, Jackson J, Melton AA et al (2008) A cell-based screen for splicing regulators identifies hnRNP LL as a distinct signal-induced repressor of CD45 variable exon 4. RNA 14(10):2038–2049
5. Stoilov P, Lin CH, Damoiseau R et al (2008) A high-throughput screening strategy identifies cardiotonic steroids as alternative splicing modulators. Proc Natl Acad Sci U S A 105(32):11218–11223
6. Warzecha CC, Sato TK, Nabet B et al (2009) ESRP1 and ESRP2 are epithelial cell-type-specific regulators of FGFR2 splicing. Mol Cell 33(5):591–601
7. Kramer MF, Coen DM (2001) Enzymatic amplification of DNA by PCR: standard procedures and optimization. Curr Protoc Mol Biol. Chapter 15: Unit 15 1
8. Struhl K (2001) Subcloning of DNA fragments. Curr Protoc Mol Biol. Chapter 3: Unit 3 16
9. Sambrook J, Russell D (2001) Molecular cloning: a laboratory manual, 3rd edn. Cold Spring Harbor Press, Cold Spring Harbor, NY
10. Anderson K, Moore MJ (1997) Bimolecular exon ligation by the human spliceosome. Science 276(5319):1712–1716

Chapter 19

Quantifying the Ratio of Spliceosome Components Assembled on Pre-mRNA

Noa Neufeld, Yehuda Brody, and Yaron Shav-Tal

Abstract

RNA processing by the splicing machinery removes intronic sequences from pre-mRNA to generate mature mRNA transcripts. Many splicing events occur co-transcriptionally when the pre-mRNA is still associated with the transcription machinery. This mechanism raises questions regarding the number of spliceosomes associated with the pre-mRNA at a given time. In this protocol, we present a quantitative FISH approach that measures the ratio of intensities between two different spliceosomal components associated on a nascent mRNA, and compares to the number of introns in the mRNA, thereby calculating the number of spliceosome complexes assembled with each transcript.

Key words RNA FISH, Immunofluorescence, Spliceosome, Intron

1 Introduction

Eukaryotic pre-mRNAs undergo several processing events that culminate in a mature mRNA molecule. One of the key steps in the maturation of newly synthesized transcripts is the mRNA splicing process, in which the noncoding intron sequences are removed and the coding exon sequences are joined in a multistep reaction, carried out by the spliceosome [1, 2]. The spliceosome can reach several megadaltons in molecular weight and is composed of many RNA and protein components, the latter contributing more than two-thirds of its mass. During the assembly of splicing factors on the pre-mRNA, a network of RNA–RNA interactions is formed, predominantly by the U snRNAs. U snRNAs are packaged to form U snRNP particles containing approximately 45 proteins, and together with the remaining non-snRNP proteins (~130) they comprise the human spliceosome [3]. Although the splicing process has been extensively studied, there are still issues that need to be addressed, such as the mechanism of spliceosome assembly in the context of multiple introns [4–6], and the location of the splicing reaction after the completion of the actual transcriptional processes [7, 8].

Klemens J. Hertel (ed.), *Spliceosomal Pre-mRNA Splicing: Methods and Protocols*, Methods in Molecular Biology, vol. 1126, DOI 10.1007/978-1-62703-980-2_19, © Springer Science+Business Media, LLC 2014

To approach these questions, we have implemented a ratio quantitative fluorescence in situ hybridization (rqFISH) method [9], which allows measuring the ratio between specific components of complexes associated with pre-mRNA transcribed on an active gene.

In this protocol, we describe how to quantify the ratio of spliceosomal components associated with pre-mRNAs containing different numbers of introns. A gene construct containing intron and exon sequences is stably transfected into a mammalian adherent cell line. The gene integrates and forms a tandem gene array and is therefore detectable using fluorescence microscopy, typically using RNA FISH with a fluorescent probe that specifically hybridizes to a unique sequence in the gene under study. The association of splicing factors with the active gene array can be visualized using immunofluorescence. The rqFISH method is then performed by marking two specific components of the complex and measuring the ratio of intensities between them when assembled on the pre-mRNA: (1) Labeling the pre-mRNA using unique fluorescence probes (by RNA FISH); and (2) Marking of either a spliceosomal protein component with a specific antibody (by immunofluorescence, I.F.) or labeling of an additional RNA component. High-resolution 3D images of the labeled cells are acquired using wide-field fluorescence microscopy, followed by deconvolution for image restoration, thereby rendering the images suitable for quantification. Subsequently, the fluorescence intensities are measured at each voxel (volumetric pixel) in the cell volumes and the ratio of intensities measured from the two channels are compared and used to calculate the proportion of the two studied molecules that are bound to the processed mRNA [9]. By comparing such ratios arising from different cells or treatments, it is possible to approach hitherto hidden stoichiometric relations between interacting moieties. In this protocol, we describe how to compare the ratio of spliceosomes assembled on a gene with three introns compared to a gene with six introns.

2 Materials

2.1 Vector Cloning

1. pSL24MS2 vector: Plasmid containing the 24 MS2 sequence repeats, for example, http://www.addgene.org/27120/.
   OR: CFP vector (for example, http://www.addgene.org/13030/).
2. Restriction enzymes.
3. 5 μ/ml T4 DNA Ligase.
4. Competent *Escherichia coli* bacteria for transformation (*see* **Note 1**).

2.2 Electroporation for Stable Integration of Gene of Interest (GOI)

1. Dulbecco's Modified Eagle's Medium (DMEM).
2. Fetal bovine serum (FBS).
3. Trypsin for detaching cells from tissue culture plate.
4. PBS solution for washing cells.
5. 1–4 μg of GOI plasmid.
6. Electroporator, e.g., Bio-Rad Gene Pulser Xcell (Bio-Rad).
7. Gene Pulser Cuvette 0.4 cm.
8. Adherent cell lines of choice. For example, U2OS or HeLa cells.
9. Antibiotics for stable selection depending on the selection marker in the gene construct.
10. Cloning cylinders (Corning).

2.3 RNA FISH

1. PBS.
2. 4 % Paraformaldehyde in PBS.
3. 70 % Ethanol.
4. 20× Saline–sodium citrate buffer (SSC): 3 M NaCl, 0.3 M *sodium citrate* $C_6H_5Na_3O_7$ at pH 7.
5. 40 % Formamide: 60 ml 4× SSC (diluted with DDW) + 40 ml 100 % formamide.
6. ssDNA/tRNA: Mix equal vol. of 10 mg/ml ssDNA (Sigma) and 10 mg/ml tRNA (Roche).
7. 10 mg/ml BSA.
8. Fluorescently labeled DNA probe to label the mRNA or snRNA. Use ~10 ng probe per coverslip (from a stock of 40 ng/ml). The fluorophore (Cy3 molecule or others such as Cy's, Alexa, ATTO, etc.) is conjugated to five amino-allyl thymidines inserted during probe synthesis at the: 5′-end, 3′-end, and three internal Ts of the probe sequence.

   For the MS2 sequence we use: 5′-TTT CTA GGC AAT TAG GTA CCT TAG GAT CTA ATG AAC CCG GGA ATA CTG CAG-3′.

   For the CFP sequence: 5′-ATA TAG ACG TTG TGG CTG ATG TAG TTG TAC TCC AGC TTG TGC CCC A-3′.

   For U snRNA probes [10] (where applicable, can use both versions to get a better signal):

   U1-v.1: 5′-CGG GAA AAC CAC CTC GTG ATC AGG TAT CTC CCC GCC AGG TAA GAT-3′

   U1-v.2: 5′-CGA ACG CAG CCC CCA CAC CAC AAA TTA GCA GTC GAG TTC CCA CAT-3′.

   U2: 5′-AGG GAC GGA GCA AGC CCT ATT CCA CTC CCT GCC CAA AAA TCC ATT-3′.

U4-v.1: 5′-AGC AAT AAC GCG CCT CGG AAA ACC TCA TGG CTA CGA TAC GCC ACT-3′.

U4-v.2: 5′-AGA CGT CAA AAA TGC CAA TGC CGA CAT ATT GCA AGC GTC ACG GCG GA-3′.

U5: 5′-GGC AAG GCT CAA AAA ATT GGG TTA AGA CTC AGA GTT GTT CCT CTC CAC GGT A-3′.

U6: 5′-CGG TCA TCC TGC GCA GGG GCC AGC TAA TCT TCC TGT ATC GTC CAA-3′.

9. Solution 1: 2.5 μl of probe (40 ng/ml stock, for ten coverslips); 3.6 μl of 20× SSC; 2 μl of 5 mg/ml of ssDNA/tRNA; 23 μl of DDW; 160 μl of 100 % formamide. Adjust to 200 μl with DDW.
10. Solution 2: 198 μl of DDW; 2 μl of BSA; 50 μl of 20× SSC.

2.4 Immunofluorescence

1. PBS.
2. 5 % BSA in PBS.
3. 0.5 % Triton in PBS.
4. Primary antibody.
5. Fluorescent secondary antibody (can use different fluorophores such as Cy's, FITC, Alexa, ATTO, etc.).

2.5 Acquisition of a 3D Image for the Detection of Transcription Sites

1. Fluorescent microscope of choice (wide field), equipped with a high-resolution CCD camera.
2. Microscope slides.
3. 13 or 18 mm coverslips.
4. Hoechst for nuclear staining.
5. Anti-fade mounting reagent.
6. Image analysis software, such as Imaris (Bitplane, Switzerland), Metamorph (Molecular Devices, Downingtown, PA), or ImageJ (NIH, Bethesda, MD; http://rsb.info.nih.gov/ij/).
7. Deconvolution software, such as Huygens Deconvolution Software (Scientific Volume Imaging, The Netherlands), AutoQuant (Media Cybernetics, Bethesda, MD), or DeltaVision (Applied Percision, Issaquah, WA).

3 Methods

3.1 Construction of an Expression Vector for Exploring Splicing in Live Cells

The first step is to build a gene construct that will express the gene of interest (GOI) to be assayed for splicing. As a specific example for the user, this protocol will make use of the study we performed to understand the dynamics of spliceosome assembly on introns. Two constructs were generated; one contained the GOI that had three exons and two introns, and as a comparison the other

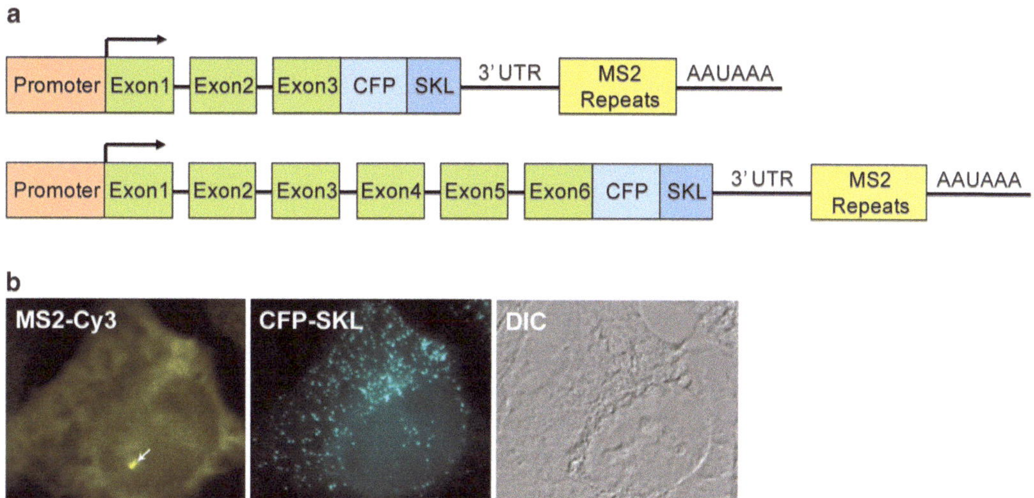

Fig. 1 Generating a cell system with comparable genes containing increasing numbers of introns for exploring splicing in live cells. (**a**) Gene constructs containing increasing numbers of introns and exons were generated. Common to all the genes are promoter (*peach*); exons (*green*) and introns (*black lines*); in-frame CFP-SKL peroxisomal protein (*pale blue*); series of MS2 repeats in the 3′ UTR (*yellow*). (**b**) Stable cell line of the construct containing the GOI with three exons and two introns. In *yellow*, the transcribed mRNA at the transcription site and throughout the cell, labeled by FISH with MS2-Cy3 probe, and in *cyan*, the CFP-SKL protein product in cytoplasmic peroxisomes. The transcription site is marked by the *white arrow*. Scale bar, 5 μm

construct harbored the same GOI but with six exons and five introns (Fig. 1a). The following steps are used to create an expression vector for exploring splicing in live cells.

1. Obtain a GOI consisting of exons and introns (*see* **Note 2**).
2. Insert the GOI into an expression vector containing an antibiotic resistance gene (*see* **Note 3**). At this point, it is important to consider that the promoter driving your gene should result in steady levels of expression. One can use, for example, viral promoters (e.g., CMV, SV2) for continuous overexpression, or endogenous promoters that usually lead to moderate and fluctuating levels of expression (*see* **Note 4**).
3. Insert a known, exclusive, sequence into the 3′ UTR of the GOI, to identify the RNA transcribed from the GOI by RNA FISH. One option is to add in-frame to the last exon, a Cyan Fluorescent Protein (CFP) coding sequence containing the peroxisomal targeting tripeptide Ser-Lys-Leu (SKL) in its C-terminus. This sequence provides a visual validation for correct translation of the GOI visualized as a cyan fluorescent cytoplasmic peroxisomes [9, 11] (*see* **Note 5**).

Another option is to insert the MS2 sequence repeats from the pSL24MS2 plasmid into the 3′ UTR of your GOI. Preferable restriction sites for the MS2 repeats in the pSL24MS2 vector are the *Bam*HI at the 5′ of the MS2 sequence repeats and *Bgl*II at the 3′ of the MS2 sequence repeats. *Bam*HI/*Bgl*II digestion will result in a 1,308 bp fragment containing the 24 MS2 sequence repeats (*see* **Notes 1** and **6**). If there are no suitable sites for the insertion of the MS2 sequence repeats fragment, an adaptor with suitable restriction sites can be added to the 3′ UTR sequence of the GOI. It is possible to add both sequences as we have performed in our study [9].

3.2 Generation of a Cell Line Stably Expressing the GOI

After creating GOI constructs, the next step is to generate cell lines. Stable integration of the constructs results in tandem arrays of the GOI, which makes detection of the transcription site easier (Fig. 1b). Also, the expression levels over many experiments will be comparable. In our above example, we generated stable cell lines with each of the two GOIs.

1. Split the cells 1 day prior to electroporation. To optimize transfection efficiency, make a single cell suspension and plate in a 10 cm tissue culture dish in fresh medium. By the next day, confluence should reach 50–80 %.
2. Day of transfection: Wash the cells with 1× PBS, trypsinze gently by adding 1–1.5 ml of trypsin to the cells and incubate for 1–5 min at 37 °C. Add medium containing 10 % FBS and transfer the cells to a 15 ml tube. Centrifuge for 5 min at 86 × *g* and aspirate the medium.
3. Suspend in 1 ml cold medium plus serum.
4. Place 200–250 μl of the cells (approx. 200,000 cells) into a sterile cuvette and add the GOI plasmid (2–10 μg of DNA per transfection) to the cells.
5. Tap gently to mix and wait for 10 min at room temperature.
6. Electroporate using machine-specific settings. The following electroporation conditions have been successfully used with the Bio-Rad Gene-Pulsar Xcell when transfecting these human cell lines: U2OS: 170 V, 950 μF; HeLa: 150 V, 500 μF; HEK-293: 300 V, 500 μF.

   For optimal transfection efficiency, specific protocols should be calibrated.
7. Plate the electroporated cells in a 10 cm plate with 10 ml fresh medium plus serum. Mix cells and medium gently, and incubate at 37 °C.
8. Next day: Add appropriate antibiotic to the medium to begin selection. Medium with antibiotics should be changed every 3 days. Selection should continue for about 2–3 weeks until single colonies develop (*see* **Note 7**).
9. Pick single colonies. Briefly wash plate with 1× PBS. Use cloning cylinders to collect colonies by placing cylinders on well-separated

colonies. Add 100 μl of trypsin into each cylinder and incubate for 1–5 min. Then add 100 μl of medium to each cylinder, gently pipette to suspend the colony, and transfer to a 24-well plate. Add 0.5 ml fresh medium plus antibiotics to each well.

10. Screen for positive colonies that have integrated the GOI. Screening is performed according to the specific construct used. For instance, using a fluorescence microscope it is possible to either detect the presence of CFP-SKL protein in the peroxisomes, or the GOI-MS2 transcript by RNA FISH with a fluorescent probe against the specific MS2 sequence.

3.3 RNA FISH

To address the biological question, fluorescent tags must be applied to the relevant moieties. In our experiments, we labeled the GOI mRNA (channel 1) with a probe against the first exon and also labeled an RNA component of the spliceosome (channel 2) with a probe against U5 snRNA (both using RNA FISH) (Fig. 2a). It is possible to combine within the FISH protocol, steps for labeling proteins by immunofluorescence (I.F.). This is included in **steps 11–16**.

1. Grow cells from a positive clone on 18 mm round coverslips in a 12-well TC dish.
2. Wash briefly in 1× PBS and fix in 4 % paraformaldehyde (PFA) in PBS for 20 min.
3. Wash briefly in 1× PBS and then add 70 % ethanol. Leave overnight at 4 °C.
4. Next day: Rinse twice prior to hybridization with 1× PBS for 10 min (with gentle shaking).
5. Wash for 10 min in 0.5 % Triton X-100 in PBS. Wash for 10 min in 1× PBS.
6. Prehybridization: Wash twice for 5 min in 40 % formamide.
7. During the rinses prepare Solution 1 and Solution 2. This volume can be used for ten 18 mm coverslips. The fluorescent probe is added directly to solution 1 (detailed above in Subheading 2.3). Just before hybridization, boil solution 1 in an Eppendorf tube for 5 min and cool on ice for 5 min. Add 200 μl of Solution 2 to 200 μl of boiled solution 1 and keep on ice.
8. Hybridization: place a 40 μl drop of the final probe solution mix in a petri dish. Gently apply the coverslip onto the drop, with the fixed cells facing down. To avoid drying, place a small reservoir with hybridization solution (40 % formamide) in the dish. Close the petri dish and seal with parafilm. Place the hybridization dish in a 37 °C incubator and hybridize for 3 h (or overnight).
9. Next day: Half an hour before rinsing, warm up the remaining 40 % formamide solution to 37 °C.

a

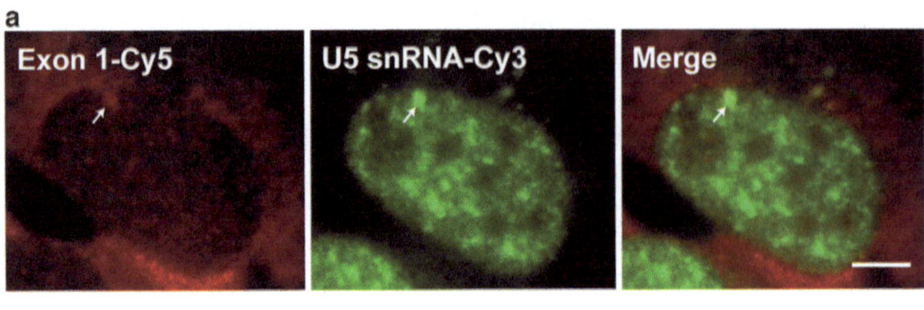

b

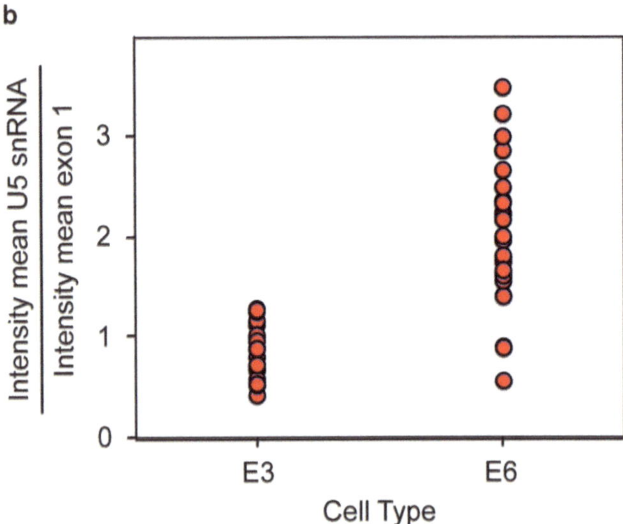

Fig. 2 Increased spliceosome recruitment in relation to increasing intron numbers shown by quantitative FISH approach. (**a**) RNA FISH was performed on a stable cell line of the construct containing the GOI with six exons and five introns. Channel 1 in *red* shows the active transcription site, tagged by a probe targeting the first exon (Cy5). Channel 2 in *green* shows the U5 snRNA tagged with a U5 snRNA-specific probe (Cy3). *White arrow* indicates the transcription site. Scale bar, 5 μm. (**b**) Quantification of the RNA FISH signal, showing the ratio between transcription site associated U5 snRNA, and exon 1 (*red dots*, each *dot* represents a transcription site ratio). The two cell lines, GOI with three exons (E3) vs. GOI with six exons (E6), are then compared side by side

10. Open the hybridization chamber and transfer the coverslips face up back into a 12-well dish already containing the pre-warmed 40 % formamide. Rinse twice in 40 % formamide for 15 min at 37 °C.

    If performing I.F. continue to step 11.

    If not performing I.F. continue to step 17.

11. Rinse for 2× 15 min in 1× PBS at RT, with gentle shaking.
12. Incubate cells with 5 % BSA in PBS for 20 min to block unspecific binding sites.
13. Primary antibody: dilute the antibody in PBS and apply to coverslips for 1 h.

14. Wash for 3× 5 min in 1× PBS.
15. Secondary antibody: dilute the antibody in PBS and apply to coverslips for 1 h.
16. Wash for 3× 5 min in 1× PBS. Go to **step 18**.
17. Rinse for 2× 1 h in 1× PBS at RT, with gentle shaking.
18. If required, perform nuclear staining for 5 min (DAPI or Hoechst).
19. Briefly wash in PBS. Mount slides in mounting solution.

3.4 Imaging and Processing

The goal is to acquire digital images from a series of focal planes and then to process these images so that the out-of-focus light (which is inherent in wide-field microscopy) from each focal plane, is quantitatively restored to the points of origin. This is done with software using a constrained deconvolution algorithm that reverses the effects of convolution on the captured images. This is followed by analysis of the acquired and restored 3D images on a specific region of interest (ROI), in this case the transcription site. The intensity of the ROIs from the two channels is then measured to calculate an internal ratio between the two channels. These internal ratio values that are calculated from the different lines (e.g., different treatments or cells with various numbers of introns, all depending on the biological question) can then be compared. In our example, we captured images of the GOI from each of the cell lines (GOI with three exons vs. GOI with six exons). After deconvolution, we calculated the internal ratios from the two channels to quantify the spliceosome component on each GOI, and then we compared between the resulting ratios of each line. This enabled us to answer the question of the relation between the spliceosome and the number of introns.

1. Acquire z-stack images of single cells using a wide-field fluorescence microscope (*see* **Note 8**). It is important to set the focus range to capture the entire cell volume. We evaluate the upper and lower borders of the cell by measuring the distance from the cell focus to the borders. The sum of the two distances represents the cell volume. This parameter is now used for all cells imaged. Cell volume varies between cell types depending on their shape. For our 100× objective with an N.A. of 1.6, we use a starting range of 15–20 μm with a z step of 0.2–0.3 μm, and 1 × 1 binning on our high-resolution CCD camera (8.6 μm pixel size), to obtain maximum resolution (*see* **Note 9**).
2. Deconvolve the z-stacks by using a deconvolution algorithm (*see* Subheading 2.5). This will also correct for bleaching. We find that 300 iterations with the Huygens Essential Deconvolution Software provides a suitable image for the quantification assay (*see* **Note 10**). For our purposes, the deconvolution will be performed on each channel separately.

3. Measure the light intensity from the selected spot in the corrected image. We use Imaris image analysis software (Bitplane) as following:
   (a) Load the 3D two channel images into the Imaris software (using "Add channel" function); select volume as initial scene and press on "Add new surface" button. Reduce time by calculating only the ROI of the transcription site area.
   (b) Estimate the diameter of the transcription site by measuring the length of a single spot in your 3D image. Mark the diameter distance by changing to "Slice" view in order to get the estimated distance in μm.
   (c) Determine the perimeter size of the area to be selected by changing the threshold function. This selection will affect the spot diameter when creating surface objects from regions in **step (d)**. Each channel can have different threshold parameters, yet the parameters chosen for each channel must be used to define all the surfaces of all cells for the same experiment for that channel.
   (d) Click on the "Spots Surfaces" and then on the "Statistics" button. After the analytical data has been calculated, press on the "Settings" button. Under "Surface Object" choose "Intensity Sum" and export the data to an Excel sheet by pressing the "Excel" button. The Intensity Sum represents the sum intensity of all pixels which make up the surface object.
   (e) Calculate the internal ratio by dividing the intensity values of one channel by the other, depending on the biological question. This must be performed on a large number of cells, at least 10, and average the resulting ratios. Then, to compare between cell lines or treatments, choose one averaged ratio to be the "normal"—divide by itself to give a value of 1—and then divide the mean ratio of the values to be compared, by this number. In our example, the intensity value of a U snRNA (obtained by the U snRNA probe—channel 1) was divided by the intensity value of an exon (obtained by the specific exon probe—channel 2) to obtain the internal ratio. This was performed on, and averaged from, multiple cells. Then, to address the biological question, this internal mean ratio value was compared to the internal mean ratio values acquired from the other cell line with a larger numbers of introns (Fig. 2b).

4 Notes

1. Bacteria tend to discard repeated sequences and therefore when cloning the MS2 repeats it is preferable to use Stbl2 competent cells that are less prone to repeat removal. Some additional steps can be taken to minimize the loss of

repeats. Clone the MS2 region as a last step into the expression vector. Then, inspect the integrity of the full 24 MS2 repeats throughout the steps of the regular cloning process, using flanking restrictions sites on either side of the MS2 sequence. It is possible to assess the actual number of repeats inserted in the vector when run in an agarose gel (also *see* **Note 6**).

2. Make sure that the RNA resulting from the GOI indeed undergoes splicing. Purify RNA from cells expressing the GOI and validate by RT-PCR.
3. It is advisable to use a vector with built in antibiotic resistance to facilitate selection for a stably expressing cell line. If this is not possible, one can perform a co-transfection of the GOI vector together with a vector containing antibiotic resistant only.
4. Promoter types. If more than one vector is to be used in the same experiment for any kind of comparison, the promoters must be the same. Additionally, it is possible to use an inducible promoter such as the Tet On/Off system, to better control the transcription and avoid any long-term overexpression issues when performing live cell imaging.
5. The end goal is to be able to label the mRNA of interest (by RNA FISH) in one fluorescent channel, in conjunction with an additional splicing related component, which can be protein or RNA depending on the biological question, in another fluorescent channel. We label the mRNA with probes to either the CPF or MS2 tag sequences, and also use probes that are specific to the exon or intron sequences in the GOI. We also labeled the splicing machinery (i.e., U snRNAs) by RNA FISH, or the RNA polymerase by immunofluorescence. Many different combinations of the above can be designed.
6. MS2 repeat number: 24 MS2 repeats produce a sufficient fluorescent signal for live cell imaging [12]. This will provide a high signal-to-noise ratio to better detect the site of active transcription over the diffuse background signal.
7. Stable cell generation: We have found that U2OS and the HeLa cells are convenient cell lines to work with for this type of microscopy analysis.
8. In order to properly quantify images using the fluorescence signal, the microscope needs to collect as much light as possible from the sample. Therefore, a series of focal planes need to be collected, and as such wide-field microscopy is desirable. Care must be taken not to under-sample the image (images acquired with sampling densities below the Nyquist Rate are said to be under-sampled). What matters for correct sampling is not only the number of slices but also the total physical volume that is actually imaged. A good image includes information of the cone of blur around it that can be large especially in

wide-field microscopes with low numerical aperture (NA) objectives. The ideal sampling rate, which depends on the NA, for normal imaging can be found using the Nyquist Calculator, but when full light scattering needs to be collected, as in this case for deconvolution purposes, over-sampling is recommended.

9. One needs to be wary of bleaching. Therefore, on the one hand, enhance the light intensity and exposure time to get a high signal to noise ratio, but on the other hand, make sure not to overexpose the images because this may lead to data distortion. These factors are dependent on the microscope parameters. The intensity initially chosen needs to be used over the entire course of the experiment, and the experiment should be completed on the same day to avoid fading. Also, it is very important to have a good FISH/IF signal because if the signal is weak, the sample will be bleached by the time the entire z-stack has been captured and the data might not be of sufficient quality.
10. To calculate the theoretical point spread function (PSF), which will enable the software to revert back from the unresolved image, to a deconvolved high-resolution image, the physical, and imaging parameters must be recorded, such as the sampling interval (*X*, *Y*, *Z*), optical information (numerical aperture, refractive indexes, lens, medium), and channel parameters (excitation/emission wavelength).

Acknowledgement

The work in the Shav-Tal lab is supported by the European Research Council (ERC).

References

1. Wahl MC, Will CL, Luhrmann R (2009) The spliceosome: design principles of a dynamic RNP machine. Cell 136(4):701–718
2. Nilsen TW (2003) The spliceosome: the most complex macromolecular machine in the cell? Bioessays 25(12):1147–1149
3. Will CL, Luhrmann R (2011) Spliceosome structure and function. Cold Spring Harb Perspect Biol 3(7)
4. Rino J, Carmo-Fonseca M (2009) The spliceosome: a self-organized macromolecular machine in the nucleus? Trends Cell Biol 19(8):375–384
5. Azubel M, Habib N, Sperling R et al (2006) Native spliceosomes assemble with pre-mRNA to form supraspliceosomes. J Mol Biol 356(4):955–966
6. Dye MJ, Gromak N, Proudfoot NJ (2006) Exon tethering in transcription by RNA polymerase II. Mol Cell 21(6):849–859
7. Girard C, Will CL, Peng J et al (2012) Post-transcriptional spliceosomes are retained in nuclear speckles until splicing completion. Nat Commun 3:994
8. Brody Y, Shav-Tal Y (2011) Transcription and splicing: when the twain meet. Transcription 2(5):216–220
9. Brody Y, Neufeld N, Bieberstein N et al (2011) The in vivo kinetics of RNA polymerase II

elongation during co-transcriptional splicing. PLoS Biol 9(1):e1000573

10. Shav-Tal Y, Blechman J, Darzacq X et al (2005) Dynamic sorting of nuclear components into distinct nucleolar caps during transcriptional inhibition. Mol Biol Cell 16(5):2395–2413

11. Tsukamoto T, Hashiguchi N, Janicki SM et al (2000) Visualization of gene activity in living cells. Nat Cell Biol 2(12):871–878

12. Janicki SM, Tsukamoto T, Salghetti SE, et al (2004) Silencing to gene expression: Real-time analysis in single cells. Cell 116(5):683–698

Chapter 20

Antisense Methods to Modulate Pre-mRNA Splicing

Joonbae Seo, Eric W. Ottesen, and Ravindra N. Singh

Abstract

The dynamic process of pre-mRNA splicing is regulated by combinatorial control exerted by overlapping *cis*-elements that are unique to every exon and its flanking intronic sequences. Splicing *cis*-elements are usually 4–8-nucleotide-long linear motifs that furnish interaction sites for specific proteins. Secondary and higher-order RNA structures exert an additional layer of control by providing accessibility to *cis*-elements. Antisense oligonucleotides (ASOs) that block splicing *cis*-elements and/or affect RNA structure have been shown to modulate alternative splicing in vivo. Consistently, ASO-based strategies have emerged as a powerful tool for therapeutic manipulation of aberrant splicing in pathological conditions. Here we describe the application of an ASO-based approach for the enhanced production of the full-length mRNA of *SMN2* in spinal muscular atrophy patient cells.

Key words Antisense oligonucleotide (ASO), Survival motor neuron (SMN), Pre-mRNA splicing, Multi-exon-skipping detection assay (MESDA), Intronic splicing silencer N1 (ISS-N1), GC-rich sequence, GM03813, Spinal muscular atrophy (SMA), Phosphorothioate, 2′-*O*-methyl modification, Transfection

1 Introduction

Pre-mRNA splicing is modulated by a strict code of conduct in which overlapping regulatory sequences that are known as exonic or intronic splicing enhancers (ESEs or ISEs) and silencers (ESSs or ISSs) play an important role [1–3]. While enhancer and silencer motifs promote or suppress splice-site (ss) selection, respectively, they do so under the influence of unique contexts furnished by varying sizes of introns and exons. RNA structure provides an additional layer of control by positioning and/or sequestering splicing *cis*-elements [4–6]. Current methods do not reliably predict the functional significance of every splicing *cis*-element in the context of endogenous gene. This is in part due to coupling of pre-mRNA splicing with transcription, which is in turn controlled by DNA modifying and chromatin remodeling factors [7]. The antisense oligonucleotide (ASO)-based approach is one of a few strategies that could be applied to assess the impact of a splicing

Klemens J. Hertel (ed.), *Spliceosomal Pre-mRNA Splicing: Methods and Protocols*, Methods in Molecular Biology, vol. 1126, DOI 10.1007/978-1-62703-980-2_20, © Springer Science+Business Media, LLC 2014

cis-element in the context of the endogenous gene [8–11]. The ability of an ASO-based approach to selectively remove or incorporate a particular exon provides a promising means to enrich a splice variant from a specific gene. Therefore, manipulation of protein levels through ASO-based splicing modulation has been considered as potential therapy for several human diseases [12]. The ASO-based strategy is also useful for uncovering the position-specific role of residues associated with long-distance RNA:RNA interactions [13].

Here we describe an example of ASO-mediated splicing correction in spinal muscular atrophy (SMA), a leading genetic cause of infant mortality [14]. SMA is caused by the loss of *Survival Motor Neuron 1* (*SMN1*) gene. *SMN2*, a nearly identical copy of *SMN1*, fails to compensate for the loss of *SMN1* due to predominant skipping of *SMN2* exon 7. Our earlier finding of intronic splicing silencer N1 (ISS-N1) has recently emerged as the leading therapeutic target for an ASO-mediated restoration of *SMN2* exon 7 inclusion in SMA [8, 15–17]. Interestingly, an 8-nucleotide (8-nt) long GC-rich sequence that partially overlaps with ISS-N1 has turned out to be the shortest ASO target for the restoration of *SMN2* exon 7 inclusion (Fig 1; 9). Here we focus exclusively on the 8-nt GC-rich target for an ASO-mediated splicing correction in SMA patient cells (Fig 1; 9). We also describe how this method could be adapted for an ASO-based splicing modulation in HeLa cells as well as in neuronal SH-SY5Y cells. Complementing this method, several recent reports describe ASO-based methods of splicing correction in mouse models of SMA [15–17]. To maintain in vivo stability, we have used RNA ASOs with phosphorothioate backbone and 2′-*O*-methyl modifications [13]. These propriety-free modifications remain one of the most frequently used oligonucleotide chemistries for in vivo applications. Of note, the methods described here can be applied to any gene, given that proper *cis*-element targets can be identified.

2 Materials

2.1 Cell Culture Components

1. GM03813 primary fibroblasts (Coriell Cell Repositories).
2. SH-SY5Y neuroblastoma cells (ATCC).
3. HeLa cells (ATCC).
4. Minimum Essential Medium (MEM) (with Non-essential Amino Acids, without Glutamine).
5. Dulbecco's Modified Eagle's Medium (DMEM), High Glucose.
6. Minimum Essential Medium (MEM) (with L-Glutamine).
7. Ham's F12 Nutrient Mixture.
8. GlutaMAX-I (Invitrogen) or equivalent.

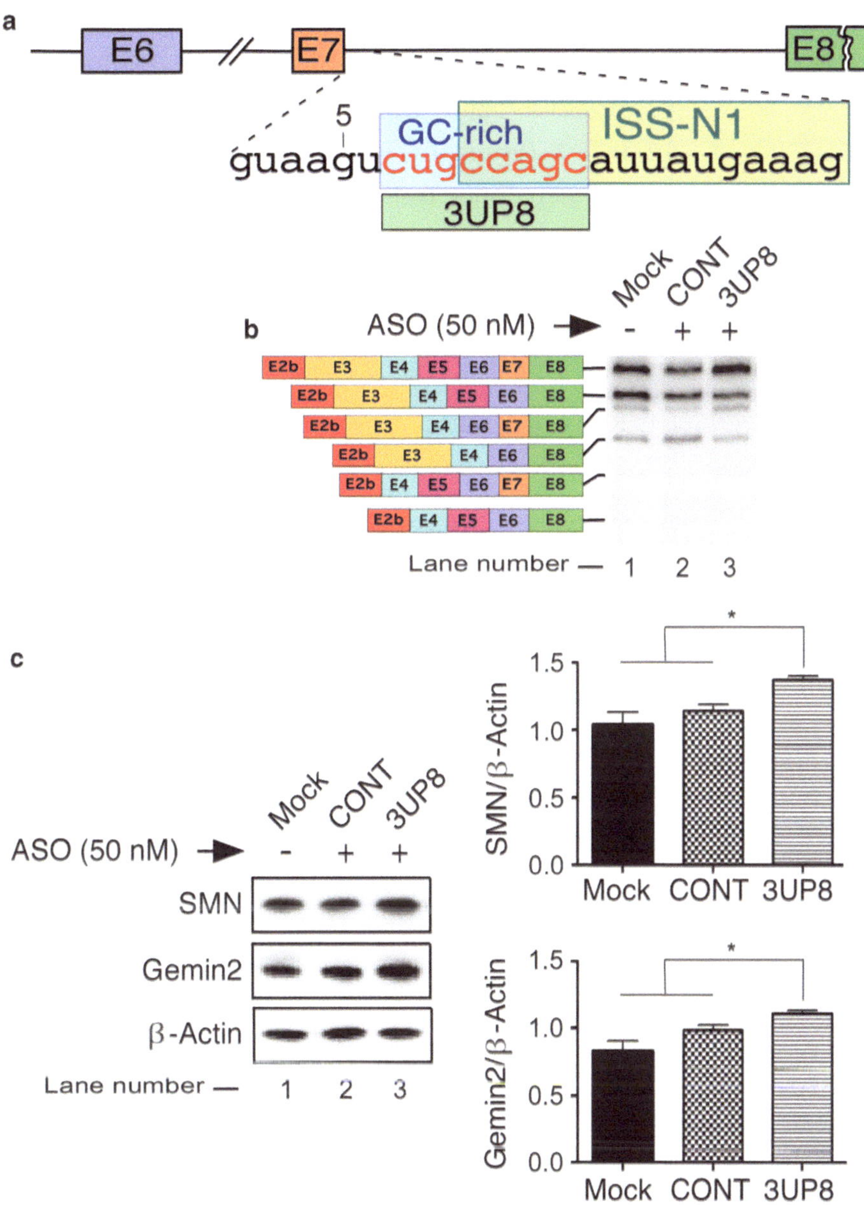

Fig. 1 ASO-based modulation of pre-mRNA splicing of *SMN2* exon 7. (**a**) Diagrammatic representation of ASO target within intron 7 of *SMN2*. Part of the *SMN2* gene is shown. Numbering starts from the first position of *SMN2* intron 7. ISS-N1 overlaps with the GC-rich sequence. 3UP8 is an ASO that targets the GC-rich sequence. (**b**) Splicing pattern of *SMN2* in GM03813 cells transfected with 3UP8 or a control ASO. We performed MESDA to capture all known splice variants of *SMN2* [11]. Bands lacking various exons are shown. Top band corresponds to the full-length transcript. 3UP8 restores *SMN2* exon 7 inclusion and increases the levels of the full-length transcript. (**c**) Levels of SMN and Gemin2 proteins in GM03813 cells transfected with various ASOs as shown in (**b**). Statistical analysis was performed using the GraphPad Prism 6.0 software. Data are expressed as mean ± standard error of the mean. We used unpaired Student's *t*-test for analysis. *P* values are two-tailed and the level of statistical significance is set at $^{*}P < 0.05$. *E* exon, *ISS-N1* intronic splicing silencer N1, *GC-rich* GC-rich intronic splicing silencer

9. Fetal Bovine Serum (FBS).
10. Opti-MEM I Reduced Serum Medium (Opti-MEM).
11. Dulbecco's phosphate-buffered saline (DPBS).
12. 0.25 % Trypsin–EDTA solution.
13. 100 mm dishes, tissue culture-treated.
14. 6-Well tissue culture-treated plates.
15. Hemocytometer.
16. NAPCO Series 8000 WJ CO_2 Cell Incubator (Thermo Scientific) or equivalent.

2.2 Antisense Transfection Components

1. RNase-free water.
2. 2′-*O*-methyl modified and phosphorothioate backbone containing Antisense Oligonucleotides (ASOs) (Dharmacon or TriLink) (*see* **Note 1**).
3. Lipofectamine 2000 (Invitrogen).
4. Cell Lifter.
5. Vortex Mixer.
6. Microcentrifuge.
7. 1.5 mL Microcentrifuge tubes.
8. 15 mL Centrifuge tubes.

2.3 RNA Isolation Components

1. TRIzol (Invitrogen) (*see* **Note 2**).
2. Chloroform (*see* **Note 2**).
3. Isopropanol.
4. Ethanol.
5. RQ1 RNase-free DNase (Promega).
6. Phenol:chloroform (1:1), Tris–EDTA buffer saturated premixed with Isoamyl Alcohol 25:24:1 (*see* **Note 2**).
7. Mini centrifuge (PHENIX Quick Spin).
8. Dry ice.
9. 15 mg/mL glycogen.
10. BioMate 3 (Thermo Scientific) or similar spectrophotometer for nucleic acid (DNA/RNA) concentration measurement.

2.4 RT-PCR and Imaging Components

1. 0.2 mL thin-walled tubes.
2. SuperScript III Reverse transcriptase (Invitrogen), 5× SuperScript III RT buffer, 0.1 M DTT, Oligo$(dT)_{12-18}$ Primer, dNTP mixture (10 mM each), RNaseOUT (Invitrogen), *Taq* DNA polymerase (5 U/μl), 10× Standard Taq (Mg-free) Reaction buffer, 25 mM $MgCl_2$.
3. Primers 5′hSMN-E2b (forward, 5′-GAATACTGCAGCTTC CTTACAACAG-3′) and P2-2 (reverse, 5′-CTTCCTTTTT CTTTCCCAACAC-3′) (*see* **Note 3**).

4. [γ-^{32}P]-ATP (3,000 Ci/mmol).
5. Micro Bio-Spin Columns with Bio-Gel P-30 in Tris Buffer (Bio-Rad).
6. 10× Tris-borate EDTA (TBE) electrophoresis buffer (*see* **Note 4**). Dilute tenfold to make working solution (1× TBE).
7. 6× DNA loading buffer (*see* **Note 5**).
8. 40 % Acrylamide/bis solution (29:1), ammonium persulfate (APS) and *N*,*N*,*N*′,*N*′-tetramethylethylenediamine (TEMED), Sodium Dodecyl Sulfate (SDS).
9. A 19.7 cm × 19.7 cm × 3 mm gel-casting cassette and an electrophoresis system.
10. Native acrylamide gel with dimensions of 17.3 cm × 16 cm × 1.5 mm (*see* **Note 6**).
11. Chromatography Paper 3MM Chr.
12. Saran wrap.
13. AB15 pH meter (Fisher Scientific) or equivalent.
14. Model 583 Gel Dryer (Bio-Rad) or equivalent.
15. FLA-5000 Image Reader (Phosphorimager) with Multi Gauge Software (Fuji Photo Film Inc) or similar system.
16. Phosphorimager screen (cassette).

2.5 Western Blotting Components

1. RIPA buffer supplemented with Halt Protease Inhibitor Single-Use cocktail (Thermo Scientific) (*see* **Note 7**).
2. BCA protein assay kit (Thermo Scientific).
3. Microplate Reader Spectrophotometer (Spectra Max) or similar instrument for measuring protein concentration.
4. 1.5 M Tris–HCl (pH 8.8), 1 M Tris–HCl (pH 7.6), 1 M Tris–HCl (pH 6.8), 5 M NaCl, Tween 20.
5. 10× Tris–Glycine Sodium Dodecyl Sulfate (SDS) Running Buffer: 250 mM Tris base, 1.92 M glycine, 1 % SDS, pH 8.3. Dilute tenfold to make working solution (1× Tris–Glycine Sodium SDS).
6. 2× Laemmli Sample Buffer (*see* **Note 8**).
7. A 10.1 cm × 8.2 cm × 1.5 mm gel-casting cassette and an electrophoresis system.
8. Minigel casting system.
9. SDS-acrylamide gel with dimensions of 8 cm × 7.3 cm × 1.5 mm (*see* **Note 9**).
10. Plastic trays.
11. 10× Transfer buffer: 0.25 M Tris base, 1.92 M glycine, pH 8.4. Dilute tenfold to make 1× Transfer buffer supplemented with 10 % methanol (working solution).
12. 10× Tris-buffered saline with Tween 20 (TBST): mix 195 mL water, 500 mL 1 M Tris–HCl (pH 7.6), 300 mL 5 M NaCl,

and 5 mL Tween 20. Dilute tenfold to make 1× TBST (working solution).

13. Western blot blocking solution: 5 % (w/v) nonfat milk in 1× TBST (*see* **Note 10**).
14. Polyvinylidene fluoride membrane (PVDF).
15. Extra thick blot paper, mini blot size.
16. Transfer-Blot SD Semi-Dry Transfer Cell (Bio-Rad).
17. Primary antibodies: mouse monoclonal anti-SMN antibody (BD Transduction Laboratories), mouse monoclonal anti-Gemin2 antibody (Sigma-Aldrich), and rabbit polyclonal anti-actin antibody (Sigma-Aldrich).
18. Secondary antibodies: horseradish-peroxidase (HRP)-conjugated anti-mouse goat antibody (Jackson ImmunoResearch) and HRP-conjugated anti-rabbit donkey antibody (GE Healthcare).
19. Restore™ Western Blot Stripping Buffer (Thermo Scientific).
20. SuperSignal West Dura Extended Duration Substrate (Thermo Scientific) or SuperSignal west Femto Maximum Sensitivity Substrate (Thermo Scientific).
21. UVP BioSpectrum AC imaging System (UVP).

3 Method

3.1 Cell Culture and Transfection of GM03813 Cells with ASOs

(Described for a single transfection in a 100-mm size dish)

1. Suspend ~8×10^5 GM03813 cells in 8 mL MEM supplemented with GlutaMax-I and FBS (*see* **Notes 11** and **12**), and then plate in a 100-mm culture dish (*see* **Note 13**). Incubate cells at 37 °C in a CO_2 incubator (set at 5 % CO_2) until they are ~80 % confluent. This takes ~24 h.
2. Transfect cells (from Subheading 3.1, **step 1**) using the following procedure (*see* **Note 14**). Prepare an ASO suspension by mixing 1.1 μL of 0.5 mM ASO with 1.5 mL of Opti-MEM I in a 15 mL tube. In a separate 15 mL tube prepare Lipofectamine suspension by mixing 30 μL Lipofectamine 2000 with 1.5 mL Opti-MEM I. Incubate Lipofectamine suspension at room temperature for 5 min. Make ASO-Lipofectamine complex by adding Lipofectamine suspension to ASO suspension. Mix gently and incubate ASO-Lipofectamine complex for 20 min at room temperature. Add entire ASO-Lipofectamine complex drop wise to a 100-mm dish containing GM03813 cells and 8 mL medium (from Subheading 3.1, **step 1**). The final ASO concentration becomes 50 nM. Mix gently by rocking the plate back and forth. Incubate the cells at 37 °C in a CO_2 incubator for 24 h (similarly as in Subheading 3.1, **step 1**). Perform

parallel transfection experiments with a control ASO as well as with no ASO (mock control, replace ASO with water).

3. Remove the culture media by aspiration at 24 h post transfection and add 10 mL fresh MEM to the culture dish. Continue to incubate the cells at 37 °C in a CO_2 incubator for an additional 24 h (similarly as in Subheading 3.1, **step 1**).
4. At 48 h post transfection remove the culture media by aspiration and wash the cells three times with 10 mL ice-cold DPBS. Following washes, add 2 mL ice-cold DPBS directly to each dish. Collect cells by scraping and make appropriate aliquots for immediate use in Subheadings 3.3, **step 1** and/or 3.4, **step 1** (*see* **Note 15** for later use).

3.2 Cell Culture and Transfection of HeLa and SH-SY5Y Cells with ASOs

(Described for a single transfection in a well of a 6-well plate)

1. Suspend ~4×10^5 HeLa cells in 2 mL DMEM or ~3×10^5 SH-SY5Y cells in 2 mL SH-SY5Y medium (*see* **Notes 11** and **12**) and then plate in a well of a 6-well tissue culture plate (*see* **Note 13**). Incubate cells at 37 °C in a CO_2 incubator (set at 5 % CO_2) until they are 70–80 % confluent (similarly as in Subheading 3.1, **step 1**). This takes ~24 h.
2. Transfect cells using following procedure (*see* **Note 14**). Prepare an ASO suspension by mixing 2.5 μL of 50 μM ASO (diluted 1:10 from 0.5 mM stock solution) with 250 μL of Opti-MEM I in a 1.5 mL tube. In a separate 1.5 mL tube prepare Lipofectamine suspension by mixing 5 μL Lipofectamine 2000 with 250 μL Opti-MEM I. Incubate Lipofectamine suspension at room temperature for 5 min. Make ASO-Lipofectamine complex by combining ASO suspension and Lipofectamine suspension in a 1.5 mL tube. Mix ASO–Lipofectamine complex gently, and incubate at room temperature for 20 min. Add entire ASO–Lipofectamine complex drop wise to the well of 6-well plate containing HeLa or SH-SY5Y cells and 2 mL of medium (from Subheading 3.2, **step 1**). The final ASO concentration becomes 50 nM. Mix gently by rocking the plate back and forth. Incubate the cells at 37 °C in a CO_2 incubator (similarly as in Subheading 3.1, **step 1**). Perform parallel transfection experiments with a control ASO as well as without an ASO.
3. Remove the culture media by aspiration at 6 h post transfection and add 2 mL fresh medium (DMEM for HeLa and SH-SY5Y medium for SH-SY5Y cells) (*see* **Note 11**) to the culture dish. Incubate the cells at 37 °C in a CO_2 incubator.
4. At 48 h post transfection remove the culture media by aspiration and wash the cells with 2 mL of DPBS. Remove DPBS by aspiration and add 0.25 mL of 0.25 % Trypsin–EDTA. Incubate cells at 37 °C for 3 min in a CO_2 incubator. Add

1 mL of cell growth medium and lift the cells by repeated pipetting. Total volume of suspension becomes 1.25 mL. Set aside 400 μL of cell suspension for RNA isolation (*see* Subheading 3.3). Transfer the remainder of the cell suspension (about 800 μL) to a 1.5 mL microcentrifuge tube.

5. Centrifuge cells at 3,500 × *g* for 1 min at 4 °C. Aspirate supernatant without disturbing pellet. Resuspend cell pellet in 1 mL of ice-cold DPBS to wash.
6. Repeat **step 5** for a total of two washes in ice-cold DPBS. Proceed to Subheading 3.4 for protein isolation.

3.3 RNA Isolation, RT-PCR, and Imaging

1. Take 0.5 mL cells from Subheading 3.1, **step 4** or 0.4 mL cells from Subheading 3.2, **step 4** and spin at 3,500 × *g* for 1 min at 4 °C. Aspirate supernatant without disturbing cell pellet. Resuspend pellet in 1 mL TRIzol Reagent and incubate at room temperature for 5 min. Addition of TRIzol Reagent will lyse the cells. Add 200 μL chloroform to the cell lysate and shake vigorously by hand for 15 s. Incubate cell lysate at room temperature for 3 min. Spin at 12,000 × *g* for 15 min at 4 °C. Transfer ~450 μL upper aqueous phase to a fresh 1.5 mL tube and mix with 500 μL isopropanol. Incubate the mixed suspension at room temperature for 10 min. To pellet RNA, spin the suspension at 12,000 × *g* for 10 min at 4 °C. Wash the pellet by adding 1 mL of 75 % (v/v) ethanol and spinning at maximum speed for 10 min at room temperature. Discard the supernatant and briefly spin the sample. Pipette out the residual ethanol and air dry the pellet at room temperature for 2–3 min. Dissolve the pellet in 20 μL of RNase-free water, measure concentration using spectrophotometer, and perform DNase treatment to remove any traces of DNA.
2. Use entire RNA sample for DNase treatment in a 100 μL reaction volume as follows: combine sample with 10 μL of 10× RQ1 DNase buffer, an appropriate amount of RQ1 RNase-free DNAse (*see* **Note 16**), and RNase-free water up to a final volume of 100 μL. Mix and incubate the reaction mixture at 37 °C for 30 min. Add 100 μL of phenol–chloroform solution and vortex to inactivate the enzyme. Spin the mixture at 16,500 × *g* in microcentrifuge for 5 min at room temperature. This will partition RNA into the aqueous phase at the top. Collect 90 μL of aqueous phase in a fresh 1.5 mL tube.
3. Recover RNA from the aqueous phase (obtained from Subheading 3.3, **step 2**) by mixing it with 1/10th volume of 3 M sodium acetate (pH 5.2), 0.25 μL of 15 mg/mL glycogen (GlycoBlue), and 2.5× volume 100 % ethanol. Mixture will turn into a cloudy suspension. Chill the suspension on dry ice for 5 min or at −20 °C overnight. Spin the suspension at 16,500 × *g* for 30 min at 4 °C to pellet the RNA.

Wash the pellet with 1 mL of 70 % (v/v) ethanol as described in Subheading 3.3, **step 1**. Dissolve the pellet in 10 μL RNase-free water. Determine RNA concentration using a spectrophotometer.

4. Assemble a 10-μL reverse-transcription reaction to generate first-strand cDNA. First combine 1.2 μg total RNA and RNase-free water to a total volume of 5.5 μL. Add 0.5 μL of Oligo$(dT)_{12-18}$ and 0.5 μL 10 mM dNTP. Mix and heat at 65 °C for 5 min. Place immediately on ice to quickly chill samples. Add 2 μL of 5× first-strand buffer, 0.5 μL of 0.1 M DTT, 0.5 μL of RNase OUT, and 0.5 μL SuperScript III RT, mix. The final reaction volume is 10 μL. Incubate the reaction at 50 °C for 1 h. Inactivate the reaction by heating at 70 °C for 15 min. Store at −20 °C until ready to use.
5. Radiolabel the 5′ primer for multi-exon-skipping detection assay (MESDA) to capture all known *SMN* transcripts (11, *see* **Note 17**). For radioactive labeling of the 5′ primer, set up a 50 μL T4 Polynucleotide Kinase (PNK) reaction. First combine 25.5 μL water, 5 μL 10× T4 PNK buffer, 3 μL 10 μM 5′ primer (5′hSMN-E2b), and 15 μL [γ-^{32}P]-ATP to a total volume of 48.5 μL. Incubate the reaction mixture at 94 °C for 2 min. Snap cool the reaction and keep it on ice for 5 min. Add 1.5 μL PNK, mix and incubate the reaction at 37 °C for 1 h. Inactivate the enzyme by heating at 65 °C for 20 min.
6. Clean up and purify the end-labeled primer by using a Micro Bio-Spin column. Start with spinning a column at 1,000 ×*g* for 2 min at room temperature to remove the packing buffer. Add 500 μL RNase-free water and spin the column at 1,000 ×*g* for 1 min at room temperature. Repeat the process three times to completely exchange the column packing buffer with water. Add 50 μL water and spin the column for 4 min at 1,000 ×*g*. Repeat this step until the volume of eluate is 50 μL. Place the column in a clean 1.5 mL tube. Carefully apply 50 μL of the end-labeled primer (from Subheading 3.3, **step 5**) directly onto the top center of the gel bed of the column. Spin the column at 1,000 ×*g* for 4 min at room temperature to collect the 5′-labeled primer in the 1.5 mL tube.
7. Assemble a 25-μL PCR as follows. Combine water (to a total volume of 25 μL), 2.5 μL 10× *Taq* reaction buffer, 1.5 μL 25 mM $MgCl_2$, 0.5 μL 10 mM dNTPs, 0.5 μL 10 μM P2-2, 0.16 μL 10 μM 5′hSMN-E2b, 5.6 μL ^{32}P-labeled 5′hSMN-E2b (from Subheading 3.3, **step 6**), 2 μL cDNA (from Subheading 3.3, **step 4**), and 0.125 μL *Taq* DNA polymerase.
8. Perform PCR using following profile: initial denaturation at 95 °C for 5 min, 20-cycle amplification (denaturation at 95 °C for 30 s, annealing at 58 °C for 30 s, and extension at

68 °C for 1 min) and a final extension at 68 °C for 7 min. Following PCR, add 5 μL of 6× DNA loading buffer to the PCR mixture and mix. Spin and analyze sample on a native polyacrylamide gel.

9. Load 3 μL of PCR products (from Subheading 3.3, **step 8**) on a 6 % native polyacrylamide gel. Run gel electrophoresis in 1× TBE running buffer at 200 V for 2.5 h at room temperature. Upon electrophoresis, transfer the gel to chromatography paper and cover the gel with Saran wrap.
10. Dry the gel on chromatography paper using Gel Dryer apparatus at 80 °C. Expose the dried gels to a phosphorimager screen.
11. Analyze and quantify the results using an Image Reader FLA-5000 and Multi Gauge software.

3.4 Western Blotting

1. Spin 1.5 mL cells from Subheading 3.1, **step 4** or 1 mL of cells from Subheading 3.2, **step 5** at 3,500 × *g* for 1 min to collect cells. Add RIPA buffer supplemented with 1× Halt Protease Inhibitor Single-Use cocktail (~2× volumes of cell pellet) and mix to resuspend cells. Incubate on ice for 30 min with occasional mixing. Spin at 15,000 × *g* for 10 min at 4 °C. Collect supernatant. Determine protein concentrations using BCA protein assay kit.
2. Resolve protein samples on a 12 % SDS–polyacrylamide gel and equilibrate gel in 1× transfer buffer for 30 min (*see* **Note 18**).
3. Transfer proteins from the gel to a PVDF membrane using Transfer-Blot SD Semi-Dry for 45 min at 75 mA (*see* **Note 19**).
4. Block the PVDF membrane in 50 mL of blocking solution at room temperature for 1 h or at 4 °C overnight (use a plastic tray).
5. Transfer the membrane into the primary antibody (mouse monoclonal anti-SMN, 1:2,000 dilution). Incubate the membrane with shaking at room temperature for 1 h.
6. Wash the membrane three times with 1× TBST for 15 min each (*see* **Note 20**).
7. Incubate the membrane with secondary antibody (anti-mouse horseradish-peroxidase-conjugated, 1:4,000 dilution) with shaking at room temperature for 1 h.
8. Wash the membrane three times with 1× TBST for 15 min each (*see* **Note 20**).
9. Detect the protein signal by developing the blot with SuperSignal West Dura Extended Duration Substrate or SuperSignal West Femto Maximum Sensitivity Substrate at room temperature for 5 min.
10. Scan the membrane using a UVP BioSpectrum AC Imaging System.

11. For reprobing, strip the membrane at room temperature for 15 min using Western Blot Stripping Buffer.
12. Wash the membrane four times with 1× TBST for 5 min each (*see* **Note 20**).
13. Block and reprobe with mouse monoclonal anti-Gemin2 (1:400 dilution) or rabbit polyclonal anti-actin (1:2,000 dilution). For reprobing, follow **steps 4** through **10**.

4 Notes

1. ASOs are dissolved at 0.5 mM in RNase-free water, aliquotted, and stored at −20 °C. We have observed batch-to-batch variations in efficiency of ASOs. Use HPLC purified ASOs for optimal results. ASO concentration should be confirmed by measuring its absorbance at 260 nm.
2. When working with the hazardous reagents, use fume hood. Use lab coats, gloves, and eye protection. When using radioactive material, use proper shielding recommended by your institution.
3. Primers 5′hSMN-E2b and P2-2 are used for amplifying *SMN2* transcripts; the product size is 774 bp when exon 7 is included, or 720 bp when exon 7 is skipped [11].
4. To make 10× TBE, dissolve 108 g Trizma base, 55 g boric acid, and 9.3 g EDTA in water (total volume 1 L). Confirm that pH is equal to 8.3.
5. To make 6× DNA loading buffer add 0.25 % (w/v) bromophenol blue, 0.25 % (w/v) xylene cyanol FF, and 30 % (v/v) glycerol in RNase-free water.
6. To make a 6 % native gel, prepare a 60 mL solution as follows. Combine 44.34 mL RNase-free water, 9 mL of 40 % (w/v) acrylamide/bis solution (29:1), 6 mL of 10× TBE. Filter the solution to remove any solid particles. Then add 600 μL of 10 % (w/v) APS and 60 μL TEMED to initiate polymerization. Pour the mixture into a gel-casting cassette, and allow the gel to polymerize for at least 30 min.
7. Add 100 μL of 100× Halt Protease Inhibitor Single-Use cocktail to 10 mL RIPA buffer immediately before usage.
8. 2× Laemmli Sample Buffer is supplemented with β-mercaptoethanol (50 μL per 950 μL sample buffer) immediately before usage.
9. To make a SDS-polyacrylamide gel, prepare a 10 mL resolving solution. For 12 % resolving gel, combine 4.3 mL RNase-free water, 3 mL of 40 % (w/v) acrylamide/bis solution (29:1), 2.5 mL of 1.5 M Tris–HCl (pH 8.8), 100 μL of 10 % (w/v)

SDS, 100 μL of 10 % (w/v) APS, and 5 μL TEMED; mix the solution by swirling gently. Pour the mixture into a gel-casting cassette, and allow the gel to polymerize for at least 30 min. For 5 % stacking gels, combine 2.92 mL RNase-free water, 500 μL of 40 % (w/v) acrylamide/bis solution (29:1), 500 μL of 1 M Tris–HCl (pH 6.8), 40 μL of 10 % (w/v) SDS, 40 μL of 10 % (w/v) APS, and 5 μL TEMED; mix the solution by swirling gently. Pour the mixture into the gel-casting cassette, and allow the gel to polymerize for at least 30 min.

10. It is best to prepare this before usage and can be stored at 4 °C for 1 week.
11. To prepare MEM media for GM03813 fibroblasts, mix 500 mL of MEM (with Non-Essential Amino Acids, without Glutamine) with 75 mL of FBS and 5 mL of 100× GlutaMAX-I. To prepare DMEM media for HeLa cells, mix 500 mL of DMEM High Glucose with 50 mL of FBS. To prepare media for SH-SY5Y cells, mix 500 mL of MEM (with L-Glutamine) with 500 mL of Ham's F12 nutrient mixture. Add 100 mL of FBS and mix.
12. The MEM media used for GM03813 and SH-SY5Y culture are different formulations. Refer to the product description to ensure that the correct MEM is used for each cell type.
13. Use a hemocytometer to count number of cells.
14. Monitor transfection efficiency by simultaneously transfecting with a fluorescent-labeled control ASO. The 5′ or 3′ end labeling of ASOs by Cy3 or Cy5 is commercially available. GM03813 cells have large nuclei. Therefore, cells could be readily identified by fluorescing nuclei several hours after transfection.
15. For using cells later, pellet the cell suspension (from Subheading 3.1, **step 4**) by centrifuging at 3,500 × *g* for 1 min at room temperature. Aspirate supernatant and store the pellet at −80 °C. For protein isolation resuspend pellet in 2× volume of RIPA + Halt Protease inhibitor. For RNA isolation resuspend pellet in 1 mL TRIzol reagent.
16. Adjust DNase concentration based on nucleic acid (DNA + RNA) concentration in the sample. We generally use 1 unit of DNase for 1 μg nucleic acid.
17. MESDA is a new PCR-based method to capture all known splice variants of *SMN* [11]. For detection of *SMN2* exon 7 skipped products only, 5′ primer annealing to exon 6 could be used [9].
18. Load ~20 μg protein per well on a 12 % SDS–polyacrylamide gel.

19. Refer to manufacturer's recommendation for set up procedure on Transfer-Blot SD Semi-Dry apparatus.
20. Use 200 mL 1× TBST for washing.

Acknowledgments

This work was supported by grants from United States National Institutes of Health (R01 NS055925, R21 NS072259, and R21 NS080294) and Salsbury Endowment (Iowa State University, Ames, IA, USA) to RNS. Authors acknowledge Dr. Natalia Singh for providing critical comments on the manuscript.

References

1. Wang Z, Burge CB (2008) Splicing regulation: from a parts list of regulatory elements to an integrated splicing code. RNA 14:802–813
2. Ke S, Shang S, Kalachikov SM et al (2011) Quantitative evaluation of all hexamers as exonic splicing elements. Genome Res 21: 1360–1374
3. Busch A, Hertel KJ (2012) HEXEvent: a database of Human EXon splicing Events. Nucleic Acids Res 41:D118–D124
4. Singh NN, Singh RN, Androphy EJ (2007) Modulating role of RNA structure in alternative splicing of a critical exon in the spinal muscular atrophy genes. Nucleic Acids Res 35: 371–389
5. Warf MB, Berglund JA (2010) Role of RNA structure in regulating pre mRNA splicing. Trends Biochem Sci 35:169–178
6. McManus CJ, Graveley BR (2011) RNA structure and the mechanisms of alternative splicing. Curr Opin Genet Dev 21:373–379
7. Irimia M, Blencowe BJ (2012) Alternative splicing: decoding an expansive regulatory layer. Curr Opin Cell Biol 24:323–332
8. Singh NK, Singh NN, Androphy EJ et al (2006) Splicing of a critical exon of human Survival Motor Neuron is regulated by a unique silencer element located in the last intron. Mol Cell Biol 26:1333–1346
9. Singh NN, Shishimorova M, Cao LC et al (2009) A short antisense oligonucleotide masking a unique intronic motif prevents skipping of a critical exon in spinal muscular atrophy. RNA Biol 6:341–350
10. Hammond SM, Wood MJ (2010) PRO-051, an antisense oligonucleotide for the potential treatment of Duchenne muscular dystrophy. Curr Opin Mol Ther 12:478–486
11. Singh NN, Seo J, Rahn S et al (2012) A multi-exon-skipping detection assay reveals surprising diversity of splice isoforms of spinal muscular atrophy genes. PLoS One 7:e49595
12. Kole R, Krainer AR, Altman S (2012) RNA therapeutics: beyond RNA interference and antisense oligonucleotides. Nat Rev Drug Discov 11:125–140
13. Singh NN, Hollinger K, Bhattacharya D et al (2010) An antisense microwalk reveals critical role of an intronic position linked to a unique long-distance interaction in pre-mRNA splicing. RNA 16:1167–1181
14. Markowitz JA, Singh P, Darras BT (2012) Spinal muscular atrophy: a clinical and research update. Pediatr Neurol 46:1–12
15. Hua Y, Sahashi K, Rigo F et al (2011) Peripheral SMN restoration is essential for long-term rescue of a severe spinal muscular atrophy mouse model. Nature 478:123–126
16. Osman EY, Yen PF, Lorson CL (2012) Bifunctional RNAs targeting the intronic splicing silencer N1 increase SMN levels and reduce disease severity in an animal model of spinal muscular atrophy. Mol Ther 20:119–126
17. Porensky PN, Mitrpant C, McGovern VL et al (2012) A single administration of morpholino antisense oligomer rescues spinal muscular atrophy in mouse. Hum Mol Genet 21: 1625–1638

Chapter 21

Using Yeast Genetics to Study Splicing Mechanisms

Munshi Azad Hossain and Tracy L. Johnson

Abstract

Pre-mRNA splicing is a critical step in eukaryotic gene expression, which involves removal of noncoding intron sequences from pre-mRNA and ligation of the remaining exon sequences to make a mature message. Splicing is carried out by a large ribonucleoprotein complex called the spliceosome. Since the first description of the pre-mRNA splicing reaction in the 1970s, elegant genetic and biochemical studies have revealed that the enzyme that catalyzes the reaction, the spliceosome, is an exquisitely dynamic macromolecular machine, and its RNA and protein components undergo highly ordered, tightly coordinated rearrangements in order to carry out intron recognition and splicing catalysis. Studies using the genetically tractable unicellular eukaryote budding yeast (*Saccharomyces cerevisiae*) have played an instrumental role in deciphering splicing mechanisms. In this chapter, we discuss how yeast genetics has been used to deepen our understanding of the mechanism of splicing and explore the potential for future mechanistic insights using *S. cerevisiae* as an experimental tool.

Key words Pre-mRNA Splicing, *Saccharomyces cerevisiae*, Yeast genetics, Synthetic lethality, Temperature-sensitive (ts) screening, Suppressor screening, DExD/H-box protein, SGA analysis, E-MAP

1 Introduction

Eukaryotic genes are often interrupted by noncoding intron sequences. In order to achieve proper gene expression, introns are removed from the pre-mRNA and the remaining exon sequences are ligated to produce a mature messenger RNA. This process, "pre-messenger RNA splicing" is carried out by an evolutionary conserved, ~3 MDa ribonucleoprotein complex called the spliceosome which is composed of 5 snRNAs and over 100 associated proteins [1, 2]. As is suggested by the functional conservation of the spliceosome, splicing is a crucial aspect of gene expression for all eukaryotic cells—from the unicellular eukaryote *S. cerevisiae* to mammalian cells. For example, it is estimated that 90 % of human genes undergo splicing, and although introns are less prevalent in *S. cerevisiae* (found in ~6 % of genes) >30 % of the total mature messages in yeast are derived from intron containing genes [3].

Klemens J. Hertel (ed.), *Spliceosomal Pre-mRNA Splicing: Methods and Protocols*, Methods in Molecular Biology, vol. 1126, DOI 10.1007/978-1-62703-980-2_21, © Springer Science+Business Media, LLC 2014

The spliceosome is a dynamic ribonucleoprotein machine. It assembles in a stepwise manner onto the nascent transcript, recognizes splice site sequences in the RNA via RNA–RNA and RNA–protein interactions, and configures into a catalytically active structure. The dynamic, ATP-driven rearrangements of the spliceosome are intricately coordinated to ensure precise cleavage and ligation of exons. Characterization of the precise nature and timing of these spliceosomal rearrangements and the proteins that direct them have been central challenges for researchers. In light of the strong functional conservation of the spliceosome, classical yeast genetics using the experimentally tractable model eukaryote *S. cerevisiae* has proven to be a powerful tool for identifying the components of the splicing machinery and elucidating their mechanisms of action. The approaches employed include a variety of screens, e.g., temperature-sensitive (ts)/cold-sensitive (cs), enhancer (e.g., synthetic lethality), and suppressor screens, all of which have led the way to identification of genes and characterization of proteins that are involved in splicing.

In this chapter, we discuss how *S. cerevisiae* has been used to study pre-mRNA splicing. We describe how temperature-sensitive mutant screens have revealed components of the splicing machinery. We also describe how suppressor screens have allowed a detailed characterization of RNA and protein interactions that guide intron recognition and catalysis. Finally, we describe low- and high-throughput methods such as Synthetic Genetic Array (SGA) and Epistatic MiniArray Profile (E-MAP) analyses used to identify functional interactions between splicing components and discuss how such data are interpreted.

1.1 Pre-mRNA Splicing and the "Awesome Power of Yeast Genetics"

Genetic manipulation of *S. cerevisiae* has been used with great effect to understand the roles of conserved genomic sequences as well as the functional relationships among genes or sets of genes. There are numerous reasons why yeast has become a favorite model organism for genetic analyses. For one, despite being a eukaryote, yeast share the technical advantages with bacteria of rapid growth, ease of mutagenesis, and ease of long-term archival storage by freezing. Moreover, transformed DNA can be integrated into the genome via homologous recombination, thus allowing efficient gene knockout and mutation. An important feature of *S. cerevisiae* that underlies its genetic tractability is the fact that it exists stably as both haploid and diploid cells, and the haploid product of meiosis can be isolated via microdissection of a tetrad ascus. Finally, yeast serves as an extremely useful model organism for understanding the basic mechanisms of pre-mRNA splicing because much of the molecular machinery involved in gene expression can be generalized to multicellular eukaryotic organisms. Genetic strategies such as mutation, deletion, or genetic depletion of factors associated with splicing have substantially contributed to understanding the mechanism of splicing, and the insights gleaned about pre-mRNA splicing beautifully illustrate the often alluded to "awesome power of yeast genetics [4]."

1.2 Identification of Temperature-Sensitive Mutations in Pre-mRNA Processing Factors

The first temperature-sensitive mutant screening study in *S. cerevisiae* was performed by Leland Hartwell in 1967 [5]. Hartwell took advantage of the small genome of *S. cerevisiae* and its ability to exist as both a haploid and a diploid cell to study the dominance or recessiveness of mutations and their complementation. Cells were exposed to mutagen, evaluated for their ability to grow at 23 °C, but not at 36 °C, and then analyzed by their abilities to produce RNA. A set of ts mutants screened in this study fell into ten complementation groups and were named *RNA2–RNA11*, as these mutants showed inhibited production of ribosomal protein gene mRNA [6] and turned out to be defective in splicing [7]. Almost 20 years later, these mutants were renamed *prp* (pre-RNA processing) mutants, and studies from John Abelson's laboratory showed that many of these *PRP*-encoded (*PRP2–PRP11*) gene products were involved in and essential for pre-mRNA splicing in vitro [8]. Additional ts *prp* mutants were isolated and screened by Northern blot analysis using an *ACT1* intron probe, which allowed analysis of the levels of actin pre-mRNA, the intron lariat intermediate, and the excised lariat product [9]. Subsequent work from Christine Guthrie's lab identified cold-sensitive (*cs*) mutants involved in splicing [10]. All of these early studies laid the groundwork for genetic analysis of yeast splicing [10]. Despite the progress that has been made toward identifying the genes and their products involved in pre-mRNA splicing, there remain many questions about the roles of these splicing factors, which can be addressed genetically. In particular, conditional alleles of essential genes can provide insights into the functions of essential components of the spliceosome, their interactions within the spliceosome, and interactions with other gene expression machineries. Moreover, mutagenesis of cells containing known gene mutations or deletions can be used to screen for functional interactions between components of the splicing machinery or between splicing factors and proteins involved in other gene expression processes, as will be discussed below.

2 Materials

Haploid *S. cerevisiae* strains (e.g., W303, S288C, or one of the BY strains derived from S288C [11]; markers may be any that are suitable for further genetics and biochemical study, such as auxotrophic or drug-resistance markers).

YPD liquid media.

YPD plates.

Sodium phosphate buffer pH 7.0.

Ethyl methanesulfonate (EMS).

5 % sodium thiosulfate (autoclaved).

3 Methods

3.1 EMS Mutagenesis: Generating Temperature-Sensitive Mutants to Study Splicing (See Note 1)

1. Grow the wild-type haploid strains to stationary phase in 50 ml of YPD medium.
2. Centrifuge cells and resuspend the pellet in 0.1 M sodium phosphate buffer pH 7.0 at a density of about 10^8 cells/ml (OD_{600} of 1.0 is ~3×10^7 cells/ml). Transfer cells to a glass culture tube.
3. At this point, save an aliquot of cells without EMS to compare cell survival (*see* **Note 2**). Then treat the cells with 3 % EMS at 30 °C for 60 min with agitation.
4. Pellet cells and remove EMS. Be sure to dispose of this in a designated EMS waste container.
5. Dilute the mutagenized cells 40-fold into 5 % sodium thiosulfate to inactivate the EMS. Spin down cells, remove supernatant, and repeat the inactivation step.
6. Wash the cells twice with sterile water. If the cells are clumpy, a brief vortexing will help facilitate uniform spreading.
7. Spread the cells onto YPD plates and incubate at the permissive temperature (23 °C) for a few days until there are ~200 colonies per plate (*see* **Note 3**).
8. Replica-plate the cells from the petri plates grown under permissive conditions onto fresh YPD plates and incubate at 37 °C. Store the 23 °C control plates.
9. Compare the original 23 °C plate and the 37 °C replicate-plate to identify colonies which show poor or no growth at 37 °C. The colonies that are identified should be restreaked onto fresh YPD plates and grown at 23 °C and 37 °C to retest growth (and decrease the likelihood of false positives).
10. Carry out an initial phenotypic analysis. To study the pre-mRNA splicing defects in these mutants a variety of functional assays have been used including Northern blots, primer extension, or in vitro splicing of a prototypical intron-containing gene, such as *ACT1*.
11. As mutant analysis can be complicated by the presence of multiple mutations, it is important to backcross mutants with an appropriate untreated WT strain of opposite mating type.

12. Check the temperature sensitivity of diploid strains by plating on the YPD plates and incubate at 23 and 37 °C, respectively (*see* **Note 4**).

For a detailed discussion of screen saturation, i.e., knowing when "enough is enough," *see* ref. 12.

3.2 Genetic Screens Identify Key Splice Site Sequences and the Proteins That Recognize Them

The discovery of introns in the late 1970s immediately raised the question of what sequence elements led to their removal. To address this question, a chimeric construct was made in which the *S. cerevisiae ACT1* intron sequence was subjected to random mutagenesis and fused upstream of the *HIS4* gene sequence. As a consequence, the His4 protein product was only generated through precise splicing of the actin intron. This construct was transformed into cells deleted for the endogenous *HIS4* gene so that splicing of this actin-*HIS4* construct was required for the cells to grow on media containing the histidine precursor histidinol [13, 14]. Colonies that showed defective growth in histidinol-containing media were further analyzed to identify the sequences in the intron that were required for different steps of the splicing reaction [13]. For example, a mutant actin-*HIS4* construct in which the branchpoint (BP) was mutated from TACTAAC to TACTACC [15] caused accumulation of the unspliced pre-mRNA in vivo, which was not sufficient for growth in media containing histidinol [13, 14].

The use of the chimeric *ACT1-HIS4* construct proved to be a powerful tool for identifying RNAs and proteins that recognize introns—both directly and indirectly. For example, it was shown that compensatory mutations in U2 snRNA that restored base pairing between the snRNA and the TACTAAC sequence in the *ACT1* intron could also restore growth on histidinol and splicing [16], thus demonstrating that base pairing between U2 snRNA and the branchpoint was critical for splicing. Similar compensatory mutation experiments demonstrated U1 base pairing with the 5′ splice site [17]. Around the same time, a trans-acting factor involved in branchpoint recognition was also identified. Growth of cells harboring the *ACT1-HIS4* construct that was mutated at the branchpoint was assessed to identify spontaneous suppressors that had acquired the ability to grow on media containing histidinol. This led to the identification of an extragenic suppressor that improved the splicing of the mutant actin-*HIS4* construct but decreased the splicing efficiency of the wild-type intron [15]. This suppressor was named *rna16-1*, and was later characterized as *prp16-1*, an allele of the *PRP16* gene which encodes an essential splicing factor Prp16 and contains ATPase activity [18]. Shortly thereafter numerous mutant branch site suppressors were identified that all mapped to the region of Prp16 responsible for its ATPase activity [19].

3.3 Suppressor Screens Reveal the Fundamental Role of DExD/H-Box Proteins in Splicing

A remarkable feature of splicing is that no new phosphodiester bonds are formed in the two catalytic reactions; nonetheless, splicing is an ATP-dependent reaction. In recent years, there has been a growing appreciation for the role of a family of proteins found to be important in numerous gene expression reactions—the DExD/H-box family of proteins, so named for the presence of a conserved motif in the protein (D (Asp)-E (Glu)-A (Ala)-D (Asp)). Eight such DExD/H-box proteins have been shown to play roles throughout the splicing cycle [20]. Whereas these proteins show RNA binding activity, RNA-dependent ATPase activity, and, similar to the DNA helicases, some nucleic acid unwinding activity, the substrates of these proteins and/or their mechanisms of action have remained elusive. Nonetheless, some of the strongest indications of their roles in splicing have come from yeast genetics.

For example, important insights into the activity of one of these DExD/H proteins, Prp16 (introduced above), were gleaned in a screen to identify suppressors of a cold-sensitive allele *prp16-302*. This genetic screen revealed that deletion of the gene encoding a component of the Prp19p-associated complex (NTC), *ISY1*, suppressed the growth defect associated with the ATPase-deficient *prp16-302* mutant [21]. This work also demonstrated that the reduced fidelity of branch site recognition seen in the *prp16-302* mutant could be suppressed by an *ISY1* deletion. These observations are consistent with a growing appreciation for the role of DExD/H-box proteins in maintaining the fidelity of splicing. Specifically, mutations in the DExD/H-box proteins and/or interacting partners are able to affect the use of nonconsensus splice sites, (i.e., splicing fidelity), indicating that the ATP-dependent activities of DExD/H-box proteins are required for proper spliceosome rearrangements that are required to maintain splicing fidelity [22, 23].

Each of the DExD/H-box proteins is encoded by an essential gene. Hence, temperature-sensitive or cold-sensitive alleles have been isolated for each—*PRP16*, *PRP5*, *SUB2*, *PRP28*, *BRR2*, *PRP2*, *PRP22*, and *PRP43*—in order to analyze their protein functions [20]. Moreover, genetic screens have identified suppressors of cs and ts mutants of each of these, which have greatly informed our understanding of the functions of DExD/H-box proteins in splicing, including crucial roles in the fidelity of splicing [20, 22, 23].

3.4 Enhancer and Suppressor Screens Identify Functional Interactions Between Components of the Spliceosome

Genetic screens such as those described above highlight the power of yeast genetics to identify functional interactions between proteins and/or protein and RNA. Applying basic yeast genetics, deletion (or mutation) of one gene can be combined with deletion (or mutation) of another gene such that the double mutant cells lead to a phenotypic modulation (Fig. 1). The degree of phenotypic modulation may allow one to predict or understand the functionality of the gene(s).

To assess genetic interactions, two haploid mutants (such as nonessential deletions) of opposite mating types can be mated to generate a diploid strain, and each mutation can be followed using a selectable marker. For example, through the yeast deletion project, a near complete collection of gene-deletion mutants has been generated (available through Open Biosystems) such that for each strain one gene has been precisely deleted [24]. Each gene deletion is marked by the *kanMX* gene, and the knockout can be followed by growth on media containing the aminoglycoside antibiotic geneticin. Diploid yeast cells undergo meiosis and unlinked genes will assort independently. When the strain is induced to sporulate by growth under nitrogen and carbon deprivation conditions, a 4-spore tetrad ascus is formed that can be dissected (Fig. 1a, b) using a micromanipulator. In this way, individual spores, representing each of the four products of meiosis can form colonies. Then the double mutants can be identified and then compared to each parent and a wild-type strain. This approach has been effectively used to directly query the relationship between two specific genes in "directed" genetic interactions studies.

When the spores containing double mutants show an enhanced negative phenotype, such as becoming more sick than the parents or inviable, this phenotype is referred to as a negative genetic interaction: a synthetic growth defect or synthetic lethality, respectively (Fig. 1b, c). In general, synthetic lethality reflects an interaction that is essential for viability, and synthetic sickness represents an interaction that is *important* for viability. Since a remarkable 80 % of yeast genes can be individually deleted in haploid cells and the cells remain viable, it is possible to analyze synthetic interactions in null alleles. Furthermore, conditional and hypomorphic alleles can be used to assess genetic interactions involving essential genes. The observation of synthetic lethality, particularly between null alleles, is generally interpreted to mean that the products of the two genes may be involved in synergistic functional pathways, may contribute to a complex, or may have activities that, in WT cells, buffer one another [25–27]. In other words, the presence of one gene allows the cells to tolerate loss of function of another gene that is essential or important for viability. The use of conditional and hypomorphic alleles has provided a powerful tool for identifying genes encoding products involved in the same pathway. An example of this is when combined mutations in two components of a complex weaken interactions within the complex sufficiently to diminish its function below the threshold for viability. Analysis of synthetic lethality has revealed important interactions between proteins involved in the same pathway such as translocation to the golgi, as well as those involved in functionally redundant or overlapping pathways such as DNA replication and DNA repair [25].

Alternatively, the double mutant products of a genetic cross could have less severe effect on growth than the single mutants

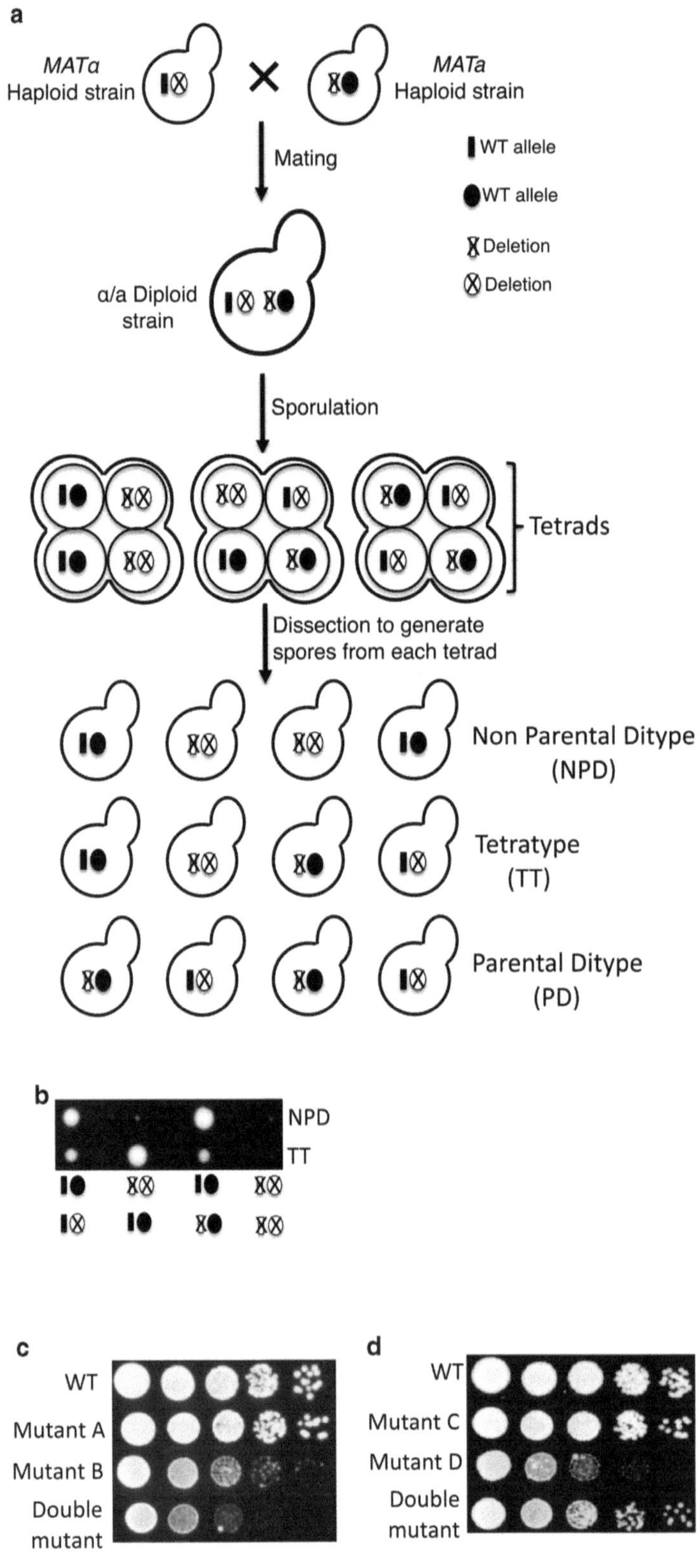

a

MATα Haploid strain × *MATa* Haploid strain

Mating

WT allele
WT allele
Deletion
Deletion

α/a Diploid strain

Sporulation

Tetrads

Dissection to generate spores from each tetrad

Non Parental Ditype (NPD)

Tetratype (TT)

Parental Ditype (PD)

b

NPD
TT

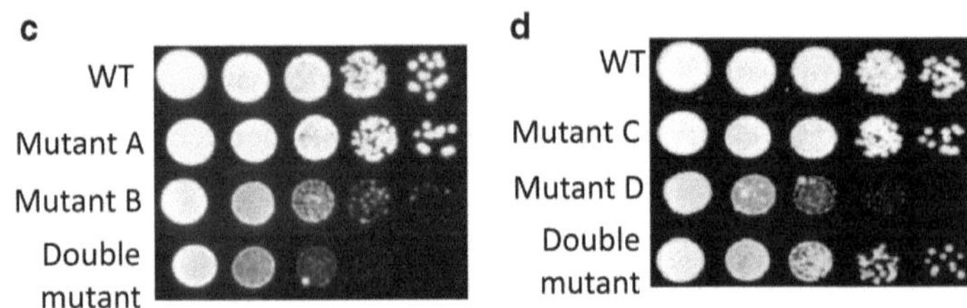

alone, and this phenotype is referred to as a positive interaction or genetic suppression (Fig. 1d). Such a phenotype in the double mutant suggests several interesting possibilities that can be further assessed experimentally. A positive genetic interaction often identifies genes acting antagonistically in the same pathway [26, 27]. For example, deletion of the U2 snRNP protein Cus2 suppresses the lethality associated with mutation of the ATP binding domain of the DExD/H protein Prp5. Cus2 is thought to negatively regulate the formation of Stem IIa of the snRNA to ensure proper timing of this spliceosomal rearrangement and U2 snRNP interaction with the branchpoint. The data suggest that one of the functions of the Prp5 ATPase activity is to displace Cus2, thus allowing the U2 snRNA to adopt the IIa conformation [28, 29]. Positive genetic interactions may also be observed when two gene products physically interact. In this case, a mutant form of the gene may generate a product with decreased functionality, but the protein produced by a suppressor mutation in another gene can associate with and correct the function. Finally, the phenotype caused by a mutation could allow a cell to bypass some defect caused by the first mutation. In such a case the genes are likely to be in separate pathways.

Although the classical, directed genetic interaction studies can uncover important functional relationships that can lead to testable hypotheses, it is important to note that further molecular and biochemical experiments are usually necessary to decipher the molecular functions of the genes. An instructive example of this is provided by the analysis of positive and negative genetic interactions between the yeast cap binding complex and the histone H2B deubiquitylation machinery [30].

Fig. 1 Genetic interaction analysis in the yeast *Saccharomyces cerevisiae*. (**a**) Schematic diagram shows how classical genetic interaction studies are performed between yeast mutant strains. The first is the mating of two haploid strains in which the gene of interest is deleted and replaced by a selectable marker. The mating will generate a heterozygous diploid strain, and sporulation of the diploid strain will generate haploid spores clustered in a 4-spore ascus. Microdissection of each tetrad generates four haploid cells that can grow to form colonies and can be genotyped to identify the combination of WT and mutant alleles. Segregation of mutant alleles leads to three classes of tetrads: nonparental ditype (NPD), tetratype (TT), and parental ditype (PD). (**b**) Representative picture of a tetrad dissection plate. Each *row* represents a tetrad, which shows growth from each spore. The genotype of each spore is depicted *below* the panel. The genotype of the spores can be determined by growing the cells under selective conditions to identify the presence of selectable markers associated with the mutations (e.g., geneticin-resistance). Alternatively, genomic DNA can be isolated from the cells, and PCR can be performed using primers that specifically identify the mutations of interest. (**c** and **d**) Show representative pictures of a dilution growth assay comparing WT, single mutants, and double mutants. Cells were grown at 30 °C to the same OD_{600} ~0.5 and then tenfold serial dilutions were spotted on YPD plates. Panel (**c**) shows the synthetic growth defect phenotype of double mutant cells (*bottom row*), which grow more slowly than either of the single mutants (*middle two rows*). The dilution assay in panel (**d**) shows the suppressor phenotype of the double mutant cells (*bottom row*), which grow better than single mutants (*middle two rows*)

3.5 High-Throughput Methods to Identify Genetic Interactions

Classical genetic screens and "low-throughput" genetic interaction studies have proven to be extremely useful for understanding the organization of molecular pathways in yeast. Furthermore, the availability of yeast deletion and mutant collections, in which each nonessential gene in the genome is deleted (with molecular barcodes at either end of the deletion cassette to allow its identification) or in which essential genes are modified to alter expression, e.g., Decreased Abundance by mRNA Perturbation or DAmP [31], has made it possible to understand, at a global level, the cellular functions of gene(s) in the context of a biological network. Large-scale genetic interaction studies have been particularly powerful for identifying and analyzing genes that encode multifunctional proteins and act in multiple cellular pathways [27]. Moreover, these methods utilizing yeast strain collections allow a systematic, quantitative assessment of interactions within and between networks. Here we describe two such tools for global analysis of genetic interactions—SGA analysis [32, 33] and quantitative interactions mapping via E-MAP [27].

3.6 SGA and E-MAP Analyses in Yeast (Saccharomyces cerevisiae)

SGA and E-MAP approaches allow systematic, unbiased, quantitative, and comprehensive methodologies for constructing a predictive network of genes, which are functionally related or distinct [27, 33, 34]. Focusing first on SGA analysis, this approach enables the systematic generation of double mutants in order to reveal genome-wide synthetic genetic interactions by using a combination of genetic methodologies and robotic devices. A detailed description of the tools and reagents used in such analysis are nicely described elsewhere [32]. Briefly a "query" strain can be crossed to an ordered array of ~5,000 viable gene-deletion mutants and ~1,000 essential genes with conditional mutations of the opposite mating type [32, 33]. After selection for diploid cells, sporulation, and selection of meiotic progeny, the double mutant cells in the arrays can be transferred to selective plates and photographed using high-resolution digital imaging. The yeast colony sizes of the two individual mutants and the double mutants are compared in order to obtain measures of fitness and genetic interactions. Using this approach, Charles Boone's lab performed SGA analysis in *S. cerevisiae* on a genome-wide scale which involved ~1,700 "query" mutants and generated ~170,000 genetic interactions [35, 36].

SGA analysis produces a large "genetic landscape" of the cell, and data from this network can be used to predict the function of the particular gene depending on the genetic interactions it shows. The interactions can then be clustered to reveal functionally related genes that exist in the same protein complex or pathway, much like the clustering of genes in a microarray. If a gene of unknown function shows similar genetic interactions with genes that make up a functional "cluster" this can provide an indication of the cellular activity of the gene's product. Moreover, genetic interactions between two clusters can reveal how they are functionally related [35, 36]. It is important to note that one limitation to interpreting the published genetic interactions data sets is the lack of validation for most of the reported interactions.

Moreover, while extremely powerful, global analysis of genetic interactions among randomly chosen gene pairs, as is the case with SGA analysis, yields mostly neutral interactions. In fact, only 0.5 % queried interactions show negative or positive interactions between the genes [35, 36]. With this in mind, E-MAP analysis was designed to explore genetic networks among genes that are likely to be involved in the same or similar functions [27]. Like SGA, E-MAP explores large-scale genetic interactions and involves genetic methodologies, robotic tools, and quantitative analysis of double mutants, and similar to SGA, E-MAP also generates a large amount of genetic interaction network data. Initially, E-MAPs were designed to measure pairwise interactions between rationally selected sets of genes increased the frequency of detecting genetic interactions, thus providing a deeper data set for analysis of specific pathways. Importantly, the computational analyses employed in E-MAP studies allowed the identification of positive interactions not strong enough to be observed in the early SGA studies (although with time enhanced computational tools have increased the sensitivity of SGA analyses as well). E-MAP analyses have been performed to study genetic interaction between genes involved in the same pathway such as chromatin assembly pathways [27], between kinases and their substrates [37], and components of the secretory pathway [38]. Applying high-density, targeted, pathway analysis via quantitative E-MAP can uncover the function of an unknown gene or known genes with unknown function. Nonetheless, it is important that the genetic interactions identified through these approaches be validated.

Interestingly, E-MAP analysis of genes acting in RNA processing pathways have revealed genetic interactions between the components of the complexes involved in RNA processing as well as significant genetic "crosstalk" between complexes. For example, positive interactions have been shown between cytoplasmic RNA biogenesis and mitochondrial RNA biogenesis, whereas negative interactions have been shown between genes involved in mRNA splicing, mRNA export, and the nuclear export. The E-MAP approach has suggested a new role for 19S proteasome subunit, Sem1/Dss1 in mRNA splicing and mRNA export [39]. E-MAP analysis also suggested that the SR-like protein Npl3 interacts with both the splicing machinery and the histone H2B deubiquitylation machinery [40, 41]. Subsequent experiments have supported this dual role for Npl3 in pre-mRNA splicing and the coupling of RNA splicing with histone H2B ubiquitylation [41].

3.7 Using Yeast Genetics to Study Complex Networks: Insights into Coordination of Gene Expression Processes

There is growing evidence that spliceosome assembly occurs co-transcriptionally (Merkhofer and Johnson, Chapter 6). As both the processes of transcription and splicing are multicomponent assembly processes, genetic interaction studies play an extremely useful role in uncovering the coordination between these two reactions. High-throughput and directed genetic studies have been performed between transcription factors and splicing factors to understand the crosstalk between transcription and splicing, for example [30, 39, 41–44]. One of the first such examples of how genetic analyses can inform understanding of mechanism comes from directed genetic studies showing synthetic interactions between the chromatin modifying enzyme, Gcn5, and components of the U2 snRNP [42]. Subsequent studies revealed that the dynamics of histone acetylation affect the recruitment of the U2 snRNP to pre-mRNA [43]. Large-scale genetic analysis using SGA analysis and E-MAP will almost certainly continue to provide important insights into the interconnections between RNA processing, transcription, and chromatin modification.

4 Notes

1. Caution: EMS is a strong mutagen and must be used in a fume hood. All the glassware must be rinsed with 5 % sodium thiosulfate to inactivate EMS.
2. Calibrate the survival efficiency by treating cells with EMS for varying amounts of time, keeping all other parameters same, to achieve approximately 10–30 % survival.
3. Adding fiduciary marks to this plate and the empty plates onto which the cells will be replica-plated will allow easier alignment of the colonies following replica-plating.
4. Diploid strains that do not show temperature sensitivity indicate recessive mutations, while diploid strains that show temperature sensitivity indicate dominant mutations.

Acknowledgments

We would like to thank Drs. Kristin Patrick (UCSF) and Lorraine Pillus (UCSD) for critical reading of the manuscript. Funding was provided by the National Institutes of General Medical Sciences (GM085764) and the National Science Foundation (MCB-1051921).

References

1. Jurica MS, Moore MJ (2003) Pre-mRNA splicing: awash in a sea of proteins. Mol Cell 12(1):5–14
2. Wahl MC, Will CL, Luhrmann R (2009) The spliceosome: design principles of a dynamic RNP machine. Cell 136(4):701–718
3. Ares M Jr, Grate L, Pauling MH (1999) A handful of intron-containing genes produces the lion's share of yeast mRNA. RNA 5(9): 1138–1139
4. Sherman F (1997) An introduction to the genetics and molecular biology of the yeast Saccharomyces cerevisiae. In: Meyers RA (ed) The encyclopedia of molecular biology and molecular medicine. VCH Publisher, Weinheim, Germany, pp 302–325
5. Hartwell LH (1967) Macromolecule synthesis in temperature-sensitive mutants of yeast. J Bacteriol 93(5):1662–1670
6. Hartwell LH, McLaughlin CS, Warner JR (1970) Identification of ten genes that control ribosome formation in yeast. Mol Gen Genet 109(1):42–56
7. Rosbash M, Harris PK, Woolford JL Jr et al (1981) The effect of temperature-sensitive RNA mutants on the transcription products from cloned ribosomal protein genes of yeast. Cell 24(3):679–686
8. Lustig AJ, Lin RJ, Abelson J (1986) The yeast RNA gene products are essential for mRNA splicing in vitro. Cell 47(6):953–963
9. Vijayraghavan U, Company M, Abelson J (1989) Isolation and characterization of pre-mRNA splicing mutants of Saccharomyces cerevisiae. Genes Dev 3(8):1206–1216
10. Noble SM, Guthrie C (1996) Identification of novel genes required for yeast pre mRNA splicing by means of cold-sensitive mutations. Genetics 143(1):67–80
11. Brachmann CB, Davies A, Cost GJ et al (1998) Designer deletion strains derived from Saccharomyces cerevisiae S288C: a useful set of strains and plasmids for PCR-mediated gene disruption and other applications. Yeast 14(2): 115–132
12. Hawley RS, Walker MY (2003) Advanced genetic analysis: finding meaning in a genome. Blackwell Publishing, Malden, MA
13. Parker R, Guthrie C (1985) A point mutation in the conserved hexanucleotide at a yeast 5′ splice junction uncouples recognition, cleavage, and ligation. Cell 41(1):107–118
14. Vijayraghavan U, Parker R, Tamm J et al (1986) Mutations in conserved intron sequences affect multiple steps in the yeast splicing pathway, particularly assembly of the spliceosome. EMBO J 5(7):1683–1695
15. Couto JR, Tamm J, Parker R et al (1987) A trans-acting suppressor restores splicing of a yeast intron with a branch point mutation. Genes Dev 1(5):445–455
16. Parker R, Siliciano PG, Guthrie C (1987) Recognition of the TACTAAC box during mRNA splicing in yeast involves base pairing to the U2-like snRNA. Cell 49(2):229–239
17. Siliciano PG, Guthrie C (1988) 5′ splice site selection in yeast: genetic alterations in base-pairing with U1 reveal additional requirements. Genes Dev 2(10):1258–1267
18. Schwer B, Guthrie C (1991) PRP16 is an RNA-dependent ATPase that interacts transiently with the spliceosome. Nature 349(6309): 494–499
19. Burgess SM, Guthrie C (1993) A mechanism to enhance mRNA splicing fidelity: the RNA-dependent ATPase Prp16 governs usage of a discard pathway for aberrant lariat intermediates. Cell 73(7):1377–1391
20. Chang TH, Tung L, Yeh FL et al (2013) Functions of the DExD/H-box proteins in nuclear pre-mRNA splicing. Biochim Biophys Acta 1829(8):764–774
21. Villa T, Guthrie C (2005) The Isy1p component of the NineTeen complex interacts with the ATPase Prp16p to regulate the fidelity of pre-mRNA splicing. Genes Dev 19(16):1894–1904
22. Semlow DR, Staley JP (2012) Staying on message: ensuring fidelity in pre-mRNA splicing. Trends Biochem Sci 37(7):263–273
23. Koodathingal P, Staley JP (2013) Splicing fidelity: DEAD/H-box ATPases as molecular clocks. RNA Biol 10(7)
24. Winzeler EA, Shoemaker DD, Astromoff A et al (1999) Functional characterization of the S. cerevisiae genome by gene deletion and parallel analysis. Science 285(5429):901–906
25. Hartman JL, Garvik B, Hartwell L (2001) Principles for the buffering of genetic variation. Science 291(5506):1001–1004
26. Boone C, Bussey H, Andrews BJ (2007) Exploring genetic interactions and networks with yeast. Nat Rev Genet 8(6):437–449
27. Collins SR, Roguev A, Krogan NJ (2010) Quantitative genetic interaction mapping using

the E-MAP approach. Methods Enzymol 470:205–231

28. Perriman R, Ares M Jr (2000) ATP can be dispensable for prespliceosome formation in yeast. Genes Dev 14(1):97–107
29. Perriman R, Barta I, Voeltz GK et al (2003) ATP requirement for Prp5p function is determined by Cus2p and the structure of U2 small nuclear RNA. Proc Natl Acad Sci USA 100(24):13857–13862
30. Hossain MA, Claggett JM, Nguyen T et al (2009) The cap binding complex influences H2B ubiquitination by facilitating splicing of the SUS1 pre-mRNA. RNA 15(8):1515–1527
31. Breslow DK, Cameron DM, Collins SR et al (2008) A comprehensive strategy enabling high-resolution functional analysis of the yeast genome. Nat Methods 5(8):711–718
32. Tong AH, Boone C (2006) Synthetic genetic array analysis in Saccharomyces cerevisiae. Methods Mol Biol 313:171–192
33. Baryshnikova A, Costanzo M, Dixon S et al (2010) Synthetic genetic array (SGA) analysis in Saccharomyces cerevisiae and Schizosaccharomyces pombe. Methods Enzymol 470:145–179
34. Tong AH, Evangelista M, Parsons AB et al (2001) Systematic genetic analysis with ordered arrays of yeast deletion mutants. Science 294(5550):2364–2368
35. Tong AH, Lesage G, Bader GD et al (2004) Global mapping of the yeast genetic interaction network. Science 303(5659):808–813
36. Baryshnikova A, Costanzo M, Kim Y et al (2010) Quantitative analysis of fitness and genetic interactions in yeast on a genome scale. Nat Methods 7(12):1017–1024
37. Fiedler D, Braberg H, Mehta M et al (2009) Functional organization of the S. cerevisiae phosphorylation network. Cell 136(5):952–963
38. Schuldiner M, Collins SR, Thompson NJ et al (2005) Exploration of the function and organization of the yeast early secretory pathway through an epistatic miniarray profile. Cell 123(3):507–519
39. Wilmes GM, Bergkessel M, Bandyopadhyay S et al (2008) A genetic interaction map of RNA-processing factors reveals links between Sem1/Dss1-containing complexes and mRNA export and splicing. Mol Cell 32(5):735–746
40. Kress TL, Krogan NJ, Guthrie C (2008) A single SR-like protein, Npl3, promotes pre-mRNA splicing in budding yeast. Mol Cell 32(5):727–734
41. Moehle EA, Ryan CJ, Krogan NJ et al (2012) The yeast SR-like protein Npl3 links chromatin modification to mRNA processing. PLoS Genet 8(11):e1003101
42. Gunderson FQ, Johnson TL (2009) Acetylation by the transcriptional coactivator Gcn5 plays a novel role in co-transcriptional spliceosome assembly. PLoS Genet 5(10): e1000682
43. Gunderson FQ, Merkhofer EC, Johnson TL (2011) Dynamic histone acetylation is critical for cotranscriptional spliceosome assembly and spliceosomal rearrangements. Proc Natl Acad Sci USA 108(5):2004–2009
44. Hossain MA, Chung C, Pradhan SK et al (2013) The yeast cap binding complex modulates transcription factor recruitment and establishes proper histone H3K36 trimethylation during active transcription. Mol Cell Biol 33(4):785–799

Chapter 22

Medium Throughput Analysis of Alternative Splicing by Fluorescently Labeled RT-PCR

Ryan Percifield, Daniel Murphy, and Peter Stoilov

Abstract

Reverse transcription-PCR (RT-PCR) is a core technique for detecting and quantifying alternative pre-mRNA splicing. RT-PCR is multistep process involving RNA isolation, reverse transcription, and PCR that is often performed using radiolabeled primers. As a result RT-PCR analysis of alternative splicing is a laborious technique that quickly becomes prohibitively expensive when applied to large numbers of samples. Here, we describe an RT-PCR approach for detecting alternative splicing in multi-well plates that can be applied to effortlessly quantify exon inclusion levels in large number of samples. The procedures outlined here can also be automated on standard liquid handling equipment to produce medium throughput assay capable of handling thousands of samples per day.

Key words Alternative splicing, RNA isolation, 96-well plate format, RT-PCR, Fluorescent primers, Capillary electrophoresis

1 Introduction

Pre-mRNA splicing has emerged as major mechanism for regulation of gene expression and protein function [1]. In higher eukaryotes alternative splicing generates astonishing protein diversity from a relatively limited number of gene [2, 3]. Perturbations in constitutive and alternative pre-mRNA splicing are a frequently cause of disease. Estimated 15 % of disease causing mutations disrupting canonical splice sites and another 20–30 % disrupting splicing regulatory sequences located within the exons [4, 5]. As a result there has been a significant interest in developing increased throughput approaches to screen for chemical and genetic modulators of alternative splicing. In the past in vivo luciferase and fluorescent protein reporters have successfully been used in high-throughput screens to identify modulators of alternative splicing [6–9]. However these approaches suffer from significant false discovery rates and require secondary validations assays to reliably identify the positive hits. Reverse transcription-PCR (RT-PCR) is the method of choice for

Klemens J. Hertel (ed.), *Spliceosomal Pre-mRNA Splicing: Methods and Protocols*, Methods in Molecular Biology, vol. 1126, DOI 10.1007/978-1-62703-980-2_22, © Springer Science+Business Media, LLC 2014

such secondary assays. However, the throughput of RT-PCR has been limited due to the relatively high cost of the necessary consumables and reagents. Here we describe a medium throughput procedure for RNA isolation and RT-PCR in multi-well plates that uses low cost consumables. The protocol outlined below can be applied effortlessly in most laboratories to process 192–384 samples per day. This throughput is sufficient to directly screen targeted compound and siRNA libraries such as the InhibitorSelect and ON-TARGETplus collections offered by EMD biosciences and Dharmacon. Furthermore, all steps of the protocol can be fully automated using standard liquid handling equipment to create a medium throughput assay capable of handling thousands of samples per day in 96- and 384-well formats.

2 Materials

Prepare all solutions with ultrapure (18 mΩ) nuclease-free water and store them as indicated in the instructions. As some waste products produced during these protocols can be harmful to the environment, please refer to your local regulations and procedures when disposing of waste.

2.1 RNA Extraction Components

1. A centrifuge with deep swing-out buckets capable of spinning two deep-well plates in each adapter at a speed of 1,500 × *g*.
2. Low volume spectrophotometer: NanoDrop (Thermo Scientific) or equivalent.
3. 96-well tissue culture plates.
4. Nuclease-free liquid troughs.
5. 96-well, 400 μl, 0.45 μm hydrophilic PVDF filter plates (Seahorse Biosciences part # 200943-100, *see* **Note 1**).
6. 96-well 1 ml deep-well plates.
7. 96-well nuclease-free PCR plates.
8. Aluminum plate sealing film.
9. DNase I, RNase free.
10. RNA Lysis Buffer: 6 M Lithium chloride, 5 % Triton X-100, 5 % DGME (Di-ethylene glycol mono-ethyl ether), 10 mM EDTA pH 8.0, 100 mM Tris–HCl pH 8.8. Filter through 0.45 μM filter and store at room temperature. Just before use add 2 % β-mercaptoethanol.
11. RNA Wash Solution I: 5 M Lithium chloride, 55 % Ethanol. Filter and store at room temperature (*see* **Note 2**).
12. RNA Wash Solution II: 30 mM Tris–HCl pH 7.6, 70 % Ethanol. Filter and store at room temperature.
13. Phosphate-Buffered Saline (PBS).

2.2 cDNA Synthesis Components

14. 96-well nuclease-free PCR plate.
15. Plate sealing film.
16. Ultrapure water.
17. dNTP mix, 10 mM each.
18. Primer mix: 10 μM Anchored oligo dT ($dT_{24}VN$) and 50 μM random hexamers.
19. 10× Reverse transcriptase buffer: 500 mM Tris–HCl (pH 8.3), 750 mM KCl, 30 mM $MgCl_2$.
20. RNase H(−) reverse transcriptase (*see* **Note 3**).

2.3 PCR Components

1. 96-well nuclease-free PCR plate.
2. Plate sealing film.
3. Ultrapure water.
4. dNTP mix, 10 mM each.
5. Forward and reverse PCR primer mix, 10 μM each. One of the primers needs to be fluorescently labeled.
6. 10× Taq buffer 500 mM KCl, 100 mM Tris–HCl (pH 9.0 at 25 °C), 15 mM $MgCl_2$, and 1 % Triton X-100.
7. Taq polymerase at 15 U/μl (*see* **Note 3**).

2.4 Acrylamide Gel Electrophoresis Components

1. Vertical gel electrophoresis apparatus (Labrepco model V16 or equivalent).
2. High voltage power supply.
3. PCR tube strips and caps (8- or 12-tube).
4. Sigmacote (Sigma Aldrich) or equivalent siliconizing reagent.
5. 10 % weight/volume Ammonium Persulfate (APS) solution in water.
6. Tetramethylenediamine (TEMED).
7. 1× Tris–Borate EDTA Buffer (TBE): 89 mM Tris, 89 mM Boric acid, 2 mM EDTA. The buffer can be made as a 5× stock solution and diluted before use.
8. Acrylamide gel solution: 4 % Acrylamide/Bis-acrylamide (19:1 crosslink ratio), 1× TBE, 7.5 M Urea. Filter solution through 0.45 μM filter and store in a dark bottle at 4 °C.
9. Clear formamide loading buffer: Deionized formamide, 2 mM EDTA.
10. Formamide loading buffer with tracking dyes: Deionized formamide, 2 mM EDTA 0.25 % (w/v) bromophenol blue, and 0.25 % (w/v) xylene cyanol FF.
11. Fluorescently labeled size standards: Life technologies/ABI GeneScan 1000 Rox or GeneScan 1200 LIZ. Alternatively custom size standards can be prepared by a simple PCR amplification with an ROX-labeled primer [10].

2.5 Capillary Electrophoresis Components

1. ABI capillary sequencer. Access to this equipment is typically available as part of sequencing core facility or commercial service.
2. 96-well half skirt PCR plate compatible with ABI sequencers (*see* **Note 4**).
3. Clear formamide loading buffer: Deionized formamide, 2 mM EDTA 0.25 % (w/v).
4. Fluorescently labeled size standard.

3 Methods

This protocol involves procedures for RNA isolation, cDNA synthesis, PCR and capillary electrophoresis that are carried out in 96-well plates. The RNA isolation procedure is adapted from Bair et al. and uses high concentrations of LiCl, which has long been known to efficiently strip the proteins from the RNA [11, 12]. The RNA is then bound to solid support, and after washing away the contaminants, eluted in water. In our hands PVDF membranes proved superior to silica or glass-fiber support that is typically used in nucleic acid purification procedures. In particular the PVDF membranes unlike glass-fiber filters did not bind the detergents used to lyse the cells and produced RNA free of contaminants (Fig. 1).

cDNA synthesis and PCR amplification procedures follow standard protocols. A key feature of the approach described here is the use of fluorescently labeled primer in the PCR amplification. The fluorescent label allows the amplification products to be subsequently quantified by capillary electrophoresis. Substituting standard gel electrophoresis procedures for automated capillary electrophoresis significantly decreases the labor involved and increases the throughput of the assay. Although we also describe

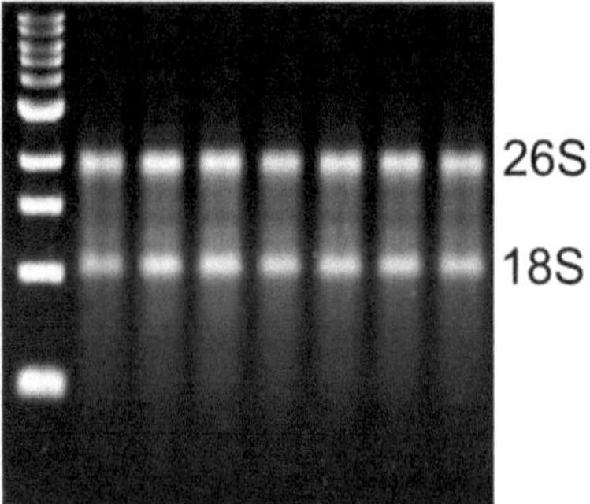

Fig. 1 Agarose gel electrophoresis of RNA extracted from cells grown in 96-well plates. The first lane contains the size standard. The positions of the 18S and 26S ribosomal RNAs are indicated on the side

the gel electrophoresis procedures, we recommend using it only on a limited subset of samples as a quality control and troubleshooting tool.

3.1 RNA Isolation from Adherent Cells in 96-Well Plates

Just prior to beginning the extraction procedure add 20 μl/ml β-mercaptoethanol to the RNA Lysis Buffer (*see* **Note 5**).

1. Grow the cells in 96-well tissue culture plates.
2. Before lysing the cells stack a 96-well PVDF filter plate on top of a deep-well plate.

Steps 3 *through* 6 *describe the lysis procedure for adherent cells. If working with suspension cultures or very loosely adherent cells skip to* step 7.

3. If working with adherent cells invert the tissue culture plate and shake off the media. Tap the plate on a stack of paper towels to remove the excess liquid.
4. Wash cells once with 200 μl of PBS. If the cells are adhering loosely to the plate, skip this step as it can result in washing the cells away.
5. Add 200 μl RNA Lysis Buffer to each well (*see* **Note 6**).
6. Using a multichannel pipette transfer the lysates into the filter plate. Proceed to **step 11** of the protocol.
7. If working with suspension cells resuspend the cells in the culture media by pipetting up and down and transfer the suspension to the filter plate.
8. Spin the plate at 1,500 × *g* for 3 min to remove the media.
9. Discard the liquid accumulated in the deep-well plate, tap the plate face down on stack of paper towels to remove excess liquid, and reuse the plate.
10. Add 200 μl of lysis buffer to each well and incubate for 3 min at room temperature.
11. Spin the plate at 1,500 × *g* for 3 min.
12. Add 200 μl RNA Wash Solution I and spin at 1,500 × *g* for 3 min.
13. Discard the liquid accumulated in the deep-well plate as in **step 9**.
14. Add 200 μl RNA Wash Solution II and spin at 1,500 × *g* for 3 min.
15. Add 20 μl of DNase solution (0.1 U/μl in 1× DNase buffer) to the membrane, seal the plate, and incubate at 37 °C for 20 min.
16. Add 200 μl RNA Lysis Buffer and incubate at room temperature for 2 min.
17. Spin at 1,500 × *g* for 2 min.

18. Discard the liquid accumulated in the deep-well plate as in **step 9**.
19. Add 200 μl RNA Wash Solution I and spin at 1,500×*g* for 2 min.
20. Add 200 μl RNA Wash Solution II and spin at 1,500×*g* for 2 min.
21. Discard the liquid accumulated in the deep-well plate as in **step 9**.
22. Repeat **step 11**, this time extending the spin to 5 min.
23. Transfer the filter plate to a full-skirt PCR plate.
24. Add 16–25 μl of water to the membrane, incubate for 5 min at room temperature, and spin at 1,500×*g* for 5 min. The filter plate retention volume is typically 1 μl per well resulting in RNA solution volume of 15–24 μl. If all RNA is going to be used for cDNA synthesis, **step 25** can be omitted. In this case the RNA can be eluted using 16 μl of water into a plate containing 5 μl of reverse transcription master mix to perform the first-strand synthesis (see below).
25. Check the RNA quality in a subset of wells by agarose gel electrophoresis of 5 μl of the sample (*see* **Note 7**). If necessary determine the RNA concentration using a NanoDrop spectrophotometer. At this point the plates can be sealed using aluminum sealing film and stored at −80 °C, or used for first-strand DNA synthesis.

3.2 First-Strand cDNA Synthesis

1. For each 96-well plate prepare a reverse transcription master mix containing 220 μl 10× reverse transcriptase buffer, 110 μl 10 mM dNTPs, 110 μl oligo dT/random hexamer mix, 55 μl of reverse transcriptase, and 55 μl of water.
2. Dispense 5 μl of the master mix in each well of a 96-well PCR plate.
3. Add 15 μl of RNA solution to each well of the plate containing the reverse transcription mix. Alternatively use the plate containing the reverse transcription mix to catch the RNA eluted with 16 μl of water in **step 24** of the RNA isolation procedure above.
4. Spin down the plate briefly to collect any drops and purge air bubbles trapped at the bottom of the wells.
5. Cover the plate with sealing film and run in a thermal cycler under the following conditions: step 1 –25 °C for 5 min, step 2 –43 °C for 40 min, step 3 –75 °C for 15 min, followed by a 10 °C hold until ready to remove the plate.
6. Remove the plate from the thermal cycler.
7. Dilute the reactions with 20 μl of water, reseal the plate, and store at −20 °C until needed.

3.3 PCR Primer Design

PCR primers for measuring exon inclusion levels can be designed using primer3. The primers are placed in the constitutive exons that flank the alternatively spliced region and should have melting temperatures of approximately 60 °C. Typically, the primers will be located in the exons immediately adjacent to the alternatively spliced region (Fig. 2a). In cases where the size or nucleotide composition of the constitutive exons place constrains on the primer design, the primers can be moved further away (Fig. 2b). This placement will produce PCR products of different size corresponding to the exon included and exon skipped mRNA isoforms. Optimally, the primers should be designed so that the shortest (skipped) product should be between 150 and 250 nt. The longest PCR product should not exceed 800 nt in size. This size limit is dictated by the lower amplification efficiency of large fragments which leads to under representation of the product derived from the exon included isoform and inaccurate quantification of the exon inclusion rates. The inclusion rate of such large alternative exons can be assessed using a set of three primers that includes a shared forward primer and two reverse primers placed in the downstream exons (Fig. 2c). The same approach using a shared forward primer and two reverse primers can also be used to detect mutually exclusive exons, which typically have the same size, or alternative 3′ terminal exons (Fig. 2d).

One of the primers in the set is synthesized with fluorescent tag at the 5′ end. In the cases where a set of three primers is used the label should be placed on the shared primer. The fluorescent tag needs to be compatible with the capillary electrophoresis equipment that will be used to separate and quantify the PCR products. The tag also needs to be different from the fluorescent label of the size standard. ABI capillary sequencers can use both ROX- and LIZ-labeled size standards. We recommend using FAM or HEX to label the primers. Bot tags are compatible with the ABI equipment (Table 1) and are commonly available as an inexpensive 5′ modification option from a number of oligonucleotide synthesis service companies.

3.4 PCR Amplification

1. For each 96-well plate prepare a PCR master mix containing 165 μl 10× Taq PCR buffer 33 μl dNTPs, 33 μl 10 μM Primer mix, 17 μl of Taq DNA polymerase, and 1,182 μl of water.
2. Dispense 13 μl of the master mix in each well of a 96-well PCR plate.
3. Transfer 2 μl of the first-strand synthesis reactions to the plate containing the PCR mix. Spin down the plate briefly to collect any drops and purge air bubbles trapped at the bottom of the wells.

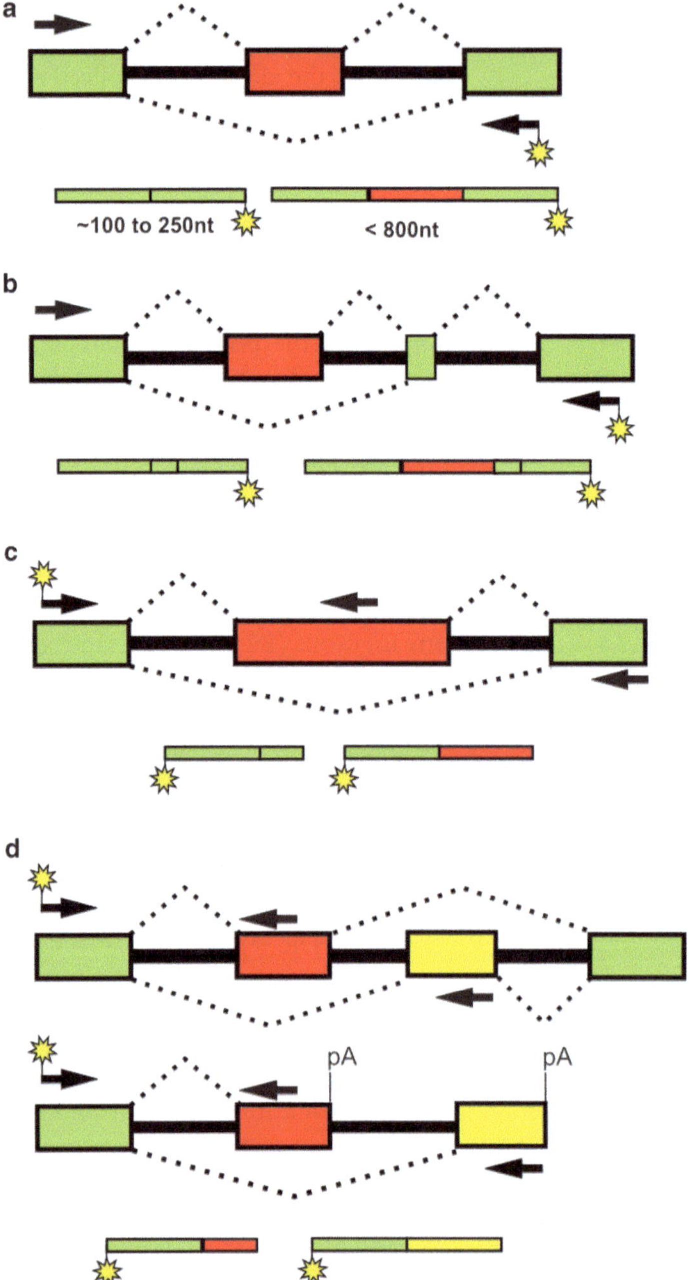

Fig. 2 Primer placement for detecting alternatively splicing events. Each panel shows a stylized gene structure (*top*) and the expected PCR products (*bottom*). The *star* indicates the label position in the primers and the PCR products. Primers are typically placed in the constitutive exons flanking the alternatively spliced regions ((**a**) and (**b**)). A combination of a shared forward primer and two reverse primers can be used to detect large cassettes (more than 700 bp), mutually exclusive or alternative 3′ exons ((**c**) and (**d**))

Table 1
Typhoon phosphorimager excitation/emission combinations and ABI capillary electrophoresis filter sets for detecting commonly used fluorescent labels

Label	Typhoon excitation laser	Typhoon emission filter	ABI dye filter set
FAM	480 nm (Blue laser) 532 nm (Green laser)	520 nm band pass 40 526 nm short pass	A, D, F, G5, C, S
HEX	532 nm (Green laser)	555 nm band pass 20	D (*see* **Note 15**)
ROX	532 nm (Green laser)	610 nm band pass 30	A, D, F
LIZ	633 nm (Red laser)	670 nm band pass 30	G5, S

4. Cover the plate with sealing film and amplify the templates using the following conditions:

   Initial denaturation: 94 °C for 4 min.

   20–35 amplification cycles: 94 °C for 30 s; 60 °C for 30 s; 72 °C for 60 s (*see* **Note 8**).

   Final extension: 72 °C for 5 min.

   Hold at 10 °C until ready to remove the plate.

3.5 Electrophoresis and Quantification of Exon Inclusion

The PCR amplicons can be visualized and quantified either by gel or capillary electrophoresis. Gel electrophoresis is significantly more laborious. However it is indispensable as a tool to control the quality of the samples and troubleshoot problems. In particular we recommend analyzing 12–16 samples by gel electrophoresis to ensure that the PCR reactions did not fail and to estimate if a dilution of the samples may be necessary prior to submitting the full sample set for capillary electrophoresis. The fluorescently labeled PCR amplicons separated by gel electrophoresis can be imaged directly on a Typhoon Phorsphorimager (GE) and quantified either by the ImageQuant software that accompanies the instrument or by the freely available ImageJ software (Fig. 3).

Capillary electrophoresis instruments offer single nucleotide resolution over a wider range of fragment sizes, increased sensitivity, and significantly higher throughput compared to gel electrophoresis.

3.5.1 Fragment Analysis by Gel Electrophoresis

Acrylamide gel electrophoresis

1. Clean the sides of each glass plate with absolute ethanol and then dry with paper towels (*see* **Note 9**).
2. If the glass plates have not been siliconized before, apply Sigmacote (*see* **Note 10**) to the ethanol cleaned sides and spread/dry with a paper towel. Clean again the plates with ethanol as described in **step 1**.

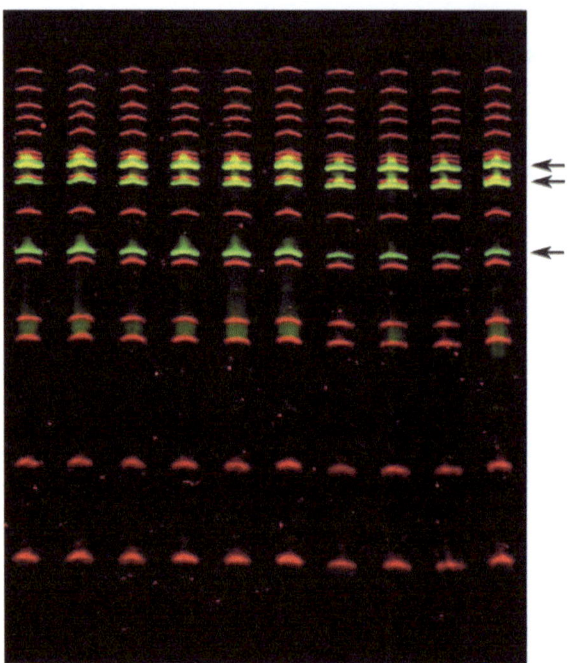

Fig. 3 Gel electrophoresis of alternatively spliced products imaged on Typhoon phosphorimager. The PCR amplicons derived from three alternative isoforms are labeled by FAM (*green bands* indicated by *arrows*). The custom size standard (75–800 nt) is labeled with ROX (*red bands*)

3. Place the spacers on the larger of the two glass plates (two side spacers with the foam dam toward the top and the bottom spacer across the bottom edge) then take the smaller of the two plates and place it on top with the cleaned side facing the other plate, thus making a plate-spacer-plate sandwich.
4. Clip the sandwich together with binder clips and set aside.
5. Assemble as many gels as needed as described above.
6. In a clean flask mix acrylamide solution (25 ml/gel) with 1/100 volume 10 % APS (250 μl/gel) and 1/1,000 volume TEMED (25 μl/gel).
7. Holding the plate sandwich at approximately 15–20° from horizontal, pour the gel solution in a steady stream along one of the side spacers allowing it to flow smoothly between the two glass plates while ensuring that no air bubbles are formed. Once filled to the top, place the gel horizontally to insert the comb. Leave the gel in this position until the gel polymerizes (approximately 20–30 min).
8. After the gel has solidified remove the clips, the bottom gel spacer, and the well comb.

9. Place the sandwich into the running apparatus with the larger glass plate facing out. Use two clips on each side of the gel to clip the sandwich to the gel.
10. Immediately rinse the wells using an 18 G needle on a 50 ml syringe filled with 1× TBE running buffer.
11. Fill the upper and lower reservoirs with 1× TBE buffer ensuring that the gel is covered and that there is no air trapped at the bottom.
12. Apply a piece of clear adhesive tape such as Scotch tape, to the outside glass plate directly under the wells. On the tape, use a Sharpie pen to label each well with a number that corresponds to the sample that will be loaded into the well (*see* **Note 11**).
13. Attach the cover of the gel apparatus and pre-run the gel for 30–50 min at 450 V. While the gel is preheating prepare the PCR amplicons for loading as described below.
14. Prepare a loading buffer mix containing 10 μl clear formamide loading buffer and 0.3 μl fluorescent size standard for each sample to be loaded on the gel.
15. Depending on the number of samples being analyzed place one or more PCR tube strips on a PCR tube rack.
16. Dispense 10 μl of the loading buffer mix to each tube.
17. Transfer 2 μl of the PCR amplicon to each of the tubes containing the loading buffer mix.
18. Seal the tubes and incubate in a thermal cycler at 95 °C for 5 min to denature the samples. Place the tubes on ice.
19. Turn off the gel power supply. Rinse again the wells as described in **step 11** to remove accumulated urea. Failing to remove the urea will interfere with loading the samples and distort the bands.
20. Load 10 μl of the denatured samples in each well.
21. Optionally load 1–2 μl of the gel loading buffer containing tracking dyes to and empty well at least one lane apart from the nearest sample (*see* **Note 12**).
22. Run the gel at 450 V for 55 min or until the bromophenol blue dye moves out of the gel then turn off the power supply.
23. Remove the gel sandwich from the electrophoresis apparatus.
24. Remove the side spacers and the adhesive tape. Do not disassemble the gel sandwich!
25. Clean the plates with deionized water to remove any dried acrylamide or urea attached to the outside of the plates.
26. Clean and dry the plates with ethanol soaked paper towels to remove any remaining dirt and dry the plates (*see* **Note 9**).

Phosphorimager visualization

1. Clean and dry the surface of the phosphorimager with ethanol soaked paper towels (*see* **Note 9**).
2. Place the gel sandwich(es) on the phosphorimager glass plate.
3. In the Typhoon control software select the scan area, then close the imager lid.
4. In the Typhoon control software set the phosphorimager to fluorescence mode.
5. Select the appropriate combinations of excitation lasers and band pass emission filters depending on labels present in the samples and the size standards. The settings for the most common labels are listed in Table 1.
6. Set the focal plane to +3 mm (this adjusts the focal point to 3 mm above the surface of the phosphorimager to account for the width of the glass plate of the gel).
7. Choose the appropriate orientation for your output image.
8. Scan the gel. While scanning make sure that there are no saturated pixels (marked in red on the preview window). If there are saturated pixels, rescan the gel after lowering the photomultiplier (PMT) voltage for the appropriate channel.

3.5.2 Fragment Analysis by Capillary Electrophoresis

1. Dilute the PCR amplicons with water. The dilution factor depends on the signal strength and can vary from 2 to 100-fold. The approximate dilution factor can be determined from the gel electrophoresis analysis. We recommend running a pilot experiment to determine the relationship between the signal strengths detected by the phosphorimager and capillary electrophoresis equipment.
2. Prepare loading buffer mix containing 1 ml clear formamide loading buffer and 30 μl for each 96-well plate.
3. Dispense 10 μl of the loading buffer mix in each well of a half skirt 96-well PCR plate.
4. Transfer 2 μl of the diluted PCR amplicons to the plate containing the loading buffer mix.
5. Seal the plate and incubate in a thermal cycler at 95 °C for 5 min to denature the samples.
6. Place the plates on ice and bring them to the facility operating the ABI capillary electrophoresis equipment to perform fragment analysis. The denatured plates can be stored frozen at −20 °C for several days.
7. The electrophoretograms generated by the capillary electrophoresis equipment can be analyzed using the PeakScanner software to determine the peak sizes and intensities. Follow the PeakScanner manual for detailed procedures (*see* **Note 13**).

8. Export the peak size and area data from PeakScanner as comma or tab delimited text file.
9. Import the peak data in spreadsheet software (Microsoft Office Excel; Libre Office Calc) and calculate the relative exon inclusion levels.
10. The relative exon inclusion rate is calculated as the amount of the bands that contain the exon normalized to the total amount of DNA in all bands (*see* **Note 14**).

4 Notes

1. In the United States hydrophilic PVDF plates from Seahorse Biosciences (part # 200943-100) are sold by Phenix Research Products (catalog # MPF-011) and ISC Bioexpress (catalog # T-3180-7). We have not tested the performance of hydrophilic PVDF plates from other manufacturers in this protocol.
2. Dissolving LiCl in water is extremely exothermic reaction. Allow the solution to cool before adding the remaining components.
3. Taq polymerase and RNase H(−) MMLV reverse transcriptase can be obtained from a number of vendors. Enzyme costs can be substantially reduced by expressing and purifying recombinant enzymes in *E. coli* following published protocols [13, 14]. We use 6× His-tagged MMLV clone containing the following mutations: (1) D524N—to eliminate the RNase H activity; (2) Q84A—to improve processivity; (3) Δ1–23—deletion of the first 23 amino acids to improve solubility [13, 15, 16].
4. ABI sequencer compatible plates can be obtained from a number of manufacturers.
5. For example, one 96 well plate will require 38.4 ml of RNA Lysis Buffer (19.2 ml at **step 2** and another 19.2 ml at **step 16**). So, it is practical to make up 40 ml total to ensure extra for ease of pipetting from the liquid troughs. After adding 40 ml of the RNA Lysis Buffer to an RNase-free conical tube, a total of 800 μl of β-mercaptoethanol is added and thoroughly mixed.
6. High numbers of cells, for example plates that contain densely seeded HEK 293 cells, may not lyse efficiently in 200 μl of lysis buffer and subsequently clog the filter plate. In such cases increase the volume of the lysis buffer to 300 μl and apply only 100 μl to the filter plate in **step 6** of the RNA extraction protocol.
7. RNA concentrations are typically 30–50 ng/μl (0.9–1.5 μg total) for 90 % confluent well of HEK293 cells (50,000 cells); 10–20 ng/μl (250–500 ng total) for 90 % confluent fibroblasts or MDA-MB-231 cells. 260/280 ratio is typically 1.9–2.0.

8. The number of cycles depends on the copy number of the template and the number of cells in the starting material. We recommend determining it experimentally for each template. Moderately expressed transcripts are easily detectable at 25–30 cycles.
9. It is critical that all surfaces are clean and free of dust as dirt and dust particles are often highly fluorescent and will interfere with the fluorescent imaging.
10. Rainex is a suitable, less expensive alternative. It is sufficient to siliconize the plates once every 6–12 months depending on the frequency of use.
11. Because the amplicon/formamide mix that will be loaded into each well is clear it is easy to lose track of which well has which sample and it is very difficult to determine if a sample was loaded into a well. By numbering the wells one can keep track of the wells that have been loaded to prevent a well from being accidentally skipped or being loaded with two samples.
12. Both bromophenol blue and xylene cyanol FF are strongly fluorescent and may interfere with the signal if placed two close to the samples.
13. PeakScanner is available after registration as a free download from Life Technologies.
14. The electrophoretograms may contain bands arising from non-specific amplification. The areas of these bands should not be included when calculating the relative exon inclusion levels.
15. The HEX label can also be detected on the G5 set although this is not supported by ABI.

References

1. Nilsen TW, Graveley BR (2010) Expansion of the eukaryotic proteome by alternative splicing. Nature 463(7280):457–463
2. Wang ET, Sandberg R, Luo S et al (2008) Alternative isoform regulation in human tissue transcriptomes. Nature 456(7221):470–476
3. Pan Q, Shai O, Lee LJ et al (2008) Deep surveying of alternative splicing complexity in the human transcriptome by high-throughput sequencing. Nat Genet 40(12):1413–1415
4. Krawczak M, Thomas NST, Hundrieser B et al (2007) Single base-pair substitutions in exon–intron junctions of human genes: nature, distribution, and consequences for mRNA splicing. Hum Mutat 28(2):150–158
5. Sterne-Weiler T, Howard J, Mort M et al (2011) Loss of exon identity is a common mechanism of human inherited disease. Genome Res 21(10):1563–1571
6. Levinson N, Hinman R, Patil A et al (2006) Use of transcriptional synergy to augment sensitivity of a splicing reporter assay. RNA 12(5): 925–930
7. Younis I, Berg M, Kaida D et al (2010) Rapid-response splicing reporter screens identify differential regulators of constitutive and alternative splicing. Mol Cell Biol 30(7): 1718–1728
8. Stoilov P, Lin C-H, Damoiseaux R et al (2008) A high-throughput screening strategy identifies cardiotonic steroids as alternative splicing modulators. Proc Natl Acad Sci USA 105(32): 11218–11223
9. Newman EA, Muh SJ, Hovhannisyan RH et al (2006) Identification of RNA-binding proteins that regulate FGFR2 splicing through the use of sensitive and specific dual color fluorescence minigene assays. RNA 12(6):1129–1141

10. DeWoody JA, Schupp J, Kenefic L et al (2004) Universal method for producing ROX-labeled size standards suitable for automated genotyping. BioTechniques 37(3):348–350, 352
11. El-Baradi TTAL, Raué HA, De Regt VCHF et al (1984) Stepwise dissociation of yeast 60S ribosomal subunits by LiCl and identification of L25 as a primary 26S rRNA binding protein. Eur J Biochem 144(2):393–400
12. Bair RJ, Heath EM, Meehan H et al (2004) Compositions and methods for using a solid support to purify RNA. US Patent number 7148343, 29 Jan 2004
13. Liu S, Goff SP, Gao G (2006) Gln84 of moloney murine leukemia virus reverse transcriptase regulates the incorporation rates of ribonucleotides and deoxyribonucleotides. FEBS Lett 580(5):1497–1501
14. Lawyer FC, Stoffel S, Saiki RK et al (1993) High-level expression, purification, and enzymatic characterization of full-length Thermus aquaticus DNA polymerase and a truncated form deficient in 5′ to 3′ exonuclease activity. Genome Res 2(4):275–287
15. Blain S, Goff S (1993) Nuclease activities of Moloney murine leukemia virus reverse transcriptase. Mutants with altered substrate specificities. J Biol Chem 268(31):23585–23592
16. Das D, Georgiadis MM (2001) A directed approach to improving the solubility of Moloney murine leukemia virus reverse transcriptase. Protein Sci 10(10):1936–1941

Chapter 23

Chromatin Immunoprecipitation Approaches to Determine Co-transcriptional Nature of Splicing

Nicole I. Bieberstein, Korinna Straube, and Karla M. Neugebauer

Abstract

Chromatin immunoprecipitation (ChIP) is a common method used to determine the position along DNA where an antigen is found. The method was initially devised for protein antigens that come in direct contact with genomic DNA, such as components of the transcriptional machinery and histones. However, ChIP can also be extended to antigens that bind RNA, as demonstrated by the specific localization of spliceosomal components to particular gene regions that correlate with when and where introns and exons are transcribed. The activities of any RNA binding protein can in principle be monitored using ChIP, and RNA dependency of binding can also be assessed through RNase treatment. Combined with qPCR or high-throughput sequencing, this method allows the detection of RNA bound proteins at individual genes or genome-wide. Here, we present a detailed protocol for "splicing factor ChIP" in tissue culture cells.

Key words Chromatin immunoprecipitation (ChIP), Splicing regulatory proteins, Spliceosomal small nuclear ribonucleoprotein particles (snRNPs), Spliceosome assembly, Quantitative PCR (qPCR), ChIP-Seq

1 Introduction

Splicing—the removal of introns and ligation of exons by the spliceosome—can take place co-transcriptionally, while the pre-mRNA is still attached to chromatin via RNA polymerase II [1]. Thereby, the nascent RNP lies close to the DNA axis, allowing for interactions between the splicing machinery and chromatin (reviewed in ref. 2). The co-transcriptional binding of splicing regulatory proteins and spliceosome assembly on nascent RNA can thus be monitored by chromatin immunoprecipitation (ChIP) [3–6]. The basic principle of this technique is in vivo crosslinking followed by the immunoprecipitation of an RNA binding protein of interest and finally the isolation of the corresponding DNA fragment (Fig. 1). First, unperturbed cells are usually crosslinked together by formaldehyde, because formaldehyde is cell-permeable and efficiently forms CH_2 linkages between amino acid side chains and nearby nitrogen atoms in nucleic acids. In the resulting complex,

Klemens J. Hertel (ed.), *Spliceosomal Pre-mRNA Splicing: Methods and Protocols*, Methods in Molecular Biology, vol. 1126, DOI 10.1007/978-1-62703-980-2_23, © Springer Science+Business Media, LLC 2014

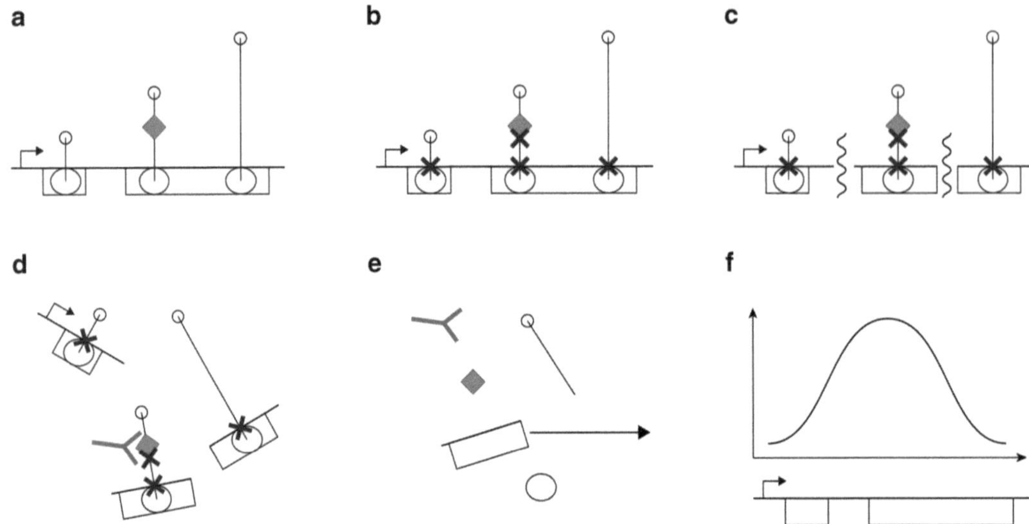

Fig. 1 Schematic of the splicing factor ChIP approach. (**a**) In vivo, splicing factors (*diamond*) can bind co-transcriptionally to nascent RNA. (**b**) Formaldehyde crosslinking covalently binds the RNA binding protein to RNA, which is attached to DNA via the polymerase (*open circle*). (**c**) After cell lysis, the DNA is sheared by sonication. (**d**) Immunoprecipitation using an antibody directed against the splicing factor of interest will isolate the DNA–RNA–protein complex to which the splicing factor was bound. (**e**) After washing and uncrosslinking, the corresponding DNA fragment is purified. (**f**) Analysis of the recovered DNA fragments by qPCR or next generation sequencing identifies the genomic region where the splicing factor was bound. The 5′ end cap on the nascent RNA is indicated by the *small open circle*

the splicing factor is covalently bound to RNA, which is in turn linked to the polymerase and to DNA. Next, the cells are lysed and the chromatin is sheared to fragments of ~200 bp by sonication. The protein–nucleic acid complex is immunoprecipitated and purified, using magnetic beads. After washing and elution, the complex is uncrosslinked by heating, and proteins are digested by Proteinase K. The DNA fragments are isolated by phenol–chloroform extraction and residual RNA is removed by the addition of RNase A. Finally, the recovered DNA is analyzed by qPCR or high-throughput sequencing (ChIP-Seq).

2 Materials

Prepare all solutions using deionized water. All reagents are stored at room temperature unless indicated otherwise.

1. 37 % Formaldehyde.
2. PBS, store at 4 °C.
3. Protease Inhibitor Cocktail 25× in PBS, store at 4 °C.

4. SDS lysis buffer: 1 % (w/v) SDS, 10 mM EDTA, 50 mM Tris–HCl pH 8.1, add 1× protease inhibitor before use.
5. Bradford reagent.
6. ChIP dilution buffer: 0.01 % (w/v) SDS, 1.1 % (v/v) Triton X-100, 1.2 mM EDTA, 16.7 mM Tris–HCl pH 8.1, 167 mM NaCl, add 1× protease inhibitor before use.
7. Dynabeads coated with protein G or A.
8. Low Salt Immune Complex wash buffer: 0.1 % (w/v) SDS, 1 % (v/v) Triton X-100, 2 mM EDTA, 20 mM Tris–HCl pH 8.1, 150 mM NaCl, store at 4 °C.
9. High Salt Immune Complex wash buffer: 0.1 % (w/v) SDS, 1 % (v/v) Triton X-100, 2 mM EDTA, 20 mM Tris–HCl pH 8.1, 500 nM NaCl, store at 4 °C.
10. LiCl Immune Complex wash buffer: 0.25 M LiCl, 1 % (v/v) NP-40, 1 % (w/v) deoxycholic acid, 1 mM EDTA, 10 mM Tris–HCl pH 8.1, store at 4 °C.
11. 1× TE: 10 mM Tris–HCl pH 8.1, 1 mM EDTA, store at 4 °C.
12. Elution buffer: 1 % (w/v) SDS, 0.1 M $NaHCO_3$, prepare freshly before use.
13. 5 M NaCl.
14. 0.5 M EDTA.
15. 1 M Tris–HCl pH 6.5.
16. 10 mg/ml Proteinase K.
17. Phenol–chloroform/isoamyl alcohol (25:24:1).
18. Chloroform/isoamyl alcohol (24:1).
19. 3 M NaOAc pH 5.4.
20. 20 mg/ml Glycogen.
21. 100 % Ethanol.
22. 70 % Ethanol.
23. Deionized water.
24. RNase A.

3 Methods

3.1 Cell Culture and Crosslinking

1. Grow cells to confluency on four 14 cm dishes. This should give cell material for four immunoprecipitations with approximately ~10^8 cells per immunoprecipitation (*see* **Note 1**).
2. Crosslinking: Add 540 μl formaldehyde (37 % solution) directly to 20 ml culture medium to a final concentration of 1 %, mix and incubate for 10 min at RT (work and incubate under the fume hood!) (*see* **Note 2**).

3.2 Harvesting the Cells

All following steps are performed on ice, if not stated otherwise.

1. Aspirate medium thoroughly. Wash cells twice using 5 ml cold PBS + protease inhibitor (1:100) (*see* **Note 3**).
2. Add 5 ml cold PBS + protease inhibitor (1:100), scrape cells using a plastic cell scraper and transfer the cells to a 50 ml tube. Repeat the scraping with 5 ml cold PBS + protease inhibitor. Pool the cells from four plates (final volume 40 ml).
3. Pellet cells for 5 min at 1,500 × *g* at 4 °C.
4. Cell pellets can be frozen in liquid nitrogen and stored at −80 °C or processed directly (*see* **Note 4**).

3.3 Preparing the Lysate

1. Resuspend the cell pellet in 1 ml of SDS lysis buffer + 1× protease inhibitor, pipette up and down to homogenize the lysate, transfer the lysate to a 15 ml tube and incubate for 10 min on ice (*see* **Note 5**).
2. Sonicate the lysate to shear the DNA to lengths between 200 and 500 bp. Keep the samples on ice/ethanol bath. Recommended sonication conditions: 30 % amplitude, 14 × 10 s pulses, 20 s pauses (*see* **Note 6**).
3. Centrifuge the lysate for 10 min at 20,000 × *g* at 4 °C and transfer the supernatant to a 1.5 ml tube. Keep the cleared lysate on ice (*see* **Note** 7).

3.4 Bradford Assay to Determine Protein Concentration (See Note 8)

1. Prepare a standard curve with 10, 5, 2.5, and 1.25 μg/ml BSA in 800 μl ddH_2O. Use 800 μl ddH_2O without BSA as a blank.
2. Dilute the ChIP lysate 1:1,000, 1:5,000, and 1:10,000 in 800 μl ddH_2O.
3. Add 200 μl Bradford reagent, mix and incubate 5 min at RT.
4. Measure absorbance at 595 nm and calculate the amount of total protein in the lysate.

3.5 Immunoprecipitation

1. Dilute 3 mg total protein ChIP lysate in 2 ml ChIP dilution buffer + 1× protease inhibitor. Prepare one 2.0 ml tube for each immunoprecipitation (*see* **Note 9**).
2. Freeze one aliquot as input (1/4 of IP volume, i.e., 0.75 mg total protein) to be used in Subheading 3.8.
3. Add 5 μg of the immunoprecipitating antibody and precipitate overnight at 4 °C with rotation. Include a mock IP as control (*see* **Notes 10** and **11**).
4. Add 18 μl of Dynabeads for 1 h at 4 °C with rotation to collect the antibody–protein complex (*see* **Notes 12–15**).

3.6 Washing the Beads

1. Capture the beads with a magnetic rack. Carefully remove the supernatant that contains unbound, nonspecific DNA.

2. Resuspend the beads in 1 ml Low Salt Immune Complex wash buffer and transfer the beads to a new 1.5 ml tube (*see* **Note 16**).
3. Wash the bead/antibody/protein complex for 4 min on a rotary shaker.
4. Capture the beads with a magnetic rack and remove the supernatant.
5. Repeat the washing with 1 ml of each of the buffers in order as listed below (*see* **Note 17**):

   1× High Salt Immune Complex wash buffer.

   1× LiCl Immune Complex wash buffer.

   1× TE.

3.7 Elution

All following steps are performed at room temperature, if not stated otherwise.

1. Freshly prepare elution buffer.
2. Capture the beads in a magnetic rack and remove the supernatant from the last washing step.
3. Elute the protein complex from the antibody by adding 250 μl elution buffer. Vortex briefly to mix and incubate at room temperature for 15 min with rotation.
4. Capture the beads, and carefully transfer the supernatant fraction (eluate) to a 1.5 ml tube and repeat elution with 250 μl fresh elution buffer.
5. Combine eluates (total volume 500 μl).

3.8 Uncrosslinking and Proteinase K Treatment

1. Take the frozen input (from Subheading 3.5) and add ChIP dilution buffer to a total volume of 500 μl. Uncrosslink the input together with the immunoprecipitation samples. This sample is considered to be the input/starting material for all the immunoprecipitations done with this extract and serves as a background control (*see* **Note 11**).
2. To all samples add 20 μl of 5 M NaCl, 10 μl of 0.5 M EDTA, 20 μl of 1 M Tris–HCl, pH 6.5 and 10 μl Proteinase K (10 mg/ml) incubate for 6 h at 65 °C.

3.9 Recover DNA by Phenol–Chloroform Extraction

1. Add 560 μl (=1 volume) phenol–chloroform/isoamyl alcohol (25:24:1) pH 8, vortex and incubate 2–3 min at RT.
2. Centrifuge for 15 min at 20,000 × *g* and 4 °C.
3. Transfer the upper aqueous phase to a new 1.5 ml tube.
4. Add 560 μl (=1 volume) chloroform/isoamyl alcohol, (24:1), vortex.
5. Centrifuge for 15 min at 20,000 × *g* and 4 °C.

6. Transfer the upper aqueous phase to a new 1.5 ml tube.
7. Precipitate the DNA by adding 1 ml 100 % EtOH, 50 μl 3 M NaAC pH 5.4, and 1 μl Glycogen (20 mg/ml). Vortex and incubate overnight at −80 °C.
8. Centrifuge for 30 min at 20,000 × *g* and 4 °C.
9. Discard the supernatant and wash the pellet with 750 μl 70 % EtOH.
10. Centrifuge for 20 min at 20,000 × *g* and 4 °C.
11. Discard the supernatant, dry the pellet, and resuspend the DNA in ddH_2O or TE + 50 μg/ml RNase A (*see* **Notes 18** and **19**).

4 Notes

1. A large number of cells is required as starting material for ChIP of spliceosomal proteins and splicing regulators. We recommend using at least 10^8 cells for each immunoprecipitation and each control. In the case of HeLa cells, one 14 cm cell culture dish provides enough material for one IP. However, due to differences in cell density and also in chromatin content between different cell lines (HeLa cells are polyploid), the optimal number of cells has to be determined experimentally for each cell line (*see* also **Notes 5–9** for optimization).
2. Formaldehyde crosslinking will covalently link proteins and nucleic acids. Thus, it is thought that splicing factors bound to RNA are crosslinked to DNA and chromatin via RNA polymerase II. However, the exact size of these protein–nucleic acid complexes is not known. The RNA binding protein of interest might be able to crosslink to a window of several nucleotides of DNA, therefore limiting the nucleotide resolution of this method. Keep that in mind, when interpreting your results. Moreover, the dependency of such a ChIP signal on bridging RNA molecules can be tested by treating the initial lysate with RNase A [6].

   The crosslinking reaction can be quenched with glycine (125 mM final concentration). However, we obtained good results by directly removing the formaldehyde containing medium and washing the cells with cold PBS + protease inhibitor.
3. Ice-cold PBS is required for washing the cells after crosslinking to preserve the crosslinked protein–nucleic acid complexes. We therefore recommend to store PBS for ChIP at 4 °C.
4. Pellets of crosslinked cells can either be lysed directly or shock frozen in liquid nitrogen and stored at −80 °C for future use. However, we recommend being consistent and treating the samples in exactly the same way for each experiment, i.e., always

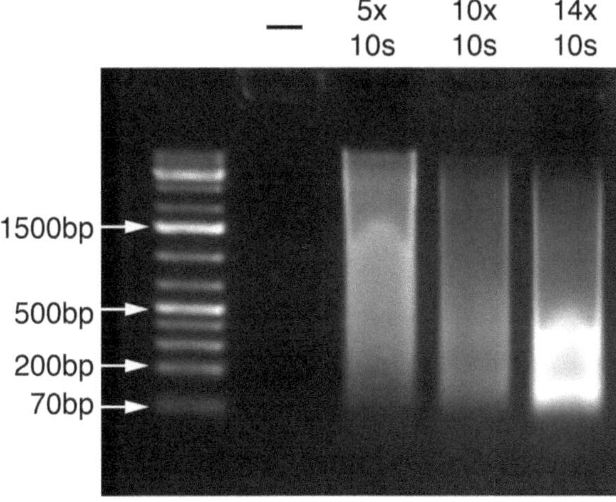

Fig. 2 DNA fragmentation by Sonication. Cells were harvested, lysed in SDS lysis buffer and sonicated for 5× 10 s, 10× 10 s, or 14× 10 s pulse with 20 s pause at an amplitude of 30 %. After centrifugation, the supernatant was uncrosslinked and treated with proteinase K before the DNA was purified by phenol–chloroform extraction. Fragmentation was assessed on a 1 % agarose gel. With increasing number of pulses, the fragment size is reduced to 100–400 bp in the last lane, which represents the desired fragment distribution around 200 bp

proceed directly OR always freeze the pellets. In our experience, freezing the cell pellets is safer than freezing and reusing lysates.

5. Pipette slowly up and down to resuspend the pellet and lyse the cells. Avoid foam! You might have to optimize the ratio of cells to SDS lysis buffer (*see* also **Notes 1**, **8**, and **9**) to obtain a homogenous and concentrated lysate.
6. The sonication conditions vary between cell types and sonicators and have to be optimized experimentally to yield fragments of approximately 200 bp. Sonication quality and fragment size can be assessed by agarose gel electrophoresis (Fig. 2). The size of DNA fragments determines the positional resolution of the ChIP assay.
7. Centrifugation should yield a small, white pellet of cell debris and a clear or milky supernatant containing the fragmented chromatin. A rather big and dense pellet indicates incomplete lysis and sonication.
8. Any assay to determine protein concentration could be used. However, due to the high SDS concentration in the lysate (SDS lysis buffer), the samples have to be diluted accordingly. We recommend Bradford or the Amido Black assay, which is insensitive to SDS, and a sample dilution of 1:1,000 to 1:10,000 in ddH_2O.
9. The amount of total protein used as staring material for each IP depends on the abundance of the protein of interest.

For splicing factor ChIP with HeLa cells, we recommend 3–4 mg of total protein per IP. For ChIP of RNA polymerase II or histones, 2 mg of total protein per IP are sufficient. The exact amount might vary between cell lines (*see* **Note 1**).

At maximum 200 μl of lysate can be diluted in a total volume of 2 ml ChIP dilution buffer, otherwise the high SDS concentration will affect the IP. You might have to optimize the lysis buffer to cells ratio, the lysis itself and sonicating conditions in order to obtain a concentrated lysate yielding ~3 mg total protein in 200 μl (*see* **Notes 1**, **5–8**).

10. The success of a ChIP experiment largely depends on the quality of the antibody. In general, polyclonal antibodies are preferred over monoclonal, as individual epitopes might not be accessible in the crosslinked state. Our lab previously showed that using a GFP-tag in combination with an anti-GFP antibody, can enhance the ChIP enrichment and signal to noise ratio for splicing factor ChIPs [6]. The GFP-tag should not be engaged in protein–protein or protein–nucleic acid interactions and thus protrude from the crosslinked complex, providing highest accessibility. Furthermore, the same anti-GFP antibody can be used for ChIPs of multiple tagged proteins of interest, thus increasing comparability. If tagging is not an option, we recommend using specific ChIP-grade antibodies. The amount of antibody per IP has to be determined experimentally; we recommend 5 μg per IP as starting point.
11. The required controls depend on the downstream analysis, i.e., qPCR or next generation sequencing. In any case, an input control is required. The input represents the total fragmented genomic DNA, which did not undergo an IP. For qPCR, the ChIP enrichment is calculated relative to Input. For ChIP-Seq, the input is required to assess the background distribution of reads due to sonication bias. Nucleosome occupancy and GC content influence the fragmentation by sonication. Thus, input samples will not yield a homogenous distribution of reads, but rather indicate hot spots of DNA shearing. It is therefore recommended to compare the ChIP-Seq signals of IP and input to test, if the enrichment is significant or simply reflects sonication bias.

    In addition, a mock IP using IgG is recommended for qPCR. This sample passes through all IP steps except for the precipitation of any chromatin complexes. The IgG control can therefore be used to determine the background of nonspecific material that was carried through the procedure by sticking to tubes, beads, pipettes, etc. Comparing the ChIP enrichment of the specific IP to the mock IP will determine, if the enrichment is significant.
12. We recommend using magnetic beads, however, we also have good experiences with sepharose beads (e.g., GammaBind

beads). The advantage of magnetic beads is that the washing is more stringent. The supernatant can be removed more completely when the magnetic beads are held back by a magnetic rack, compared to sepharose beads that were pelleted by centrifugation. In our hands, using magnetic beads greatly improved the signal to noise ratio by reducing background.

13. Whether to use protein A or protein G coupled beads depends on the Ig origin of the antibodies (*see* supplier information for more details).
14. Beads have to be washed twice in ChIP dilution buffer before use.
15. An alternative approach is to pre-couple the antibody to beads before adding the ChIP lysate. Aliquot 18 μl beads into a 2 ml tube, add 500 μl ChIP dilution buffer, and 5 μg antibody. Incubate for 2 h at 4 °C with rotation. Wash the beads twice in ChIP dilution buffer to remove excess antibody. Add the diluted ChIP lysate to the pre-couples beads, incubate overnight at 4 °C with rotation and proceed with Subheading 3.6.
16. Chromatin can also stick to plastic tubes and thus be carried through the whole procedure increasing the nonspecific background. One option is using non-sticky tubes, or alternatively, transferring the beads to a new tube during the washing step.
17. We recommend using washing buffers with increasing stringency.
18. How the DNA is finally resuspended depends on the downstream analysis. 50 μg/ml RNase A should be added to remove residual RNA.
19. The relative amount of DNA recovered can be determined by qPCR. The ChIP signal is calculated as enrichment over input using the following equation: $\Delta Ct = 2^{(CtInput - CtIP)}$, where Ct_{Input} is the threshold cycle of the input sample and Ct_{IP} that of the specific IP. The ChIP enrichment is further normalized to a control primer pair such as an intergenic gene desert region as $\Delta\Delta Ct = \Delta Ct^{experiment} / \Delta Ct^{control}$.

References

1. Brugiolo M, Herzel L, Neugebauer KM (2013) Counting on Co-transcriptional Splicing. F1000Prime Reports 5:9–15
2. Carrillo Oesterreich F, Bieberstein N, Neugebauer KM (2011) Pause locally, splice globally. Trends Cell Biol 21(6):328–335
3. Görnemann J, Kotovic KM, Hujer K et al (2005) Cotranscriptional spliceosome assembly occurs in a stepwise fashion and requires the cap binding complex. Mol Cell 19:53
4. Listerman I, Sapra A, Neugebauer K (2006) Cotranscriptional coupling of splicing factor recruitment and precursor messenger RNA splicing in mammalian cells. Nat Struct Mol Biol 13:815
5. Lacadie S, Rosbash M (2005) Cotranscriptional spliceosome assembly dynamics and the role of U1 snRNA:5′ ss base pairing in yeast. Mol Cell 19:65
6. Sapra A, Ankö ML, Grishina L et al (2009) SR protein family members display diverse activities in the formation of nascent and mature mRNPs in vivo. Mol Cell 34:179

Chapter 24

Computational Approaches to Mine Publicly Available Databases

Rodger B. Voelker, William A. Cresko, and J. Andrew Berglund

Abstract

Publicly available sequence annotation data is a vital resource for researchers. Many types of information are available, including structural annotations (i.e., the locations and identities of genomic features) and functional annotations (e.g., gene expression and protein interactions). Annotation data is especially useful for interrogating Next-Gen sequencing data (e.g., identifying genomic features that are associated with mapped reads). Additionally, the vast amount of data that is available offers researchers the opportunity to mine existing data sets and make new discoveries. The ability to efficiently obtain, manipulate, and interrogate this data is a valuable and empowering skill. In this chapter, we introduce several primary data repositories and describe the most commonly encountered file formats. In order to highlight some of the key concepts, operations, and utilities that are involved in working with annotation data we provide a fully worked example of using annotations to answer some basic questions about a particular CHIP-seq data set.

Key words Sequence annotation, Bioinformatics, BED format, UCSC genome browser, Genomic interval operations

1 Introduction

The amount of publicly available biological sequence data has grown nearly exponentially since the establishment of the first public databases in the 1980s. Today more than one hundred eukaryotic genomes and several thousand bacterial genomes have been sequenced and are publicly available. However, raw genomic sequence data by itself is of little direct value. One of the central goals of molecular biology is to decode this information. Decoding a genome entails determining the locations, identities, and roles of the various functional elements. Associating this type of information to specific genomic sequences is known as annotation [1].

Numerous types of annotation data are publicly available. Such data includes structural annotations that are concerned with identifying the genomic locations of functional elements such as

Klemens J. Hertel (ed.), *Spliceosomal Pre-mRNA Splicing: Methods and Protocols*, Methods in Molecular Biology, vol. 1126, DOI 10.1007/978-1-62703-980-2_24, © Springer Science+Business Media, LLC 2014

transcription units, CDSs, UTRs, poly-adenylation sites, regulatory elements, SNPs, and more. Another type, often referred to as functional annotation, focuses on associating more dynamic processes such as gene expression, protein binding, and methylation states with genomic features. Collectively the enormous wealth of publicly available sequence and annotation data forms a critically valuable resource for biological research.

The ability to efficiently obtain, manipulate, and interrogate this type of data is a valuable and empowering skill. The shear magnitude and complexity of genomic data poses challenges for interpretation. Until relatively recently experiments were typically focused on characterizing individual genes and interpretation often involved visual inspection of bands on gels. However, the advent of next-generation sequencing (NGS) inspired the creation of new techniques that make it possible to obtain genome-wide portraits of cellular processes. In contrast to more traditional techniques, it is not possible to interpret NGS data by visually inspecting the millions of short reads that are obtained. Instead interpretation requires the use of sophisticated computational methods that rely upon collating one's own data with publicly available sequence and annotation data. For instance, in order to use RNA-seq to identify differentially regulated genes the NGS reads must be mapped to a reference gene model. Additionally, the vast amount of data that is available offers researchers the opportunity to mine existing data sets and make new discoveries.

Annotation data can be used to answer many types of questions. Some examples are as follows: Which genes do RNA-seq reads map to? Which exons are highly conserved? Which introns do CLIP-seq reads map to? Which SNPs are located in an exon? What diseases are associated with a set of genes? Although these are diverse questions, the procedures for answering them are surprisingly similar, and in each case the answers can be obtained by performing various set operations (e.g., intersection, union, or difference) between two or more sets of annotated genomic intervals.

Another common type of question involves data aggregation. Examples include: What is the average CHIP-read density across promoters? What is the average conservation for intronic regions flanking alternatively spliced exons? How are CLIP-reads distributed relative to 5′ splice sites? Again, these questions can be answered using a similar set of operations.

In this chapter, we describe the concepts and processes necessary to answer these types of questions. We introduce primary data repositories, describe the most commonly encountered file formats, and describe some software utilities and packages that are useful for working with large biological data sets. In Subheading 3, we work through an example that highlights some of the most common procedures.

2 Materials

The material presented here is designed for researchers who have little or no programming skills, but have basic knowledge of Unix and in using the Unix command-line. A Unix-based operating system such as Linux or Mac OS X is required. In addition we will use the following packages:

BEDtools [2] (available at: http://code.google.com/p/bedtools).

GenomicTools [3] (available at: http://code.google.com/p/ibm-cbc-genomic-tools).

SAMtools [4] (available at: http://samtools.sourceforge.net).

The liftOver utility (pre-compiled versions are available at: http://hgdownload.cse.ucsc.edu/admin/exe).

Instructions for compiling and using these tools are included in their distributions and at the sites listed above (Mac OS X users should *see* **Note 1** before attempting to install these tools).

3 Methods

Sequence annotations can be used to answer many types of questions. Here we will use publicly available data to answer several questions about the relationship between a specific type of histone modification (H3K36me3) and exons. Several studies have shown that nucleosome occupancy and the H3K36me3 modification are enriched over exon bodies [5–7]. We will explore this relationship further by asking whether the association between H3K36me3 and exons is dependent upon where within the genome the exon is located. In particular we will ask whether the association differs for exons located in different types of isochores (*see* **Note 2**).

Sequence and annotation data are available at many online sites. The Journal of Nucleic Acids Research hosts a helpful database of biological databases (*see* **Note 3**). As of 2012, 1,380 databases are listed [8]. Three that are especially useful are NCBI, Ensembl, and the UCSC Genome Browser (for more information on these sites *see* **Notes 4–6**). For this example we will use the UCSC site.

A note on font conventions: In the following text we will use Courier font to indicate commands that are entered in the Unix command-line. However, many of the commands that should be entered on a single line are wrapped to fit within the page margins. In the following text individual commands begin with a "$" character, and wrapped lines are indented.

3.1 Create a BED Table Containing the Coordinates for All Human Exons

First we need to create a file containing the coordinates for all human exons. Generally, it is not possible to simply retrieve a table containing just the data that one wants in the format one desires. Exonic annotations are typically made available in much larger files containing other structural annotations. There are several ways to create a table of specific structural features. Here we will use the UCSC Table Browser, which is a utility that can be used to perform simple queries of the UCSC databases. Alternative approaches are to use the Ensembl BioMart (*see* **Note 7**) or to download and parse the larger parent table locally (*see* **Note 8**).

We must also consider the format for the final table. Several formats are commonly used (for descriptions of the most common *see* **Notes 9–11**). Here we will use the BED format (*described in* **Note 9**), which has been widely adopted for representing simple structural annotations. The BED format is concise and is recognized by a wide range of tools that can be used to manipulate and analyze annotation data.

Using the UCSC Table Browser (available at: http://genome.ucsc.edu/cgi-bin/hgTables), we will retrieve a BED file containing the coordinates of all human exons annotated in the latest ENCODE data [9] (*see* **Note 12**). In the Table Browser select the following:

```
Clade= Mammal
Genome= Human
Assembly= Feb. 2009 (GRch37/hg19)
Group= Genes and Gene Prediction Tracks
Track= GENCODE Genes V12
Table= Basic (wgEncodeGencodeBasicV12)
Region= genome
Output format= BED
Output file= GENCODE_V12_exons.bed
File type returned= gzip compressed
```

Click "get output." This will take you to a new page where you can choose formatting options. In this page select: Exons plus = 0. Then click "get BED" to download the file. The first few lines are shown below (Note: your data may look slightly different):

```
chr1   11869 12227 DDX11L1      0      +
chr1   12613 12721 DDX11L1      0      +
chr1   13221 14409 DDX11L1      0      +
```

In the file above, each line contains information for a single exon. However, users need to be aware of several issues. The first is that different file formats use different coordinate systems. The BED format uses a "space-based" format (also known as "0-based half-open") (for a discussion of the coordinate systems *see* **Note 13**). The second is that tables that are returned from the Table Browser often contain redundant and/or overlapping features. Redundant features most likely indicate that the table was generated from a parent table of all transcripts. In our case there are also overlapping entries that result from exons having alternative 5′ or 3′ splice sites. One can use the Unix "sort" and "uniq" commands to remove redundant entries. However, in this case we also want to merge overlapping entries. Both processes can be performed using the "mergeBed" utility from BedTools. The command for this is as follows:

```
$ mergeBed -s -nms -n -i gencodeV12_exons.bed >
   gencodeV12_exonsNR.bed
```

The "-s" option causes the strand to be ignored. The "-nms" and "-n" options specify that the names of the merged features should be merged and the number of features that were merged will be reported in the resulting file. These last two options are not necessary but they are often useful. If we use the Unix "wc" command to count the lines in original and merged files we see that the original file had 1,188,378 entries and the merged file has only 306,266 entries.

3.2 Retrieve the BAM File Containing the Histone-CHIP Data

Next we will retrieve a file containing data from a H3K36me3 CHIP-seq experiment. We will use one of the datasets generated from K562 cells. This data represents just a small fraction of the ENCODE data that is available. The Unix "wget" command can be used to retrieve the file directly from UCSC. On the command-line issue the following (note this command should be issued on one line and the return characters in the URL should be removed):

```
$ wget http://hgdownload.cse.ucsc.edu/
  goldenPath/hg19/encodeDCC/wgEncodeUwHistone/
  wgEncodeUwHistoneK562H3k36me3StdAlnRep1. bam
```

Next we will use the "samview" utility from samtools to convert this BAM file to SAM format (*see* **Note 14**). We can use the following command to both convert the file and rename the SAM file with a more user-friendly name:

```
$ samtools view
  wgEncodeUwHistoneK562H3k36me3StdAlnRep1.bam >
  hk36.sam
```

3.3 Retrieve and Convert the Isochore Coordinates to BED Format

We will use the isochore definitions that are available in isobase which is an online database that contains isochore profiles for several organisms and genome assemblies [10]. It is available at: http://www.geneinfo.eu:8080/isobase/. Since we are using the hg19 build we could directly choose the hg19 isochore profiles. However, for the sake of this exercise, we will choose the hg18 build data so we can demonstrate how to convert between build coordinates using the UCSC "liftOver" tool. A web interface to liftOver is available at the UCSC site, but we will use a locally installed version (*see* Subheading 2). The first several lines of the isobase file are shown below:

```
chr  start   end     gc         isotype  conf
1    10000   60000   40.813999    L2     2.0
1    60000   100000  39.702499    H1     2.25
1    100000  110000  35.139999    L1     2.0
```

Before we can use this file we need to convert it to BED format. Since all we need are the coordinates (the first three columns) and the isotype data we use the BED4 format. Custom conversions like this are common and can usually be performed using simple Perl or Awk scripts. We could break the processes into individual steps (for a demonstration *see* **Note 15**). However, here we will use a Perl "one-liner" to remove the header, alter the chromosome name, move the isotype data to column 4, and adjust the coordinate system. We can do this by issuing the following, and if you are unfamiliar with Perl, somewhat cryptic, command (again the entire command should be issued on one line):

```
$ perl -lane 'if($.>1){$F[1] -= 1; print "chr",
   join("\t",@F[0..2,4])}'
   isobase_hg18_consensus.txt >
   isobase_hg18.bed
```

The phrase "-lane" is a series of directives to the Perl interpreter. The "l" tells Perl to add a line terminator. The "a" directive causes Perl to split the data on tab-characters. The "n" argument tells the interpreter to apply the code to each line in the file, and "e" tells the interpreter to execute the code that follows. As the interpreter reads lines it stores the line numbers in the special variable "$." The "if" statement therefore skips the header line. The default delimiter for "a" is the tab-character, and when Perl splits lines it puts the data into a special array known as "@F". Perl arrays begin at 0; therefore the statement "$F[1] - = 1" adjusts the first coordinate (*see* **Note 13**). The print statement prints "chr," and then uses several array operators to print the information contained in columns 1, 2, 3, and 5. The first few lines of the new file now look like:

```
chr1     9999     60000    L2
chr1     59999    100000    H1
chr1     99999    110000    L1
```

Now we can use the liftOver tool to convert from hg18 to hg19 coordinates. This utility needs a "chain" file containing the data necessary for the conversion. These are available at: http://hgdownload.cse.ucsc.edu/goldenPath/hg19/liftOver/. The command is as follows:

```
$ liftOver isobase_hg18.bed
   hg18ToHg19.over.chain isobase_hg19.bed
   unmapped.txt
```

The converted data is contained in the file "isobase_hg19.bed." Typically however, some coordinates cannot be converted. These will be listed in the "unmapped.txt" file. It is always a good idea to examine this file to make sure that there were not too many problems.

3.4 Separate Exons According to Their Isochore Type

Ultimately we want to know whether exons located in different types of isochores have different CHIP-read densities. Now that we have a BED file with exon coordinates and a BED file with isochore coordinates we can perform an interval intersect operation to split the exons according to which type of isochore they are located in. First we will split the isochore BED file into separate files each containing entries for just one type of isochore. Since each line in the isochore BED file contains an annotation indicating the type of isochore (column 4) we can use the Unix "grep" command to create separate files for each type of isochore.

```
$ grep L1 isobase_hg19.bed > iso19_L1.bed
$ grep L2 isobase_hg19.bed > iso19_L2.bed
$ grep H1 isobase_hg19.bed > iso19_H1.bed
$ grep H2 isobase_hg19.bed > iso19_H2.bed
$ grep H3 isobase_hg19.bed > iso19_H3.bed
```

Next we can use the "intersectBed" utility from BEDtools to recover the exons that are located in each type of isochore. The command for recovering exons located in L1 isochores is shown below:

```
$ intersectBed -a gencodeV12_exonsNR.bed -b
   iso19_L1.bed > exons_L1.bed
```

3.5 Create CHIP-seq Density Profiles Across Exon/Intron Boundaries

Our goal is to produce a histogram that is a composite of the CHIP-read densities over all exon/intron flanks in each of the five sets of exons. To achieve this we will use several tools from the GenomicTools package. We will have to perform several operations. GenomicTools are designed to allow data from one operation to be piped directly into another using the Unix "|" operator so we will break the operations into three discrete sets of commands.

First we need to use the exon coordinates to create a new set of equally sized intervals that represent a fixed number of bases upstream and downstream of the donor splice sites. These new coordinates essentially represent an alignment of donor splice sites. We will use the Unix "cat" command to read the exon files into the "genomic_regions" tool from GenomicTools. This tool has several functions. First we will use the genomic_regions "pos" function to get just the 3′ coordinate for each exon (i.e., the coordinate adjacent to the 5′ splice site). These will be piped into the "shiftp" function to get the coordinates that correspond to 100 bases upstream and 500 bases downstream. Then we will use the "fix" utility to remove any invalid coordinates (e.g., if an exon was located less than 100 bases from the beginning of the reference the shiftp operation would produce a negative number). The command for exons located in L1 isochores is as follows:

```
$ cat exons_L1.bed|genomic_regions pos -op 3p |
   genomic_regions shiftp -5p -100 -3p +500 |
   genomic_regions fix > tempL1_regions.bed
```

Next we can use the "offset" function in the "genomic_overlaps" program (also from GenomicTools) to calculate the offsets between the 5′ end of each of the intervals obtained above and any CHIP-seq reads that overlap the interval. The command for the L1 data is shown below:

```
$ cat hk36.sam | genomic_overlaps offset -v -i
   -op 5p -a tempL1_regions.bed > tempL1_offsets.txt
```

The "-op 5" option specifies that the offset will be calculated relative to the 5′ end of the reads, the "-i" option specifies that the strand will be ignored, and the "-a" option causes the offset to be calculated as the fractional distance from the 5′ end of the interval. The first few lines of the output are shown below:

```
RP5-1057J7.6 0.968386 1.026622
RP5-1057J7.6 0.916805 0.975042
RP5-1057J7.6 0.750416 0.808652
```

One line is generated for each feature overlap. The output contains three columns with the following information: the name of the exonic interval, the distance between the 5′ end of the interval to the 5′ end of the overlapping read, and the distance between the 5′ end of the interval and the 3′ end of the overlapping read. Users should be aware that the first and second columns are separated by a tab character, while the second and third are separated by a space character.

The profile that we aim to generate is essentially the distribution of the data contained in the second column of the offsets file. We could calculate this in several ways, here we will use the "vectors" program from GenomicTools. The vectors program has a wide variety of functions for analyzing and manipulating data vectors. We can use the Unix "cut" command to extract the second column and pipe this into the vectors program as follows:

```
$ cut -d' ' -f1 tempL1_offsets.txt | vectors
  -hist -n 6 -b 150 > profileL1.txt
```

The "-hist" option tells vectors to generate a distributional histogram, "-n 6" specifies that the data will be output to 6 decimal places, and "-b 150" specifies that the data will be divided into 150 equally sized bins. The data range can also be specified using the "-min" and "-max" options, but these default to 0–1, which is the range for our data. The output contains three columns of data: the bin start, the distribution as frequency, and the distribution as

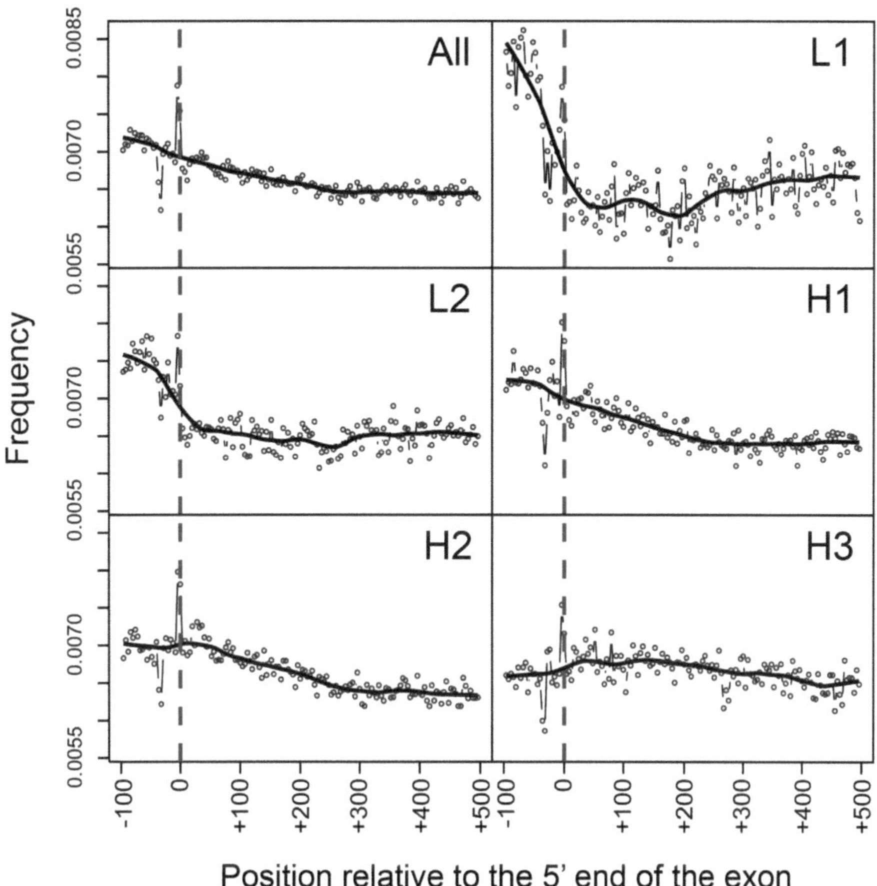

Fig. 1 H3K36me3 CHIP-read density profiles across donor-side exon/intron boundaries. The *vertical dashed line* represents the exon/intron boundary. The isochoric region is indicated in the *upper right*. The *y*-axis is the distributional frequency, and the *x*-axis is the distance relative to the 5′ exonic boundary

number of items. By plotting the data in the first column relative to the data in the second column we can generate the profile that we are interested in. This could be done using Excel or any other plotting program.

After processing the other isochores in the manner described above we generated Fig. 1 using R (*see* **Note 16**). The results are interesting and reveal an interesting phenomenon. Previously it was demonstrated that the H3K36me3 mark is generally associated with exons, and our results confirm this (Fig. 1, "All"). However, our analysis also reveals that the association between exons and H3K36me3 is more complicated. It appears to be highly dependent upon which type of isochore that the exon is located in. Exons in the most AT-rich regions (Fig. 1, L1 and L2) show a profile similar to that seen previously. In sharp contrast, the H3K36me3-mark is dramatically decreased and even shifts to the flanking intronic region for exons located in the most GC-rich isochores (Fig. 1, H2 and H3). Although the biological relevance for this phenomenon is currently unclear, this analysis demonstrates that researchers can make new observations in existing data by using a small number of freely available tools and freely available data.

4 Notes

1. Mac OS X users need to be aware that although OS X is built on Unix, Apple doesn't preinstall the tools that are needed to compile and install C/C++ source code. The easiest way to get these is to install the Apple Developer Tools (available for free from Apple). Apple also doesn't preinstall many of the most commonly used libraries. Before attempting to compile the packages listed in Subheading 3, Mac OS X users also need to install the GNU Scientific Library (GSL) (available at: http://www.gnu.org/software/gsl) and the zlib library (http://www.zlib.net) after installing Developer Tools.
2. Mammalian genomes are highly heterogeneous. Large segments of chromosomes have very high %GC content and other regions have relatively low %GC content (AT-rich) [11]. Computational methods have been used to categorize genomic segments into five types known as isochores: L1, L2, H1, H2, and H3 (listed in order of increasing %GC) [10]. The evolutionary origins and implications of this compositional heterogeneity are currently unclear. However GC-rich isochores are associated with several genomic processes such as DNA replication timing [12], recombination rates [13], and G/C-biased gene conversion [14].
3. *NAR database.* The journal Nucleic Acids Research hosts a web site that describes and links to all of the databases that have been covered in their annual reviews of biological databases. It is available at: http://www.oxfordjournals.org/nar/database/c/.
4. *NCBI.* The National Center for Biotechnology Information (NCBI) hosts GenBank and provides access to many types of biological data (http://www.ncbi.nlm.nih.gov). For a full list of resources that are available at NCBI see: http://www.ncbi.nlm.nih.gov/guide/all. The amount of data that is available can make it difficult to locate a specific set of data. To help overcome this NCBI hosts a search tool known as "Entrez" which uses its own query language (for more information on using Entrez *see* ref. 15). NCBI tends to organize data in a gene/locus centric manner.
5. *Ensembl.* The Ensembl web site is a joint project between EMBL-EBI and the Wellcome Trust Sanger Institute. Ensembl contains genomic data for many organisms and they provide a

genome browser capable of mapping various types of annotation data back to a reference genome. Data can be accessed in several ways. Complete sequence and structural annotation files can be directly downloaded, or users can retrieve customized data sets using the BioMart interface (*see* **Note 7**). For more sophisticated and automated queries users can utilize the Perl API. Information regarding each of these resources is available at the Ensembl web site. The main site (http://www.ensembl.org) focuses on model organisms and vertebrates. Mirrors are available at: http://uswest.ensembl.org, http://useast.ensembl.org, and http://asia.ensembl.org. Genomic data for other organisms including fungi, plants, protozoans, and other metazoans can be found at: http://www.ensemblgenomes.org.

6. *UCSC Genome Browser.* The UCSC Genome Browser provides interactive access to genomic sequences for many vertebrate and non-vertebrate organisms [16]. It also does an excellent job of integrating a large variety of annotation data. These data (referred to as tracks within the browser) include structural annotations such as gene models, ESTs, mRNAs, conservation profiles, SNPs, and repeat sequences. There are also tracks representing functional annotations such as gene expression, transcription factor binding sites, DNA methylation profiles, histone modification, and DNA replication profiles. One of the most valuable features of the UCSC site is the ability to download the annotation files that are used to draw the tracks. Users can also upload their own custom tracks for viewing in the Browser. In order to view custom tracks the data must be formatted in one of several recognized data formats (*see* **Notes 9–11**). An excellent introduction to the Genome Browser can be found at: http://www.nature.com/scitable/ebooks/guide-to-the-ucsc-genome-browser-16569863. Another useful resource is the book *Genomes, Browsers, and Databases* [17].

7. BioMart is an online utility for performing custom queries of the Ensembl databases (available at: http://www.ensembl.org/biomart/martview/). It is easy to use and can perform rather complex queries. Only limited format options are available for the output. However tab-delimited format is an option and is easily parsed using simple Perl or Awk scripts.

8. *Parsing GFF/GTF files.* Whole genome structural annotations are available from Ensembl (*see* **Note 5**) in GTF format (*see* **Note 10**). These files are easy to parse and specific feature types can be easily recovered using "grep." Users should be aware, however, that depending upon the feature type this may result in redundant entries (as defined simply by feature coordinates). The resulting files can be converted to BED format using simple Perl scripts, but users must remember to convert

to the BED coordinate system (*see* **Note 13**). The script "gff2bed.py" is a nice utility for converting GFF/GTF files to BED format. It is available as part of the "bedops" package [18] (http://code.google.com/p/bedops/).

9. *BED format*. The BED format was developed to display tracks in the UCSC Genome Browser. The BED format is concise and easily parsed, and it has become widely adopted as a general format for representing simple annotations. Each line in a BED file contains information for a single feature. Up to 12 tab-delimited fields are defined (*see* Fig. 2), but only the first three are required and contain, in order, a sequence identifier, the feature start-coordinate, and the feature end-coordinate (defined using the space coordinate system discussed above). The next nine fields were designed for displaying data in the browser, but can be used to contain user-defined data. It should be noted that the order of the optional fields is fixed, and lower-numbered fields must always be populated if higher-numbered fields are used. More information regarding the BED format can be found at: http://genome.ucsc.edu/FAQ/FAQformat.html.

   The full 12-field BED format (often called BED12) can accommodate features that are composed of multiple non-overlapping sequence intervals (e.g., exons in a gene). But since users often do not need all 12 fields several abbreviated variants are commonly encountered. The two most common are referred to as BED3 and BED6. BED3 refers to files containing just the coordinates (i.e., first three fields). This format is commonly used when all of the features in a file have the same attribute (e.g., exon, RNA-seq read, SNP) and all that is needed is to specify the genomic intervals having this attribute. The BED6 format uses the first six fields, which allows one to associate additional attributes such as a categorical label (field 4), a numerical value (field 5), and the strand (field 6).

   More information about the BED format and the binary version "bigBED" are available at: http://genome.ucsc.edu/FAQ/FAQformat.html.

10. *GFF/GTF formats* The BED format was originally developed for the specific purpose of displaying data in the UCSC Browser. The Gene Feature Format (GFF), however, was devised to be a generic way to encode structural annotations. It has several commonly used versions including GFF2, GTF (also known as GFF2.5), and GFF3. The GFF format defines 9 tab-separated fields (Fig. 2) all of which are required. The primary difference between GFF versions is in the ninth field, which is used to define hierarchical relationships. GFF feature coordinates use the "base coordinate" system (*see* **Note 13**).

BED

	Field	Value type	Description	Example 1 BED12	Example 2 BED6
1	Sequence Name	Text	Name of the chromosome, scaffold, or contig	chr22	chr1
2	Start	Integer	The starting coordinate for the feature. Note that the first base is always numbered as 0.	1000	35721
3	End	Integer	The coordinate for the base just after the feature.	5000	35736
4	Name	Text	An optional name or tag that is associated with the feature.	cloneA	FAM138A
5	Score	Integer	A score between 0 and 1000. This is used by the browser to determine the level of gray to use to display the feature.	960	.
6	Strand	One of '+' or '-'	Strand for the feature.	+	-
7	Thick Start	Integer	The starting position at which the feature is drawn using a thick line.	1000	
8	Thick End	Integer	The ending position for drawing the feature using a thick line.	5000	
9	RGB value	Comma delimited set of values	Set of comma delimited RGB values used for displaying the feature.	0	
10	Block Count	Integer	The number of blocks or intervals that are associated with the feature (e.g. number of exons).	2	
11	Block Sizes	Comma delimited set of integers	A comma-delimited list of interval lengths (e.g. lengths of each exon).	567,488,	
12	Block Starts	Comma delimited set of integers	A comma-delimited list of the starting coordinates for each of the associated intervals (relative to field 2).	0,3512	

GFF

	Field	Value type	Description	Example1	Example2
1	Sequence Name	Text	Name of sequence (e.g. chromosome) that contains the feature	ctg123	Chr1
2	Source	Text (or '.' if not relevant)	Source of this feature.	.	RefSeq
3	Type	Text	The type of feature. Preferrably this should correspond to types defined within the Sequence Ontology (see www.sequenceontology.org)	gene	exon
4	Start	Integer	The starting coordinate for the feature (where the first nucleotide of the parent sequence = 1).	1000	7000
5	End	Integer	The ending coordinate for the feature.	9000	9000
6	Score	Floating point (or '.' if not relevant)	Can be any value that is meaningfully related to the feature (e.g. blast E-value, conservation score)	0.015	.
7	Strand	One of: '+', '-', '.', '?'	Indication of the strand for the feature.	+	+
8	Phase	One of: 0,1,2, or '.'	Sense phase if the feature is coding.	.	.
9	Attribute	Semi-colon delimited list of feature attributes defined as tag=value pairs.	Tags are predefined and described at www.sequenceontolgy.org	ID=gene0000 1; Name=EDEN	Parent=mR NA00001, mRNA0000 2, mRNA0000 3

Fig. 2 Field definitions for BED and GFF files

More information about the GFF format can be found at: http://gmod.org/wiki/GFF, http://www.sanger.ac.uk/resources/software/gff/, and http://www.sequenceontology.org/resources/gff3.html.

11. Other commonly used annotation formats are described at: http://genome.ucsc.edu/FAQ/FAQformat.html.
12. The ENCODE project is a world-wide multi-consortium project devoted to developing functional annotations including

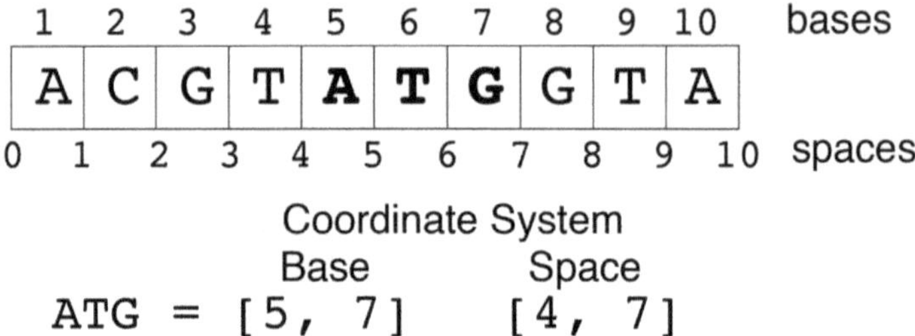

Fig. 3 Schematic representation of the two coordinate systems that are commonly used

transcription factor binding sites, chromatin structure, and histone modification in the human genome [9]. Both raw data and processed annotations related to the ENCODE experiments are available at: http://genome.ucsc.edu/ENCODE.

13. *Coordinate systems.* Sequence annotations are an association between categorical or numerical data and individual nucleotides or ranges of nucleotides in a reference sequence. Since the reference sequence is an ordered set of nucleic acids then annotations can be thought of as ordered intervals located along the reference genome, and their locations can be defined using a three-coordinate system composed of a sequence identifier, a starting coordinate, and an ending coordinate. The sequence identifier is simply the identifier contained in the sequence header, the starting coordinate refers to the start of the feature, and the ending coordinate refers to the end of the feature. For example an exon that is composed of the tenth through twentieth nucleotides of chromosome 1 could be indicated as chr1:10–20. However, readers must be aware that two different coordinate systems are in common use. These systems differ both in terms of the numbering system used for the reference and in how the annotation intervals are defined. In the first system (which we will refer to as the "base coordinate system"), the first nucleotide is numbered 1 and the coordinates for the interval are simply the positions of the first and last nucleotides of the feature (*see* Fig. 3). This is the system used by most annotation formats including GenBank, EMBL, GFF/GTF, SAM, and WIG. In the second system (referred to as the "space coordinate system") the coordinates can be thought of as the junctions or spaces between nucleotides and the position before the first base is referred to as 0 (*see* Fig. 3). In this system an interval is defined by the spaces that bound the nucleotides that actually compose the feature. For instance in Fig. 3, the interval [4, 7] refers to the bases at positions 5–7 (*see* Fig. 3). Although this system may be less intuitive, it simplifies calculating the length of an interval and easily accommodates annotating features that exist between nucleotides

(e.g., splice-junctions). This system is used primarily in the BED format. In reality it is easy to inter-convert between the two systems. For example to convert the coordinates in a BED file to a GFF file just add 1 to the first coordinate, and to convert GFF coordinates to BED coordinates just subtract 1 from the first coordinate.

14. *SAM/BAM format.* Several short-read aligners have been developed to map NGS reads back to a reference genome. These mappings are essentially a type of categorical mapping since they associate an individual read to one or more genomic intervals. The sequence alignment format (SAM) has become the standard way report these alignments. Each SAM entry specifies the coordinates of the alignment (relative to the reference) as well as information regarding the quality of the alignment and gap-structure. A full description of the SAM format is available at: http://samtools.sourceforge.net/. Considering that a typical NGS experiment may consist of millions of reads SAM files are often quite large and are slow to parse. This inspired the introduction of an indexed binary version of the SAM format (known as BAM). BAM files are significantly smaller than their SAM counterparts, and although they cannot be parsed using simple Unix commands such as "grep" and "wc" they can be quickly parsed and manipulated using tools that are available in software packages such as SAMtools [4], bamtools [19], and BEDtools [2].

15. Reformatting the isobase file can be broken down in several discrete steps. We could use a single one-liner (as shown in Subheading 3) or we could concatenate a series of operations using the Unix pipe command "|". For instance we could break it down as follows:

    (a) Remove the header line using either of the following:

    ```
    $ cat myfile | perl -ne 'print if($.>1)' >
    mynewfile
    $ cat myfile | awk 'NR > 1' > t.tmp
    ```

    (b) Append "chr" to the start of the line using either of the following:

    ```
    $ cat myfile | perl -ne 'print "chr",$_' >
    mynewfile
    $ cat myfile | awk '{print "chr" $0}' >
    mynewfile
    ```

    (c) Remove and reorder columns

    ```
    $ cat myfile | cut -f 1-3,5 > mynewfile
    ```

    (d) Change the coordinate for the first column:

    ```
    $ cat myfile | perl -lane '$F[1] -= 1;
    print join("\t",@F)' >
      mynewfile
    ```

Finally, these commands could all be strung together as follows (remember in reality it would be on one line):

```
$ cat myfile | awk 'NR > 1' | perl -ne 'print
"chr",$_' |
  cut -f 1-3,5 | perl -lane '$F[1] -= 1; print
join("\t",@F)' >
  mynewfle
```

16. R is a very powerful open-source data analysis package, and we feel it is well worth the effort to learn. It is available at: http://www.r-project.org/.

References

1. Stein L (2001) Genome annotation: from sequence to biology. Nat Rev Genet 2:493–503
2. Quinlan AR, Hall IM (2010) BEDTools: a flexible suite of utilities for comparing genomic features. Bioinformatics (Oxford, England) 26:841–842
3. Tsirigos A, Haiminen N, Bilal E et al (2012) GenomicTools: a computational platform for developing high-throughput analytics in genomics. Bioinformatics (Oxford, England) 28:282–283
4. Li H, Handsaker B, Wysoker A et al (2009) The sequence alignment/map format and SAMtools. Bioinformatics (Oxford, England) 25:2078–2079
5. Spies N, Nielsen CB, Padgett RA et al (2009) Biased chromatin signatures around polyadenylation sites and exons. Mol Cell 36: 245–254
6. Schwartz S, Meshorer E, Ast G (2009) Chromatin organization marks exon-intron structure. Nat Struct Mol Biol 16: 990–995
7. Andersson R, Enroth S, Rada-Iglesias A et al (2009) Nucleosomes are well positioned in exons and carry characteristic histone modifications. Genome Res 19:1732–1741
8. Galperin MY, Fernandez-Suarez XM (2011) The 2012 nucleic acids research database issue and the online molecular biology database collection. Nucleic Acids Res 40:D1–D8
9. ENCODE Project Consortium, Bernstein BE, Birney E et al (2012) An integrated encyclopedia of DNA elements in the human genome. Nature 489:57–74
10. Schmidt T, Frishman D (2008) Assignment of isochores for all completely sequenced vertebrate genomes using a consensus. Genome Biol 9:R104
11. Costantini M, Clay O, Auletta F et al (2006) An isochore map of human chromosomes. Genome Res 16:536–541
12. Costantini M, Bernardi G (2008) Replication timing, chromosomal bands, and isochores. Proc Natl Acad Sci USA 105:3433–3437
13. Fullerton SM, Bernardo Carvalho A, Clark AG (2001) Local rates of recombination are positively correlated with GC content in the human genome. Molecular Biol Evol 18:1139–1142
14. Duret L, Galtier N (2009) Biased gene conversion and the evolution of mammalian genomic landscapes. Annu Rev Genomics Hum Genet 10:285–311
15. Gibney G, Baxevanis AD (2002) Current protocols in bioinformatics. Wiley, Hoboken, NJ
16. Kent WJ, Sugnet CW, Furey TS et al (2002) The human genome browser at UCSC. Genome Res 12:996–1006
17. Schattner P (2008) Genomes, browsers and databases: data-mining tools for integrated genomic databases. Cambridge University Press, Cambridge
18. Neph S, Kuehn MS, Reynolds AP et al (2012) BEDOPS: high-performance genomic feature operations. Bioinformatics (Oxford, England) 28:1919–1920
19. Barnett DW, Garrison EK, Quinlan AR et al (2011) BamTools: a C++ API and toolkit for analyzing and managing BAM files. Bioinformatics (Oxford, England) 27:1691–1692

Chapter 25

Approaches to Link RNA Secondary Structures with Splicing Regulation

Mireya Plass and Eduardo Eyras

Abstract

In higher eukaryotes, alternative splicing is usually regulated by protein factors, which bind to the pre-mRNA and affect the recognition of splicing signals. There is recent evidence that the secondary structure of the pre-mRNA may also play an important role in this process, either by facilitating or hindering the interaction with factors and small nuclear ribonucleoproteins (snRNPs) that regulate splicing. Moreover, the secondary structure could play a fundamental role in the splicing of yeast species, which lack many of the regulatory splicing factors present in metazoans. This chapter describes the steps in the analysis of the secondary structure of the pre-mRNA and its possible relation to splicing. As a working example, we use the case of yeast and the problem of the recognition of the 3′ splice site (3′ss).

Key words RNA, Secondary structure, Splicing, Bioinformatics, Posttranscriptional regulation, Yeast

1 Introduction

Splicing is the mechanism by which introns are removed from the pre-mRNA to create the mature transcript. In higher eukaryotes this process involves, apart from the core machinery of the spliceosome, many auxiliary factors, e.g., SR proteins or hnRNPs, which can enhance or block the recognition of splicing signals [1]. These factors allow the modulation of the splicing reaction and thus, the existence of alternative splicing.

During transcription, the synthesized RNA can fold [2]. Accordingly, secondary structures adopted by the pre-mRNA may influence splicing regulation. RNA structures can hinder the recognition of splicing signals by occluding them and preventing their recognition by spliceosome components. Alternatively, they could expose signals necessary for regulation. Interestingly, predicted secondary structures have been identified to aid the computational prediction of splice sites [3, 4], and genome-wide analyses have shown that conserved RNA secondary structures overlapping splice sites are related to alternative splicing [5]. Besides, these

Klemens J. Hertel (ed.), *Spliceosomal Pre-mRNA Splicing: Methods and Protocols*, Methods in Molecular Biology, vol. 1126, DOI 10.1007/978-1-62703-980-2_25, © Springer Science+Business Media, LLC 2014

pre-mRNA structures can facilitate the recognition of splicing signals by shortening the distance between them [6, 7]. In other cases, RNA structures can regulate complex splicing patterns, as shown in *Drosophila melanogaster* [8, 9] and human [10].

All these examples indicate that the secondary structure adopted by the pre-mRNA modulates splicing. However, this may be a transient process, since RNA folds co-transcriptionally and the structure may change as more RNA gets produced [11, 12]. Furthermore, these structures can be altered by temperature, transcription, or other factors that prevent their formation or stabilize them [2, 7, 13], thus providing more possibilities for splicing regulation. It is still unclear to which extent secondary structures play a role in splicing in human and in general, in metazoans. However, studies in single cell eukaryotes have provided some insights. In contrast to what happens in higher eukaryotes, yeast species do not have as many of the splicing auxiliary factors [14, 15], which reduces dramatically the number of regulatory mechanisms and makes splicing more dependent on *cis*-acting elements.

Recent works have suggested that RNA structures could be a general mechanism to explain 3′ss selection in yeast [16, 17], expanding previous observations suggesting a role of RNA structures in splicing regulation in yeast [18–22]. This proposed mechanism could resolve, in particular, those cases where a scanning mechanism from the BS onwards [23] could not explain splice site selection. Furthermore, secondary structures have been shown to explain some cases of alternative splicing in yeast, in which changes in temperature affect the stability of the RNA structure and thus, produces altered splicing patterns [17, 24].

In this chapter, we provide the resources and steps to obtain information on the secondary structure of the RNA in relation to splicing, which may serve as starting point for an integrative analysis with multiple other features, for instance, using Machine Learning methodologies [24]. In particular, we describe how to calculate optimal and suboptimal secondary structures, how to calculate the effective distance and the accessibility, and how to predict conserved secondary structures affecting splicing. As a practical example, we use the case of the RNA secondary structure in introns that has been shown to be relevant for 3′ss selection in yeast and that could be a general splicing regulatory mechanism [16, 17, 24].

2 Materials

In this chapter, we describe the use of several online tools and databases to retrieve and analyze data. Furthermore, we illustrate the use of some available programs and simple Perl programs on a unix terminal to perform data analysis such as prediction of RNA structures, calculation of effective distance, and prediction of accessibility.

Therefore, a computer with a Unix terminal and Perl programming language installed is required. Other websites and resources used along this chapter are listed below:

2.1 Online Databases and Tools

Saccharomyces genome database: http://www.yeastgenome.org/

Ensembl database: www.ensembl.org/

UCSC genome browser: http://genome.ucsc.edu/

Galaxy: http://galaxyproject.org/

Alignment format converter: http://biotechvana.uv.es/bioinformatics/

2.2 Other Software

Vienna RNA package: http://www.tbi.univie.ac.at/RNA/

Perl: http://www.perl.org/

3 Methods

3.1 Retrieving Sequence Datasets

As splicing often occurs co-transcriptionally [2], we expect that the RNA structures involved in splicing regulation are going to be short and dynamic, i.e., they will not be very stable and may change as the amount of pre-mRNA sequence transcribed increases. Furthermore, we have to consider the scenario in which RNA structures compete with RNA binding proteins (RBPs) or small nuclear ribonucleoproteins (snRNPs). Therefore, to predict secondary structure that may affect splicing we will use short sequences around splicing signals (or any other elements of interest such SR protein binding sites). Accordingly, we will need to have some prior knowledge about RBPs or snRNPs that may be involved in the process to limit the amount of sequence to be used. For instance, in the example proposed in this chapter, we will use pre-mRNA sequences spanning from the BS to the region downstream of the 3′ss.

The sequence of introns and exons from *S. cerevisiae* can be obtained from several online resources such as Saccharomyces Genome Database (http://www.yeastgenome.org/) [25], Ensembl (www.ensembl.org/) [26], UCSC (http://genome.ucsc.edu/) [27] or Galaxy (http://galaxyproject.org/) [28]. These resources provide tools to facilitate sequence retrieval for genes and genomic regions; hence, we will not go over this process. As an example we will use the gene SCN1 from yeast. In Fig. 1 you can see the sequence of SCN1 pre-mRNA, with the exons in lower case and the intron in upper case. The sequence of the BS and the 3′ss are highlighted in boldface. We will use this sequence to illustrate the analyses described below.

```
>SNC1(YAL030W)
augucgucaucuacucccuuugacccuuaugcucuauccgagcacgaugaagaacgaccc
cagaauguacagucuaagucaaggacugcggaacuacaagcuGUAAGUACAGAAAGCCAC
AGAGUACCAUCUAGGAAAUUAACAUUAUACUAACUUUCUACAUCGUUGAUACUUAUGCGU
AUACAUUCAUAUACGUUCUUCGUGUUUAUUUUUAGgaaauugaugauaccgugggaauaa
ugagagauaacauaaauaaaguagcagaaagaggugaaagauuaacguccauugaagaua
aagccgauaaccuagcggucucagcccaaggcuuuaagaggggugccaauaggucagaa
aagccaugugguacaaggaucuaaaaaugaagaugugucuggcuuuaguaaucaucauau
ugcuuguuguaaucaucguccccauugcuguucacuuuagucgauag
```

Fig. 1 Sequence of the SNC1 gene in Fasta format. Fasta format consist of a header line starting with ">" and additional lines with the sequence data, generally split in blocks of 60 residues. In the figure, the exon sequence is shown in *grey lower case letters*, whereas the intron sequence is shown in *black upper case letters*. The branch site (BS) sequence (UACUAACUU) and the 3′ss (UAG) are highlighted in *bold* with the BS A colored in *red*

```
>scn1_bs_3ss
AUCGUUGAUACUUAUGCGUAUACAUUCAUAUACGUUCUUCGUGUUUAUUUU
```

Fig. 2 Intronic sequence between the BS and the 3′ss, discarding the 3′ss sequence and 8 nt downstream of the BS A

3.2 Secondary Structure Prediction

RNA structure prediction generally involves the search for configurations of maximum base pairing or of minimum free energy (MFE). As this search entails the exploration of an enormous RNA configuration space, different computation methods propose different strategies to arrive at a result. Besides, these methods must also rely on the availability of correct free energy estimates for the base pairings. There are many methods for RNA structure prediction, e.g., *mfold* [29], *RNAsubopt* [30], *RNAfold* [31]. There are also methods that calculate the secondary structure using information from multiple sequences, either from an alignment or performing the alignment simultaneously to the structure prediction, like *RNAalifold* [32], *evofold* [33], *RNAz 2.0* [34], or *Locarna* [35].

1. We will use *RNAfold* (http://rna.tbi.univie.ac.at/) [31] to make RNA secondary structure predictions using the command line. To make a simple prediction, first we need to get a sequence in Fasta format. From the SCN1 gene, we extract the sequence between the BS and the 3′ss, discarding the first 8 nt downstream from the BS A and the 3′ss sequence (*see* **Note 1**). We save this sequence in Fasta format as shown in Fig. 2. The RNA secondary structure for this sequence can be predicted using the program *RNAfold* (*see* **Note 2**):

   ```
   RNAfold < seq.fa > rna_struct.txt
   ```

2. As can be seen in Fig. 3a, the file `rna_struct.txt` contains the sequence and the MFE structure prediction in bracket notation, labeled as (1) and (2), respectively. Furthermore, we

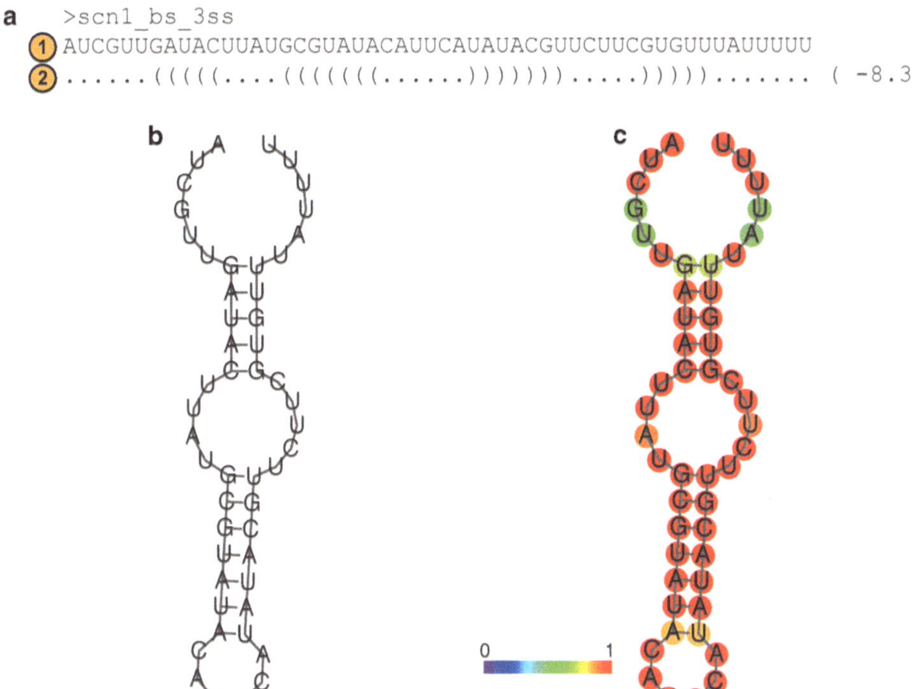

Fig. 3 (**a**) MFE structure prediction output by RNAfold. In the output we get (*1*) the nucleotide sequence given as input and (*2*) the MFE secondary structure prediction in bracket notation. In this format, "(" and ")" denote positions that are forming a base pair, whereas "." correspond to unpaired nucleotides. The free energy of the structure, expressed in kcal/mol is provided between brackets. (**b**) Graphical representation of the predicted MFE structure. (**c**) Graphical representation of the MFE structure showing the pair probabilities of the nucleotides in the MFE structure. For nucleotides outside the secondary structure (i.e., in bulges, loops, or unstructured), the color represents the probability of not being in a base pair for the MFE structure in the same scale. The color scale goes from *purple*, which represent the lowest pair probability to *red*, which represents the highest probability

also obtain the free energy of the predicted structure, in this case, −8.30 kcal/mol. This command produces an additional file, `scn1_bs_3ss_ss.ps`, which contains the drawing of the MFE structure predicted (Fig. 3b). In this structure, base pairings between the nucleotides are shown as lines connecting nucleotides in different parts of the sequence. Nucleotides that are not in any base pair are shown as loops and bulges.

3. We can obtain further information about the stability of MFE structure by calculating the pair probabilities of the base pairs in the MFE structure. Nucleotide pairs with a high pair probability represent very stable base pairs. In contrast, low pair probabilities suggest that a particular base pair in the structure is not very likely to occur and thus, in the majority of the cases, it will not happen. We can calculate the RNA secondary structure and the base pair probabilities of the structure using the option `-p`:

```
RNAfold -p < seq.fa > rna_struct.txt
```

4. In this case, we obtain another additional file, `scn1_bs_3ss_dp.ps`, which contains the pair probabilities of all possible base pairs. We can use this last file to redraw the predicted secondary structure (Fig. 3b), adding the probability of the base pairs in the structure, using the program `relplot.pl` from the Vienna RNA package:

```
relplot.pl -p scn1_bs_3ss_ss.ps scn1_bs_3ss_
dp.ps> scn1_bs_3ss_rss.ps
```

The structure displaying the pair probabilities, `scn1_bs_3ss_rss.ps`, is shown in Fig. 3c. In this case, the nucleotides in the structure are colored according to their probability in the MFE structure.

3.2.1 Suboptimal Structure Prediction

To do a more accurate analysis of the possible secondary structures, we can calculate suboptimal structures that are similar to the MFE but not as probable. Assuming that the structures involved in splicing regulation are transient and unstable, e.g., by occurring along a short time span during transcription, it is plausible that the effect of the RNA secondary structure on splicing is the effect not from a single optimal structure but also from other suboptimal but nearly identical structures. To assess this possibility, one can predict suboptimal structures whose free energy are close to that of the optimal secondary structure using the program *RNAsubopt* [22, 30]. The relation between the stability of a structure and its probability is given by

$$P(S_i) = \frac{1}{Z} e^{\frac{\Delta G(S_i)}{RT}}$$

where $\Delta G(S_i)$ is the free energy of the structure S_i for sequence S,

$$Z = \sum_{S_i} e^{-\frac{\Delta G(S_i)}{RT}}$$

is the partition function of all possible secondary structures S_k of sequence S, R is the physical gas constant, and T is the temperature. This equation determines that the lower the free energy, the higher its probability. Accordingly, structures with energies close to the MFE can still be highly probable. The method *RNAsubopt* calculates a sample of the possible secondary structure space within a given variation of the MFE. Using these suboptimal structures, one can for instance calculate the distribution of effective distances for each of the introns analyzed. This allows determining the effect of the variability of the secondary structure.

1. In our example, we will generate a random sample of 1,000 suboptimal structures drawn with probabilities equal to their

Boltzmann weights (-p 1000) and whose energy does not vary more than 5 % from the MFE structure (-ep 5):

```
RNAsubopt -ep 5 -p 1000 < seq.fa > subop_rna_
structs.txt
```

In this case, the resulting file, subop_rna_structs.txt, contains only the secondary structures predicted in bracket notation.

3.3 Linking RNA Structures to Splicing Regulation

The two main mechanisms by which a secondary structure can hinder splicing is by (1) affecting the distance between splicing signals (i.e., the BS and the 3′ss), which will alter splicing efficiency or (2) blocking the recognition of splicing signals, i.e., changing splicing signal accessibility [17]. These two effects can be measured by calculating the effective distance and the nucleotide accessibility.

3.3.1 Effective Distance

The effective distance is defined as the linear distance in nucleotides (nt) after removing the secondary structure. More specifically, removing all the bases that are part of a structured region and keeping the two bases corresponding to the beginning and the end of the structured region. The simplest way of calculating the effective distance between two signals in the RNA (i.e., the BS and the 3′ss) is to predict the MFE structure and calculate the distance between them after discarding the positions included within the secondary structure. To calculate the effective distance we can use a small program in Perl, effective_distance.pl, which will parse the information contained in the RNA structure predicted in bracket notation and will return the effective distance calculated in nucleotides:

```
perl effective_distance.pl < rna_struct.txt >
effective_dist.txt
```

The program effective_distance.pl could look like this:

```
#!/usr/bin/perl -w
use strict;
my $effective_length = 0;
my $open = 0;
my $close = 0;
while (<STDIN>) {
    next if ($_=~m/>/|| $_=~m/^[AUGC]/);
    chomp;
    my $effective_length = 0;
    my @line = split;
    my @structure = split (//, $line[0]);
    foreach my $i (0..$#structure){
       if ($structure[$i] eq "." && $open ==
$close){
          $effective_length++;
       }
```

```
                elsif ($structure[$i] eq "("){
                    $open++;
                }
                elsif ($structure[$i] eq ")"){
                    $close++;
                    if ($open == $close){
                        $effective_length += 2;
                    }
                }
            }
            $effective_length = $effective_length+8+3;
            print $effective_length, "\n";
        }
        close (STDIN);
```

The output given by this program is a number, which represents the effective distance in nucleotides between the BS and the 3′ss. This number, also considers the 8 nt discarded downstream of the BS A at the beginning of the sequence and the 3 nt of the 3′ss, which should be considered to calculate the effective distance [17].

As before, besides the MFE structure, we can also incorporate suboptimal structures to the calculation of the effective distance. In this case, we can run the program `effective_distance.pl` using the suboptimal structures predicted before with *RNAsubopt*:

```
perl effective_distance.pl < subop_rna_structs.
txt > effective_dist_subopt.txt
```

The output file contains the effective distance of each of the 1,000 suboptimal structures predicted before. Given that the structures predicted are a weighted sample of all possible structures, we can use this data to calculate the mean effective distance of the 3′ss analyzed. In Fig. 4 we see the distribution of effective distances calculated for the suboptimal structures. For comparison, we have colored in red the bar for the effective distance obtained from the MFE structure. We observe that the distribution of effective distances is bimodal. Furthermore, the most frequent effective distance in the suboptimal structures predicted (22 nt) differs from that of the optimal structure (28 nt; red bar). Therefore, using only the MFE structure may result in a wrong estimate of the effective distance.

3.3.2 Accessibility of Splicing Signals

When secondary structures are placed overlapping *cis* elements in the sequence, they can hinder the recognition of these elements by other proteins or RNAs. Therefore, when measuring the ability to recognize a signal in an RNA molecule such as a splice site, we will have to measure its accessibility, i.e., whether the signal will be available to other proteins or will be hidden by an RNA structure.

Even though the MFE structure may be the most frequent, we have already shown that suboptimal structures are important to

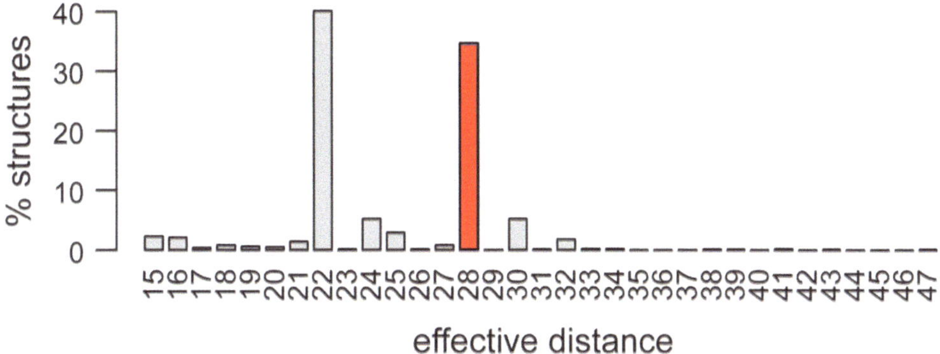

Fig. 4 Barplot showing the distribution of effective distances (in nucleotides) for the 1,000 suboptimal structures predicted. The *x*-axis shows the effective distances measured in nucleotides. The *y*-axis shows the % of structures with a given effective distance. The value corresponding to the effective distance of the MFE is indicated with a *red bar*

understand the effect of RNA structures in splicing regulation. The pair probability, defined above, can also be calculated considering the contribution from all possible structures. In this way, we will be able to determine a local effect of all structures on the recognition of a splicing signal. Moreover, the pair probability over all possible structures also allows describing the probability of not being in a base pair, i.e., the accessibility. This accessibility is what will actually give us information about the likelihood that a signal in the RNA is accessible to a protein factor to bind, or on the contrary, is likely to be "hidden" inside a secondary structure.

For our present example, we will include the sequence upstream of the 3′ss till the BS and also some nucleotides downstream of the 3′ss, as they can also be included in secondary structures affecting its recognition. In other cases, such as in the case of the 5′ss, we will be interested in selecting the sequence in a different way, as only some nucleotides upstream and downstream of the 5′ss may affect its recognition. The pair probability of a given position can be calculated using the program *RNAfold* [31]. From the result given by *RNAfold*, we will calculate the accessibility of the nucleotides from the 3′ss.

1. From the original Fasta sequence, extract the sequence between the BS and the 3′ss, discarding the first 8 nt downstream from the BS A. In this case, we will include the 3′ss and 5 nt downstream of the 3′ss, as we will want to quantify the probability of the 3′ss being included in different secondary structures. We will save this secondary structure in Fasta format, `seq_ext.fa`, as described above (Fig. 5).
2. For each of the sequences, predict the RNA secondary structure with *RNAfold* as described in Subheading 3.2. In this case, we will use the option `-noPS`, which avoids producing the postscript figure of the MFE structure:

```
RNAfold -p -noPS < seq_ext.fa > rna_struct_ext.txt
```

```
>scn1_ext
AUCGUUGAUACUUAUGCGUAUACAUUCAUAUACGUUCUUCGUGUUUAUUUUUAGGAAAU
```

Fig. 5 Intronic sequence between the BS and 5 nt downstream of the 3′ss, discarding the 8 nt downstream of the BS A. The 3′ss sequence is shown in *bold*

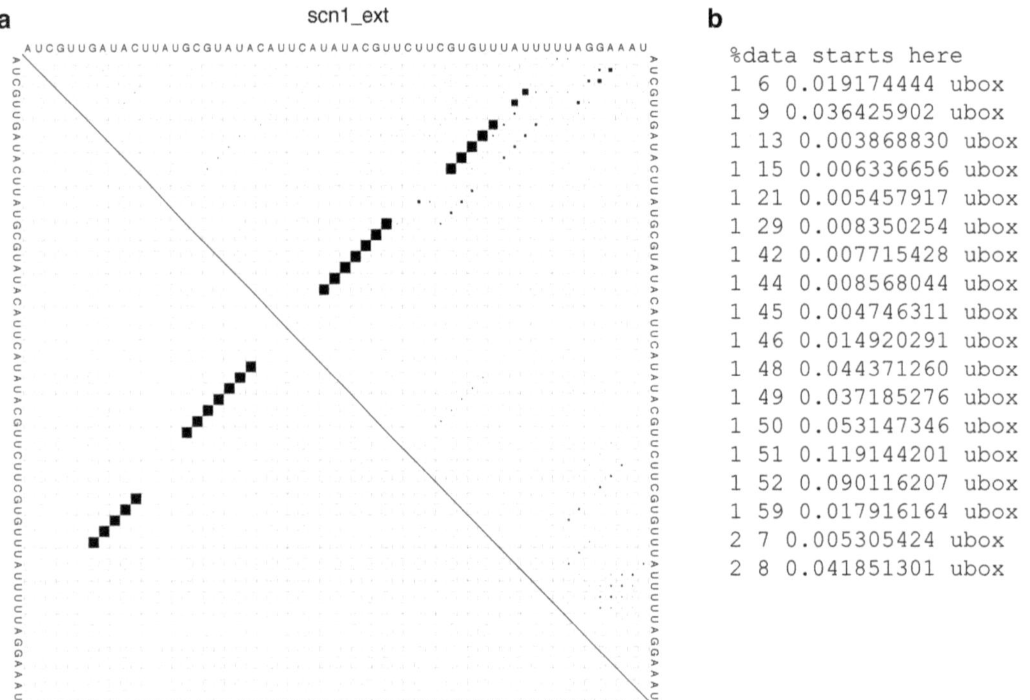

Fig. 6 (**a**) Dot plot showing the base pair probabilities. The input sequence is shown at both sides of the matrix. For each pair of nucleotides, *i* and *j*, we have a *black square* whose size is proportional to the probability of *i* and *j* being in a base pair. The elements above the diagonal (ubox) represent the base pair probabilities calculated from all structures for each pair *i* and *j*. The elements below the diagonal represent the base pair probabilities in the MFE structure for each pair *i* and *j*. Only probabilities larger than 10^{-6} are shown. (**b**) The dot plot postscript file also includes the probability of each pair of nucleotides *i j* to be in a base pair in the form: *i*, *j*, *square root of the probability*, *ubox*

As before, the option `-p` will produce a file called `scn1_ext_dp.ps`, which is a dot plot that contains for each pair of nucleotides in the sequence the probability of them being in a base pair. Graphically, the file shows a matrix. Each position in the matrix is represented by a black square whose size is proportional to the probability that a given pair of nucleotides is in a base pair (Fig. 6a). The probability of a pair of nucleotides being in a base pair is also provided inside of the dot plot file in multiple lines (Fig. 6b), each line of the form

```
i j sqrt(p) ubox
```

where `i` and `j` are the nucleotides evaluated, `sqrt(p)` is the square root of the pair probability of the base pair between `i` and `j`, and `ubox` indicates that these are the elements above the diagonal, i.e., representing the pair probabilities from all possible structures. The label `lbox` is used for the matrix elements below the diagonal, which represent the pair probabilities of the optimal structure.

3. We will use another small program, `accessibility.pl`, to parse the information inside the dot plot file and calculate the average accessibility of the 3′ss:

```
perl accessibility.pl < scn1_ext_dp.ps >
accessibility.txt
```

the program `accessibility.pl` looks like this:

```
#!/usr/bin/perl -w
use strict;
my $seq="";
my @pair_probability;
my $seq_flag = 0;
while (<STDIN>) {
     chomp;
     if ($_=~m/^\/sequence\s+\{/){
       $seq_flag =1;
     }
     elsif ($seq_flag == 1){
       if ($_=~m/^\)\s+\}\s+def/){
        $seq_flag=0;
     @pair_probability = split (//,0 x length
($seq));
      }
      else{
           $seq .= $_;
           $seq =~s/\\//g;
      }
     }
     elsif  ($_=~m/(\d+)\s+(\d+)\s+([0-9.Ee-
]+)\s+ubox/){
   my ($i, $j, $probability) = ($1, $2, $3);
   $probability *=$probability;
   $pair_probability[$i] += $probability;
   $pair_probability[$j] += $probability;
     }
}
```

```
close (STDIN);
my @ss = splice (@pair_probability,-8,3);
my $average_pp = ($ss[0]+$ss[1]+$ss [2])/3;
my $average_accessibility = 1 - $average_pp;
print $average_accessibility, "\n";
```

This will return the average accessibility of the 3′ss of interest, which will be saved in the file `accessibility.txt`.

If we want to use the accessibility of a signal to understand if a 3′ss could be functional or not, what we can do is to compare the accessibility of a candidate 3′ss to that of the annotated 3′ss. If we find any candidate 3′ss that have an accessibility similar or higher than a nearby annotated 3′ss and it is in range, i.e., the effective distance between the BS and the 3′ss is not too big, this candidate could be a possible alternative 3′ss. Furthermore, we can also compute the accessibility using sequences of different length, which allows estimating the fact that splicing and transcription are coupled.

3.4 Conserved Secondary Structures

Another aspect in which we can be interested is in the identification of conserved secondary structures, which may be indicative of a mechanism conserved across different species. In human, it has been shown that conserved secondary structures overlapping a splice site are more frequent in alternative exons than in constitutive ones [5], suggesting that structure could actually be a mechanism of splicing regulation conserved across eukaryotes. In this case, we will do an RNA prediction based on a sequence alignment. This prediction can be done with programs such as *RNAalifold* [32] or *evofold* [33], to which we will have to input an alignment in Clustal format (*see* **Note 3**) to make the prediction.

1. First, we get the homologous sequences to the one used before to make the prediction. If we know the genomic coordinates of our sequence (in this case, ChrI:87447-87500) we can extract the homologous region from the genomic alignments in UCSC using Galaxy [28] (for more details on how to perform this, see the available screen casts and demos from Galaxy at http://wiki.g2.bx.psu.edu/Learn/Screencasts#Getting_Started).
2. Using Galaxy we can convert the original alignment format from MAF to Fasta using the *Convert Formats* tool. Additionally, the resulting file, `yeast_all.fa`, should be converted into Clustal format, `yeasts_all.aln`, which can be done with tools like the *Alignment format converter* (http://biotechvana.uv.es/bioinformatics) [36].
3. Using the aligned sequences, we predict the RNA secondary structure with *RNAalifold*.

```
RNAalifold < yeasts_all.aln > yeast_all.txt
```

As before, we can use the output file of the prediction, `yeast_all.txt`, to calculate the effective distance between the BS and the 3′ss using the program `effective_distance.pl`.

4. If we run *RNAalifold* with the option `-p` and include the 3′ss sequence plus 5 nt downstream (as done before), we will produce a file called `alidot.ps` that could be used to measure the accessibility of the 3′ss according to the conserved secondary structure.

3.5 Significance of Results

In general, the longer the sequence and the higher its GC content, the more likely it is to predict a secondary structure computationally. Accordingly, we must evaluate the significance of our analyses taking into consideration these and other possible biases. One of the most effective ways to assess significance is consider a control set, which would represent the null hypothesis. For the analysis of secondary structures, we can generally consider two types of control sets: randomized sequences and a negative control set. Randomized sequences are obtained from the original set by shuffling nucleotides. Within intron regions, shuffling single nucleotides could be enough, but shuffling while keeping dinucleotide frequencies can help controlling for more subtle structural biases. For exonic regions, the nucleotide shuffling should be done such that the encoded amino acid sequence, codon usage, and base composition of the RNA are preserved [37]. By construction, this control set maintains the sequence content and length distribution. On the other hand, when performing an analysis using a multiple alignment, we can consider a different form of shuffling: vertical shuffling. In this method, each column of the alignment is shuffled vertically. In this way, the sequence conservation is preserved, but possible structural dependencies within each sequence are broken. This can also be extended to di- or trinucleotides (*see* ref. 38 for an example).

A control set can also be built by extracting random genomic regions that resemble the regions being analyzed, but that are known to be nonfunctional to some extent. For instance, a control set for exons could consist of intronic regions flanked by motifs similar to splice sites, but have no evidence of being expressed (*see* ref. 39 for an example). Likewise, a control set for intronic regions could be extracted from random intergenic regions of the same sizes, known to be devoid of any expression evidence and selected such that they have a similar sequence content bias (*see* ref. 17 for an example). Significance is then assessed by performing the structure prediction analysis on the control set, exactly in the same way as we did before on our input data set. Properties from both sets, e.g., effective distance, accessibility, frequency for structures per length, can then be compared to obtain a measure of significance, for instance, by using false discovery rate or any other statistical test [40].

4 Notes

1. We discard these nucleotides downstream of the BS as it has been shown that they are not generally included in a secondary structure [17].
2. The chapter describes how to use the programs *RNAfold*, *RNAsubopt*, and *RNAalifold* from the command line. However, these and other programs from the Vienna package can also be executed online (http://rna.tbi.univie.ac.at/).
3. A file in CLUSTAL format is a plain text file with a header starting with the Word "CLUSTAL" followed by information of the version. Multiple alignment programs generate alignments in this format, possibly adding extra information. The alignment is generally represented in blocks of 60 residues, where each block starts with a sequence identifier. Additionally, the end of each line may include the number of residues in that line of the alignment. Below each block, the symbol "*" indicates whether the position in the alignment is identical for all sequences (see http://www.clustal.org/ for more details). In the case of amino acid alignments, the symbols ":" and "." indicate conserved or semi-conserved substitutions. Below, we show the example of the multiple sequence alignment used for the prediction of the conserved secondary structure using *RNAalifold* (Fig. 7).

```
CLUSTAL X (1.81) multiple sequence alignment

sacCer3     ATCGTTGATACTTATGCGTATAC-ATTCATATACG-TTCTTCGTGTTTAT-TTTTAG
sacPar      GTCATTGATATATATACGTATAC-ATACGTGTACG-TATGCCGTGTTTAT-TTTTAG
sacMik      GTCGTTAATGTTTTTACGTATAT-GTATGTATACG-TATATCACGTTATT-TTACAG
sacKud      GACATTGATGTACATACGCATACGGTGTATGTACATTTTTTCATGTTTTTCTTCCAG
sacBay      GACATTACTGTATATACGTATAC-GTTTATGTATG-T------CGTTATCTTCATAG
sacKlu      ---------------------------------------------TTTTT-TAACAG
                                                         **    *   **
```

Fig. 7 Nucleotide sequence alignment in Clustal format. The alignment has been extracted from the 7-way genome alignment from UCSC for yeast species, for the region between the BS and the 3′ss (excluding the BS signal). The species included in the alignment are *S. cerevisiae* (sacCer3), *S. paradoxus* (sacPar), *S. mikatae* (sacMik), *S. kudriavzevii* (sacKud), *S. bayanus* (sacBay), and *S. Kluyveri* (sacKlu)

References

1. Jurica MS, Moore MJ (2003) Pre-mRNA splicing: awash in a sea of proteins. Mol Cell 12:5–14
2. Pan T, Sosnick T (2006) RNA folding during transcription. Annu Rev Biophys Biomol Struct 35:161–175
3. Patterson DJ, Yasuhara K, Ruzzo WL (2002) Pre-mRNA secondary structure prediction aids splice site prediction. Pac Symp Biocomput 2002:223–234
4. Marashi SA, Eslahchi C, Pezeshk H et al (2006) Impact of RNA structure on the prediction of donor and acceptor splice sites. BMC Bioinformatics 7:297
5. Shepard PJ, Hertel KJ (2008) Conserved RNA secondary structures promote alternative splicing. RNA 14:1463–1469
6. Buratti E, Baralle FE (2004) Influence of RNA secondary structure on the pre-mRNA splicing process. Mol Cell Biol 24:10505–10514
7. Warf MB, Berglund JA (2010) Role of RNA structure in regulating pre-mRNA splicing. Trends Biochem Sci 35:169–178
8. Graveley BR (2005) Mutually exclusive splicing of the insect Dscam pre-mRNA directed by competing intronic RNA secondary structures. Cell 123:65–73
9. Raker VA, Mironov AA, Gelfand MS et al (2009) Modulation of alternative splicing by long-range RNA structures in Drosophila. Nucleic Acids Res 37(14):4533–4544
10. Pervouchine DD, Khrameeva EE, Pichugina MY et al (2012) Evidence for widespread association of mammalian splicing and conserved long-range RNA structures. RNA 18(1):1–15
11. Bevilacqua PC, Blose JM (2008) Structures, kinetics, thermodynamics, and biological functions of RNA hairpins. Annu Rev Phys Chem 59:79–103
12. Mahen EM, Watson PY, Cottrell JW et al (2010) mRNA secondary structures fold sequentially but exchange rapidly in vivo. PLoS Biol 8(2):e1000307
13. Chen SJ (2008) RNA folding: conformational statistics, folding kinetics, and ion electrostatics. Annu Rev Biophys 37:197–214
14. Plass M, Agirre E, Reyes D et al (2008) Co-evolution of the branch site and SR proteins in eukaryotes. Trends Genet 24:590–594
15. Schwartz SH, Silva J, Burstein D et al (2008) Large-scale comparative analysis of splicing signals and their corresponding splicing factors in eukaryotes. Genome Res 18:88–103
16. Gahura O, Hammann C, Valentova A et al (2011) Secondary structure is required for 3′ splice site recognition in yeast. Nucleic Acids Res 39(22):9759–9767
17. Meyer M, Plass M, Pérez-Valle J et al (2011) Deciphering 3′ss selection in the yeast genome reveals an RNA thermosensor that mediates alternative splicing. Mol Cell 43(6):1033–1039
18. Deshler JO, Rossi JJ (1991) Unexpected point mutations activate cryptic 3′ splice sites by perturbing a natural secondary structure within a yeast intron. Genes Dev 5:1252–1263
19. Goguel V, Rosbash M (1993) Splice site choice and splicing efficiency are positively influenced by pre-mRNA intramolecular base pairing in yeast. Cell 72:893–901
20. Goguel V, Wang Y, Rosbash M (1993) Short artificial hairpins sequester splicing signals and inhibit yeast pre-mRNA splicing. Mol Cell Biol 13:6841–6848
21. Charpentier B, Rosbash M (1996) Intramolecular structure in yeast introns aids the early steps of in vitro spliceosome assembly. RNA 2:509–522
22. Rogic S, Montpetit B, Hoos HH et al (2008) Correlation between the secondary structure of pre-mRNA introns and the efficiency of splicing in Saccharomyces cerevisiae. BMC Genomics 9:355
23. Smith CW, Chu TT, Nadal-Ginard B (1993) Scanning and competition between AGs are involved in 3′ splice site selection in mammalian introns. Mol Cell Biol 13:4939–4952
24. Plass M, Codony-Servat C, Ferreira PG et al (2012) RNA secondary structure mediates alternative 3′ss selection in Saccharomyces cerevisiae. RNA 18(6):1103–1115
25. Cherry JM, Hong EL, Amundsen C et al (2012) Saccharomyces Genome Database: the genomics resource of budding yeast. Nucleic Acids Res 40((Database issue)):D700–D705
26. Flicek P, Amode MR, Barrell D et al (2012) Ensembl 2012. Nucleic Acids Res 40((Database issue)):D84–D90
27. Kuhn RM, Haussler D, Kent WJ (2012) The UCSC genome browser and associated tools. Brief Bioinform 14(2):144–161
28. Hillman-Jackson J, Clements D, Blankenberg D et al (2012) Using Galaxy to perform large-scale interactive data analyses. Curr Protoc Bioinformatics. Chapter 10:Unit10.5
29. Zuker M, Mathews DH, Turner DH (1999) Algorithms and thermodynamics for RNA secondary structure prediction: a practical guide. In: Barciszewski J, Clark BFC (eds) RNA biochemistry and biotechnology, NATO ASI series. Kluwer Academic Publishers, Dordrecht, NL
30. Wuchty S, Fontana W, Hofacker IL et al (1999) Complete suboptimal folding of RNA and the

stability of secondary structures. Biopolymers 49(2):145–165
31. Hofacker IL (2009) RNA secondary structure analysis using the Vienna RNA package. Curr Protoc Bioinformatics. Chapter 12:Unit12.2
32. Hofacker IL (2003) Vienna RNA secondary structure server. Nucleic Acids Res 31(13):3429–3431
33. Pedersen JS, Bejerano G, Siepel A et al (2006) Identification and classification of conserved RNA secondary structures in the human genome. PLoS Comput Biol 2(4):e33
34. Gruber AR, Findeiß S, Washietl S et al (2010) Rnaz 2.0: improved noncoding RNA detection. Pac Symp Biocomput 2010:69–79
35. Will S, Reiche K, Hofacker IL et al (2007) Inferring noncoding RNA families and classes by means of genome-scale structure-based clustering. PLoS Comput Biol 3(4):e65
36. Llorens C, Futami R, Vicente-Ripolles M et al (2008) The alignment format converter server 1.0. In: Biotechvana Bioinformatics 2008. Biotechvana, Valencia. SCR: AFC
37. Katz L, Burge CB (2003) Widespread selection for local RNA secondary structure in coding regions of bacterial genes. Genome Res 13:2042–2051
38. Fernández N, Fernandez-Miragall O, Ramajo J et al (2011) Structural basis for the biological relevance of the invariant apical stem in IRES-mediated translation. Nucleic Acids Res 39(19):8572–8585
39. Corvelo A, Eyras E (2008) Exon creation and establishment in human genes. Genome Biol 9(9):R141
40. Noble WS (2009) How does multiple testing correction work? Nat Biotechnol 27(12):1135–1137

Chapter 26

Methods to Study Splicing from High-Throughput RNA Sequencing Data

Gael P. Alamancos, Eneritz Agirre, and Eduardo Eyras

Abstract

The development of novel high-throughput sequencing (HTS) methods for RNA (RNA-Seq) has provided a very powerful mean to study splicing under multiple conditions at unprecedented depth. However, the complexity of the information to be analyzed has turned this into a challenging task. In the last few years, a plethora of tools have been developed, allowing researchers to process RNA-Seq data to study the expression of isoforms and splicing events, and their relative changes under different conditions. We provide an overview of the methods available to study splicing from short RNA-Seq data, which could serve as an entry point for users who need to decide on a suitable tool for a specific analysis. We also attempt to propose a classification of the tools according to the operations they do, to facilitate the comparison and choice of methods.

Key words RNA-Seq, Splicing, Alternative splicing, Isoform, Quantification, Reconstruction

1 Introduction

The development of novel high-throughput sequencing (HTS) methods for RNA (RNA-Seq) has facilitated the discovery of many novel transcribed regions and splicing isoforms [1] and has provided evidence that a large fraction of the transcribed RNA in human cells undergo alternative splicing [2, 3]. RNA-Seq thus represents a very powerful tool to study alternative splicing under multiple conditions at unprecedented depth. However, the large datasets produced and the complexity of the information to be analyzed has turned this into a challenging task. In the last few years, a plethora of tools have been developed (Fig. 1), allowing researchers to process RNA-Seq data to study the expression of isoforms and splicing events, and their relative changes under different conditions. In this chapter, we provide an overview of the methods available to study alternative splicing from short RNA-Seq data.

Klemens J. Hertel (ed.), *Spliceosomal Pre-mRNA Splicing: Methods and Protocols*, Methods in Molecular Biology, vol. 1126, DOI 10.1007/978-1-62703-980-2_26, © Springer Science+Business Media, LLC 2014

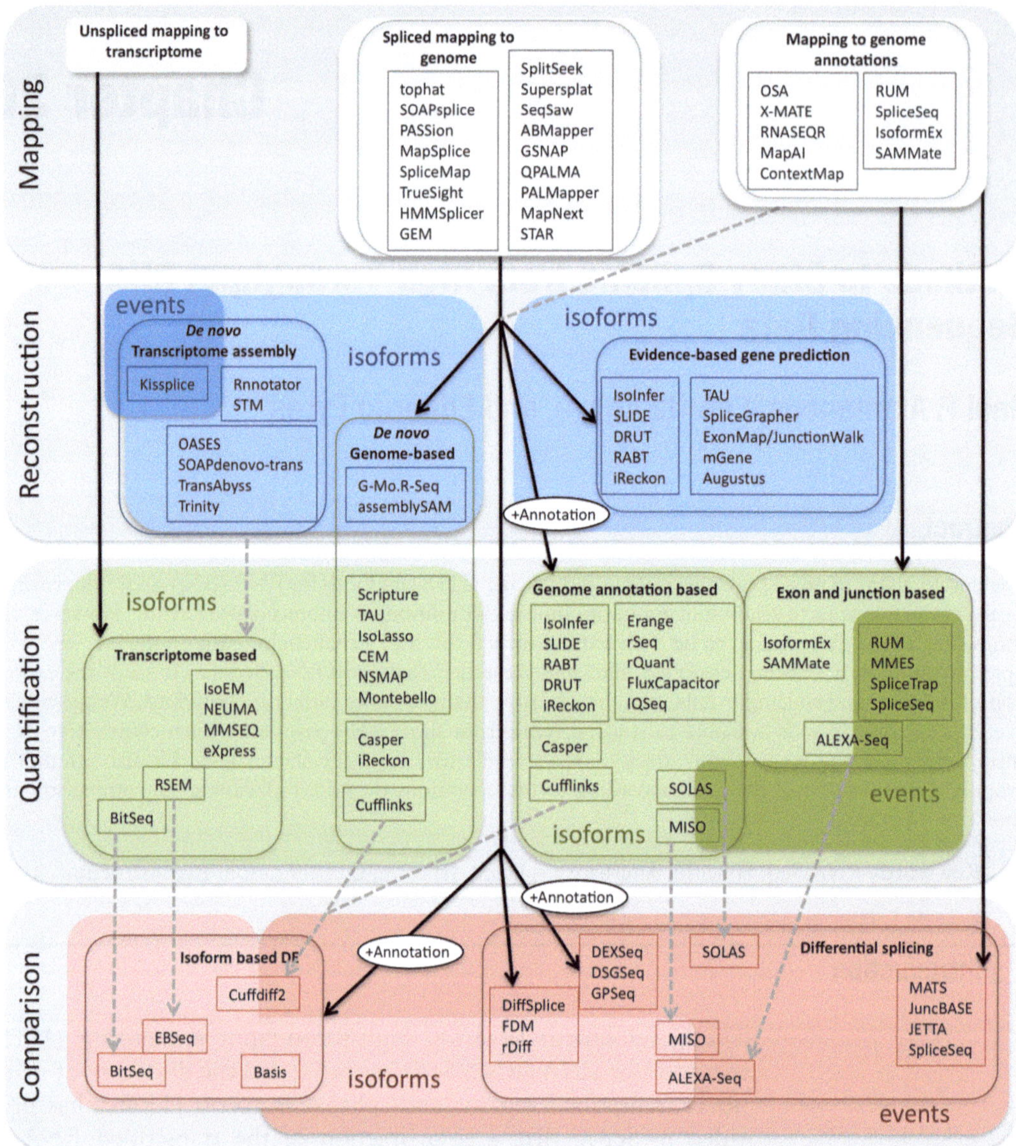

Fig. 1 Graphical representation of methods to study splicing from RNA-Seq. Methods are divided according to whether they perform mapping, reconstruction of events/isoforms, quantification of events/isoforms and whether they can perform a comparison between two or more conditions of event/isoform relative abundances, or of isoform expression. We only list the mapping methods that are spliced-mappers or the ones that use some heuristics to map to known exons and junctions. Mapping methods that also perform quantification are repeated in both levels. Methods for reconstruction (*blue*), quantification (*green*), and comparison (*red*) are divided according to whether they work with isoforms (*lighter color*) or with events (*darker color*); when they work at both levels, events and isoforms, they are overlapped by the two color tones, darker and lighter, respectively. Methods are also grouped by rounded rectangles according to the tables in Subheading 2. Some methods perform reconstruction and quantification and are grouped with those that only perform reconstruction. Methods that require an annotation are indicated. Quantification methods that work with or without annotation are in different groups. *Solid arrows* connect Mapping methods to the tools in the other three levels; since, in principle, any mapping method producing BAM as output could be fed to methods reading BAM as input. Some methods perform mapping and quantification or mapping and differential splicing and are connected with a *solid arrow* too. We indicate with *dashed gray arrows* those cases when a comparison method can use the output from a quantification method

We will group the methods according to the different questions they address:

1. Assignment of the sequencing reads to their likely gene of origin. This is addressed by methods that map reads to the genome and/or to the available gene annotations (Subheading 2.1).
2. Quantification of events and isoforms. Either using an annotation (Subheadings 2.2 and 2.3) or after reconstructing transcripts (Subheading 2.4), many methods estimate the expression level or the relative usage of isoforms and/or events.
3. Recovering the sequence of splicing events and isoforms. This is addressed by transcript reconstruction and de novo assembly methods (Subheadings 2.4, 2.5, and 2.6).
4. Providing an isoform or event view of differential splicing or expression. These include methods that compare relative event/isoform abundance or isoform expression across two or more conditions (Subheadings 2.7 and 2.8).
5. Visualizing splicing regulation. Various tools facilitate the visualization of the RNA-Seq data in the context of alternative splicing (Subheading 2.9).

In this review, we use transcript or isoform to refer to a distinct RNA molecule transcribed from a gene locus. We use gene to refer to the set of isoforms transcribed from the same genomic region and the same strand, sharing some exonic sequence; and a gene locus refers to this genomic region. A splicing event refers to the exonic region of a gene that shows variability across two or more of its isoforms. Splicing events generally include exon skipping (or cassette exon), alternative 5′ and 3′ splice-sites, mutually exclusive exons, retained introns, alternative first exons and alternative last exons (*see* for example [4]), although other events may occur as a combination of two or more of these ones. In this review, we do not enter into the details of the specific mathematical models behind each method; for a comparative analysis of the mathematical models behind many of these methods *see* ref. 5. Our aim is rather to provide an overview that could serve as an entry point for users who need to decide on a suitable tool for a specific analysis. We also attempt to propose a classification of the tools according to the operations they do, to facilitate the comparison and choice of methods.

2 Materials

This section includes the list of methods described in subsequent sections.

2.1 Spliced-Mappers

In Table 1, we provide a list of mapping tools that are able to locate exon–intron boundaries. Some of the methods use annotation information for mapping (OSA, X-MATE, SAMMate, IsoformEx,

Table 1
A list of mapping tools that are able to locate exon–intron boundaries

Method	Type	Uses annotation	Paired-end reads	Splice site model	Reference	Web site
TopHat	Exon-first	Optional	Yes	Exact match to GT/C-AG	[6]	http://tophat.cbcb.umd.edu/
SOAPsplice	Exon-first	No	Yes	Exact match to GT-AG, GC-AG, AT-AC	[7]	http://soap.genomics.org.cn/soapsplice.html
PASSion	Exon-first	No	Only paired-end	Exact match to GT-AG, GC-AG, AT-AC	[8]	https://trac.nbic.nl/passion
MapSplice	Exon-first. Seed-and-extend for spliced reads	No	Yes	Unbiased	[9]	http://www.netlab.uky.edu/p/bioinfo/MapSplice
SpliceMap	Exon-first. Seed-and-extend for spliced reads	No	Yes	Exact match to GT-AG, GC-AG, AT-AC	[10]	http://www.stanford.edu/group/wonglab/SpliceMap/
HMMSplicer	Exon-first. Seed-and-extend for spliced reads	No	Yes	Hidden Markov Model	[11]	http://derisilab.ucsf.edu/index.php?software=105
TrueSight	Exon-first. Seed-and-extend for spliced reads	No	Yes	Exact match to GT-AG, GC-AG, AT-AC	[12]	http://bioen-compbio.bioen.illinois.edu/TrueSight/
GEM	Exon-first. Seed-and-extend for splice reads	Optional	Yes	User defined regular expression and known junctions (optional)	[13]	http://algorithms.cnag.cat/wiki/The_GEM_library
SplitSeek	Seed-and-extend	No	Yes	Unbiased	[14]	http://solidsoftwaretools.com/gf/project/splitseek
Supersplat	Seed-and-extend	No	No	Unbiased	[15]	https://github.com/mocklerlab/supersplat

SeqSaw	Seed-and-extend	No	Yes	Unbiased	[16]	http://bioinfo.au.tsinghua.edu.cn/software/seqsaw
ABMapper	Seed-and-extend	No	Yes	Exact match to GT-AG, GC-AG, AT-AC	[17]	http://abmapper.sourceforge.net/
MapNext	Seed-and-extend	Optional	No	Known-junctions and GT-AG for novel ones	[18]	http://evolution.sysu.edu.cn/english/software/mapnext.htm
STAR	Seed-and-extend	Optional	Yes	Exact match to GT-AG, GC-AG, AT-AC and unbiased	[19]	http://gingeraslab.cshl.edu/STAR/
GSNAP	Seed-and-extend	No	Yes	Exact match to GT-AG, GC-AG, AT-AC	[20]	http://research-pub.gene.com/gmap/
QPALMA	Seed-and-extend	No	No	SVM model for splice-sites	[21]	http://www.raetschlab.org/suppl/qpalma
PALMapper	GenomeMapper + QPalma	No	Yes	Qpalma model	[22]	http://galaxy.raetschlab.org/
CRAC	Seed-and-extend	No	No	Unbiased	[23]	http://crac.gforge.inria.fr/
OLEgo	Multi-seed	No	Yes	Combined model of splice-site sequence and intron length	[24]	http://zhanglab.c2b2.columbia.edu/index.php/OLego
Subread	Multi-seed	No	Yes	Exact match to GT-AG	[25]	http://bioconductor.org/packages/release/bioc/html/Rsubread.html
OSA	Seed-and-extend	Yes	Yes	Known and splice-sites and exact match to GT-AG, GC-AG, AT-AC	[26]	http://omicsoft.com/osa/

(continued)

Table 1
(continued)

Method	Type	Uses annotation	Paired-end reads	Splice site model	Reference	Web site
X-MATE	Recursive mapping to genome and junctions	Yes	No	Known splice-sites	[27]	http://grimmond.imb.uq.edu.au/X-MATE/
RNASEQR	Bowtie and BLAT on transcripts and genome	Yes	Yes	Known splice-sites and BLAT model	[28]	https://github.com/rnaseqr/RNASEQR
MapAI	Bowtie alignments to transcripts	Yes	No	Known splice-sites	[29]	http://www.bioinf.boku.ac.at/pub/MapAl/
SAMMate	Bowtie to exons and junctions	Yes	Yes	Known splice-sites	[30]	http://sammate.sourceforge.net/
IsoformEx	Bowtie to exons and junctions	Yes	No	Known splice-sites	[31]	http://bioinformatics.wistar.upenn.edu/isoformex
RUM	Bowtie and BLAT on transcripts and genome	Yes	Yes	Known splice-sites and BLAT model	[32]	http://www.cbil.upenn.edu/RUM/userguide.php
SpliceSeq	Bowtie alignments to Splicing graphs	Yes	Yes	Known splice-sites	[33]	http://bioinformatics.mdanderson.org/main/SpliceSeq:Overview
PASTA	Bowtie alignment of read fragments	No	Yes	Logistic-regression model for splice-sites	[34]	http://genome.ufl.edu/rivalab/PASTA
ContextMap	Genome alignments from other methods	No	No	Unbiased	[35]	http://www.bio.ifi.lmu.de/softwareservices/contextmap

RNASEQR, RUM, SpliceSeq, MapAI), some can use annotation as an option (GEM, MapNext, STAR, TopHat), and others (the rest) work directly with the genome reference. Additionally, some methods perform quantification (Subheading 2.2) (SAMMate, IsoformEx, RUM, SpliceSeq) and are included here since they provide an independent method for mapping. We also indicate whether the method can map paired-end reads, the type of splice-site model used, the reference where the method is described and the URL where the software is available.

2.2 Genome-Based Quantification of Known Events and Isoforms

In Table 2, we give a list of methods that can be used to quantify known splicing events (RUM, SpliceSeq, MMES, SpliceTrap), known isoforms (SAMMate, IsoformEx, Erange, rSeq, rQuant, FluxCapacitor, IQSeq, Cufflinks, Casper, CEM, IsoInfer, SLIDE, RABT, DRUT, iReckon), or both (MISO, ALEXA-Seq, SOLAS) when a genome-based annotation is available. Some include the mapping step (RUM, SpliceSeq, SAMMate, IsoformEx). Some isoform-based methods can quantify known and novel isoforms simultaneously (IsoInfer, SLIDE, RABT, DRUT, iReckon) or choose between quantifying known or novel isoforms (Cufflinks, Casper, CEM, IsoLasso). We indicate the type of input used by each method, whether they exploit paired-end read information in the calculation and what type of quantification is given. We also provide the reference where the method is described, and the URL (or email) where the software is available.

2.3 Isoform Quantification Guided by a Transcriptome

Table 3 includes methods that quantify isoforms using a transcriptome annotation and reads mapped with a non-spliced mapper. All the methods listed used bowtie to map reads to transcripts in the original publication. Although they generally work with reads mapped to a transcriptome, some methods (RSEM, MMSEQ) can work with reads mapped to a genome. We indicate the type of input used by the method, whether they exploit paired-end read information in the calculation and what type of isoform quantification is given. We also provide the reference where the method is described, and the URL where the software is available.

2.4 Genome-Based Reconstruction and Quantification Without Annotation

Table 4 includes methods to reconstruct (all methods) and to quantify (all methods except for G-Mo.R-Se and assemblySAM) multiple isoforms from genome-mapped reads without using any gene annotation. Some methods can also be run with annotations for quantification (Cufflinks, IsoLasso, Casper, CEM). Some perform simultaneously the reconstruction and quantification of novel isoforms (NSMAP, Montebello, IsoLasso). We indicate the type of input used by each method, whether they exploit paired-end read information in the calculation and what type of isoform quantification is given. We also provide the reference where the method is described and the URL or email where the software is available.

Table 2
A list of methods that can be used to quantify known splicing events (RUM, SpliceSeq, MMES, SpliceTrap), known isoforms (SAMMate, IsoformEx, Erange, rSeq, rQuant, FluxCapacitor, IQSeq, Cufflinks, Casper, CEM, IsoInfer, SLIDE, RABT, DRUT, iReckon), or both (MISO, ALEXA-Seq, SOLAS) when a genome-based annotation is available

Method	Type	Input used in publication	Uses paired-end reads	Quantification	Reference	Web site
RUM	Exon and junction quantification	Bowtie and BLAT on transcripts and genome	Yes	Read counts and RPKM of exons and junctions	[32]	http://www.cbil.upenn.edu/RUM/userguide.php
SpliceSeq	Exon and junction quantification	Bowtie alignments to Splicing graphs	Yes	Inclusion level of exons and junctions	[33]	http://bioinformatics.mdanderson.org/main/SpliceSeq:Overview
MMES	Junction quantification	SOAP alignments to junctions	No	Junction scores	[36]	Email to Wang.Liguo@mayo.edu
SpliceTrap	Exon and junction quantification	Bowtie on inclusion/skipping events	Yes (models insert sizes)	Exon inclusion level	[37]	http://rulai.cshl.edu/splicetrap/
SAMMate	Isoform quantification	Bowtie on genome and junctions	Yes	RPKM/FPKM	[30]	http://sammate.sourceforge.net/
IsoformEx	Isoform quantification	Bowtie on genome and junctions	No	Isoform expression (~RPKM)	[31]	http://bioinformatics.wistar.upenn.edu/isoformex
MISO	Event and isoform quantification	Bowtie on genome and junctions	Yes	Isoform PSI value	[38]	http://genes.mit.edu/burgelab/miso/

ALEXA-Seq	Event and isoform quantification	Reads mapped to genome and junctions	Yes	Event and isoform expression level	[39]	http://www.alexaplatform.org/alexa_seq/
SOLAS	Event and isoform quantification	Reads mapped to genome	No	Isoform expression (~RPKM)	[40]	http://cmb.molgen.mpg.de/2ndGenerationSequencing/Solas/
Erange	Isoform quantification	Bowtie on genome and junctions	No	Isoform RPKM	[41]	http://woldlab.caltech.edu/rnaseq
rSeq	Isoform quantification	SeqMap alignments to exons and exon–exon junctions	Yes (in latest version)	Isoform RPKM	[42]	http://www.personal.umich.edu/~jianghui/rseq/
rQuant	Isoform quantification	Reads mapped to genome	No	Isoform average read coverage and RPKM	[43]	http://galaxy.raetschlab.org/
FluxCapacitor	Isoform quantification	Reads mapped to genome	Yes	Isoform relative abundance (~PSI)	[44]	http://flux.sammeth.net/capacitor.html
IQSeq	Isoform quantification	GFF/MRF/BED	Yes	Isoform RPKM	[45]	http://archive.gersteinlab.org/proj/rnaseq/IQSeq/
Cufflinks	Known or novel Isoform quantification	TopHat alignments	Yes	FPKM	[46]	http://cufflinks.cbcb.umd.edu/
Casper	Known or novel Isoform quantification	TopHat alignments	Yes	Isoform PSI value	[47]	https://sites.google.com/site/rosselldavid/software
CEM	Known or novel Isoform quantification	TopHat alignments	Yes	Isoform expression	[48]	http://alumni.cs.ucr.edu/~liw/cem.html

(continued)

Table 2
(continued)

Method	Type	Input used in publication	Uses paired-end reads	Quantification	Reference	Web site
IsoLasso	Known or novel Isoform quantification	TopHat alignments	Yes	RPKM	[49]	http://alumni.cs.ucr.edu/~liw/isolasso.html
IsoInfer	Known and novel isoform quantification	TopHat alignments	Yes	Isoform RPKM	[50]	http://www.cs.ucr.edu/~jianxing/IsoInfer.html
SLIDE	Known and novel isoform quantification	modEncode spliced mappings	Yes	Isoform RPKM	[51]	https://sites.google.com/site/jingyijli/SLIDE.zip
RABT	Known and novel isoform quantification	TopHat alignments	Yes	Isoform FPKM	[52]	http://cufflinks.cbcb.umd.edu/
DRUT	Known and novel isoform quantification	Bowtie/TopHat alignments to transcriptome/genome	No	FPKM (computed by IsoEM)	[53]	http://www.cs.gsu.edu/~serghei/?q=drut
iReckon	Known and novel isoform quantification	TopHat alignments	Yes	Isoform RPKM	[54]	http://compbio.cs.toronto.edu/ireckon/

Table 3
Methods that quantify isoforms using a transcriptome annotation and reads mapped with a non-spliced mapper

Method	Input reads format	Uses paired-end reads	Isoform quantification	Reference	Web site
RSEM	BAM/SAM	Yes (models insert size)	"Expected number of fragments per isoform" and "fraction of transcripts represented by the isoform"	[55]	https://github.com/bli25wisc/RSEM/
IsoEM	SAM	Yes (models insert size)	Isoform expression	[56]	http://dna.engr.uconn.edu/?page_id=105
NEUMA	Fastq/Fasta mapped with Bowtie	Yes	FVKM (fragments per virtual kilobase per million sequenced reads)	[57]	http://neuma.kobic.re.kr
BitSeq	SAM	Yes (models insert size)	Isoform expression	[58]	http://www.bioconductor.org/packages/2.11/bioc/html/BitSeq.html
MMSEQ	Sorted BAM	Yes	Haplotype and isoform-specific expression	[59]	http://bgx.org.uk/software/mmseq.html
eXpress	BAM	Yes	FPKM, estimated counts	[60]	http://bio.math.berkeley.edu/eXpress/

Table 4
Methods to reconstruct (all methods) and to quantify (all methods except for G-Mo.R-Se and assemblySAM) multiple isoforms from genome-mapped reads without using any gene annotation

Method	Type	Input used in publication	Uses paired-end reads	Isoform quantification	Reference	Web site
G-Mo.R-Se	De novo isoform reconstruction	SOAP alignments	No	No	[61]	http://www.genoscope.cns.fr/externe/gmorse/
assemblySAM	De novo isoform reconstruction	Own heuristics for read-mapping using Bowtie	Yes	No	[62]	http://sammate.sourceforge.net/assemblysam.html
TAU	De novo isoform reconstruction and quantification	Supersplat alignments	No	Average per-base sequencing depth	[63]	Email to HPriest@danforthcenter.org
Scripture	De novo isoform reconstruction and quantification	TopHat alignments	Yes (models insert size)	RPKM	[64]	http://www.broadinstitute.org/software/Scripture/
Cufflinks	Known or novel isoform quantification	TopHat alignments	Yes (models insert size)	FPKM	[46]	http://cufflinks.cbcb.umd.edu/
Casper	Known or novel Isoform quantification	TopHat alignments	Yes	Isoform PSI value	[47]	https://sites.google.com/site/rosselldavid/software

CEM	Known or novel isoform quantification	TopHat alignments	Yes	Isoform expression	[48]	http://alumni.cs.ucr.edu/~liw/cem.html
IsoLasso	Known or novel isoform quantification	TopHat alignments	Yes	RPKM	[49]	http://alumni.cs.ucr.edu/~liw/isolasso.html
Montebello	Novel isoform reconstruction and quantification	SpliceMap alignments	Yes	Isoform expression	[65]	http://www.stanford.edu/group/wonglab/Montebello/Montebello_0.8.tar.gz
NSMAP	Novel isoform reconstruction and quantification	TopHat alignments	Yes (models insert size)	RPKM	[66]	https://sites.google.com/site/nsmapfornaseq

2.5 Evidence-Based Alternatively Spliced Gene Prediction

Table 5 includes methods that could be used to perform alternatively spliced gene prediction from RNA-Seq data. Besides the de novo reconstruction and quantification methods from Subheading 2.4 and those from Subheading 2.2 that can predict novel and known isoforms simultaneously (IsoInfer, SLIDE, RABT, DRUT, iReckon), we also include methods that can use various sources of evidence to predict alternatively spliced genes (TAU, SpliceGrapher, ExonMap/JunctionWalk) and methods that predict alternatively spliced protein-coding genes from multiple evidences (Augustus, mGene). We also include classical protein-coding gene prediction methods that could potentially use RNA-Seq as evidence (Gaze, JigSaw, EVM, Evigan). For each method, we indicate the type of input used, whether they exploit paired-end read information in the calculation or provide any isoform quantification. We also give the reference where the method is described and the URL or email where the software is available.

2.6 De Novo Transcriptome Assembly

Table 6 includes methods for de novo transcriptome assembly. Some of these methods produce multiple isoforms per assembled gene (OASES, SOAPdenovo-trans, TransAbyss, Trinity) and only two quantify the alternative isoforms (TransAbyss, Trinity). Nonetheless, these methods could be coupled with transcriptome-based quantification methods (Subheading 2.3). KisSplice assembles alternatively spliced events rather than isoforms and quantifies the read coverage of these events. We indicate whether they exploit paired-end read information in the calculation, the *k*-mer approach (single/multiple), whether they detect multiple isoforms per gene and whether they perform isoform quantification. We also provide the reference where the method is described and the URL (or email) where the software is available.

2.7 Differential Splicing

Table 7 includes methods that measure relative changes in inclusion/usage between two or more conditions at the exon level (DEXSeq, DSGSeq, GPSeq, SOLAS), event level (MATS, JuncBASE, JETTA, SpliceSeq), and isoform region level (DiffSplice, SplicingCompass, FDM, rDiff) or at both, isoform and event levels (MISO, ALEXA-Seq). We indicate whether the methods perform any quantification per sample, the measure of differential splicing provided, whether they exploit paired-end read information in the calculation, the reference where the method is described and the URL where the software is available.

2.8 Isoform-Based Differential Expression

Table 8 includes methods that measure differential expression at the transcript level between two or more conditions, allowing multiple transcripts per gene. Cuffdiff2, additionally, can calculate significant changes in the relative abundance of isoforms. For each method, we indicate the quantification performed per sample, whether it exploits paired-end read information in the calculation,

Table 5
Methods to perform alternatively spliced gene prediction from RNA-Seq data

Method	Type	Input used in publication	Uses paired-end reads	Isoform quantification	Reference	Web site
IsoInfer	Known and novel isoform quantification	TopHat alignments	Yes	Isoform RPKM	[50]	http://www.cs.ucr.edu/~jianxing/IsoInfer.html
SLIDE	Known and novel isoform quantification	modEncode spliced mappings	Yes	Isoform RPKM	[51]	https://sites.google.com/site/jingyijli/SLIDE.zip
RABT	Known and novel isoform quantification	TopHat alignments	Yes	Isoform FPKM	[52]	http://cufflinks.cbcb.umd.edu/
DRUT	Known and novel isoform quantification	Bowtie (TopHat) alignments to transcriptome/genome	No	FPKM (computed by IsoEM)	[53]	http://www.cs.gsu.edu/~serghei/?q=drut
iReckon	Known and novel isoform quantification	TopHat alignments	Yes	Isoform RPKM	[54]	http://compbio.cs.toronto.edu/ireckon/
TAU	Evidence-based isoform reconstruction and quantification	Supersplat alignments	No	Average per-base sequencing depth	[63]	Email to hpriest@danforthcenter.org
SpliceGrapher	Evidence-based isoform reconstruction	TopHat alignments	Yes	No	[67]	http://SpliceGrapher.sf.net

(continued)

Table 5 (continued)

Method	Type	Input used in publication	Uses paired-end reads	Isoform quantification	Reference	Web site
ExonMap/ JunctionWalk	Evidence-based isoform reconstruction	Reads mapped to exons and junctions	Handled by SpliceMap	No	[68]	http://gluegrant1.stanford.edu/~DIC/ RNASeqArray/TranscriptConstruction.html
mGene	Evidence-based alternatively spliced gene prediction	Reads mapped to genome	Yes	No	[69]	http://mgene.org/
Augustus	Evidence-based alternatively spliced gene prediction	Spliced evidences	No	No	[70]	http://bioinf.uni-greifswald.de/augustus/
Gaze	Evidence-based gene prediction	Evidence in GFF format	No	No	[71]	http://www.sanger.ac.uk/resources/software/ gaze/
JigSaw	Evidence-based gene prediction	Spliced evidences	No	No	[72]	http://www.cbcb.umd.edu/software/jigsaw/
EVM	Evidence-based gene prediction	PASA alignments	No	No	[73]	http://evidencemodeler.sourceforge.net/
Evigan	Evidence-based gene prediction	Spliced evidences	No	No	[74]	http://www.seas.upenn.edu/~strctlrn/evigan/ evigan.html

Table 6
Methods for de novo transcriptome assembly

Method	Uses paired-end reads	Graph approach	Detects alternative isoforms	Isoform quantification	Reference	Web site
Rnnotator	Yes	Variable *k*-mer	No	No	[75]	Email to vtdelapuente@lbl.gov
STM	Yes	Variable *k*-mer	No	No	[76]	http://www.surget-groba.ch/downloads/stm.tar.gz
OASES	Yes	Variable *k*-mer	Yes	No	[77]	http://www.ebi.ac.uk/~zerbino/oases/
SOAPdenovo-trans	Yes	Variable *k*-mer	Yes	No	[78]	http://sourceforge.net/projects/soapdenovotrans/
TransAbyss	Yes	Variable *k*-mer	Yes	Isoform read coverage	[79]	http://www.bcgsc.ca/platform/bioinfo/software/
Trinity	Yes	Single *k*-mer	Yes	Yes (uses RSEM)	[80]	http://TrinityRNASeq.sourceforge.net
KisSplice	No	Single *k*-mer	Events only	Event read coverage	[81]	http://alcovna.genouest.org/kissplice/

Table 7
Methods that measure relative changes in inclusion/usage between two or more conditions at the exon level (DEXSeq, DSGSeq, GPSeq, SOLAS), event level (MATS, JuncBASE, JETTA, SpliceSeq), and isoform region level (DiffSplice, SplicingCompass, FDM, rDiff) or at both isoform and event levels (MISO, ALEXA-Seq)

Method	Type	Quantification	Uses paired-end reads	Models biological variability	Differential quantification	Reference	Web site
DEXSeq	Exon level	No	No	Yes	Differential exon inclusion	[82]	http://www.bioconductor.org/packages/release/bioc/html/DEXSeq.html
DSGSeq	Exon level	Isoform relative abundances	No	Yes	Differential exon inclusion	[83]	http://bioinfo.au.tsinghua.edu.cn/software/DSGseq
GPSeq	Exon level	Exon splicing ratio	No	Yes	Differential exon splicing index	[84]	http://cran.r-project.org/web/packages/GPseq/index.html
SOLAS	Exon level	Event and isoform inclusion	No	No	Differential exon inclusion	[40]	http://cmb.molgen.mpg.de/2ndGenerationSequencing/Solas/
MATS	Event level	Event inclusion	Handled by mapping method	Yes	Differential event inclusion	[85]	http://rnaseq-mats.sourceforge.net/
JuncBASE	Event level	Event inclusion	Yes	No	Differential event inclusion	[86]	http://compbio.berkeley.edu/proj/juncbase/Home.html
JETTA	Event level	SeqMap alignments	Handled by mapping method	No	Differential event inclusion	[87]	http://igenomed.stanford.edu/~junhee/JETTA/rnaseq.html

SpliceSeq	Event level	Inclusion level of exons and junctions	Yes	No	Differential event inclusion	[33]	http://bioinformatics.mdanderson.org/main/SpliceSeq:Overview
MISO	Event and isoform levels	PSI	Yes	No	Differential event/isoform PSI	[38]	http://genes.mit.edu/burgelab/miso/
Alexa-Seq	Event and isoform levels	Gene, transcript, and event expression levels	Yes	No	Differential relative event/isoform expression	[39]	http://www.alexaplatform.org/alexa_seq/
SplicingCompass	Isoform region level	Normalized exon density	Handled by mapping method	No	Differential relative isoform abundance	[88]	http://www.ichip.de/software/SplicingCompass.html
DiffSplice	Isoform region level	Expression of "Alternative Splicing Modules"	Yes	Yes	Differential Expression of "Alternative Splicing Modules"	[89]	http://www.netlab.uky.edu/p/bioinfo/DiffSplice
FDM	Isoform region level	Isoform region relative expression	No	Yes	Differential relative isoform abundance	[90]	http://csbio-linux001.cs.unc.edu/nextgen/software/FDM
rDiff	Isoform region level	Isoform region relative expression	Yes	Yes	Differential relative isoform abundance	[91]	http://bioweb.me/rdiff

Table 8
Methods that measure differential expression at the transcript level between two or more conditions, allowing multiple transcripts per gene

Method	Quantification	Uses paired-end reads	Models biological variability	Differential quantification	Reference	Web site
BitSeq	Isoform expression	Yes	Yes	Differential isoform expression	[58]	http://www.bioconductor.org/packages/2.11/bioc/html/BitSeq.html
BASIS	Isoform relative expression	No	No	Differential isoform expression	[92]	http://www.rcf.usc.edu/~liangche/software.html
Cuffdiff2	Isoform expression	Yes	Yes	Differential isoform expression	[93]	http://cufflinks.cbcb.umd.edu/
EBSeq	Isoform expression quantified by input method	Handled by quantification method	Yes	Differential isoform expression	[94]	http://www.biostat.wisc.edu/~kendzior/EBSEQ/

the measure of differential expression provided, the reference where the method is described and the URL where the software is available.

2.9 Visualization of Alternative Splicing

Table 9 includes some of the available tools for the visualization of alternative splicing using RNA-Seq data. Some of them can be used as command line tools that are included in the distribution of the analysis tools (RSEM, SpliceGrapher, DiffSplice, DEXSeq, SplicingCompass) or provided separately (Sashimi Plots), whereas others are Graphical User Interfaces (Savant, ALEXA-Seq, SpliceSeq).

3 Methods

3.1 Spliced-Mapping Short Reads

Event and Isoform quantification are very much dependent on the correct assignment of RNA-Seq reads to the molecule of origin. Accordingly, we will start by reviewing some of the read mappers that are splice-site aware, and therefore, can be used to detect exon–intron boundaries and connections between exons. This alignment problem has been addressed in the past by tools that combine fast heuristics for sequence matching with a model for splice-sites, for example, Exonerate [97], BLAT [98], or GMAP [99]. These methods, however, are not competitive enough to map all reads from a sequencing run in a reasonable time. In the last few years, a myriad of methods have been developed for mapping short reads to a reference genome [100]. Those that are splice-site aware and incorporate intron-like gaps are generally called spliced-mappers, split-mappers, or spliced aligners. Their main challenge is that reads must be split into shorter pieces, which may be harder to map unambiguously; and although introns are marked by splice-site signals, these occur frequently by chance in the genome.

Spliced-mappers have been classified previously into two main classes [101], *exon-first* and *seed-and-extend* (Subheading 2.1). *Exon-first* methods map reads first to the genome using an unspliced approach to find read-clusters; unmapped reads are then used to find connections between these read-clusters. These methods include TopHat [6], SOAPsplice [7], PASSion [8], MapSplice [9], SpliceMap [10], HMMsplicer [11], TrueSight [12], and GEM [13]. *Seed-and-extend* methods generally start by mapping part of the reads as *k*-mers or substrings; candidate matches are then extended using different algorithms and potential splice-sites are located. These methods include SplitSeek [14], Supersplat [15], SeqSaw [16], ABMapper [17], MapNext [18], STAR [19], GSNAP [20], and PALMapper [22]. A generalization of seed-and-extend methods is represented by the multi-seed methods, like CRAC [23],

Table 9
Some of the available tools for the visualization of alternative splicing using RNA-Seq data

Method	Type	Used with	Input data	Visualization	Reference	Web site
Sashimi Plots	Command line tool	MISO	GFF3	Splicing events and read coverage	[38]	http://genes.mit.edu/burgelab/miso/docs/sashimi.html
RSEM	Command line tool	RSEM	Transcript BAM file	Read profiles (WIG)	[55]	https://github.com/bli25wisc/RSEM/
SpliceGrapher	Command line tool	SpliceGrapher	GFF files	Isoforms	[67]	http://SpliceGrapher.sf.net
DEXSeq	Command line tool	DEXSeq	DEXSeq results	Differential exon usage	[82]	http://www.bioconductor.org/packages/release/bioc/html/DEXSeq.html
SplicingCompass	Command line tool	SplicingCompass	SplicingCompass results	Differential exon usage	[88]	http://www.ichip.de/software/SplicingCompass.html
DiffSplice	Command line tool	DiffSplice	GTF (graphs)	Isoforms	[89]	http://www.netlab.uky.edu/p/bioinfo/DiffSplice
SpliceSeq	GUI	SpliceSeq	SpliceSeq processed data	Isoforms and alternatively spliced events	[33]	http://bioinformatics.mdanderson.org/main/SpliceSeq:Overview
ALEXA-Seq viewer	GUI	ALEXA-Seq	Alexa-seq database	Splicing events and expression levels	[39]	http://www.alexaplatform.org/alexa_seq/
Savant Browser	GUI	iReckon	GFF	Isoforms	[95]	http://genomesavant.com/savant/
SplicingViewer	GUI	Splicing Viewer	BAM	Splicing events coverage	[96]	http://bioinformatics.zj.cn/splicingviewer/

OLego [24], and Subread [25], which consider multiple subreads within each read. Similarly, ABMapper consider multiple read-splits for mapping. Some methods actually use a hybrid strategy, following an exon-first approach for unspliced reads, and then using seed-and-extend approach for spliced reads, like MapSplice, SpliceMap, HMMSplicer, TrueSight, GEM, and PALMapper; the latter being a combination of GenomeMapper [102] and QPalma [21] for spliced reads. *Exon-first* methods depend strongly on sufficient coverage on potential exons to incorporate spliced reads, but are generally faster than *seed-and-extend* methods. On the other hand, *seed-and-extend* methods tend to be less dependent on recovering exon-like read-clusters and may recover more novel splice-sites. However, the storage of *k*-mers for long reads requires sufficient computer memory for large *k*, and the mapping has limited accuracy for small *k* [7].

There is also a different class of tools, which use the annotation and/or some heuristics to map reads. These include OSA [26], X-Mate [27], RNASEQR [28], MapAI [29], SAMMate [30], IsoformEx [31], RUM [32], SpliceSeq [33], and PASTA [34]. RNASEQR and RUM use Bowtie [103] to map reads to the transcriptome and genome; and then identify novel junctions from the unmapped reads using BLAT [98]. Similarly, SAMMate and IsoformEx use Bowtie to locate reads in exons and junctions, whereas SpliceSeq uses Bowtie to map reads to a graph representation of the annotation; X-Mate uses its own heuristics to trim and map reads recursively to locate reads on exons and junctions. On the other hand, PASTA does not use any gene annotation; it uses Bowtie and a splice-site model to locate read fragments on exon junctions. Among these methods, SAMMate, IsoformEx, RUM, and SpliceSeq also provide some level of quantification for exons, events, or isoforms (Subheading 2.2) (Fig. 1), which makes them convenient as a pipeline tool. OSA is actually a seed-and-extend mapping method but relies on an annotation. OSA avoids splitting reads into subreads which helps improving speed; and like other annotation-guided methods, also split-maps reads that are not located in the provided annotation using the seed-and-extend approach. Finally, unlike the other methods, MapAI and ContextMap use reads already mapped to a reference genome. MapAI uses reads mapped to a transcriptome to assign them to their genomic positions, whereas ContextMap refines the genome mappings using the read context, extending to all reads the context approach used by methods like MapSplice or GEM for spliced reads. In the newest version, ContextMap can also be used as a standalone read-mapping tool. Annotation-guided mapping methods are possibly the best option to accurately assign reads to gene annotations, whereas de novo mapping tools are convenient for finding new splicing junctions.

Besides the differences in the mapping procedure, de novo mapping tools detect splice-sites using a variety of approaches, which may determine the reliability of the splice-sites detected and the possibility of obtaining novel ones. Most tools search for an exact match of the flanking intronic dinucleotides to the canonical splice-sites GT-AG, GC-AG, AT-AC (*see* Subheading 2.1). Tools like MapNext and Tophat use a two-step approach, first mapping to the known junctions and then locating novel ones with GT-AG dinucleotides, whereas tools like MapSplice, Supersplat, SpliceMap, and HMMSplicer use a gapped-alignment approach that allows the detection of junctions regardless of the exon coverage. HMMSplice, QPalma, PASTA, and OLego use a more complex representation for splice-sites. HMMSplice is based on a hidden Markov model, QPalma on a Support Vector Machine, PASTA on a logistic regression, and OLego in the combined logistic modeling of sequence bias and intron-size; all of which are trained on known splice-sites. In contrast, MapSplice, SeqSaw, STAR, SplitSeek, and CRAC can do an unbiased search of splice-junctions, not necessarily looking for the splice-site motif and generally using support from multiple reads; hence, they can potentially recover noncanonical splice-sites. Annotation-guided methods will accurately assign reads to known splice-sites, but will miss novel ones, unless they use some heuristics for novel junctions like RUM and RNASEQR. Mapping methods like STAR, GEM, MapNext, and TopHat accept annotations as optional input, which will guide the initial mapping of reads. Other parameters may be important too, like the search range of intron lengths. Most models impose restrictions in the minimum and maximum intron lengths, but methods like MapSplice does not impose any restriction and OSA has a specific search for novel exons using distal fragments. The decision of which tool to use depends very much on whether the aim is to assign reads to known annotations or to find novel splice-sites.

3.2 Definition and Quantification of Events and Isoforms

First reports using RNA-Seq to quantify splicing followed an approach analogous to splicing junction arrays [104]. They were based on the analysis of junctions built from known gene annotations [2, 3, 105–108]. In these and later methods, reads aligning to candidate alternative exons and its junctions are considered as inclusion reads, whereas reads mapping to flanking exons and to junctions skipping the candidate alternative exon are considered as skipping or exclusion reads. These reads are then used to provide an estimate of the relative inclusion of the regulated exon [109], generally called inclusion level. This approach has shown a reasonable agreement with microarrays and can be modified to include exon-body reads and variable exon lengths [2, 109]

An alternative measure, "percent spliced in" (PSI), has been defined as the number of isoforms that include the exon over the total isoforms [110], or equivalently, as the fraction of mRNAs that represent the inclusion isoform [38]. If the PSI value is calculated for a particular splicing event, it can be considered equivalent to the inclusion level. Isoform quantification can be expressed in terms of either a global measure of expression [58], which may provide a global ranking comparable across genes in one sample, or in terms of a relative measure of expression, which is normalized per gene locus and comparable across conditions. The global measure is generally given in terms of RPKM or FPKM (Reads or Fragments Per Kilobase of transcript sequence per Millions mapped reads); and the relative measure is given in terms of a PSI value or a similar value.

Besides the original approaches [2, 3, 105–108], various tools have been developed recently to quantify events and isoforms. These range from simply quantifying the inclusion of events to the reconstruction and quantification of novel isoforms. Some of the tools that reconstruct isoforms also estimate their quantification, and some tools may quantify either known isoforms or novel ones, or both simultaneously. Accordingly, we classify the methods depending on whether they use annotation or not and on the type of input and output:

1. Event/isoform quantification using known (genome-based) gene annotations (Subheading 2.2).
2. Isoform quantification using a transcriptome annotation (Subheading 2.3).
3. De novo isoform reconstruction with a genome reference, either purely focused on reconstruction or also providing isoform quantification (Subheading 2.4).
4. Isoform reconstruction and quantification guided by annotation. These methods use a gene annotation as a guide and can complete the annotation with new exons, new isoforms, or even with some new gene loci (Subheading 2.5).
5. Finally, some of the de novo transcript assembly methods also quantify isoforms (Subheading 2.6).

3.2.1 Event and Isoform Quantification Guided by Gene Annotation

Various tools have been developed for event quantification from a single condition (Subheading 2.2) (Fig. 1): RUM [32], SpliceSeq [33]. MMES [36], SpliceTrap [37], MISO [38], ALEXA-Seq [39], and SOLAS [40]. RUM provides quantification of genes, exons, and junctions in terms of read-counts and RPKM (reads per kilobase per million mapped reads), whereas SpliceTrap and MMES use the reads mapped to junctions and employ a statistical model to calculate exon inclusion levels and junction scores, RUM and MMES also provide the mapping step. RUM has its own heuristics (Subheading 2.1), whereas MMES maps reads to exon–exon junctions using SOAP [111]. Similarly, SpliceSeq maps reads to a splicing-graph to obtain exon and junction inclusion levels. MISO and

ALEXA-Seq use reads on exons and junctions, whereas SOLAS uses only reads on exons. MISO provides PSI values, while ALEXA-Seq and SOLAS event and isoform expression levels. MISO, ALEXA-Seq, and SOLAS can also estimate isoform relative abundances and can be further used for differential splicing (Subheading 2.7).

Quantification of isoforms is more complicated than that of events, as it requires the correct assignment of reads to isoforms sharing part of their sequence. One of the first attempts to do this was Erange [41], where reads mapped to the genome and known junctions were distributed in isoforms according to the coverage of the genomic context, and isoform expression was defined in terms of RPKM. However, the uncertainty in the assignment of reads shared by two or more isoforms must be appropriately modeled. Accordingly, a number of methodologies have been proposed to address this issue (Subheading 2.2): SAMMate [30], IsoformEx [31], MISO [38], ALEXA-Seq [39], SOLAS [40], rSeq [42], rQuant [43], FluxCapacitor [44], IQSeq [45], Cufflinks [46], Casper [47], CEM [48], IsoLasso [49], IsoInfer [50], SLIDE [51], RABT [52], DRUT [53], and iReckon [54]. Isoform quantification is generally given in terms of RPKM, FPKM, some equivalent *isoform expression level* value, PSI, or an equivalent *relative expression* value.

SAMMate and IsoformEx use the reads mapped to exons and junctions by their own methods to quantify gene and isoform expression in terms of RPKM values. SAMMate incorporates two quantification methods, one that is not sensitive to coverage, so it can be used on early sequencing platforms [112] and a recent one that is aimed for deeper coverage and uses a filtering of non-expressed transcripts [113]. SOLAS and rSeq use reads on exons to estimate isoform expression levels, whereas rQuant uses the position-wise density of mapped reads to calculate two abundance estimates: the RPKM and the estimated average read coverage for each transcript. IQSeq provides a statistical model that facilitates the incorporation of data from multiple technologies; and FluxCapacitor, unlike other methods, does not account for the mapping variability across isoforms and directly solves the constraints derived from distributing the reads in isoforms according to the splicing graph built from the read evidence.

IsoInfer, SLIDE, RABT, DRUT, and iReckon can quantify the known annotation and at the same time predict and quantify novel isoforms in known gene loci. RABT quantifies known and novel isoforms, taking into account existing gene annotations and using the same graph assembly algorithm of Cufflinks, combining the sequencing reads with reads obtained by fragmenting known transcripts. RABT is part of the Cufflinks distribution, but here we distinguish it from the original Cufflinks, which quantifies abundances of either only annotated or only novel isoforms [46, [52]. Similar to RABT, SLIDE uses RNA-Seq data and existing gene annotation to discover novel isoforms and to estimate the abundance of known and new isoforms. Additionally, it can use other sources of evidence, like RACE, CAGE, and EST, or even the output from other isoform reconstruction algorithms. IsoInfer uses the transcript start and end sites, plus exon–intron boundaries to enumerate all possible isoforms, estimate their expression levels and then choose the subset of isoforms that best explain the observed reads, predicting novel isoforms from the existing exon data. On the other hand, iReckon can work with just transcript start and end sites or with full annotations; it models multimapped reads, intron-retention and unspliced pre-mRNAs and performs reconstruction and quantification simultaneously. DRUT uses a modified version of the IsoEM algorithm [56] in combination with a de novo reconstruction method similar to Cufflinks to complete partial existing annotations as well as to estimate isoform frequencies. Casper, similar to Cufflinks, estimates abundances of known or novel isoforms separately, but unlike other methods, uses information of the connectivity of more than two exons. Generally, known isoform quantification methods show a high level of agreement with experimental validation [54] and can be improved using annotation-guided methods for read mapping [29].

3.2.2 Isoform Quantification Guided by a Transcriptome

A number of methods consider reads mapped to a transcriptome for isoform quantification (Subheading 2.3); these include RSEM [55], IsoEM [56], NEUMA [57], BitSeq [58], MMSEQ [59], and eXpress [60]. Although these methods depend on a transcriptome annotation, they can use a standard (non-spliced) mapper to obtain the input data. Additionally, they can work also with predicted isoforms from transcript assembly methods (Fig. 1). All of them provide a measure of global isoform expression, similar to RPKM. Moreover, RSEM also calculates the fraction of transcripts represented by the isoform, equivalent to PSI. RSEM and IsoEM use both an Expectation–Maximization algorithm and model paired-end fragment size. RSEM models the mapping uncertainty to transcripts and provides confidence intervals of the abundance estimates. IsoEM uses the fragment-size information to disambiguate the assignment of reads to isoforms. BitSeq is based on a Bayesian approach, incorporates the mapping step to the transcriptome, models the nonuniformity of reads, and provides an expression value per isoform. BitSeq can also be used for differential isoform expression (see below). MMSEQ also takes into account the nonuniform read distribution and deconvolutes the mapping to isoforms to estimate isoform-expression and haplotype-specific isoform-expression. The method eXpress is in fact a general tool for quantifying abundances of a set of sequences in a generic experiment and can be used with a reference genome or transcriptome. For RNA-Seq reads mapped to a transcriptome, eXpress provides isoform quantification in terms of FPKM. Finally, NEUMA is different from the other methods, as it does not use any probabilistic description and assumes uniformity of the reads along transcripts. NEUMA labels reads according to whether they are isoform or gene specific and calculates a measure of isoform quantification defined as the number of fragments per virtual kilobase per million reads (FVKM). Transcript-based methods can be generally applied to the transcripts obtained from genome annotations, so that the correspondence of transcripts to gene loci is maintained. Additionally, they can be used in combination with de novo transcript assembly methods (see below) to estimate isoform abundance in genomes without a reference.

3.2.3 Genome-Based Transcript Reconstruction and Quantification Without Annotation

These methods use the reads mapped to the genome to reconstruct isoforms de novo. They are generally based on previous approaches to transcript reconstruction from ESTs [114–117]. As for ESTs [118], accuracy is limited by the lengths of the input reads; hence, the use of paired-end sequencing becomes crucial. Additionally, as RNA abundance spans a wide range of values, the correct recovery of lowly expressed isoforms requires sufficient sequencing coverage. Although these methods work independently of the mapping procedure, they strongly rely on the accuracy of the spliced-mapper.

Purely reconstruction methods, without isoform quantification, include G-Mo.R-Se [61] and assemblySAM [62]. Methods that reconstruct isoforms as well as estimate their abundances include Cufflinks [46], Casper [47], CEM [48], IsoLasso [49], TAU [63], Scripture [64], Montebello [65], and NSMAP [66]. G-Mor.R-Se, Scripture, and TAU proceed in a similar way by first obtaining candidate exons from read-clusters and then connecting them using reads spanning exon–exon junctions. Subsequently, all possible isoforms from the graph of connected exons are computed. As they explore all possible connections between potential exons, they ensure a high sensitivity but at the cost of a high false-positive rate. In contrast, Cufflinks first connects predicted exons trying to identify the minimum number of possible isoforms using a graph generated from the reads; expression levels are then calculated using a statistical model [42]. IsoLasso also tries to obtain the minimal set of isoforms from predicted exons, but maximizing the number of reads included in each isoform. CEM model takes into account positional, sequencing and mappability biases of the RNA-Seq. Casper follows a heuristics similar to Cufflinks but exploiting the reads that connect more than 2 exon. Some of these methods use paired-end reads and/or model the insert-size distribution, which improve the reconstruction accuracy [119]. NSMAP, IsoLasso, and Montebello perform identification of the exonic structures and estimation of the isoform expression levels simultaneously in a single probabilistic model; iReckon does so too, but was not included in this section as it needs at least the transcript start and end positions. The rest of methods perform reconstruction and quantification independently.

Although reasonable overlap among methods has been reported [120], there are still many predictions unique to each method. Interestingly, given a fixed number of sequenced bases, sequencing longer reads does not seem to lead to more accurate quantifications [55, 56], although exonic structures may be better predicted [48]. These de novo reconstruction and quantification methods seem a good option for finding novel isoforms [64], alternatively spliced genes in a genome with partial annotation [61] and for quantifying isoforms under various conditions [46]. However, they depend much on coverage. Accordingly, if the aim is to obtain the expression of known isoforms, gene-based methods may be a better choice. Alternatively, for protein-coding gene finding there are other options available, as discussed next.

3.2.4 Evidence-Based Alternatively Spliced Gene Prediction

The methods described above are mainly focused on isoform quantification, using available annotation, or on the de novo reconstruction and quantification of isoforms, using reads mapped to the genome. Quantification methods based solely on gene annotations could miss many novel genes and isoforms, whereas de novo approaches not using annotations may produce many false positives. Combined approaches that discover novel isoforms in known and new loci and, at the same time, quantify them, could help improving the gene annotation. Some of the annotation-based quantification methods can also reconstruct and quantify new isoforms in known gene loci (Subheading 2.5): IsoInfer [50], SLIDE [51], RABT [52], DRUT [53], and iReckon [54]. Some of these methods can work with partial evidence, like iReckon. However, they do not predict new isoforms in new gene loci. To this end, a number of methods can use RNA-Seq and other sources of evidence to predict the exon–intron structures of isoforms, or even to predict full protein-coding gene structures. These methods include (Subheading 2.5) TAU [63], SpliceGrapher [67], mGene [69], and the method described in ref. 68. The method mGene is an SVM-based gene predictor (*see*, e.g., [121]) that first reconstructs a high-quality gene set, which then uses to train a gene model that is applied using RNA-Seq data in addition to the previously determined genomic signal predictors. In contrast, SpliceGrapher and TAU incorporate into the same graph model information from ESTs and RNA-Seq reads to complete known gene annotations and produce novel variants. ExonMap/JunctionWalk proposed in ref. [68] combine SpliceMap [10] alignments with known annotations to complete known isoforms and obtain novel ones without quantification (Fig. 1).

Some of these methodologies are reminiscent of the evidence-based gene prediction methods. These are generally based on probabilistic models of protein-coding genes, which can incorporate external spliced evidence like ESTs and cDNAs into the model to guide the prediction of the exon–intron structure, and some of which can predict multiple isoforms in a gene locus. Accordingly, evidence-based gene prediction methods could still be useful for splicing analysis from RNA-Seq. In particular, Augustus [70] is an evidence-based protein-coding gene prediction method, capable of finding multiple isoforms per gene, which has been shown to be highly accurate using a blind test set [122, 123]. Other evidence-based prediction methods include (Subheading 2.5) GAZE [71], JigSaw [72], EVM [73], and Evigan [74]. Although these four methods do not explicitly model alternative isoforms, they can still produce multiple transcripts in a locus.

Evidence-based gene prediction methods can take as input transcripts reconstructed by other methods and generate protein-coding isoforms. They do not provide a quantification of isoforms, but in combination with quantification methods (Subheadings 2.2 and 2.3) they could be a powerful approach to annotate and quantify alternatively spliced protein-coding genes from newly sequenced genomes using RNA-Seq data.

3.2.5 *De Novo Transcript Assembly*

De novo transcript assemblers put together reads into transcriptional units without mapping the reads to a genome reference, similar to building Unigene clusters from ESTs prior to having a genome reference [124]. A transcriptional unit can be defined as the set of RNA sequences that are transcribed from the same genome locus and share some sequence, i.e., the set of RNA isoforms from the same gene. This is generally represented as a sequence-based graph, where paths along the graph potentially resolve the different isoforms. Methods for transcript assembly include (Subheading 2.6) Rnnotator [75], STM [76], OASES [77], SOAPdenovo-trans [78], TransAbyss [79], Trinity [80], and Kissplice [81]. Although KisSplice focuses on recovering alternative splicing events, we include it here as it follows a similar approach to the other methods. *See* ref. 125 for a recent comparison between some of these methods.

The main challenge of these methods is not only to distinguish sequence errors from polymorphisms but also to distinguish close paralogues from alternative isoforms, which requires correctly capturing the exonic variability. All these methods are based on a graph built from *k*-mer overlaps between read sequences. The choice of *k*-mer length affects the assembly, being more sensitive at low values of *k* and more specific at high values. Accordingly, some use a variable *k*-mer approach. Isoforms are recovered as paths through the graph with sufficient read coverage. Not all methods can provide multiple isoforms from the same gene (Subheading 2.6).

Genome-independent methods are useful when there is no genome reference sequence available, and could also be valuable when the RNA is expected to contain much variation, like in a cancer cell with many copy number alterations, mutations and genome rearrangements compared to the reference genome. De novo assembly methods tend to be more sensitive to sequencing errors and low coverage, and generally require more computational resources, although full parallelization of the graph algorithms can alleviate this issue [126]. Some of the methods also consider the comparison to reference sets of DNA or protein sequences [76]. In fact, mapping assembled transcripts to a reference genome, even from a related species, seems to improve accuracy in transcript quantification [127]. KisSplice is explicitly designed to obtain and quantify de novo alternative splicing events, which may potentially be coupled with other methods to study differential splicing. On the other hand, OASES, TransAbyss, Trinity, and SOAPdenovo-trans can produce multiple isoforms, but only TransAbyss and Trinity perform quantification. Nonetheless, multiple assembled isoforms can be quantified with transcript-based methods (Subheading 2.3) or further processed with isoform-based differential expression methods (Subheading 2.8).

3.3 Comparing Splicing Across Samples

The comparison of events and isoforms across two or more conditions provide valuable information to understand the regulation of alternative splicing. However, it is important to distinguish differential isoform relative abundance, from differential isoform expression. Changes in relative abundance of isoforms, regardless of the expression change, indicate a splicing-related mechanism. On the other hand, there can be measurable changes in the expression of isoforms across samples, without necessarily changing the relative abundance, which possibly indicates a transcription-related mechanism. With this in mind, we can consider two types of methods, those that measure relative event or isoform usage (Subheading 2.7) and those that measure isoform-based changes in expression (Subheading 2.8).

3.3.1 Differential Splicing

Most of these methods are focused on splicing events, thereby summarizing the isoform relative abundance into two possible splicing outcomes in a local region of the gene (Fig. 1). They use a predetermined set of splicing events, generally calculated from gene annotations and additional EST and cDNA data; hence, they are suitable for studying splicing variation in well-annotated genomes. They all consider exon-skipping events (cassette exons), and some also include alternative 5′ and 3′ splice-sites, mutually exclusive exons and retained introns; and in very few cases, multiple-cassette exons, alternative first exons and alternative last exons [38]. Potential novel events are sometimes built by considering hypothetical exon–exon junctions from the annotation [85].

Methods that calculate differential relative abundance of events or exons under at least two conditions include (Subheading 2.7) SpliceSeq [33], MISO [38], ALEXA-Seq [39], SOLAS [40], DEXSeq [82], DSGSeq [83], GPSeq [84], MATS [85], JuncBase [86], JETTA [87], SplicingCompass [88], DiffSplice [89], FDM [90] rDiff [91, 128], and the methods from ref. 129. ALEXA-Seq estimates inclusion levels on a set of pre-calculated events using only unambiguous reads, i.e., reads that map to one unique event, and calculates various measures of differential expression, including the splicing index, i.e., a measure of change in expression of an event between two conditions relative to the change in expression of the entire gene locus between the same two conditions. On the other hand, SOLAS uses single-reads and only takes into account those mapping within exons, disregarding reads spanning exon–exon junctions, to detect differentially spliced events between two conditions. DEXSeq, DSGSeq, and GPSeq use read counts on exons to calculate those genes with differential splicing between two conditions. They do not provide any event or isoform information and report the exons with significant change (Fig. 1). MATS and MISO use both a Bayesian approach to calculate the differential inclusion of splicing events between two samples, using reads that map to exons and to the inclusion and skipping exon junctions. JuncBASE also uses reads mapped to exon junctions and uses a Fisher exact test to compare the read count in the inclusion and exclusion forms in two conditions. JETTA estimates the differential inclusion between two conditions from pre-calculated expression values for genes, exons, and junctions, which the authors obtain using SeqMap [130] and rSeq [49]. SpliceSeq calculates read coverage along genes, exons, and junctions for each sample, which are then compared to identify significant changes in splicing across samples. SpliceSeq also includes the evaluation of the impact of alternative splicing on protein products and a visualization of the events (see below). Besides all these methods, various methods were proposed in ref. 129 based on reads over exon junctions to find robust estimates of PSI, taking into account the positional bias of reads relative to the junction.

Some of these methods can also measure the change in the relative abundance of isoforms (Fig. 1): MISO can measure changes in isoform relative abundances from previously calculated isoform PSI values; ALEXA-Seq uses the events that are differentially expressed to infer isoform abundance differences between two conditions. Finally, rDiff, FDM, and DiffSplice are methods that work with a more general definition of event and that can operate without an annotation. FDM and DiffSplice are graph-based methods and both identify regions of differential abundance of transcripts between two samples using the variability of reads that define a splicing graph. Similarly, rDiff uses a Maximum Mean Discrepancy test [131] to estimate regions that have a significant distance between the read distributions in the two conditions. Alternatively, rDiff can work with an annotation; it considers reads in exonic regions that are not in all isoforms and groups those regions according to whether they occur in the same set of isoforms. Finally, SplicingCompass uses a geometric approach to detect differentially spliced genes and quantifies relative exon usage.

In summary, these methods test whether events, isoforms, or genic regions, change their relative abundances between two or more conditions, and so directly address the question of differential splicing.

When comparing two or more conditions, biological variability becomes an important issue, which has been shown to be relevant for studying expression [132] and splicing [82] from RNA-Seq data. However, not all methods take this into account. From the methods described here, DEXSeq, DSGSeq, GPSeq, DiffSplice, FDM, rDiff, and a newer version of MATS accept multiple replicates and model biological variability in different ways. In contrast, the initial methods for calculating splicing changes from RNA-Seq data [2, 3, 105], as well as MISO, ALEXA-Seq, JETTA, SpliceSeq, SOLAS, and SplicingCompass, do not work with multiple replicates. On the other hand, JuncBASE can work with replicated data but does not seem to model variability. As the cost of sequencing continue to decrease, it will be more common to include replicates in the differential splicing analysis, which will prove relevant to discern actual regulatory changes from biological variability.

3.3.2 Isoform-Based Differential Expression

Current methods to study differential splicing at the event level show a high validation rate [2, 85]. However, their agreement with microarray-based methods is not as high as one may expect [2]. This limitation could be due to the simplification of considering only events, rather than full RNA isoforms. An improvement in this direction would be to quantify changes in isoform expression. A possible approach is to combine methods that quantify isoforms with methods for differential gene expression. However, as previously pointed out [5, 90, 93], this may be problematic, since tools for differential gene expression analysis do not generally take into account the uncertainty of mapping reads to isoforms. We will not discuss here the many methods that have been proposed to study differential gene expression analysis from RNA-Seq data; for a recent review *see* refs. 5, 133.

A number of methods have been proposed to detect expression changes at the isoform level (Subheading 2.8): BitSeq [58], BASIS [92], Cuffdiff2 [93], and EBSeq [94]. Cuffdiff2, BitSeq, and EBSeq take into account the read-mapping uncertainty, accept multiple replicates and model biological variability. BASIS does not accept replicates, but it models variability along genes. Cuffdiff2 and BitSeq provide quantification and differential expression of isoforms from genome-mapped and transcriptome-mapped reads, respectively. Cuffdiff2 can use reads directly mapped to the genome or can use the results from Cufflinks on two conditions after using cuffcompare [46] (Fig. 1), which gives equivalent transcripts in both conditions. On the other hand, EBSeq relies on the isoform quantification from other methods, like RSEM or Cufflinks, and is actually included in the current release of RSEM;

whereas BASIS uses coverage over exon regions that are isoform-specific to calculate differential expression of isoforms. These methods rely on an annotation, either genome-based (Cuffdiff2, BASIS, and EBSeq) or transcriptome-based (BitSeq and EBSeq). Except for Cuffdiff2, these methods do not explicitly address the question of whether the relative abundance of these isoforms change across samples (Fig. 1). Accordingly, if there is an increase of transcription but the relative abundance of isoforms remain constant, they can detect changes in isoform expression, even though there might not be an actual change in splicing. On the other hand, if there are changes in the relative abundance of isoforms, they may possibly detect expression changes, but they will not provide information about the change of the relative abundances, and therefore do not directly address the question of differential splicing.

3.4 Visualizing Alternative Splicing

Being able to visualize the complexity of alternative splicing is an important aspect of the analysis. In the past, there have been multiple efforts to store and visualize alternative isoforms from ESTs and cDNAs [134, 135]. Visualization for RNA-Seq requires specialized tools that can efficiently process large amount of data from multiple samples. This has triggered the development of specialized tools to visualize alternative isoforms and events from RNA-Seq data (Subheading 2.9). Perhaps the simplest way to visualize isoforms and events is to generate track files for a genome browser. For instance, RSEM produces WIG files that can be viewed as tracks in the UCSC browser [136]. Similarly, SpliceGrapher and DiffSplice produce files in GFF-like formats (http://gmod.org/wiki/GFF), which can be uploaded into visualization tools like GBrowse [137] or Apollo [138]. On the other hand, SpliceGrapher and Alexa-Seq have their own visualization utilities. Other tools have been developed independently from the analysis method. For instance, the Sashimi plot toolkit to visualize isoforms and events and their relative coverage was used with MISO but can be used with the results from other tools (Subheading 2.8). Similarly, the browser Savant [95] has been used in conjunction with iReckon, but can be used independently for multiple HTS data formats. Finally, SpliceSeq [33] and SplicingViewer [96] are stand-alone tools that, besides mapping reads and quantifying events, also provide a visualization of results.

4 Conclusions and Outlook

The rapid development of short-read RNA sequencing technologies has triggered the development of new methods for data analysis. In this review, we have tried to provide an overview of methods applicable to the study of alternative splicing. These provide a way to detect and quantify exon–exon junctions, transcript isoforms, and differential splicing. Despite the many tools available not all are necessarily applicable to every purpose. For instance, for genomes with good annotation coverage, like human, the expression of known isoforms and possibly their changes under several conditions might be more accurately assessed using annotation-guided methods. Similarly, if sufficient annotation is available, there are also hybrid methods that can quantify known isoforms and predict novel ones simultaneously. For newly sequenced genomes, there are effective methods to perform de novo reconstruction and quantification of isoforms. However, if one is specifically interested in protein-coding genes, there are also evidence-based gene prediction methods available, which can be quite effective for isoform prediction.

One can identify some open questions and areas of improvement. For instance, not all of the de novo transcript assembly methods describe multiple isoforms per gene and only few actually quantify them. These are still two hard problems to solve, as incompleteness or absence of transcriptomes can lead to many reconstruction and quantification errors [139]. There are different approaches to improve these questions, either by a combination of methods and homology searches [140] or by using error correction of sequencing reads before assembly [141]. These tools are of great relevance for non-model organisms and we will probably see substantial improvements in the near future. Accurate reconstruction and quantification of isoforms is crucial for downstream analysis and in particular, for differential analysis of isoform abundances. Methods to estimate differential splicing at the event level seem to provide accurate measures as shown by experimental validation. However, differential expression at the isoform level is still an active area of development.

Extending de novo transcriptome assembly methods to calculate differential expression of isoforms between two or more conditions could facilitate the analysis of isoform expression for non-model organisms. Although this may be done currently with a combination of methods, a tool that integrates all these could provide a powerful approach to study expression and splicing in tumor samples, where multiple genome rearrangements and copy number alterations are expected to have occurred. On a different direction, considering that a reference genome sequence does not represent all DNA that can be possibly transcribed in a cell, unmapped RNA reads may come from functional RNAs not represented in the genome annotation. Tools that map reads to a genome reference and simultaneously attempt to perform transcript assembly will be also quite useful to perform systematic analyzes of RNA in cancer samples as well as in genomes that are partly assembled.

Besides the technical improvements, there is probably also a need to improve the comparison and evaluation of current methods. Transcript reconstruction methods should be evaluated using manual gene annotation sets, as proposed previously for gene prediction methods [123] and currently by RGASP for RNA-Seq based methods (http://www.gencodegenes.org/rgasp). Additionally, these comparisons should use measures that take into account alternative splicing [123, 142]. Similarly, there is the need to develop an experimental gold standard dataset for isoform quantification and differential isoform expression [143].

As a final question, we may ask for how long some of these methods will be needed. There are new technologies for single-molecule sequencing that soon will be used to probe the transcriptome. This may preclude the need to perform reconstruction of isoforms. Nonetheless, short-read RNA-Seq may still be necessary for efficient quantification. On the other hand, single-molecule sequencing technologies will open up a whole new set of problems, like that of reconciling new cell-specific RNA sequences with the information available for the genome sequence and its annotation. In fact, we will be in the position to quantify multiple transcriptomes and to revisit previous studies of differential splicing and expression in cancer, as the DNA and transcription complexity of the tumor cell is fully revealed.

With this review, we have aimed to provide an overview of the different tools to study different aspects of alternative splicing from RNA-Seq data, organized such that it is useful for the end user to navigate through the list of methods. All of them have their advantages and disadvantages, but are certainly useful to answer specific questions. We also hope that this review makes it easier to identify the tools that are still missing in order to improve the study of splicing with RNA-Seq.

Acknowledgements

We thank Y. Xing, K. Hertel, J.R. González, M. Kreitzman, and P. Drewe for comments and suggestions. This work was supported by the Spanish Ministry of Science with grants BIO2011-23920 and CSD2009-00080 and by Sandra Ibarra Foundation for Cancer with grant FSI 2011-035.

References

1. Djebali S, Davis CA, Merkel A et al (2012) Landscape of transcription in human cells. Nature 489(7414):101–108
2. Wang ET, Sandberg R, Luo S et al (2008) Alternative isoform regulation in human tissue transcriptomes. Nature 456(7221):470–476
3. Pan Q, Shai O, Lee LJ et al (2008) Deep surveying of alternative splicing complexity in the human transcriptome by high-throughput sequencing. Nat Genet 40(12): 1413–1415
4. Chen L (2011) Statistical and computational studies on alternative splicing. In: Horng-Shing Lu H et al (eds) Handbook of statistical bioinformatics. Springer, New York. doi:10.1007/978-3-642-16345-6_2

5. Pachter L (2011) Models for transcript quantification from RNA-Seq. arXiv:1104.3889v2 (http://arxiv.org/abs/1104.3889)
6. Trapnell C, Pachter L, Salzberg SL (2009) TopHat: discovering splice junctions with RNA-Seq. Bioinformatics 25(9):1105–1111
7. Huang S, Zhang J, Li R et al (2011) SOAPsplice: genome-wide ab initio detection of splice junctions from RNA-Seq data. Front Genet 2(July):46
8. Zhang Y, Lameijer EW, 't Hoen PA et al (2012) PASSion: a pattern growth algorithm-based pipeline for splice junction detection in paired-end RNA-Seq data. Bioinformatics 28(4):479–486
9. Wang K, Singh D, Zeng Z et al (2010) MapSplice: accurate mapping of RNA-seq reads for splice junction discovery. Nucleic Acids Res 38(18):e178
10. Au KF, Jiang H, Lin L et al (2010) Detection of splice junctions from paired-end RNA seq data by SpliceMap. Nucleic Acids Res 38(14):4570–4578
11. Dimon MT, Sorber K, DeRisi JL (2010) HMMSplicer: a tool for efficient and sensitive discovery of known and novel splice junctions in RNA-Seq data. PloS one 5(11):e13875
12. Li Y, Li-Byarlay H, Burns P et al (2013) TrueSight: a new algorithm for splice junction detection using RNA-seq. Nucleic Acids Res 41(4):e51
13. Marco-Sola S, Sammeth M, Guigó R et al (2012) The GEM mapper: fast, accurate and versatile alignment by filtration. Nat Methods 9(12):1185–1188
14. Ameur A, Wetterbom A, Feuk L et al (2010) Global and unbiased detection of splice junctions from RNA-seq data. Genome Biol 11(3):R34
15. Bryant DW, Shen R, Priest HD et al (2010) Supersplat– spliced RNA-seq alignment. Bioinformatics 26(12):1500–1505
16. Wang L, Wang X, Wang X et al (2011) Observations on novel splice junctions from RNA sequencing data. Biochem Biophys Res Commun 409(2):299–303
17. Lou SK, Ni B, Lo LY et al (2011) ABMapper: a suffix array-based tool for multi-location searching and splice-junction mapping. Bioinformatics 27(3):421–422
18. Bao H, Xiong Y, Guo H et al (2009) MapNext: a software tool for spliced and unspliced alignments and SNP detection of short sequence reads. BMC Genomics 10(Suppl 3):S13
19. Dobin A, Davis CA, Schlesinger F et al (2013) STAR: ultrafast universal RNA-seq aligner. Bioinformatics 29(1):15–21
20. Wu TD, Nacu S (2010) Fast and SNP-tolerant detection of complex variants and splicing in short reads. Bioinformatics 26(7):873–881
21. De Bona F, Ossowski S, Schneeberger K et al (2008) Optimal spliced alignments of short sequence reads. Bioinformatics 24(16): i174–i180
22. Jean G, Kahles A, Sreedharan VT et al. (2010) RNA-Seq read alignments with PALMapper. Curr Protoc Bioinformat Chapter 11:Unit 11.6
23. Philippe N, Salson M, Commes T et al (2013) CRAC: an integrated approach to the analysis of RNA-seq reads. Genome Biol 14(3):R30
24. Wu J, Anczuków O, Krainer AR et al (2013) OLego: fast and sensitive mapping of spliced mRNA-Seq reads using small seeds. Nucl Acids Res 41(10):5149–5163
25. Liao Y, Smyth GK, Shi W (2013) The Subread aligner: fast, accurate and scalable read mapping by seed-and-vote. Nucleic Acids Res 41(10):e108
26. Hu J, Ge H, Newman M, Liu K (2012) OSA: a fast and accurate alignment tool for RNA-Seq. Bioinformatics 28(14):1933–1934
27. Wood DL, Xu Q, Pearson JV et al (2011) X-MATE: a flexible system for mapping short read data. Bioinformatics 27(4):580–581
28. Chen LY, Wei KC, Huang AC et al (2012) RNASEQR—a streamlined and accurate RNA-seq sequence analysis program. Nucleic Acids Res 40(6):e42
29. Labaj PP, Linggi BE, Wiley HS et al (2012) Improving RNA-Seq Precision with MapAl. Front Genet 3:28
30. Xu G, Deng N, Zhao Z et al (2011) SAMMate: a GUI tool for processing short read alignments in SAM/BAM format. Source Code Biol Med 6(1):2
31. Kim H, Bi Y, Pal S et al (2011) IsoformEx: isoform level gene expression estimation using weighted non-negative least squares from mRNA-seq data. BMC Bioinforma 12:305
32. Grant GR, Farkas MH, Pizarro AD et al (2011) Comparative analysis of RNA-Seq alignment algorithms and the RNA-Seq unified mapper (RUM). Bioinformatics 27(18): 2518–2528
33. Ryan MC, Cleland J, Kim R et al (2012) SpliceSeq: a resource for analysis and visualization of RNA-Seq data on alternative splicing and its functional impacts. Bioinformatics 28(18):2385–2387
34. Tang S, Riva A (2013) PASTA: splice junction identification from RNA-Sequencing data. BMC Bioinforma 14(1):116
35. Bonfert T, Csaba G, Zimmer R et al (2012) A context-based approach to identify the most likely mapping for RNA-seq experiments. BMC Bioinforma 13(Suppl 6):S9
36. Wang L, Xi Y, Yu J et al (2010) A statistical method for the detection of alternative splicing using RNA-seq. PLoS one 5(1):e8529

37. Wu J, Akerman M, Sun S et al (2011) SpliceTrap: a method to quantify alternative splicing under single cellular conditions. Bioinformatics 27:3010–3016
38. Katz Y, Wang ET, Airoldi EM et al (2010) Analysis and design of RNA sequencing experiments for identifying isoform regulation. Nat Methods 7(12):1009–1015
39. Griffith M, Griffith OL, Mwenifumbo J et al (2010) Alternative expression analysis by RNA sequencing. Nat Methods 7(10):843–847
40. Richard H, Schulz MH, Sultan M et al (2010) Prediction of alternative isoforms from exon expression levels in RNA-Seq experiments. Nucl Acids Res 38(10):e112
41. Mortazavi A, Williams BA, Mccue K et al (2008) Mapping and quantifying mammalian transcriptomes by RNA-Seq. Nat Methods 5(7):1–8
42. Jiang H, Wong WH (2009) Statistical inferences for isoform expression in RNA-Seq. Bioinformatics 25(8):1026–1032
43. Bohnert R, Behr J, Rätsch G (2009) Transcript quantification with RNA-Seq data. BMC Bioinforma 10(Suppl 13):P5
44. Montgomery SB, Sammeth M, Gutierrez-Arcelus M et al (2010) Transcriptome genetics using second generation sequencing in a Caucasian population. Nature 464(7289): 773–777
45. Du J, Leng J, Habegger L et al (2012) IQSeq: integrated isoform quantification analysis based on next-generation sequencing. PLoS One 7(1):e29175
46. Trapnell C, Williams BA, Pertea G et al (2010) Transcript assembly and quantification by RNA-Seq reveals unannotated transcripts and isoform switching during cell differentiation. Nat Biotechnol 28(5):511–515
47. Rossell D, Attolini CSO, Kroiss M et al. (2012) Quantifying alternative splicing from paired-end RNA-sequencing data. COBRA Preprint Series. Working Paper 97 http://biostats.bepress.com/cobra/art97
48. Li W, Jiang T (2012) Transcriptome assembly and isoform expression level estimation from biased RNA-Seq reads. Bioinformatics 28(22):2914–2921
49. Li W, Feng J, Jiang T (2011) IsoLasso: a LASSO regression approach to RNA-Seq based transcriptome assembly. J Comput Biol 18(11):1693–1707
50. Feng J, Li W, Jiang T (2010) Inference of isoforms from short sequence reads. In: Berger B (ed) Research in computational molecular biology, lecture notes in computer science, vol 6044. Springer, Heidelberg, pp 138–157
51. Li JJ, Jiang CR, Brown JB et al (2011) Sparse linear modeling of next-generation mRNA sequencing (RNA-Seq) data for isoform discovery and abundance estimation. PNAS 108(50):19867–19872
52. Roberts A, Pimentel H, Trapnell C et al (2011) Identification of novel transcripts in annotated genomes using RNA-Seq. Bioinformatics 27(17):2325–2329
53. Mangul S, Caciula A, Glebova O et al (2012) Improved transcriptome quantification and reconstruction from RNA-Seq reads using partial annotations. Silico Biol 11(5): 251–261
54. Mezlini AM, Smith EJ, Fiume M et al (2013) iReckon: simultaneous isoform discovery and abundance estimation from RNA-seq data. Genome Res 23(3):519–529
55. Li B, Dewey CN (2011) RSEM: accurate transcript quantification from RNA-Seq data with or without a reference genome. BMC Bioinforma 12:323
56. Nicolae N, Mangul S, Mandoiu I et al (2011) Estimation of alternative splicing isoform frequencies from RNA-seq data. Algorithms Mol Biol 6:9
57. Lee S, Seo CH, Lim B et al (2011) Accurate quantification of transcriptome from RNA-seq data by effective length normalization. Nucleic Acids Res 39(2):e9
58. Glaus P, Honkela A, Rattray M (2012) Identifying differentially expressed transcripts from RNA-seq data with biological variation. Bioinformatics 28(13):1721–1728
59. Turro E, Su SY, Gonçalves Â et al (2011) Haplotype and isoform specific expression estimation using multi-mapping RNA-seq reads. Genome Biol 12(2):R13
60. Roberts A, Pachter L (2013) Streaming fragment assignment for real-time analysis of sequencing experiments. Nat Methods 10(1): 71–73
61. Denoeud F, Aury JM, Da Silva C et al (2008) Annotating genomes with massive-scale RNA sequencing. Genome Biol 9(12):R175
62. Zhao Z, Nguyen T, Deng N et al. (2011) SPATA: a seeding and patching algorithm for de novo transcriptome assembly. 2011 IEEE International Conference on Bioinformatics and Biomedicine Workshop (IEEE BIBMW'11) pp. 26–33
63. Filichkin S, Priest H, Givan S et al (2010) Genome-wide mapping of alternative splicing in Arabidopsis thaliana. Genome Res 20(1): 45–58
64. Guttman M, Garber M, Levin JZ et al (2010) Ab initio reconstruction of cell type-specific transcriptomes in mouse reveals the conserved multi-exonic structure of lincRNAs. Nat Biotechnol 28(5):503–510
65. Hiller D, Wong WH (2012) Simultaneous isoform discovery and quantification from RNA-Seq. Stat Biosci 5(1):100–118

66. Xia Z, Wen J, Chang CC et al (2011) NSMAP: a method for spliced isoforms identification and quantification from RNA-Seq. BMC Bioinforma 12:162
67. Rogers MF, Thomas J, Reddy AS et al (2012) SpliceGrapher: detecting patterns of alternative splicing from RNA-Seq data in the context of gene models and EST data. Genome Biol 13(1):R4
68. Seok J, Xu W, Jiang H et al (2012) Knowledge-based reconstruction of mRNA transcripts with short sequencing reads for transcriptome research. PLoS ONE 7(2):e31440
69. Behr J, Bohnert R, Zeller G et al (2010) Next generation genome annotation with mGene.ngs. BMC Bioinforma 11(Suppl 10):O8
70. Stanke M, Schöffmann O, Morgenstern B et al (2006) Gene prediction in eukaryotes with a generalized hidden Markov model that uses hints from external sources. BMC Bioinforma 7:62
71. Howe KL, Chothia T, Durbin R (2002) GAZE: a generic framework for the integration of gene-prediction data by dynamic programming. Genome Res 12(9):1418–1427
72. Allen JE, Salzberg SL (2005) JIGSAW: integration of multiple sources of evidence for gene prediction. Bioinformatics 21(18): 3596–3603
73. Haas BJ, Salzberg SL, Zhu W et al (2008) Automated eukaryotic gene structure annotation using EVidenceModeler and the Program to Assemble Spliced Alignments. Genome Biol 9(1):R7
74. Liu Q, Mackey AJ, Roos DS et al (2008) Evigan: a hidden variable model for integrating gene evidence for eukaryotic gene prediction. Bioinformatics 24(5):597–605
75. Martin J, Bruno VM, Fang Z et al (2010) Rnnotator: an automated de novo transcriptome assembly pipeline from stranded RNA-Seq reads. BMC Genomics 11:663
76. Surget-Groba Y, Montoya-Burgos J (2010) Optimization of de novo transcriptome assembly from next-generation sequencing data. Genome Res 20(10):1432–1440
77. Schulz MH, Zerbino DR, Vingron M et al (2012) Oases: robust de novo RNA-seq assembly across the dynamic range of expression levels. Bioinformatics 28(8):1086–1092
78. Xie Y, Wu G, Tang J et al. (2013) SOAPdenovo-Trans: De novo transcriptome assembly with short RNA-Seq reads. arXiv:1305.6760 [q-bio.GN] (http://arxiv.org/abs/1305.6760)
79. Robertson G, Schein J, Chiu R et al (2010) De novo assembly and analysis of RNA-seq data. Nat Methods 7(11):909–912
80. Grabherr MG, Haas BJ, Yassour M et al (2011) Full-length transcriptome assembly from RNA-Seq data without a reference genome. Nat Biotechnol 29(7):644–652
81. Sacomoto GA, Kielbassa J, Chikhi R et al (2012) KISSPLICE: de-novo calling alternative splicing events from RNA-seq data. BMC Bioinforma 13(Suppl 6):S5
82. Anders S, Reyes A, Huber W (2012) Detecting differential usage of exons from RNA-seq data. Genome Res 22(10): 2008–2017
83. Wang W, Qin Z, Feng Z et al (2013) Identifying differentially spliced genes from two groups of RNA-seq samples. Gene 518(1):164–170
84. Srivastava S, Chen L (2010) A two-parameter generalized Poisson model to improve the analysis of RNA-seq data. Nucleic Acids Res 38(17):e170
85. Shen S, Park JW, Huang J et al (2012) MATS: a Bayesian framework for flexible detection of differential alternative splicing from RNA-Seq data. Nucleic Acids Res 40(8):e61
86. Brooks AN, Yang L, Duff MO et al (2011) Conservation of an RNA regulatory map between Drosophila and mammals. Genome Res 21(2):193–202
87. Seok J, Xu W, Gao H et al (2012) JETTA: junction and exon toolkits for transcriptome analysis. Bioinformatics 28(9):1274–1275
88. Aschoff M, Hotz-Wagenblatt A, Glatting KH et al (2013) SplicingCompass: differential splicing detection using RNA-Seq data. Bioinformatics 29(9):1141–1148
89. Hu Y, Huang Y, Du Y et al (2013) DiffSplice: the genome-wide detection of differential splicing events with RNA-seq. Nucleic Acids Res 41(2):e39
90. Singh D, Orellana CF, Hu Y et al (2011) FDM: a graph-based statistical method to detect differential transcription using RNA-seq data. Bioinformatics 27(19):2633–2640
91. Drewe P, Stegle O, Hartmann L et al (2013) Accurate detection of differential RNA processing. Nucl Acids Res 41(10):5189–5198
92. Zheng S, Chen L (2009) A hierarchical Bayesian model for comparing transcriptomes at the individual transcript isoform level. Nucleic Acids Res 37(10):e75
93. Trapnell C, Hendrickson DG, Sauvageau M et al (2013) Differential analysis of gene regulation at transcript resolution with RNA-seq. Nat Biotechnol 31(1):46–53
94. Leng N, Dawson JA, Thomson JA et al (2013) EBSeq: an empirical Bayes hierarchical model for inference in RNA-seq experiments. Bioinformatics 29(8):1035–1043
95. Fiume M, Williams V, Brook A et al (2010) Savant: genome browser for high-throughput sequencing data. Bioinformatics 26(16): 1938–1944

96. Liu Q, Chen C, Shen E et al (2012) Detection, annotation and visualization of alternative splicing from RNA-Seq data with SplicingViewer. Genomics 99(3):178–182
97. Slater GS, Birney E (2005) Automated generation of heuristics for biological sequence comparison. BMC Bioinforma 6:31
98. Kent WJ (2002) BLAT–the BLAST-like alignment tool. Genome Res 12(4):656–664
99. Wu TD, Watanabe CK (2005) GMAP: a genomic mapping and alignment program for mRNA and EST sequences. Bioinformatics 21(9):1859–1875
100. Fonseca NA, Rung J, Brazma A et al (2012) Tools for mapping high-throughput sequencing data. Bioinformatics 28(24):3169–3177
101. Garber M, Grabherr MG, Guttman M et al (2011) Computational methods for transcriptome annotation and quantification using RNA-seq. Nat Methods 8(6):469–477
102. Schneeberger K, Hagmann J, Ossowski S et al (2009) Simultaneous alignment of short reads against multiple genomes. Genome Biol 10(9):R98
103. Langmead B, Trapnell C, Pop M et al (2009) Ultrafast and memory efficient alignment of short DNA sequences to the human genome. Genome Biol 10(3):R25
104. Clark TA, Sugnet CW, Ares M Jr (2002) Genome wide analysis of mRNA processing in yeast using splicing-specific microarrays. Science 296(5569):907–910
105. Sultan M, Schulz MH, Richard H et al (2008) A global view of gene activity and alternative splicing by deep sequencing of the human transcriptome. Science 321(5891):956–960
106. Cloonan N, Forrest ARR, Kolle G et al (2008) Stem cell transcriptome profiling via massive scale mRNA sequencing. Nat Methods 5(7):613–619
107. Cloonan N, Xu Q, Faulkner GJ et al (2009) RNA-MATE: a recursive mapping strategy for high-throughput RNA-sequencing data. Bioinformatics 25(19):2615–2616
108. Tang F, Barbacioru C, Wang Y et al (2009) mRNA-Seq whole-transcriptome analysis of a single cell. Nat Methods 6(5):377–382
109. Chen L (2012) Statistical and computational methods for high-throughput sequencing data analysis of alternative splicing. Stat Biosci 5(1):138–155
110. Venables JP, Klinck R, Bramard A et al (2008) Identification of alternative splicing markers for breast cancer. Cancer Res 68(22): 9525–9531
111. Li R, Yu C, Li Y et al (2009) SOAP2: an improved ultrafast tool for short read alignment. Bioinformatics 25(15):1966–1967
112. Deng N, Puetter A, Zhang K et al (2011) Isoform-level microRNA-155 target prediction using RNA-seq. Nucleic Acids Res 39(9):e61
113. Nguyen TC, Deng N, Zhu D (2013) SASeq: a selective and adaptive shrinkage approach to detect and quantify active transcripts using RNA-Seq. arXiv:1208.3619v2 [q-bio.QM] (http://arxiv.org/abs/1208.3619v2)
114. Heber S, Alekseyev M, Sze SH et al (2002) Splicing graphs and EST assembly problem. Bioinformatics 18(Suppl 1):S181–S188
115. Haas BJ, Delcher AL, Mount SM et al (2003) Improving the Arabidopsis genome annotation using maximal transcript alignment assemblies. Nucleic Acids Res 31:5654–5666
116. Xing Y, Resch A, Lee C (2004) The multiassembly problem: reconstructing multiple transcript isoforms from EST fragment mixtures. Genome Res 14(3):426–441
117. Xing Y, Yu T, Wu YN et al (2006) An expectation-maximization algorithm for probabilistic reconstructions of full-length isoforms from splice graphs. Nucleic Acids Res 34(10):3150–3160
118. Nagaraj SH, Gasser RB, Ranganathan S (2007) A hitchhiker's guide to expressed sequence tag (EST) analysis. Brief Bioinform 8(1):6–21
119. Salzman J, Jiang H, Wong WH (2011) Statistical modeling of RNA-Seq data. Stat Sci 26(1):62–83
120. Li B, Ruotti V, Stewart R et al (2010) RNA-Seq gene expression estimation with read mapping uncertainty. Bioinformatics 26(4):493–500
121. Sonnenburg S, Schweikert G, Philips P et al (2007) Accurate splice site prediction using support vector machines. BMC Bioinforma 8(Suppl 10):S7
122. Stanke M, Keller O, Gunduz I et al (2006) AUGUSTUS: ab initio prediction of alternative transcripts. Nucleic Acids Res 34(Web Server issue):W435–W439
123. Guigó R, Flicek P, Abril JF et al (2006) EGASP: the human ENCODE genome annotation assessment project. Genome Biol 7(Suppl 1):S2.1–31
124. Pontius JU, Wagner L, Schuler GD (2003) UniGene: a unified view of the transcriptome. In: The NCBI Handbook. Bethesda (MD): National Center for Biotechnology Information http://www.ncbi.nlm.nih.gov/books/NBK21083/
125. Zhao QY, Wang Y, Kong YM et al (2011) Optimizing de novo transcriptome assembly from short-read RNA-Seq data: a comparative study. BMC Bioinforma 12(Suppl 14):S2
126. Jackson B, Schnable P, Aluru S (2009) Parallel short sequence assembly of transcriptomes. BMC Bioinforma 10(Suppl 1):S14
127. Vijay N, Poelstra JW, Künstner A et al (2013) Challenges and strategies in transcriptome

assembly and differential gene expression quantification. A comprehensive in silico assessment of RNA-seq experiments. Mol Ecol 22(3):620–634

128. Stegle O, Drewe P, Bohnert R et al (2010) Statistical tests for detecting differential rna-transcript expression from read counts. Nat Preced. doi:10.1038/npre.2010.4437.1
129. Kakaradov B, Xiong HY, Lee LJ et al (2012) Challenges in estimating percent inclusion of alternatively spliced junctions from RNA-seq data. BMC Bioinforma 13(Suppl 6):S11
130. Jiang H, Wong WH (2008) SeqMap: mapping massive amount of oligonucleotides to the genome. Bioinformatics 24(20):2395–2396
131. Borgwardt KM, Gretton A, Rasch MJ et al (2006) Integrating structured biological data by Kernel Maximum Mean Discrepancy. Bioinformatics 22(14):e49–e57
132. Hansen KD, Wu Z, Irizarry RA et al (2011) Sequencing technology does not eliminate biological variability. Nat Biotechnol 29: 572–573
133. Oshlack A, Robinson MD, Young MD (2010) From RNA-seq reads to differential expression results. Genome Biol 11(12):220. doi:10.1186/gb-2010-11-12-220
134. Bhasi A, Philip P, Sreedharan VT et al (2009) AspAlt: A tool for inter-database, inter-genomic and user-specific comparative analysis of alternative transcription and alternative splicing in 46 eukaryotes. Genomics 94(1):48–54
135. Martelli PL, D'Antonio M, Bonizzoni P et al (2011) ASPicDB: a database of annotated transcript and protein variants generated by alternative splicing. Nucleic Acids Res 39(Database issue):D80–D85
136. Karolchik D, Hinrichs AS, Kent WJ (2012) The UCSC Genome Browser. Curr Protoc Bioinformatics Chapter 1:Unit1.4
137. Donlin MJ. (2009) Using the Generic Genome Browser (GBrowse). Curr Protoc Bioinformatics, Chapter 9:Unit 9.9
138. Lee E, Harris N, Gibson M et al (2009) Apollo: a community resource for genome annotation editing. Bioinformatics 25: 1836–1837
139. Pyrkosz AB, Cheng H, Brown CT. (2013) RNA-Seq Mapping Errors When Using Incomplete Reference Transcriptomes of Vertebrates. arXiv:1303.2411 [q-bio.GN] (http://arxiv.org/abs/1303.2411)
140. Birzele F, Schaub J, Rust W et al (2010) Into the unknown: expression profiling without genome sequence information in CHO by next generation sequencing. Nucleic Acids Res 38(12):3999–4010
141. MacManes MD, Eisen MB (2013) Improving transcriptome assembly through error correction of high-throughput sequence reads. arXiv:1304.0817 [q-bio.GN] (http://arxiv.org/abs/1304.0817) (3/April/2013)
142. Eyras E, Caccamo M, Curwen V et al (2004) ESTGenes: alternative splicing from ESTs in Ensembl. Genome Res 14(5):976–987
143. Lovén J, Orlando DA, Sigova AA et al (2012) Revisiting global gene expression analysis. Cell 151(3):476–482

Chapter 27

Global Protein–RNA Interaction Mapping at Single Nucleotide Resolution by iCLIP-Seq

Chengguo Yao, Lingjie Weng, and Yongsheng Shi

Abstract

Eukaryotic genomes encode a large number of RNA-binding proteins, which play critical roles in many aspects of gene regulation. To functionally characterize these proteins, a key step is to map their interactions with target RNAs. UV crosslinking and immunoprecipitation coupled with high-throughput sequencing has become the standard method for this purpose. Here we describe the detailed procedure that we have used to characterize the protein–RNA interactions of the mRNA 3′ processing factors.

Key words CLIP, iCLIP, UV crosslinking, RNA-binding proteins, High-throughput sequencing

1 Introduction

The human genome encodes more than 800 potential RNA-binding proteins [1], which play a wide variety of important roles in gene expression, including RNA processing, trafficking, translation, and degradation [2]. Functional characterization of these proteins remains a key task in the post-genomic era. A major challenge in this effort has been to identify the natural RNA targets of these proteins in vivo. For this purpose, early studies relied on immunoprecipitation (IP) to isolate specific proteins with their associated RNAs, which are subsequently identified through differential display or microarray analysis (RIP-chip) [3–5]. In some cases, formaldehyde crosslinking was applied prior to IP to capture transient and/or weak protein–RNA interactions [6]. These methods, however, tend to suffer from relatively high background. Additionally, although the RNA targets can be identified by RIP analysis, the specific protein-binding sites within the RNAs could not be mapped [3–6].

To overcome these limitations, the Darnell group pioneered a method called CLIP (UV crosslinking and IP) [7]. In this method, UV irradiation is used to specifically crosslink proteins and RNAs

Klemens J. Hertel (ed.), *Spliceosomal Pre-mRNA Splicing: Methods and Protocols*, Methods in Molecular Biology, vol. 1126, DOI 10.1007/978-1-62703-980-2_27, © Springer Science+Business Media, LLC 2014

that are in direct contact. Cellular RNAs are then digested into smaller sizes by RNase, and specific proteins and their directly crosslinked RNA fragments are IPed. Following gel purification and linker ligation, the crosslinked RNAs are reverse transcribed and amplified by PCR for sequencing. Compared to RIP analysis, CLIP has a number of important advantages. First, UV irradiation, which only crosslinks proteins and RNAs that are in direct contact, is much more specific than formaldehyde crosslinking. Second, as the proteins and their RNA targets are covalently linked, IP can be performed under highly stringent conditions to improve specificity. Third, gel purification enriches the RNAs that are crosslinked to the target proteins instead of other co-purified proteins, further enhancing the specificity. Finally, when coupled with high-throughput sequencing (called HITS-CLIP or CLIP-seq), CLIP allows global mapping of the protein-binding sites at a high resolution. More recently, two modified versions of CLIP, PAR-CLIP (photoactivatable ribonucleoside-enhanced CLIP) and iCLIP (individual-nucleotide resolution CLIP), were introduced that enable global protein–RNA interaction mapping at single nucleotide resolution [8, 9]. We have adopted the iCLIP method developed by the Ule group and successfully applied it to the mRNA 3′ processing factor CstF64 [10]. Here we describe the detailed procedure and offer technical advice on how to optimize it for your protein of interest.

2 Materials

2.1 Solutions

1. 1× PBS buffer: (137 mM NaCl, 2.7 mM KCl, 8 mM Na_2HPO_4, 1.46 mM KH_2PO_4, pH 7.4).
2. Cell lysis buffer (50 mM Tris–HCl, pH 7.4; 100 mM NaCl; 1 % NP-40; 0.1%SDS; 0.5 % sodium deoxycholate).
3. High-salt buffer (50 mM Tris–HCl, pH 7.4; 1 M NaCl; 1 mM EDTA; 1 % NP-40; 0.1 % SDS; 0.5 % sodium deoxycholate).
4. PNK buffer (20 mM Tris–HCl, pH 7.4; 10 mM $MgCl_2$; 0.2 % Tween-20).
5. PK buffer (50 mM Tris–HCl, pH 7.4; 50 mM NaCl; 10 mM EDTA).
6. PK-urea buffer (50 mM Tris–HCl, pH 7.4; 50 mM NaCl; 10 mM EDTA; 7 M urea).
7. TE buffer (10 mM Tris–HCl, pH 7.4; 1 mM EDTA).
8. Sodium acetate (3 M, pH 5.2).
9. 100 % Ethanol.
10. NuPAGE MOPS SDS running buffer (Life Technologies).
11. NuPAGE transfer buffer (Life Technologies).

12. RNA phenol/chloroform (MP Biomedicals).
13. 2× SDS gel-loading buffer (100 mM Tris–Cl, pH 6.8; 200 mM β-mercaptoethanol; 2.5 % SDS; 0.2 % bromophenol blue; 25 % glycerol).
14. 2× RNA gel-loading buffer (0.025 % SDS; 0.025 % bromophenol blue; 0.025 % xylene cyanol; 0.5 mM EDTA; 95 % formamide).
15. 8 % Urea-PAGE gel (1× TBE; 8 % polyacrylamide gel [acrylamide:bisacrylamide = 19:1]; 8 M urea).

2.2 Enzymes, Reagents, Equipment

1. UV Stratalinker 1800 or equivalent.
2. Protein A Dynabeads (Life Technologies).
3. CstF64 antibody (Bethyl Laboratories).
4. RNase I.
5. DNase I Turbo (Life Technologies).
6. Protease inhibitor cocktail.
7. Shrimp alkaline phosphatase.
8. RNasin.
9. T4 RNA ligase.
10. T4 PNK.
11. [γ-P32]ATP.
12. Pre-stained protein marker.
13. 4–12 % NuPAGE Bis-Tris gel (Life Technologies).
14. Novex mini-cell electrophoresis system (Life Technologies).
15. Nitrocellulose membrane.
16. Thermomixer (Eppendorf).
17. Glycogen.
18. Proteinase K.
19. Superscript III reverse transcriptase.
20. RNase A.
21. Circligase II (Epicentre).
22. FastDigest *Bam*HI (Fermentas).
23. Phusion High-Fidelity DNA polymerase.
24. PCR purification kit (Qiagen).

2.3 Primer Sequences and Linkers

1. RNA linker (Thermo Scientific, formerly Dharmacon).

   5′-Phosphate-AGAUCGGAAGAGCGGUUCAG-3′-puromycin
2. Annealing oligo harboring *Bam*HI restriction enzyme site.

   5′-GTTCAGGATCCACGACGCTCTTCAAAA

3. Reverse transcription primers with different barcodes.
4. Rclip1: 5′-phosphate-NNAACCNNNAGATCGGAAGAGC GTCGTGGATCCTGAACCGC

   Rclip2: 5′-phosphate-NNACAANNNAGATCGGAAGAGCG TCGTGGATCCTGAACCGC

   Rclip3: 5′-phosphate-NNATTGNNNAGATCGGAAGAGCG TCGTGGATCCTGAACCGC
5. PCR primers

   P5: 5′-AATGATACGGCGACCACCGAGATCTACACTCTT TCCCTACACGACGCTCTTCCGATCT

   P3: 5′-CAAGCAGAAGACGGCATACGAGATCGGTCTCG GCATTCCTGCTGAACCGCTCTTCCGATCT

3 Methods

3.1 UV Crosslinking of Cells/Tissues

For adherent cells

1. Grow cells in 100-mm dishes to 80–90 % confluence.
2. Rinse plates with 1× PBS three times and remove PBS after final wash.
3. Remove lid and place the plate on ice. Irradiate once with 400 mJ/cm^2 at 254 nm in a Stratalinker (or other UV crosslinkers).
4. Add 4 ml 1× PBS to the plate and harvest cells by scraping with a cell lifter. Transfer cell suspension to three 1.5 ml microtubes.
5. Spin down at 15,000 ×*g* for 1 min at 4 °C in a minifuge to pellet cells and remove supernatant.
6. Snap-freeze cell pellets on dry ice and store at −80 °C.

For suspension cells

1. Spin down cells at 600 ×*g* for 5 min, wash with 1× PBS three times, leave cells in 1× PBS, and transfer to the 100-mm dishes.
2. Place the dishes on ice, remove the lid, irradiate, harvest, and freeze cells as described above.

For tissues

1. Harvest tissue and rinse with cold 1× PBS.
2. Dissociate the tissue by passing through the 200 μl pipette tip.
3. Transfer tissues to 100-mm dishes, irradiate, harvest, and freeze cells as described above.

3.2 Beads and Cell Lysate Preparation

Beads preparation

1. Use 100 μl of protein A Dynabeads beads (slurry volume) per IP and wash beads twice with 600 μl cell lysis buffer.
2. Resuspend beads in 200 μl cell lysis buffer and add 4–10 μg antibody.
3. Rotate tubes at room temperature for 1 h.
4. Wash beads with 600 μl cell lysis buffer three times.

Cell lysate preparation

1. Resuspend cell pellet in 500 μl cell lysis buffer.
2. Add 5 μl DNase I (2 U/μl), 5 μl protease inhibitor cocktail (100×), and appropriate amount of RNase I (to be determined in pilot experiments).
3. Incubate for 3 min at 37 °C with shaking at 1,200 rpm in a Thermomixer.
4. Transfer to ice and leave on ice for 2 min.
5. Spin down at 15,000×*g* in 4 °C for 15 min and collect the supernatant for IP.

3.3 Immunoprecipitation

1. Remove wash buffer from beads and mix beads with cell lysate.
2. Rotate the samples overnight at 4 °C.
3. Collect the beads using a magnet and discard the supernatant.
4. Wash beads twice with 600 μl high-salt buffer.
5. Wash beads twice with 600 μl wash buffer.

3.4 Dephosphorylation of the 5′ Ends of RNAs

1. Resuspend beads in

   35 μl water

   4 μl 10× Shrimp alkaline phosphatase buffer

   1 μl Shrimp alkaline phosphatase (10 U/μl)

   Total volume of resuspension reaction: 40 μl.
2. Incubate at 37 °C for 10 min (1,200 rpm for 10 s every half a min in a Thermomixer).
3. Wash beads twice with 600 μl high-salt buffer.
4. Wash beads once with 600 μl PNK buffer.
5. Wash beads once with 50 μl 1× T4 RNA ligase buffer.

3.5 3′ Linker Ligation

1. Resuspend beads in

   4 μl PEG8000 (50 %)

   4 μl RNA linker (20 μM)

   2 μl 10× T4 RNA ligase buffer

   2 μl BSA (10 μg/μl)

7 μl water

0.5 μl RNaseOUT (40 U/μl)

0.5 μl T4 RNA ligase (10 U/μl)

Total volume of resuspension reaction: 20 μl.

2. Incubate for 21 h at 16 °C (1,200 rpm for 10 s every 3 min in a Thermomixer).
3. Wash beads with 600 μl PNK buffer twice.

3.6 RNA 5′ End Labeling

1. Resuspend beads in

   15 μl water

   2 μl 10× T4 PNK buffer

   2 μl [γ-P^{32}]ATP (10 μCi/μl)

   1 μl T4 PNK (10 U/μl)

   Total volume of resuspension reaction: 20 μl.
2. Incubate at 37 °C for 10 min (1,200 rpm for 10 s every 3 min in a Thermomixer).
3. Wash beads three times with 600 μl PNK buffer.

3.7 SDS-PAGE and Membrane Transfer

1. Add 20 μl 1× SDS gel-loading buffer to the beads and heat at 70 °C for 5 min.
2. Collect the beads on a magnet and load the supernatant on a NuPAGE gel and load a pre-stained protein marker in the next lane.
3. Run the gel in 1× MOPS running buffer at 180 V until the bromophenol blue dye reaches the bottom of the gel.
4. Transfer the gel to a nitrocellulose membrane using the Novex wet transfer apparatus (400 mA for 1 h at 4 °C).
5. Rinse membrane in 1× PBS, wrap the membrane in plastic wrap and expose it to a phosphorimager screen. To help align the gels with the image, we usually mark the protein ladder bands with a small amount of radioactivity.

3.8 RNA Isolation and Reverse Transcription

1. Based on the autoradiograph image from the last step (a typical image is shown in Fig. 1, further discussed in Subheading 4), cut out the smear band above the expected protein size (20–70 kDa above the expected protein size). Place the membrane pieces in a 1.5 ml tube.
2. Add 200 μl PK buffer and 10 μl proteinase K (20 μg/μl) to the membrane pieces, incubate at 37 °C for 20 min with shaking at 1,200 rpm in a Thermomixer.
3. Add 200 μl PK-urea buffer, incubate at 37 °C for 20 min with shaking at 1,200 rpm in a Thermomixer.

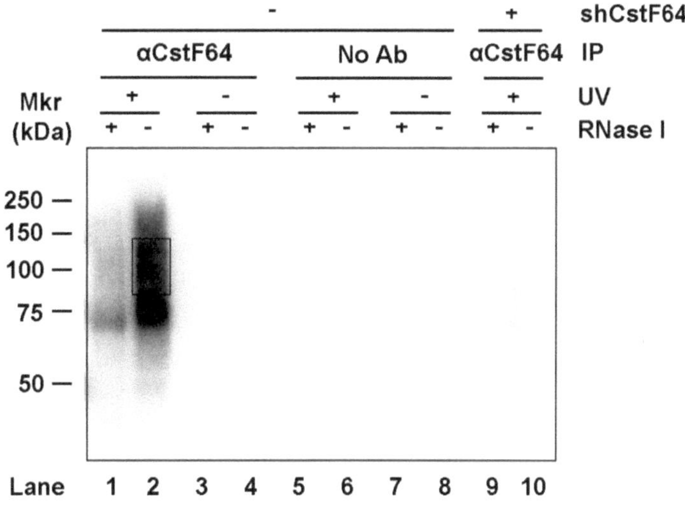

Fig. 1 *Top panel*: Autoradiography image of the 5′ ^{32}P-labeled RNA–protein complexes from IP using no antibody (No Ab) or anti-CstF64 antibodies (αCstF64) with cell lysates from control HeLa (−shCstF64) or a HeLa cell line that stably expresses shRNAs targeting CstF64 mRNA (+shCstF64). The rectangle area was cut out from the membrane for subsequent steps. *Bottom panel*: CstF64 and actin western blotting results of the lysates used for IP in the *top panel*. (Gel images are from ref. 13)

4. Add 400 μl phenol/chloroform and incubate at 37 °C for 5 min with shaking at 1,200 rpm in a Thermomixer.
5. Spin down at 15,000 × *g* for 10 min.
6. Take the supernatant and add 1 μl glycogen (5 μg/μl), 40 μl 3 M sodium acetate, and 1 ml 100 % ethanol.
7. Incubate overnight in −20 °C.
8. Spin down at 15,000 × *g* for 15 min. Wash pellet with 200 μl 70 % ethanol. Air dry the pellet and dissolve it in 11.5 μl water.
9. Prepare reverse transcription reaction as follows:

   11.5 μl RNA

   1 μl RT primer (2 μM, Rclip1, 2, or 3)

   1 μl dNTPs (10 μM)

   Total volume of resuspension reaction: 13.5 μl.
10. Incubate at 65 °C for 5 min and quickly chill on ice for 2 min.
11. Add the following:

    4 μl 5× First strand reverse transcriptase buffer

    1 μl DTT(0.1 M)

    0.5 μl RNaseOUT (40 U/μl)

    1 μl Superscript III reverse transcriptase (200 U/μl)

    Total volume of resuspension reaction: 20 μl.

12. Incubate at 42 °C for 10 min, then 50 °C for 40 min, 85 °C for 5 min, and then hold at 4 °C.
13. Add 1 µl RNase A (20 µg/µl), incubate at 37 °C for 20 min.
14. Add 80 µl TE, 1 µl glycogen (5 µg/µl), 10 µl 3 M sodium acetate, and 300 µl 100 % ethanol.
15. Incubate at −20 °C overnight.

3.9 Gel Purification and Circularization of cDNA

1. Spin down at 15,000 × *g* for 15 min.
2. Wash pellet with 70 % ethanol, air dry the pellet and resuspend it in 5 µl 1× RNA gel-loading dye.
3. Heat sample at 75 °C for 2 min and load it on a 8 % Urea-PAGE gel along with a molecular weight marker. Cut out 80–300 nt (nucleotide) gel pieces and elute the cDNAs from the gel pieces with 400 µl TE at room temperature overnight.
4. Spin down at 15,000 × *g* for 1 min, take out the supernatant, and add 40 µl 3 M sodium acetate, 1 ml 100 % ethanol, and 1 µl glycogen (5 µg/µl).
5. Incubate overnight at −20 °C.
6. Spin down at 15,000 × *g* for 15 min, wash the pellet with 70 % ethanol, dissolve the pellet in 12 µl water and add:

   1.5 µl 10× CircLigase buffer II

   0.75 µl $MnCl_2$ (50 mM)

   0.75 µl CircLigase II (100 U/µl)

   Total volume of resuspension reaction: 15 µl.

   Incubate at 60 °C for 2 h.
7. Add: 26 µl water

   5 µl 10× Fastdigest buffer

   1 µl Annealing oligo (10 µM)

   Total volume of resuspension reaction: 50 µl.

   Heat at 95 °C at 2 min.
8. Slowly cool down to room temperature.
9. Add 3 µl FastDigest *Bam*HI (20 U/µl) and incubate at 37 °C for 30 min.
10. Add 50 µl TE, 1 µl glycogen (5 µg/µl), 10 µl 3 M sodium acetate, and 300 µl 100 % ethanol. Incubate overnight at −20 °C.

3.10 PCR Amplification

1. Spin down at 15,000 × *g* for 15 min, wash the pellet, and dissolve pellet in 36.5 µl water. Then add:

   1 µl dNTPs (10 mM each)

   1 µl P5Solexa (10 µM)

   1 µl P3Solexa (10 µM)

10 μl 5× High-fidelity phusion polymerase buffer

0.5 μl High-fidelity phusion polymerase (2 U/μl)

Total volume of resuspension reaction: 50 μl.

2. Run the following PCR program on a thermocycler:
   (a) 98 °C 30 s
   (b) 98 °C 10 s
   (c) 65 °C 30 s
   (d) 72 °C 30 s
   (e) Go to (b) for 29 cycles
   (f) 72 °C for 5 min
   (g) Hold at 4 °C.
3. Purify the PCR products using the PCR purification kit.
4. Run a 1 % agarose gel to examine the size. A typical gel picture of the iCLIP library is shown in Fig. 2.
5. Measure the DNA concentration. Check with your sequencing facility about the required DNA concentration (usually 20–100 nM) and adjust your sample accordingly.

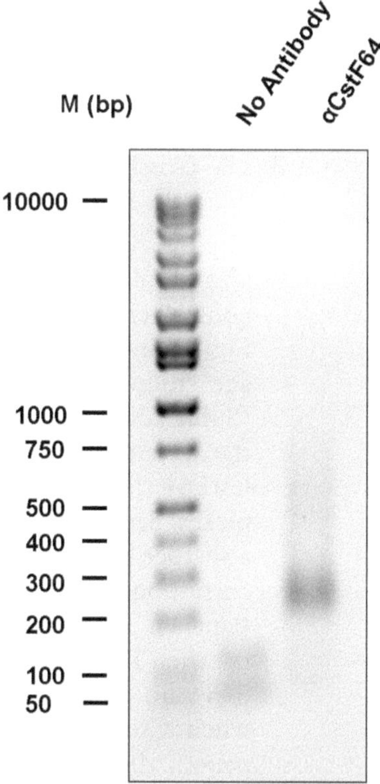

Fig. 2 SYBR staining of PCR-amplified CstF64 iCLIP cDNA library (*lane 2*) and the no antibody negative control (*lane 1*). Sizes of the DNA ladder are marked

3.11 Bioinfomatic Analysis

1. Raw reads are first demultiplexed using unique sequencing barcodes, and the random trinucleotides identifying individual cDNA molecules are clipped.
2. The remaining sequences are then filtered and mapped to the reference genome using bowtie with the setting (-n 2 -m 1 -s 1) (up to two nucleotide mismatches and one unique match to the reference genome allowed) [11].
3. After mapping, reads that truncate at the same sites and have the same barcodes are combined.
4. For each read, the base upstream of the 5′ end is marked as the crosslinking site, and the total number of reads sharing the same crosslinking site on the same strand, called "cDNA count" is calculated.
5. Crosslinking sites identified in multiple replicate libraries are considered high confidence sites and are used for further analyses.

For further details on the bioinformatics analysis, please refer to refs. 8, 10, 12.

4 Trouble-Shooting Tips

1. All the homemade solutions are prepared with Milli-Q water.
2. A Thermomixer or a similar shaker is recommended for iCLIP library construction.
3. UV crosslinking needs to be optimized. 100–400 mJ/cm^2 is generally recommended for most proteins.
4. Highly specific IP is key to the success of iCLIP-seq analysis and should be carefully optimized. An example of the quality control experiments we performed for CstF64 is shown in Figs. 1 and 2. The top panel of Fig. 1 displays a phosphorimage following gel transfer to nitrocellulose membrane (Subheading 3.7). First, a strong smeary band was observed in the IP sample that extends upwards from the expected size of CstF64 (lane 2). This corresponds to the CstF64–RNA complexes. Second, when the cell lysate was treated with RNase I, a sharper band at the expected size of CstF64 appeared, which corresponds to CstF64 crosslinked to small RNA fragments (lane 1). Third, when UV irradiation was omitted (lanes 3–4) or when IP was performed using protein A beads alone (lanes 5–8), no protein–RNA complex signal was detected. Fourth, when CstF64 is knocked down by RNAi (Fig. 1, bottom panel), the protein–RNA complex signal was proportionally reduced (lanes 9–10). Finally, specific CLIP PCR products

were only obtained in the IP sample and no specific PCR products were detected when no antibody was used (Fig. 2). These experiments demonstrate that the UV crosslinking and IP were efficient and specific. It is highly recommended that similar pilot experiments be carried out all CLIP analyses to ensure the specificity.

5. RNase digestion needs to be optimized. Several RNases have been used in CLIP analyses, including RNase A, RNase A/T1 mix, RNase I, and micrococcal nuclease (MN) [7, 8, 13, 14]. As MN activity is dependent on calcium, MN digestion can be terminated by using EGTA [13]. No matter which RNase is used, the amount of RNase and digestion time need to be carefully optimized to maximize the yield.
6. cDNA is purified on a 0.4 mm thick 8 % PAGE-urea gel. We usually cut out the band above the xylene cyanol which corresponds to ~80 nt. This step removes the free RT primers which may interfere with the following steps.
7. If the majority of the final PCR products are primer dimers (128 bp), there are two potential reasons. First, not enough protein–RNA complexes were IPed. In this case, the experiments should be further optimized or scaled up to increase the yield. Second, the RNAs or cDNAs may be lost. Carefully monitor the RNA in all steps. Following 5′ labeling, RNAs can be traced by using a Geiger counter.
8. Before submitting iCLIP libraries for high-throughput sequencing, it is recommended to clone an aliquot of the iCLIP-seq libraries into a DNA vector and sequence a few clones using Sanger sequencing. Carefully check the sequences to make sure the insert sizes are appropriate and the libraries are properly constructed.
9. As a final measure to ensure the specificity of iCLIP, it is recommended that three replicate libraries from individually processed samples be constructed and sequenced. Careful comparisons of the replicates provide valuable information on the reproducibility of the analysis.

Acknowledgement

This work was supported by an NIH grant R01GM090056 and ACS grant RSG-12-186.

References

1. Castello A, Fischer B, Eichelbaum K et al (2012) Insights into RNA biology from an atlas of mammalian mRNA-binding proteins. Cell 149:1393–1406
2. Hieronymus H, Silver PA (2004) A systems view of mRNP biology. Genes Dev 18:2845–2860
3. Trifillis P, Day N, Kiledjian M (1999) Finding the right RNA: identification of cellular mRNA substrates for RNA-binding proteins. RNA 5:1071–1082
4. Brooks SA, Rigby WF (2000) Characterization of the mRNA ligands bound by the RNA binding protein hnRNP A2 utilizing a novel in vivo technique. Nucleic Acids Res 28:E49
5. Tenenbaum SA, Carson CC, Lager PJ et al (2000) Identifying mRNA subsets in messenger ribonucleoprotein complexes by using cDNA arrays. Proc Natl Acad Sci USA 97:14085–14090
6. Gilbert C, Kristjuhan A, Winkler GS et al (2004) Elongator interactions with nascent mRNA revealed by RNA immunoprecipitation. Mol Cell 14:457–464
7. Ule J, Jensen KB, Ruggiu M et al (2003) CLIP identifies Nova-regulated RNA networks in the brain. Science 302:1212–1215
8. Konig J, Zarnack K, Rot G et al (2010) iCLIP reveals the function of hnRNP particles in splicing at individual nucleotide resolution. Nat Struct Mol Biol 17:909–915
9. Hafner M, Renwick N, Brown M et al (2011) RNA-ligase-dependent biases in miRNA representation in deep-sequenced small RNA cDNA libraries. RNA 17:1697–1712
10. Yao C, Biesinger J, Wan J et al (2012) Transcriptome-wide analyses of CstF64-RNA interactions in global regulation of mRNA alternative polyadenylation. Proc Natl Acad Sci USA 109:18773–18778
11. Langmead B, Trapnell C, Pop M et al (2009) Ultrafast and memory-efficient alignment of short DNA sequences to the human genome, Genome Biol 10:R25
12. Sugimoto Y, Konig J, Hussain S et al (2012) Analysis of CLIP and iCLIP methods for nucleotide-resolution studies of protein-RNA interactions, Genome Biol 13:R67
13. Yeo GW, Coufal NG, Liang TY et al (2009) An RNA code for the FOX2 splicing regulator revealed by mapping RNA-protein interactions in stem cells. Nat Struct Mol Biol 16:130–137
14. Sanford JR, Wang X, Mort M et al (2009) Splicing factor SFRS1 recognizes a functionally diverse landscape of RNA transcripts. Genome Res 19:381–394

Chapter 28

Predicting Alternative Splicing

Yoseph Barash and Jorge Vaquero Garcia

Abstract

Alternative splicing of pre-mRNA is a complex process whose outcome depends on elements reviewed in the previous chapters such as the core spliceosome units, how the core spliceosome units interact between themselves and with other splicing enhancers and repressors, primary sequence motifs, and local RNA secondary structure. Connections between RNA splicing, transcription, and other processes have also been reviewed in the previous chapters. Splicing is inherently a stochastic process: Some defective transcripts are produced and handled by mechanisms such as nonsense-mediated decay (NMD), and studies report high variability at the transcript level between cells supposedly in similar states. Nonetheless, splicing is obviously not a random process: Many determinants of splicing regulation have been identified, and experimental measurements detect highly robust and conserved splicing changes between developmental stages and tissues. These observations naturally lead to the following questions: Can we devise a method that predicts given a cellular context and the primary transcript what would be the splicing outcome? What can such a method tell us about the underlying mechanisms that govern alternative splicing?

This chapter describes how these questions can be framed and addressed using machine-learning methodology. We describe how to extract putative RNA regulatory features from genomic sequence of exons and proximal introns, how to define target values based on experimental measurements of exon inclusion, how to learn a simple splicing model that optimizes the prediction the observed exon inclusion levels from the identified RNA features, and how to subsequently evaluate the learned model accuracy.

Key words RNA, Alternative splicing, Machine learning, Computational biology, Posttranscriptional regulation

1 Introduction

A vast amount of knowledge has accumulated about splicing since its initial discovery in the 70′ [1, 2]. A wide variety of experimental methods reviewed in the previous chapters have been applied in order to gain that knowledge. Much of our mechanistic understanding about splicing was gained using classical techniques such as in vitro and cell-based splicing of minigenes (Chapters 11–13, 18), mass spec, and siRNA. More recently, high-throughput experiments added a complementary and much needed genome-wide view of splicing determinants. Techniques such as RNAseq

Klemens J. Hertel (ed.), *Spliceosomal Pre-mRNA Splicing: Methods and Protocols*, Methods in Molecular Biology, vol. 1126, DOI 10.1007/978-1-62703-980-2_28, © Springer Science+Business Media, LLC 2014

(Chapter 26) and CLIPseq (Chapter 27) allow us to quantify pre-mRNA and identify binding locations of splice factors at previously unparalleled resolution. Overall, the picture that has emerged through years of research about the determinants of splicing outcomes is a complex one, involving interactions between many elements.

Splicing involves several cores sequence motifs, namely the 5′ and 3′ splice site (5′ss, 3′ss), the branch point sequence (BPS), and the polypyrimidine tract. These core signals are recognized multiple times during spliceosome assembly (*see* Chapter 3) and can be represented as sequence motifs using various models [3, 4]. However, it was observed sometime ago that these sequence motifs are generally not sufficient to define where splicing occurs [5, 6].

Thousands of general splicing regulatory elements (SREs) have been identified via computational enrichment and conservation analysis of alternative and constitutively spliced exons. These SREs are usually represented using short k-mers, typically 5-6b long. SREs are generally divided to exonic splicing enhancers and silencers (ESEs and ESSs, respectively) and intronic splicing enhancers and silencers (ISEs and ISSs, respectively). A fraction of these SREs have been verified to affect splicing using specific experimental systems [7–9]. Some SREs recruit trans-acting factors that activate or suppress splice site recognition [7, 10], but overall the mechanisms by which SREs operate and interact are not well understood [3, 11].

While some splice factors are abundant in most cell types, others such as Fox1/2 or the members of the muscleblind-like (Mbnl) family of RNA-binding proteins, change their expression and protein levels across developmental stages and tissue types. These changes result in clear tissue or condition specific splicing changes in thousands of target exons [12, 13]. Dozens of splice factors have been identified so far, but the binding sites of many of those have not been well characterized. Many splice factor binding sites involve loosely defined motifs such as "CA-rich" for hnRNP-L or "CU-rich" for PTB1 and PTB2 [14, 15]. Subsequently, using such motifs as predictors for splicing outcome results in poor prediction accuracy [16].

Structural elements have also been implicated as splicing determinants. In mammalians, exons are relatively short, while introns are generally much longer and vary in length. In humans for example, exon length is sharply distributed around an average of approximately 147b. This, along with other observations, such as the dominance of exon skipping and cross-exon coevolution of splice sites, led to a "cross-exon" view of splicing in mammals, compared to an "intron centered" definition in invertebrates, plants, and fungi [17, 18]. Nonetheless, spliceosome assembly involves snRNPs interactions across the intron and changes to both exon and intron length have been demonstrated to affect alternative splicing outcome in mammals [19]. Similarly, RNA secondary structure in regions proximal to alternative exons has been shown to affect splicing outcome. For example, local structures expose or obscure recognition of sequence motifs by the core spliceosome machinery or auxiliary splicing factors (*see* Chapter 25).

Recently, new high-throughput experiments supplied valuable genome-wide information about various splicing regulatory determinants. RNAseq (Chapter 26) allows us to quantify each exon inclusion levels, while CLIP-based techniques (Chapter 27) allow us to identify binding locations of splice factors across the genome at previously unparalleled resolution. However, such experiments are still confounded with experimental noise and typically measure only the output (i.e., exon inclusion) or a single component (e.g., a known splice factor binding locations) of a complex system in a single cellular condition. For example, CLIPseq typically yields thousands of sites, many of which do not seem to exert a regulatory effect [13]. Thus, while the motifs recovered from CLIP sites generally match well computationally derived motifs, derivation of more accurate motif representation and target exons is still challenging. Contributing to this challenge is the fact that binding sites for RNA-binding proteins clearly have a context-specific effect and evoke different regulatory mechanisms. For example, CLIPseq-based "Motifs Maps" reveal that on average Fox1/2 binding upstream and downstream of an alternative exon cause an opposite effect on an alternative exon's inclusion [20, 21]. However, many of the affected exons appear to be exceptions to these "rules" as the motif maps simply marginalize the motif relative position compared to the exon but do not take into account other contextual elements. Indeed, well-studied examples of tissue-specific exon inclusion such as the N1 exon in src gene [19, 22–24], or exon 16 in daam1 [16], show complex combinations of many binding sites and structural elements that affect the observed splicing outcome.

Given the vast body of knowledge and newly derived high-throughput measurements a natural question is whether we can devise a method to combine all available data so that given a cellular context it will predict the splicing outcome of a given primary transcript. Consequently, we are also interested in what can such a method teach us about the underlying mechanisms that govern alternative splicing. In the next few paragraphs we discuss several points related to these two questions.

The first point to be made is that addressing the above two questions using computational methods can be seen as a natural extension to the experimental methods described in the previous chapters. Computational methods should allow researchers to derive specific regulatory hypotheses and perform in silico experiments such as mutating the genomic sequence to test the possible effect on splicing. Consequently, researchers should be able iterate between experimental techniques and computational modeling, increasing the yield of both.

Secondly, when referring to predictions given a "cellular context" it is important to define what the context is. In practice, we are not likely to have complete knowledge of a cellular state so the method we derive will need to train and consequently predict splicing outcome for generally defined conditions such as "normal mouse kidney cells." Depending on the input data, such a definition may average out many determinants of within cell and between cell variability, different cell types within the kidney, time-dependent changes (e.g., age, time of day), gender, and diet.

Thirdly, the ability to predict the splicing outcome of a given primary transcript can be framed in machine-learning methodology and theory as achieving generalization power: the ability of an algorithm to perform its task (i.e., splicing outcome prediction) accurately on new, unseen examples (aka test set) after learning the task from a training dataset. Real-life applications include predicting splicing outcome for a new transcript without performing the experiment, and evaluating the effect of sequence mutations or splice factor knockdowns in silico. A method that can achieve such generalization power may save on costly, labor intensive, experiments and overcome technical limitations such as read coverage depth or the inability to access tissue samples (e.g., a brain sample from a live patient).

Finally, the question of what can a predictive method for splicing teach us about underlying regulatory mechanisms concerns the interpretability of the applied method. In general, we would like a splicing prediction method to not only produce accurate predictions but also capture complex relations between known and novel regulatory elements in an easily interpretable way. Inferring meaningful causal relations from observational data or direct interference is the topic of much machine-learning research and theory and is beyond the scope of this chapter. Suffice to point out that such ability depends on the domain modeled, available data, and the type of algorithm employed. In what follows, we will consider the ability to easily interpret results as a consideration in algorithms derivation but will not focus on this issue here.

Previously, various computational works analyzed specific aspects of splicing regulation such enriched motifs around differentially spliced exons [25, 26], or built "motif maps" derived from CLIP for the binding locations of a specific splice factor [27]. Other works focused on the ability to distinguish alternative from constitutive exons [28]. Recently, we developed the first predictive splicing model for regulated alternative splicing [16, 29]. This model parsed over a thousand putative regulatory features (e.g., sequence motifs, local RNA structure) from a given genomic sequence to predict tissue-dependent inclusion of cassette exons. Many of the regulatory features identified by this model were inline with previous results about splicing regulation, and some novel elements predicted by the code were subsequently tested and verified experimentally.

In this chapter, we follow the machine learning-based approach used to derive the first splicing model. We review the process of building a predictive model as a machine-learning task, describing input feature definition and extraction, target value definition and extraction, scoring function, optimization method, and model interpretation.

2 Materials

The code used in this chapter is written in Python. Software to run the code can be downloaded from http://www.python.org/. The implementation of the machine-learning algorithms used here is freely available at http://scikit-learn.org/ [30]. All the example input/output and code files can be downloaded from http://kbc.biociphers.org

3 Methods

3.1 Retrieving Sequence Information for a Set of Cassette Exons

The first step in order to construct our dataset is to retrieve a set of putative cassette exons for which we can later extract features to train and test our algorithm. For each sample (i.e., putative cassette exon) we extract the genomic regions defined in Fig. 1. Here, we will assume the cassette exons are extracted from a set of known RefSeq transcripts. The RefSeq transcript set can be downloaded from the genome browser (http://genome.ucsc.edu/). The following simple script extracts the coordinates of the genomic regions shown in Fig. 1 from the RefSeq transcript track:

get_from_refSeq.py mm9.refSeq.txt events.names events.bed

Where mm9.refSeq.txt is the RefSeq track file, events.names is a list of events names, and events.bed is the output file. The events names serve as keys and are constructed as TRANSCRIPT.X.Y.Z (e.g., NM_133916.11.12.13), where TRANSCRIPT is the name of the RefSeq transcript and X.Y.Z are the indexes of the three exons used to define a putative cassette exon (Fig. 1).

The output .bed file can be used to extract matching genomic sequences using for example Galaxy (https://main.g2.bx.psu.edu/) [31]. Here, we will assume basic familiarity with

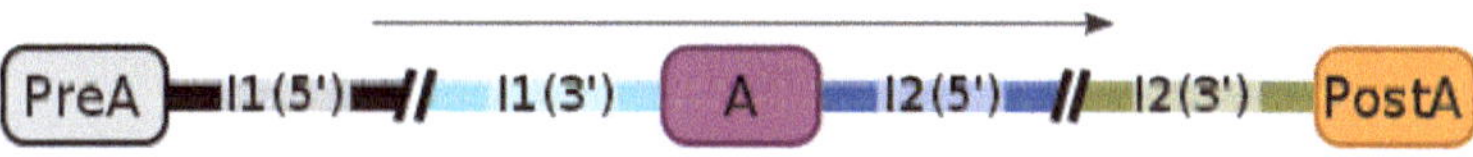

Fig. 1 Sequence regions map. The regions used to extract RNA features in order to predict alternative cassette exons

Galaxy or an equivalent tool, and basic knowledge in scripting. Thus, at the end of this processing step the sequence file downloaded from Galaxy should be parsed into a fasta format file named events.fa. Sequences in events.fa are named TRANSCRIPT.X.Y.Z_REG, where REG is any of the seven regions defined in Fig. 1.

3.2 Retrieving Putative Regulatory Features

Next, we need to define a set of putative regulatory features and extract those for the cassette exons set. In general, the feature set definition depends on the exact question we address, such as "What are the exons that are highly included in kidney but not in brain?" Since we generally do not know in advance the exact set of regulatory features and splicing is a complex process we prefer to extract an extended set of putative regulatory features and let the learning algorithm identify combinations of relevant ones.

The extracted features in our working example include sequence motifs known to bind several key splice factors, single and dinucleotide frequencies, and the length of the genomic regions considered (*see* **Notes 1–3**). The following simple script extracts these features for a given set of sequences:

```
get_motif_counts.py events.fa events.bed kmer.list diNuc.list
```

The output is a set of tab-delimited files named kmerFeatures, NucFreq, and regionLength, all sharing the given prefix (i.e., "events." in our example).

Features representing splice junctions strength supplement the above set of features. In order to compute the strength of the 3′ and 5′ splice sites we use the MaxEnt method [32] available at http://genes.mit.edu/burgelab/maxent/. The program, implemented using perl, requires as input the sequences around each splice sites. The following simple script can be used to parse those input sequences:

```
juncScore.py events.fa
```

The output .fa sequence files named junc3, junc5 share the "events." prefix and list the sequences for the 3′ and 5′ splice sites. These files can then be used to run the MaxEnt programs

```
perl Score3.pl events.junc3.fa > events.junc3.scores
perl Score5.pl events.junc5.fa > events.junc5.scores
```

Where the output of those is directed to create matching junc3.scores and junc5.scores.

3.3 Computing Target Values

Training a predictive splicing model involves supplying it with target values per sample and a loss function to optimize. The definition of these not only affects the model's accuracy, but also what the model actually captures in terms of the underlying regulatory mechanisms. For example, training on a labeled set of alternative vs. constitutive exons or training on differences in exon inclusion between muscle and kidney tissues will result in distinct models that capture different regulatory determinants.

In the example we follow here we will assume the labels are binary, representing whether the exon is alternatively spliced or not (*see* **Note 4**). Such information can be derived from EST/cDNA libraries and parsing RNAseq data. For example, the RUM pipeline [33] allows users to easily parse sequencing reads and detect reads that span across putative cassette exons' inclusion and exclusion junctions. The target values for our working example are given in the file events.labels with "1" and "0" representing alternative cassette or constitutive exons, respectively.

3.4 Learning a Predictive Splicing Model

At this stage we have created the input feature set and the target values. We can now apply various methods to achieve prediction of the target values given the input features. The original splicing model used a mixture of decision trees learned using a variant of boosting [16]. We therefore show here how to use a package for learning decision trees, freely available at http://scikit-learn.org/ [30].

Decision trees (DTs) have been used extensively for various machine-learning task [34]. Briefly, DTs are used for predicting an outcome given a sample's vector (i.e., the set of features parsed from a putative cassette exon) by applying a set of "rules" to it. Every internal node in the tree, starting from the root, contains a "rule" (e.g., "Is the alternative exon longer than 100b?"). Depending on the outcome of applying the rule to the sample, the sample is propagated to the appropriate node at the next level of the tree, until it reaches a leaf where a prediction is made. An example of such a tree for classifying alternative vs. constitutive exons is shown in Fig. 2.

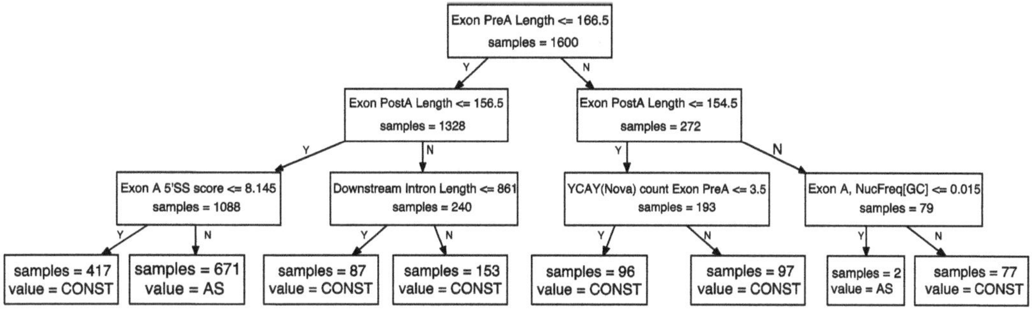

Fig. 2 A decision tree example. This tree was trained to distinguish between alternative and constitutive exons. Given a sample, i.e., a putative cassette exon feature set, rules in internal nodes are applied until the sample reaches a leaf node where a prediction is made. The number of samples marked in each node represents the total number of samples that reach that specific node from the training dataset

Trees can be used for both regression and classification tasks, with a variety of loss functions and rule definitions [34]. Some of their benefits include fast learning, avoiding the need for careful normalization and scaling when handling heterogeneous features, and interpretability of the learned model (*see* **Note 5**). Learning a decision tree involves learning the rules and the prediction functions at the leafs and much research has been dedicated to effective learning of those (*see* **Note 6**). However, as we shall see below the structure of a decision tree may not be optimal for certain learning tasks. For example, a learning task that involves sparse linear combination of the input features or a mixture of several groups of features will generally not be captured well by a DT. Specifically, splicing is known to involve different combinations of regulators and thus deriving a single DT to capture splicing outcome will likely result in subpar results.

A commonly used solution to the limitations of a single decision tree is to learn a combination or a mixture of those. Boosting has been previously introduced as a method to effectively learn models composed of a weighted mixture of decision trees for a variety of prediction tasks [35] (*see* **Notes 6** and **7**). Using the scikit-learn package this can be done using the following commands:

```
boost_tree = GradientBoostingClassifier(n_estimators =m,learning_
    rate = 1.0,max_depth = t_size)
clf= boost_tree.fit(x_train, y_train)
```

Where the first line creates the boost tree object and the second line executes the learning algorithm, returning the learned mixture of DT as the clf object. Learning a mixture of decision trees using boosting involves setting the number of boosting steps (n_estimators) that corresponds to the number of trees, the size of the trees (max_depth), and the learning rate (learning_rate). Setting these parameters is done by evaluating the performance of the learning algorithm, which is the topic of the next section.

3.5 Testing Procedure

The opening section of this chapter introduced the concept of generalization. In order to evaluate a machine-learning algorithm generalization power we require a dataset to train the algorithm ("training set"), and a dataset to test its performance ("test set"). Sometimes a third, "validation set" is required to set some of the algorithm's parameters during training. It is imperative that the test set will not include any sample from the training set. When a limited amount of samples is available, a K-fold cross-validation procedure is commonly employed: The dataset is randomly divided to K similarly sized subset. A fraction of 1/k of the data is kept "hidden." The algorithm trains on the remaining (k − 1)/k fraction of the data and is then tested on the "hidden" test set. The procedure is repeated K times for each 1/k data subset so every sample ends up appearing in the test set exactly once. Averaging the algorithm's performance by comparing predictions to the true values on all test data gives us an estimate of its generalization capabilities. The K-fold cross-validation can be repeated several times to also give an estimated variance of this performance evaluation.

The K-fold cross-validation procedure can be used with a variety of scoring metrics for evaluating performance. When dealing with binary classification a classifier will typically produce class prediction for every sample with some confidence value (e.g., the probability an exon is alternative given its observed features). Applying a specific threshold on this confidence value results in a certain false-positive rate (FPR—the fraction of false positives from the negative set) and sensitivity (fraction of correctly labeled samples from the positive set). The tradeoff between higher sensitivity and higher FPR for all possible confidence thresholds can be visualized using a receiver operation characteristic (ROC) curve. Several ROC curves for different algorithms tested on the example we follow here are shown in Fig. 3. Ideally, a classifier will reach the top left corner of the ROC curve, corresponding to 100 % sensitivity and 0 % FPR (i.e., perfect classification). Finally, in order to compare different ROC curves a commonly used summary statistic is the area under the curve (AUC). A theoretical random classifier

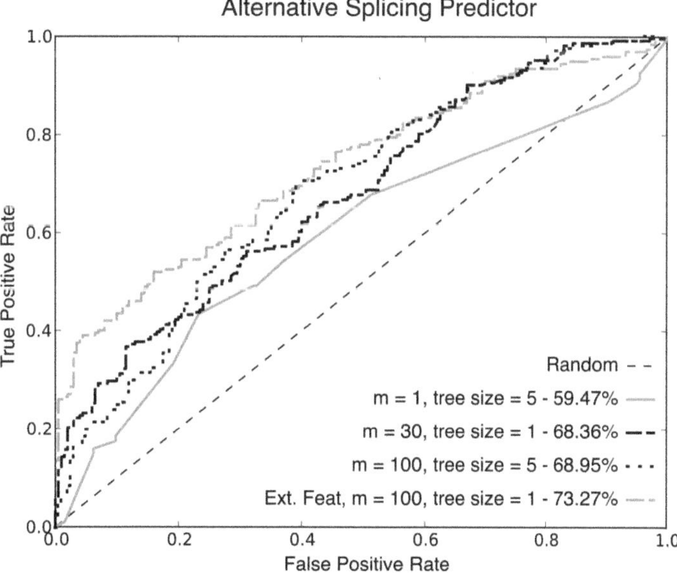

Fig. 3 Test data ROC curves for classification of alternative and consecutive exons from the sample dataset. Each curve represents a different setting of the number of decision trees (m) and the number of nodes in each tree (tree size). The extended feature set (bottom legend), including additional motifs and conservation level, helps improve prediction accuracy

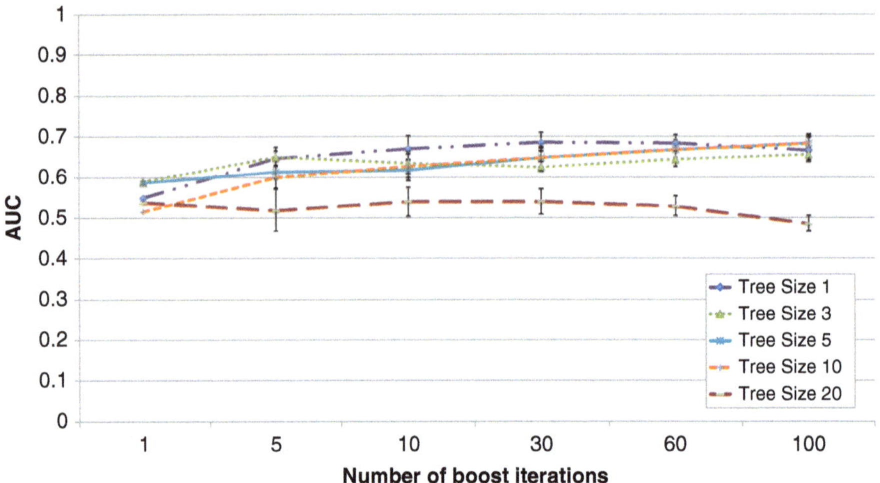

Fig. 4 AUC test performance (*y*-axis) as a function of boosting iterations (*x*-axis) for different tree sizes. AUC mean and variance were computed using 5 repetitions of fivefold cross-validation. Over-fitting can be observed for a mixture of 100 trees of size 20, leading to degradation of test performance

will produce a diagonal line with an area of 50 %, while a perfect classifier hitting the top left corner will cover the entire rectangle area with an AUC of 100 %.

Using the AUC as the test statistic we can evaluate the effect of different parameters on the learning algorithm (*see* **Note 8**). Figure 4 shows that from all the parameters tested for the sample dataset best test performance is achieved using a tree size of 10 and 100 boost iterations. Figure 3 shows a subset of ROC curves from those used in Fig. 4, along with the performance achieved using a more complex classifier trained with a larger set of features. Notably, the AUC on this specific dataset is relatively low. Recent work showed the accuracy for a much larger dataset with a more extensive set of putative regulatory features is around 91 %. Moreover, when discriminating between tissue-dependent alternative exons and constitutive exons the AUC reaches 95–98 %. It is important to note though that this and similar datasets are likely to suffer from false negatives that degrade performance as some exons deemed constitutive may be alternative exons that have not been experimentally detected yet.

4 Notes

1. Alternative methods to computing motif occurrences can be used to create putative regulatory features. For example, instead of using discrete values of occurrences of k-mers as in the example here, one can use more complex probabilistic

models that assign a real value conveying the posterior probability of a binding site occurrence given the genomic sequence or other elements [36, 37].

2. When computing motif occurrences, allowing or prohibiting motif overlap may result in a different feature value. For example, the sequence "AAAA" can be counted as having either two or three occurrences of the motif "AA," depending whether overlaps are allowed. While both are legitimate for usage as input features, they may yield different results. In any case, one must be cautious in interpreting enrichment of low complexity motifs (cf [38]).
3. For improving prediction accuracy it is recommended to add features that represent conservation in the flanking intronic regions. Since many regulatory elements controlling alternative splicing reside in the introns, increased conservation in the noncoding regions serve as excellent "clues" for detecting alternative exons.
4. Besides the gained prediction power, the biological interpretation of the inferred decision rules may be of interest. It is important to note though that the rule derivation as described here does not necessitate a causative or mechanistic relation. For example, conservation-based features described above may be highly indicative of regulation but do not offer a specific mechanism by themselves. Simple analysis of the derived rules includes testing what features where used, at what height of the trees where these features used, and how would prediction accuracy change if the features were removed.
5. Deriving a robust set of regulatory features can help improve both prediction accuracy and the confidence in the biological relevance of the features used by the model. One simple approach to derive a robust feature set is bootstrapping, with repetitive resampling of the original datasets with replacement. Each sampled dataset is used to train a model, and the resulting set of models can then be used to average test predictions. Similarly, the set of models can be used to identify "robust" rules, i.e., rules that are repeatedly selected by the trained models.
6. The choice of the loss function can have a significant effect on performance. The choice needs to fit well both the learning task and the data available [34]. For example, exponential loss functions may over penalize during training when dealing with noisy or mislabeled data. It therefore may be necessary to evaluate different loss functions using the procedures as those described above.
7. In the learning setting described here, it is important to use methods that encourage model sparseness and control for over-fitting.

The boosting learning procedure described above is an easy to implement and efficient method to achieve this. However, boosting should not be considered a solution by itself and proper evaluation using the train and test paradigm is mandatory. For example, the decision trees mixture model with 20 tree nodes and 100 trees shown in Fig. 4 clearly demonstrates over-fitting that result in diminished performance on test data.

8. The learning rate parameter, which was not discussed, can be evaluated using a similar procedure to the one described here for other parameters.

References

1. Berget SM, Moore C, Sharp PA (1977) Spliced segments at the 5′terminus of adenovirus 2 late mRNA. Proc Natl Acad Sci 74:3171–3175
2. Chow LT, Gelinas RE, Broker TR et al (1977) An amazing sequence arrangement at the 5? ends of adenovirus 2 messenger RNA. Cell 12:1–8
3. Chen M, Manley JL (2009) Mechanisms of alternative splicing regulation: insights from molecular and genomics approaches. Nat Rev Mol Cell Biol 10:741–754
4. Roca X, Krainer AR, Eperon IC (2013) Pick one, but be quick: 5′ splice sites and the problems of too many choices. Genes Dev 27:129–144
5. Lim LP, Burge CB (2001) A computational analysis of sequence features involved in recognition of short introns. Proc Natl Acad Sci 98:11193–11198
6. Black DL (1995) Finding splice sites within a wilderness of RNA. RNA 1:763–771
7. Yu Y, Maroney PA, Denker JA et al (2008) Dynamic regulation of alternative splicing by silencers that modulate 5′ splice site competition. Cell 135:1224–1236
8. Stadler M, Shomron N, Yeo GW et al (2006) Inference of splicing regulatory activities by sequence neighborhood analysis. PLoS Genet 2:e191
9. Wang Y, Xiao X, Zhang J et al (2013) A complex network of factors with overlapping affinities represses splicing through intronic elements. Nat Struct Mol Biol 20:36–45
10. Lam BJ, Hertel KJ (2002) A general role for splicing enhancers in exon definition. Rna 8:1233–1241
11. Shepard PJ, Hertel KJ (2010) Embracing the complexity of pre-mRNA splicing. Cell Res 20:866–868
12. Kalsotra A, Xiao X, Ward AJ et al (2008) A postnatal switch of CELF and MBNL proteins reprograms alternative splicing in the developing heart. Proc Natl Acad Sci 105:20333–20338
13. Wang ET, Cody NAL, Jog S et al (2012) Transcriptome-wide regulation of Pre-mRNA splicing and mRNA localization by muscleblind proteins. Cell 150:710–724
14. Ray D, Kazan H, Chan ET et al (2009) Rapid and systematic analysis of the RNA recognition specificities of RNA-binding proteins. Nat Biotechnol 27:667–670
15. Xue Y, Zhou Y, Wu T et al (2009) Genome-wide analysis of PTB-RNA interactions reveals a strategy used by the general splicing repressor to modulate exon inclusion or skipping. Mol Cell 36:996–1006
16. Barash Y, Calarco JA, Gao W et al (2010) Deciphering the splicing code. Nature 465:53–59
17. Robberson BL, Cote GJ, Berget SM (1990) Exon definition may facilitate splice site selection in RNAs with multiple exons. Mol Cell Biol 10:84–94
18. Wang Z, Burge CB (2008) Splicing regulation: From a parts list of regulatory elements to an integrated splicing code. RNA 14:802–813
19. Black DL (1991) Does steric interference between splice sites block the splicing of a short c-src neuron-specific exon in non-neuronal cells? Genes Dev 5:389–402
20. Yeo GW, Coufal NG, Liang TY et al (2009) An RNA code for the FOX2 splicing regulator revealed by mapping RNA-protein interactions in stem cells. Nat Struct Mol Biol 16:130–137
21. Zhang C, Zhang Z, Castle J et al (2008) Defining the regulatory network of the tissue-specific splicing factors Fox-1 and Fox-2. Genes Dev 22:2550–2563

22. Chan R, Black D (1997) The polypyrimidine tract binding protein binds upstream of neural cell-specific c-src exon N1 to repress the splicing of the intron downstream. Mol Cell Biol 17:4667–4676
23. Markovtsov V, Nikolic J, Goldman J et al (1992) Activation of c-src neuron-specific splicing by an unusual RNA element in vivo and in vitro. Cell 69:795–807
24. Rooke N, Markovtsov V, Cagavi E et al (2003) Roles for SR proteins and hnRNP A1 in the regulation of c-src exon N1. Mol Cell Biol 23:1874–1884
25. Yeo GW, Nostrand EL, Liang TY (2007) Discovery and analysis of evolutionarily conserved intronic splicing regulatory elements. PLoS Genet 3:e85
26. Castle JC, Zhang C, Shah JK et al (2008) Expression of 24,426 human alternative splicing events and predicted cis regulation in 48 tissues and cell lines. Nat Genet 40(12):1416–1425
27. Ule J, Stefani G, Mele A et al (2006) An RNA map predicting Nova-dependent splicing regulation. Nature 444(7119):580–586
28. Dror G, Sorek R, Shamir R (2005) Accurate identification of alternatively spliced exons using support vector machine. Bioinformatics 21:897–901
29. Xiong HY, Barash Y, Frey BJ (2011) Bayesian prediction of tissue-regulated splicing using RNA sequence and cellular context. Bioinformatics 27:2554–2562
30. Pedregosa F, Varoquaux G, Gramfort A et al (2011) Scikit-learn: machine learning in python. J Mach Learn Res 12:2825–2830
31. Giardine B, Riemer C, Hardison RC et al (2005) Galaxy: a platform for interactive large-scale genome analysis. Genome Res 15:1451–1455
32. Yeo G, Burge CB (2004) Maximum entropy modeling of short sequence motifs with applications to RNA splicing signals. J Comput Biol 11:377–394
33. Grant GR, Farkas MH, Pizarro AD et al (2011) Comparative analysis of RNA-Seq alignment algorithms and the RNA-Seq unified mapper (RUM). Bioinformatics 27:2518–2528
34. Hastie T, Tibshirani R, Friedman JH (2003) The elements of statistical learning. Springer, New York
35. Friedman JH (2001) Greedy function approximation: a gradient boosting machine. Ann Stat 29:1189–1232
36. Barash Y, Elidan G, Kaplan T et al (2005) CIS: compound importance sampling method for protein–DNA binding site p-value estimation. Bioinformatics 21:596–600
37. Barash Y, Elidan G, Friedman N et al (2003) Modeling dependencies in Protein–DNA binding sites. Proceedings of Seventh International Conference Res in Comp Mol Bio (RECOMB)
38. Sinha S, Tompa M (2000) A statistical method for finding transcription factor binding sites. Proc Int Conf Intell Syst Mol Biol 8:344–354

INDEX

A

A complex 19, 36, 38, 40, 41, 57, 60, 67, 72–74, 107, 147, 179, 291, 412, 413, 416
Affinity purification................................ 180, 181, 187, 212
Alternative splicing.............................. 18, 36, 45, 55, 86, 97, 151, 161, 193, 243, 299–312, 341, 357, 411–422
Antisense oligonucleotide (ASOs) 91, 180, 271–274, 276–278, 281, 282

B

Baculovirus 205
Basepairing......................... 14, 18, 28, 38–41, 57, 60, 61, 72, 73, 75, 141, 143, 147, 194, 195, 289, 344, 345
B complex.................................... 36–38, 40, 41, 73, 179, 180
Bioinformatics 408
Branchpoint...... 14, 15, 17, 18, 21, 24–26, 179, 180, 289, 293

C

Cassette exons 45–47, 49, 51, 57, 59, 63, 65, 66, 73, 74, 97, 246, 359, 387, 414–417
C complex 36, 38–42, 180, 181, 187
ChIP. *See* Chromatin immunoprecipitation (ChIP)
CHIP-seq.......................... 102, 107, 316, 322, 329, 331–334
Chromatin........................ 56, 71, 83–85, 87, 89–93, 97–108, 271, 295, 296, 315, 316, 320–323, 338
Chromatin immunoprecipitation (ChIP)............. 84, 85, 92, 315–323, 326, 331, 333
CLIP 55, 64, 224, 225, 326, 339, 400, 408, 409, 413, 414
Co-immunoprecipitation (Co-IP)...................................117
Convergence...23
Cotranscriptional.. 83–93, 99

D

Databases................ 4, 5, 23, 93, 223, 325–340, 342, 343, 378
Deep sequencing 48, 92, 93, 102
Drosophila................................ 5, 18, 48, 59, 60, 67, 74, 83, 85, 88, 93, 99, 102, 342

E

E complex... 36, 57, 72, 73, 179
Electrophoresis 124, 131, 153, 155, 172, 182, 184, 188, 229, 230, 244, 250, 275, 280, 301–305, 307–310, 321, 401
Epigenetics 108
ESEs. *See* Exonic splicing enhancer (ESEs)
ESSs. *See* Exonic splicing silencers (ESSs)
Eukaryotes..............13–17, 19, 21, 23, 26, 28–30, 46, 97–99, 101, 105, 123, 285, 286, 299, 341, 342, 352
Evolution 13–30, 49, 83
Exon
  definition 6, 19, 20, 29, 37, 38, 59, 60, 67, 74, 75, 98, 101, 103, 106
  skipping 20, 28, 29, 46, 50, 59, 65, 66, 68, 75, 76, 91, 305, 359, 387, 412
Exonic splicing enhancer (ESEs) 6, 24, 26, 57, 58, 60, 62, 63, 72–74, 76, 77, 271, 412
Exonic splicing silencers (ESSs) 57, 58, 70, 72, 74, 77, 271, 412

F

Fluorescence.................................... 125, 217–220, 224, 233, 234, 258, 263, 265, 267, 310

G

Gene expression........................... 9, 36, 49, 69, 91, 101, 104, 105, 151, 169, 193, 252, 283, 285–287, 290, 296, 299, 326, 335, 389, 399

H

HeLa cells 117, 119, 120, 124, 125, 152, 169–171, 189, 259, 267, 272, 277, 282, 320, 322, 405
High throughput sequencing (HTS)...................... 102, 316, 357–392, 400, 409
Histone................................ 8, 85, 89–91, 93, 100–105, 108, 293, 295, 296, 322, 327, 329, 335, 338
hnRNP........................... 59, 63, 67, 68, 72, 73, 88, 341, 412
HTS. *See* High throughput sequencing (HTS)
Human 4, 15, 38, 48, 57, 85, 97, 125, 180, 193, 224, 247, 257, 272, 285, 328, 342, 357, 399, 412

I

Immunofluorescence (IF) 258, 263, 267
Intron 3, 13, 35, 45, 56, 85, 97, 129, 151, 162, 174, 179, 194, 220, 245, 257, 271, 285, 315, 326, 341, 359, 412
  definition 19, 20, 38, 103

Klemens J. Hertel (ed.), *Spliceosomal Pre-mRNA Splicing: Methods and Protocols*, Methods in Molecular Biology, vol. 1126, DOI 10.1007/978-1-62703-980-2, © Springer Science+Business Media, LLC 2014

Intronic splicing enhancers (ISEs)................ 57, 73, 271, 412
Intronic splicing silencers (ISSs)......58, 76, 77, 271–273, 412
Intronless..4, 16, 90, 106, 108
In vitro splicing................................117, 119, 123–125, 151, 152, 155, 156, 161, 162, 164, 165, 187, 189, 205, 223, 233, 244, 288
ISEs. *See* Intronic splicing enhancers (ISEs)
Isoforms.........45, 48–52, 58, 69–71, 107, 249, 254, 305, 308, 357–359, 363, 364, 367, 368, 370, 373, 378–392
ISSs. *See* Intronic splicing silencers (ISSs)

K

Kinetics..................................56, 71, 74, 86–89, 92, 98–100, 103–108, 152, 161–168, 218, 224, 225

M

Machine learning.........................65, 342, 414, 415, 417, 418
Metazoan..........................4, 5, 9, 36, 37, 84, 87, 97, 335, 342
Microscope......................................220, 226, 227, 237, 238, 260, 263, 265, 267, 268
Minigene................88, 99, 117, 151, 152, 157, 243–255, 411
Mobility shift assays (EMSA)..117
Muscular dystrophy...76, 91

N

Nonsense-mediated decay (NMD) 27, 47, 50, 76
Nuclear extracts.......... 73, 117–120, 124, 125, 151–171, 173, 175–177, 180, 182, 185, 189
Nuclear ribonucleoproteins 36, 179, 194, 218, 343
Nucleosomes........... 60, 88, 89, 100–102, 106, 108, 322, 327

P

PhosphorImager...............................128, 130, 154, 159, 164, 165, 172, 174, 188, 197, 200, 222, 230, 254, 275, 280, 307, 308, 310, 404
Phylogenetic...................................14, 16–19, 24, 29, 46–48
Plasmid............................ 127, 129, 133, 170, 172, 182, 183, 189, 196, 198, 201, 206, 208, 209, 214, 244, 247, 248, 250, 258, 259, 261, 262
Polyadenylation 46, 117, 152, 169–177
Polypyrimidine tract..........................4–7, 14, 21, 25, 36, 37, 57, 72, 73, 180, 412
pre-mRNA........................ 3, 13, 35, 45–52, 55–77, 98, 117, 123, 138, 151–169, 179, 193, 205, 218, 243, 257–268, 271–283, 299, 341, 383, 412
pre-mRNA splicing..........3–9, 35–37, 45–52, 55–76, 84, 87, 89, 91, 98, 100, 105, 108, 117, 123–125, 130, 138, 151–169, 174, 179, 185, 193, 194, 197, 200–201, 205, 220, 243, 271–283, 286–288, 291, 296, 299
Protein expression..209–210
Protein purification 209–211, 213
Protein tag... 206, 223, 224

R

Recombinant protein...........64, 117, 205, 206, 210–215, 223
Ribonucleoproteins..................................3, 35, 36, 179, 190, 194, 218, 285, 286, 343
RNA binding proteins...............................51, 52, 55, 56, 58, 59, 63, 70, 92, 98, 103, 105, 108, 290, 315, 316, 320, 343, 399, 412, 413
RNA elements...58, 152, 161
RNA features...415
RNA isolation 251, 274, 278, 282, 300, 302–304, 404
RNA ligation.. 137–138, 140, 230
RNA oligonucleotides 137, 140, 147, 191
RNA polymerase II..............................71, 84, 85, 89, 98, 99, 315, 320, 322
RNA recognition motif (RRM) 61–63, 67, 70, 72, 75
RNA-seq 92, 93, 106, 107, 326, 336, 357–359, 370, 371, 377, 378, 380, 382–386, 389, 390, 392
RRM. *See* RNA recognition motif (RRM)
RT-PCR..55, 243, 245, 247–254, 267, 274–275, 278–280, 299–312

S

S. cerevisiae.............................. 4–6, 13, 15, 16, 18, 19, 24–26, 28, 29, 47, 57, 85, 86, 88, 92, 93, 180, 193–203, 285–287, 289, 292, 294, 343, 354
Secondary structure..........................8, 14, 39, 57–59, 61, 65, 68, 141, 142, 147, 251, 254, 341–354, 413
Sequence annotation 327, 338
Single molecule 74, 125, 217–240, 392
siRNA 104, 300, 411
Small nuclear ribonucleoproteins (snRNP)......... 6, 7, 36–42, 57, 62, 67, 70, 72–75, 85, 87, 89, 92, 105, 179, 180, 193–203, 218, 293, 296, 315, 341, 343, 412
Small nuclear RNAs (snRNA).................6, 7, 14, 18, 20, 25, 35–41, 56, 57, 61, 67, 72–75, 90, 124, 125, 138, 152, 174, 176, 193–203, 257, 259, 263, 264, 266, 267, 285, 289, 293
Species..............................4, 5, 13–18, 20, 26–29, 46, 47, 49, 50, 64, 65, 88, 93, 102, 124, 125, 187, 188, 218, 342, 352, 354, 387
Spinal muscular atrophy (SMA)...........................76, 77, 272
Splice junctions.. 48, 83, 123, 339, 380, 416
Spliceosome...........3, 13, 35, 62, 97, 124, 138, 179, 194, 205, 217-240, 257-268, 285, 315, 341, 412
Spliceosome assembly.. 6, 8, 35–38, 42, 72, 73, 84, 85, 87, 90, 92, 99, 101, 107, 108, 125, 138, 179, 180, 185, 217–240, 257, 260, 296, 315, 412
Spliceosome complex................................. 97, 180, 182–183
Splice site.................. 4, 15, 36, 45, 56, 86, 97, 123, 152, 164, 179, 194, 247, 271, 286, 299, 326, 341, 359, 412

Splicing factors..........7, 18, 19, 28, 29, 45, 51, 56, 60–62, 69, 71–73, 87, 89–91, 93, 99, 104, 105, 108, 123, 162, 205, 206, 209–212, 215, 218, 257, 258, 287, 289, 296, 316, 320, 322, 413
Splicing isoforms..........357
Splicing regulatory elements (SREs)..........76, 161, 412
Splicing regulatory proteins..........58, 61, 68, 315
Splicing silencer..........57, 58, 72, 76, 88, 271–273, 412
SR protein..........6, 19, 29, 60–62, 65, 69, 70, 72, 73, 84, 87–90, 117, 190, 241, 343
Stable transfection..........250
Synthetic lethality..........286, 291

T

Transcript..........3, 8, 20, 21, 27, 55–63, 65, 66, 71, 75, 87, 92, 98, 124, 127, 130, 176, 183–186, 188, 194, 196, 199, 200, 202, 220, 221, 229–231, 250, 263, 273, 286, 341, 359, 370, 375, 376, 378, 381–387, 391, 392, 413–415
Transcription..........7, 27, 46, 56, 84, 98, 117, 128, 138, 152, 161, 169–177, 182, 196, 205, 217, 254, 257, 271, 296, 299, 326, 341, 387, 402
Transcriptome..........48, 55, 63, 64, 97, 363, 366, 367, 370, 371, 373, 379, 381, 383, 389–392
Transesterification..........3, 7, 8, 36, 42
Transient transfection..........99, 207–208, 245–247, 252

U

UCSC Genome Browser..........327, 335, 336, 343
Untranslated regions (UTRs)..........3, 45, 326
UV crosslinking..........399, 402, 408, 409

W

Western blotting..........275–276, 280–281, 405
Whole cell extract..........117, 123–134, 218, 225

Y

Yeast..........4–7, 15, 18, 24, 47, 57, 74, 84–86, 88, 90–92, 123–134, 183, 195, 196, 199, 205, 218, 220, 223–225, 228, 231, 232, 234, 239, 285, 342, 343, 354
Yeast genetics..........35, 92, 93, 124, 223, 285–296

MIX
Papier aus verantwortungsvollen Quellen
Paper from responsible sources
FSC® C105338

If you have any concerns about our products,
you can contact us on
ProductSafety@springernature.com

In case Publisher is established outside the EU,
the EU authorized representative is:
Springer Nature Customer Service Center GmbH
Europaplatz 3, 69115 Heidelberg, Germany

Printed by Libri Plureos GmbH
in Hamburg, Germany